Theoretical, Numerical and Experimental Studies on Clean Energy and Combustion

Theoretical, Numerical and Experimental Studies on Clean Energy and Combustion

Topic Editors

Pedro Miguel Rebelo Resende
Mohsen Ayoobi
Alexandre M. Afonso

MDPI

Basel • Beijing • Wuhan • Barcelona • Belgrade • Novi Sad • Cluj • Manchester

Topic Editors

Pedro Miguel Rebelo Resende
Polytechnic Institute of Viana do Castelo
Viana do Castelo
Portugal

Mohsen Ayoobi
Division of Engineering Technology
Wayne State University
Detroit
USA

Alexandre M. Afonso
Department of Mechanical Engineering
University of Porto
Porto
Portugal

Editorial Office
MDPI AG
Grosspeteranlage 5
4052 Basel, Switzerland

This is a reprint of the Topic, published open access by the journals *Aerospace* (ISSN 2226-4310), *Applied Sciences* (ISSN 2076-3417) and *Energies* (ISSN 1996-1073), freely accessible at: https://www.mdpi.com/topics/5G0ERDO7HU.

For citation purposes, cite each article independently as indicated on the article page online and as indicated below:

Lastname, A.A.; Lastname, B.B. Article Title. *Journal Name* **Year**, *Volume Number*, Page Range.

ISBN 978-3-7258-4957-4 (Hbk)
ISBN 978-3-7258-4958-1 (PDF)
https://doi.org/10.3390/books978-3-7258-4958-1

Contents

applied sciences

MDPI

Editorial

An Overview of Theoretical, Numerical and Experimental Studies on Clean Energy and Combustion

Pedro R. Resende [1], Mohsen Ayoobi [2] and Alexandre M. Afonso [3,4,*]

1 ProMetheus—Escola Superior de Tecnologia e Gestão, Instituto Politécnico de Viana do Castelo, 4900-347 Viana do Castelo, Portugal; pedroresende@estg.ipvc.pt
2 Division of Engineering Technology, College of Engineering, Wayne State University, Detroit, MI 48202, USA; mohsen.ayoobi@wayne.edu
3 Transport Phenomena Research Center, Faculdade de Engenharia, Universidade do Porto, Rua Dr. Roberto Frias, 4200-465 Porto, Portugal
4 ALiCE, Associate Laboratory in Chemical Engineering, Faculdade de Engenharia, Universidade do Porto, Rua Dr. Roberto Frias s/n, 4200-465 Porto, Portugal
* Correspondence: aafonso@fe.up.pt

1. Introduction

This paper reviews recent theoretical, numerical and experimental studies on clean energy and combustion, focusing on advancements in combustion efficiency, emission reduction technologies and renewable energy integration. Specifically, this paper highlights key findings from 17 studies that address the modeling of combustion processes, energy conversion systems and clean energy sources. Such findings are significantly important in providing insights into improving the sustainability of energy systems while mitigating the environmental impacts of using traditional energy sources.

The need for cleaner energy sources and conversion mechanisms is increasing with a continuous increase in the global demand for energy. The aviation industry, in particular, is under pressure to reduce its carbon footprint and to explore sustainable alternatives to traditional jet fuel. Meanwhile, other industries, such as power generation, are investigating ways to optimize combustion efficiency and to reduce emissions. The studies reviewed in this paper focus on theoretical, numerical and experimental advances in clean energy and combustion technologies.

This review summarizes the key findings from 17 studies published between 2022 and 2025 in the MDPI topic "Theoretical, Numerical and Experimental Studies on Clean Energy and Combustion", focusing on alternative fuels, combustion dynamics, reduced-order modeling, thermo-acoustic and turbulent combustion and emission control strategies. By integrating these diverse investigations, this paper aims to elucidate the interconnected advances in combustion science and their implications for sustainable energy applications.

2. Alternative Fuels in Aviation and Engines

The aviation sector, concerned with the escalating fuel consumption and greenhouse gas emissions, has shown interest in exploring the use of alternative fuels, as exemplified by Villette et al. [1], who evaluated the performance of Liquefied Natural Gas (LNG) and methanol in a geared turbofan engine, demonstrating that these non-drop-in fuels can reduce post-combustion temperatures by 1–3% and increase thrust by 3–10% compared to Jet-A, while achieving substantial reductions in CO_2, NO_x and CO emissions by up to 20%, 60% and 39%, respectively.

In another work conducted by Janovec et al. [2], the authors modeled hydrogen and battery-powered aircraft concepts, revealing that the power-to-weight ratio of existing battery technologies still need to be improved before electrified propulsion becomes a contender in the light-sport aircraft segment.

Complementing these aviation-focused studies, Aljabri et al. [3] conducted a computational comparison of spark-ignited (SI) and pre-chamber (PC) hydrogen engines, finding that lean mixtures (air–fuel ratio $\lambda = 3.33$) with advanced spark timing in SI mode yield efficiencies exceeding 50% with minimal NO_x, whereas the PC configuration increases pressure rise rates and heat losses, necessitating further optimization—a challenge that echoes the trade-offs observed by Villette et al. [1], with low-energy-density fuels requiring larger tank volumes. On the ground, alternative fuels also show promise, as Garzón et al. [4] investigated straight soybean oil–diesel blends (50% and 80% v/v) in a compression ignition engine, noting that higher vegetable oil fractions delay combustion due to prolonged diffusive phases, but the oxygen content of the fuel improves combustion efficiency, a finding that parallels the emission reduction potential seen in aviation fuels.

Similarly, Wang et al. [5] explored co-firing coal with biomass syngas in a 300 MW boiler, observing that introducing 3×10^4 m^3/h of wood syngas reduces coal consumption by 5.1% and lowers the furnace exit temperatures due to diminished flame radiation, suggesting a synergy with Villette et al. [1] and Garzón et al. [4] in leveraging oxygenated fuels to curb emissions, albeit with thermodynamic adjustments to maintain system performance.

3. Combustion Dynamics and Modeling

Combustion modeling under varying conditions can provide critical insights into flame behavior, which could inform the system design requirements. Xu et al. [6] conducted thorough numerical simulations to investigate how an equivalence ratio can affect the detonation characteristics and performance of methane–oxygen combustion in rotating detonation rocket engines (RDREs). They showed that increasing the equivalent ratio reduces the number of detonation waves while boosting specific impulse, a trend that correlates with the enhanced reactivity observed by Wang et al. [7] in ammonia–air flames with higher cracking ratios.

In a different context, Gong and Tang [8] examined the high-gravity combustion characteristics of propane–air in a channel with backward-facing steps (in an ultra-compact combustor) utilizing large eddy numerical simulations.

Analyzing their simulation result, they concluded that the centrifugal forces double the flame propagation speed by triggering Rayleigh–Taylor and Kelvin–Helmholtz instabilities, which enhance mixing and reduce combustion time—an effect that resonates with the turbulence-driven flame enhancements noted by Martinez-Sanchis et al. [9] in methane–oxygen rocket flames. Zhang et al. [10] further explored the flame stabilization by simulating lean premixed methane flames piloted by rich premixed flames, showing that the pilot flame adjusts stretch and heat transfer to anchor the main flame closer to the wall, a mechanism that could inform the micro-combustion stability challenges addressed by Kutkut et al. [11].

Micro-scale combustion introduces additional complexities, as reviewed by Dias et al. [12], who highlighted phenomena such as flashback, quenching diameter and flames with repeated extinctions and ignition (FREI) cycles, which Kutkut et al. [11] substantiated through simulations of premixed methane–air combustion in micro-channels, noting that higher inlet velocities stabilize flames while lower velocities trigger FREI modes influenced by equivalence ratio and channel size.

Huang et al. [13] utilized numerical simulations to investigate the impact of thermal gas on the boundary layer flame flashback in channels. They demonstrated that thermal gas expansion globally modifies flow in narrow channels but remains localized in wider ones, challenging the model critical gradient assumptions while aligning with Gong and Tang [8] in emphasizing the role of flow–flame interactions under confinement.

Modeling efforts to reduce computational costs are advanced by Gitushi and Echekki [14], who compared different species selection methods, including the principal component analysis (PCA), directed relation graphs (DRGs), the global pathway selection (GPS) and the manifold-informed species selection methods, to reduce the complexity of the chemical mechanisms describing complex hydrocarbon fuel combustion. They suggested that these approaches effectively capture key species across the oxidation stages, with sensitivity to variance and target species as also implied in parametric sensitivities in Xu et al. [6].

Building on this, Alqahtan et al. [15] developed a hybrid chemistry framework using neural networks to model the low-temperature oxidation of n-heptane, coupling it with foundational chemistry to achieve acceptable accuracies at reduced computational expense. Such a strategy could enhance the detonation simulations of Xu et al. [6] or the micro-flame analyses of Kutkut et al. [11].

4. Thermoacoustic and Turbulent Combustion

Thermoacoustic oscillations and turbulent combustion significantly influence the stability and performance of a combustion system. Tao et al. [16] performed a theoretical analysis of these effects in a lean premixed combustion regime, revealing that extending the length of the combustion chamber from 1.2 m to 6.0 m can reduce the amplitude of the sound pressure by 34% and delay the trigger time from 0.32 s to 0.91 s, which could mitigate the pressure fluctuations noted by Gong and Tang [8] under high gravity conditions.

In a different study, Martinez-Sanchis et al. [9] simulated turbulent non-premixed methane–oxygen flames at 20 bar, observing that despite a non-premixed setup, premixed combustion occurs in lean and rich regions, with turbulence shifting products toward lean conditions, an effect that parallels the flow modifications in the study performed by Huang et al. [13] and suggests a link to the flame stabilization mechanisms explained in the work of Zhang et al. [10].

5. Catalyst and Emission Control

Emission control through catalysts and fuel modifications is pivotal for clean combustion, as Liu et al. [17] demonstrated with Zr-modified Cu-Ce/SAPO-34 catalysts for NH_3-SCR, where Zr reduces acid sites and denitration efficiency yet achieves over 99% NO removal at 400–500 °C, contrasting with the results of Wang et al. [5], who found that higher ammonia cracking ratios in turbulent flames increase NO formation despite enhanced reactivity, highlighting a trade-off between combustion intensity and emissions, akin to the findings of Villette et al. [1]. These studies collectively underscore the need for integrated fuel and catalyst strategies to balance efficiency and emission goals.

6. Conclusions

The collective works from the studies presented in the MDPI topic "Theoretical, Numerical and Experimental Studies on Clean Energy and Combustion" illuminate the multifaceted advancements in clean energy and combustion, where alternative fuels such as LNG, methanol, hydrogen and biomass syngas offer substantial emission reductions but necessitate careful engine and system design adjustments to address efficiency and stability challenges, as seen in Villette et al. [1], Janovec et al. [2] and Wang et al. [5]. Combustion dynamics, from detonation waves [2] to micro-flames [12], reveal the critical interplay of flow and heat transfer, while modeling innovations [5,6] and catalyst developments [3] provide tools to optimize these processes, suggesting a future where integrated approaches could fully realize the potential of sustainable combustion technologies.

Funding: P. R. Resende acknowledges the funding by FAPESP through Project No. 2015/26842-3. A.M. Afonso acknowledge FCT—Fundação para a Ciência e a Tecnologia for financial support through LA/P/0045 (ALiCE), UIDB/00532 and UIDP/00532(CEFT), funded by national funds through FCT/MCTES (PIDDAC).

Conflicts of Interest: The authors declare no conflicts of interest.

References

1. Villette, S.; Alexiou, A.; Aretakis, N.; Mathioudakis, K. Expanding Known Performance Capabilities of Geared Turbofan Engine When Powered by LNG and Methanol. *Aerospace* **2025**, *12*, 96. [CrossRef]
2. Janovec, M.; Babčan, V.; Kandera, B.; Šajbanová, K.; Škultéty, F.; Halvoň, Ľ. Performance and Weight Parameters Calculation for Hydrogen- and Battery-Powered Aircraft Concepts. *Aerospace* **2023**, *10*, 482. [CrossRef]
3. Aljabri, H.; Silva, M.; Houidi, M.B.; Liu, X.; Allehaibi, M.; Almatrafi, F.; AlRamadan, A.S.; Mohan, B.; Cenker, E.; Im, H.G. Comparative Study of Spark-Ignited and Pre-Chamber Hydrogen-Fueled Engine: A Computational Approach. *Energies* **2022**, *15*, 8951. [CrossRef]
4. Garzón, N.A.N.; Oliveira, A.A.M.; Bazzo, E. Heat Release Rates of Straight Soybean and Diesel Oil Blends in a Compression Ignition Engine. *Appl. Sci.* **2024**, *14*, 9215. [CrossRef]
5. Wang, J.; Yao, Q.; Jin, X.; Deng, L. The Influence of Co-Firing Coal with Biomass Syngas on the Thermodynamic Parameters of a Boiler. *Appl. Sci.* **2023**, *13*, 11477. [CrossRef]
6. Xu, X.; Han, Q.; Zhang, Y. Numerical Investigation of the Effect of Equivalent Ratio on Detonation Characteristics and Performance of CH_4/O_2 Rotating Detonation Rocket Engine. *Aerospace* **2025**, *12*, 68. [CrossRef]
7. Wang, C.; Wang, H.; Luo, K.; Fan, J. The Effects of Cracking Ratio on Ammonia/Air Non-Premixed Flames under High-Pressure Conditions Using Large Eddy Simulations. *Energies* **2023**, *16*, 6985. [CrossRef]
8. Gong, Z.; Tang, H. Numerical Study of High-g Combustion Characteristics in a Channel with Backward-Facing Steps. *Aerospace* **2024**, *11*, 767. [CrossRef]
9. Martinez-Sanchis, D.; Sternin, A.; Haidn, O.; Tajmar, M. Combustion Regimes in Turbulent Non-Premixed Flames for Space Propulsion. *Aerospace* **2023**, *10*, 671. [CrossRef]
10. Zhang, L.; Cui, Y.; Yin, P.; Mao, W.; Zhang, P. Numerical Investigation on the Flame Characteristics of Lean Premixed Methane Flame Piloted with Rich Premixed Flame. *Energies* **2024**, *17*, 3430. [CrossRef]
11. Kutkut, A.; Ayoobi, M.; Baumgardner, M.E.; Akkerman, V. Investigating the Ignition and Stability Limits of Premixed Methane/Air Combustion in Micro-Channels. *Energies* **2023**, *16*, 6752. [CrossRef]
12. Dias, D.M.; Resende, P.R.; Afonso, A.M. A Review on Micro-Combustion Flame Dynamics and Micro-Propulsion Systems. *Energies* **2024**, *17*, 1327. [CrossRef]
13. Huang, K.; Valiev, D.M.; Zhong, H.; Han, W. Numerical Study of the Influence of the Thermal Gas Expansion on the Boundary Layer Flame Flashback in Channels with Different Wall Thermal Conditions. *Energies* **2023**, *16*, 1844. [CrossRef]
14. Gitushi, K.M.; Echekki, T. Comparisons of Different Representative Species Selection Schemes for Reduced-Order Modeling and Chemistry Acceleration of Complex Hydrocarbon Fuels. *Energies* **2024**, *17*, 2604. [CrossRef]
15. Alqahtani, S.; Gitushi, K.M.; Echekki, T. A Data-Based Hybrid Chemistry Acceleration Framework for the Low-Temperature Oxidation of Complex Fuels. *Energies* **2024**, *17*, 734. [CrossRef]
16. Tao, C.; Sun, R.; Wang, Y.; Zhang, L.; Ye, J.; Liang, S. The Effects of Parameter Settings on Triggering Time and Climb Rate during Lean-Premixed Combustion Thermoacoustic Oscillations. *Appl. Sci.* **2024**, *14*, 806. [CrossRef]
17. Liu, C.; Wang, X.; Xing, L.; Cheng, X.; Zhang, X.; Li, H.; Liu, M. Effect of Zr Modification on NH_3-SCR Reaction Performance of Cu-Ce/SAPO-34 Catalysts. *Appl. Sci.* **2023**, *13*, 4763. [CrossRef]

energies

MDPI

Article

Comparative Study of Spark-Ignited and Pre-Chamber Hydrogen-Fueled Engine: A Computational Approach

Hammam Aljabri [1,*], Mickael Silva [1], Moez Ben Houidi [1], Xinlei Liu [1], Moaz Allehaibi [1,2], Fahad Almatrafi [1], Abdullah S. AlRamadan [3], Balaji Mohan [3], Emre Cenker [3] and Hong G. Im [1]

1 Clean Combustion Research Center (CCRC), King Abdullah University of Science and Technology, Thuwal 23955, Saudi Arabia
2 Mechanical Engineering Department, College of Engineering and Islamic Architecture, Umm Al-Qura University, Mecca 24382, Saudi Arabia
3 Transport Technologies Division, R&DC, Saudi Aramco, Dhahran 34466, Saudi Arabia
* Correspondence: hammam.aljabri@kaust.edu.sa

Abstract: Hydrogen is a promising future fuel to enable the transition of transportation sector toward carbon neutrality. The direct utilization of H_2 in internal combustion engines (ICEs) faces three major challenges: high NO_x emissions, severe pressure rise rates, and pre-ignition at mid to high loads. In this study, the potential of H_2 combustion in a truck-size engine operated in spark ignition (SI) and pre-chamber (PC) mode was investigated. To mitigate the high pressure rise rate with the SI configuration, the effects of three primary parameters on the engine combustion performance and NO_x emissions were evaluated, including the compression ratio (CR), the air–fuel ratio, and the spark timing. In the simulations, the severity of the pressure rise was evaluated based on the maximum pressure rise rate (MPRR). Lower compression ratios were assessed as a means to mitigate the auto-ignition while enabling a wider range of engine operation. The study showed that by lowering CR from 16.5:1 to 12.5:1, an indicated thermal efficiency of 47.5% can be achieved at 9.4 bar indicated mean effective pressure (IMEP) conditions. Aiming to restrain the auto-ignition while maintaining good efficiency, growth in λ was examined under different CRs. The simulated data suggested that higher CRs require a higher λ, and due to practical limitations of the boosting system, λ at 4.0 was set as the limit. At a fixed spark timing, using a CR of 13.5 combined with λ at 3.33 resulted in an indicated thermal efficiency of 48.6%. It was found that under such lean conditions, the exhaust losses were high. Thus, advancing the spark time was assessed as a possible solution. The results demonstrated the advantages of advancing the spark time where an indicated thermal efficiency exceeding 50% was achieved while maintaining a very low NO_x level. Finally, the optimized case in the SI mode was used to investigate the effect of using the PC. For the current design of the PC, the results indicated that even though the mixture is lean, the flame speed of H_2 is sufficiently high to burn the lean charge without using a PC. In addition, the PC design used in the current work induced a high MPRR inside the PC and MC, leading to an increased tendency to engine knock. The operation with PC also increased the heat transfer losses in the MC, leading to lower thermal efficiency compared to the SI mode. Consequently, the PC combustion mode needs further optimizations to be employed in hydrogen engine applications.

Keywords: hydrogen combustion; spark ignition; pre-chamber combustion

1. Introduction

The transportation sector is one of the main contributors to greenhouse gas (GHG) emissions, which has promoted substantial research and development. The demand for passenger and cargo transport is increasing rapidly, while the net zero emission by 2050 scenario requires a 20% decrease in the transport sector emissions by 2030 [1]. As a result, research efforts are now focused on employing renewable fuels and more efficient engines to reduce the carbon footprint. The hydrogen-powered internal combustion engine is gaining popularity due to its potential to achieve high brake thermal efficiencies of over 45% with emissions less than the EURO 6 standards without any after-treatment by utilizing lean ($\lambda > 2.5$) operation [2–11]. The high auto-ignition temperature and Research Octane Number (RON $\geq$ 130) [12] of hydrogen compared to standard fuels make it a promising candidate fuel to mitigate the knocking tendency. However, its minimum ignition energy is lower than most hydrocarbon fuels, implying that hydrogen can be easily ignited by hot spots or residues in the combustion chamber. Studies with a pressure in the range of 0.2–1 atm and temperatures varying between 273–373 K have shown that the minimum ignition energy of hydrogen–air mixtures has an inverse relation with the square of pressure and with temperature [13]. If these trends are extrapolated to higher pressure and temperature, the minimum ignition energies of hydrogen–air mixtures will likely be significantly lower than reported for normal engine temperatures and pressures.

Having low minimum ignition energy may result in pre-ignition, which is defined as combustion during the compression stroke before the desired ignition timing. This may lead to a loss of control over combustion phasing, heavy knocking, and in extreme cases, mechanical failure of the engine. Although knock and pre-ignition have a nearly indistinguishable global impact, the causes of the two phenomena are vastly different [2]. Unlike knock, pre-ignition can be prevented with the proper engine design but knock mostly depends on the maximum compression ratio that can be used with a certain fuel [14]. Another difference is that knock can be mitigated by retarding the spark timing, while pre-ignition cannot be controlled in such a way.

The minimum ignition energy for hydrogen is highly sensitive to the air–fuel ratio as the stoichiometric condition ($\lambda = 1$) is approached from the lean side ($\lambda > 1$). This tendency is responsible for the experimentally observed fact that operating an H_2-ICE at or near the stoichiometric state in the absence of pre-ignition events is extremely challenging. As a result, for practical reasons, the minimum λ and peak power output can be restricted by the pre-ignition limit. The pre-ignition limit depends mainly on the engine specifications and operational conditions, of which consistent patterns have been discovered. For instance, it has been reported that the λ limit to prevent pre-ignition grows monotonically when the compression ratio (CR) [15–17] and mixture temperature are increased [16]. Moreover, the engine speed has also been demonstrated to influence [15,17], although the trend is more complicated owing to the associated effect of the residual mass fraction.

Hydrogen has a large flammability range of $0.14 \leq \lambda \leq 10$ at 1 atm and 298 K, compared to that of gasoline in the range of $0.25 \leq \lambda \leq 1.42$, implying that a hydrogen internal combustion engine will have more stable operation in a highly diluted environment, potentially leading to lower emissions and fuel consumption. One of the most researched modes of H_2-ICE is the port-fuel injection spark ignition (PFI SI) engine. Das [18] and White et al. [2] examined the performance and control techniques of this engine mode in depth, and reported that operating the engine under ultra-lean conditions ($\lambda \geq 2$) can reduce NO_x emissions to less than 100 ppm without the need for after-treatment. When operating at supercharged intake pressures of 2.6 bar, Nagalingam et al. [19] published data indicating values below 100 ppm for $\lambda > 2.5$. Natkin et al. [20] found 90 ppm emissions at $\lambda = 4$, which is similar. Ford's supercharged engine was operated at a leaner conditions of $\lambda = 4.34$ to achieve emissions of 3–4 ppm [20]. The use of pure H_2 in SI engines have been extensively studied in literature [21–24]. Moreover, H_2 is also used to guarantee stable operation for different fuels such as, gasoline and compressed natural gas [25,26]. The H_2 addition lowers cycle-to-cycle variability and increases the lean-burn limit because of its low ignition energy and rapid flame speed. Thus, the use of pure H_2 permits high dilution rates, achieving very low NOx emissions at lean and ultra-lean mixtures while keeping acceptable combustion stability.

The use of H_2 in a homogenous charge compression ignition (HCCI) mode was investigated thoroughly by [27] where high compression ratios ranging from 20:1 up to 45:1 under various air–fuel ratios was studied. It was found that with no charge preheating, a very high compression ratio was required to reach the auto-ignition temperature of H_2. Under ultra-lean conditions ($\lambda \geq 2$), it was not possible to reach the auto-ignition temperature of H_2 using a CR < 26:1.

The use of an ultra-lean mixture is an effective strategy to further reduce emissions and improve fuel consumption. It has also been proven that lean mixtures will improve the indicated thermal efficiency by increasing the specific heat ratio, reducing heat transfer losses, and allowing for higher compression ratios to be used. Nonetheless, the ultra-lean mixture makes it challenging to ignite the charge, which could result in high incomplete combustion losses or misfires in more severe cases. In such a situation, pre-chamber combustion is considered a suitable solution to achieve stable combustion [28–30].

The pre-chamber (PC) consists of a spark plug that ignites the mixture in a small cavity, resulting in high temperature and pressure jets, which in turn travel through orifices to the main chamber leading to spatial ignition [31,32]. The flame propagation distance is shortened by the rapid penetration and dispersion of the hot turbulent jet. The main chamber (MC) combustion process is improved by the unburned intermediate products as well as the enhanced turbulence level resulting from the hot jet [33–36].

Depending on how the fuel is supplied, PCs are classified as active or passive. In the passive system, all the fuel is provided through the intake port into the MC, while in the active PC, an additional fuel injector provides a fraction of fuel into the PC to create local stratification [37–41]. Passive PCs have been used in gasoline engines in favor of simplicity, as a means to increase the combustion speed, reduce knocking tendency, and allow for higher compression ratios. However, one of the major limitations of the passive PC is that the lean burn limit cannot be significantly extended compared to conventional SI engines [38,42,43]. The active PC is thus considered to achieve the lean limit extension by the enrichment in the PC [44].

The objective of the present study was to identify the optimal operation modes to enable hydrogen-fueled engines by employing various lean-burn strategies. Starting with the SI engine mode as the baseline case, the effects of the overall air–fuel ratio, compression ratio, and spark timing on the auto-ignition tendency were numerically investigated in terms of combustion, thermal efficiency, and NO_x emissions. Subsequently, the use of pre-chamber under the same load conditions was assessed in comparison. Both active and passive PCs were used and investigated with different compression ratios, and their relative performance metrics were examined.

2. Numerical Setup and Model Validation

CONVERGE^{TM} 3.0 [45] was used to conduct three-dimensional CFD simulations. For turbulence modeling, the original formulation of the RNG k-ε model was used [46]. The O'Rourke and Amsden [47] model was used to predict the heat transfer process. A real-fluid equation of state (Redlich-Kwong-Soave) combined with the gas species' critical properties consisting of pressure, temperature, and acentric factor for precise prediction of the Joule–Thomson effect with hydrogen [48] was used.

The SAGE detailed chemical kinetic solver with a detailed kinetic mechanism for H_2 combustion. Through the literature search, we adopted the mechanism developed by Park et al. [49], which was an optimized version of the original mechanism by Burke et al. [50] to improve the prediction at ultra-lean H_2 combustion conditions, in terms of ignition delay time and laminar flame speed validated against the experiment conducted by Bradley et al. [51].

Full-cycle simulations starting from the exhaust valve opening timing were conducted. The computational domain and mesh details are seen in Figure 1. A base grid of 4 mm was adopted, with both fixed embedding and adaptive mesh refinement (AMR) in the regions of interest, with the smallest grid size of 0.125 mm. In addition, three levels of dynamic refinement were used in the main chamber based on the gradients of velocity, temperature, and H_2 species mass fraction. Additional details on mesh and grid scaling can be found in [52], including the detailed geometry of the PC that replaces the spark plug in the subsequent study. Considering the small volume, the interior of PC is sufficiently resolved.

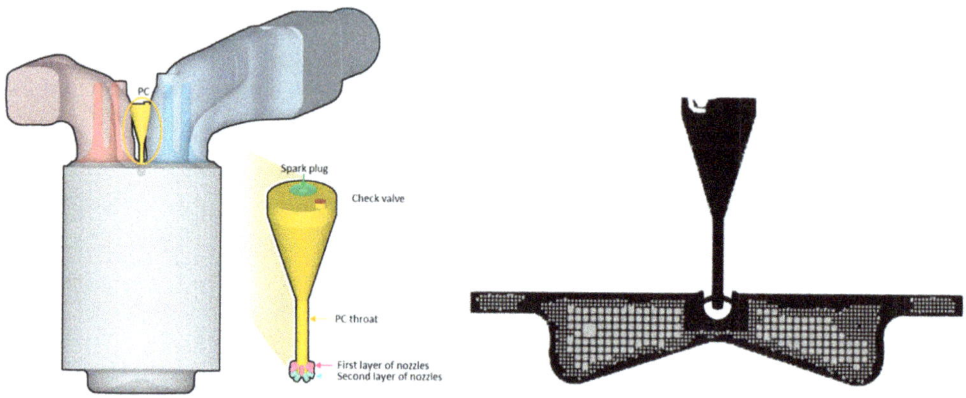

(a) (b)

Figure 1. Schematic of: (**a**) computational geometry; and (**b**) grid control technique.

The engine specifications used in the simulations are listed in Table 1. The geometrical compression ratio corresponds to the standard bowl piston mounted on the Volvo D13C500 engines; details on engine setup and specifications are found in previous works [53]. This reference configuration was considered as we aimed to assess the utilization of hydrogen fuel in this engine with minimal modification of the geometry.

Table 1. Engine specifications.

Engine Type	4-Stroke
Bore/stroke (mm)	131/158
Connecting rod length (mm)	255
Displacement volume (L)	2.13
Geometric compression ratio	17:1
Intake valve open (°aTDC)	347
Intake valve close (°aTDC)	−167
Exhaust valve open (°aTDC)	−140
Exhaust valve close (°aTDC)	352
H_2 injected mass (mg/cycle)	38.2

The fidelity of the current modeling approach was verified in previous studies using methane as a fuel with in-house data [54–56], which is briefly discussed in the current work. The selected cases refer to the equivalent experimental data with the maximum break torque. The initial and boundary conditions were further calibrated using 1D GT-Power; details on the methodology for calibration are shown in previous work [53]. The input parameters for the model have been experimentally measured, including the valve lift profiles, intake, exhaust, and pre-chamber fuel line pressure traces. For the pressure traces on both the pre-chamber and main chamber, the ensemble average of 500 experimental cycles is considered. The parameter Pre-chamber fueling ratio (PCFR) is the fraction of total fuel injected through the pre-chamber and more details can be found in Figure 2. The conditions for the validation are shown in Table 2. It should mentioned that the simulated engine could withstand an in-cylinder peak pressure up to 220 bar.

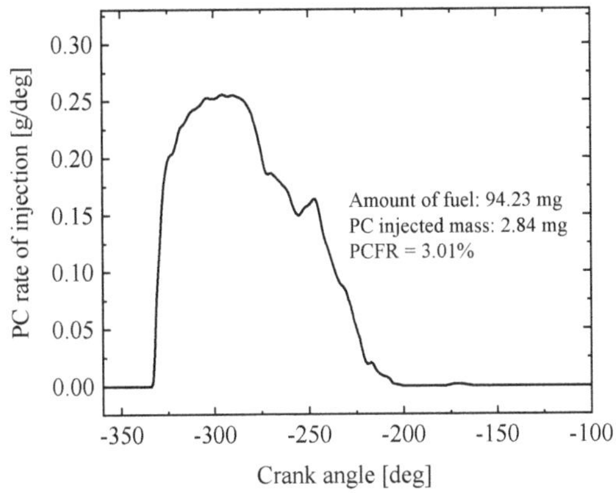

Figure 2. Pre-chamber fueling profile.

Table 2. Specifications of load and boundary conditions [57].

Load (bar IMEP)	9.5
Engine speed (RPM)	1200
Effective compression ration	11.1
Intake pressure (bar)	1.5
Intake temperature (K)	303
Exhaust pressure (bar)	1.7
Exhaust temperature (K)	800
Total fuel mass (mg)	94
Global-λ	1.8
Pre-chamber fueling ratio (PCFR) (%)	3
Spark timing (CAD aTDC)	−13

The in-cylinder trapped fuel and total gas amounts are the key quantities to be matched between 3D CFD and 1D GT-Power solutions. For the 3D CFD, the field is initialized quiescently with the major species (CO_2, CO, O_2, N_2, and H_2O), pressure, and temperature. Since the simulations ran for a full cycle starting at the exhaust valve opening, it allows for minimizing the field initialization before the combustion stroke. For the current simulations, the effective compression ratio was set to be 11.1, instead of the geometrical value of 11.5 in order to reproduce the experimental pressure trace. The PC fueling was performed during the intake stroke, as shown in Figure 2. Additional details on the injection strategy can be found in [53].

The validation of the pressure curve in Figure 3 showed satisfactory agreement between the experiment and the simulation. Further comparison of the engine performance parameters complemented the validation in Figure 3b. Small differences are inevitably caused by uncertainties in the employed physical and chemical models, but overall good prediction capability was demonstrated.

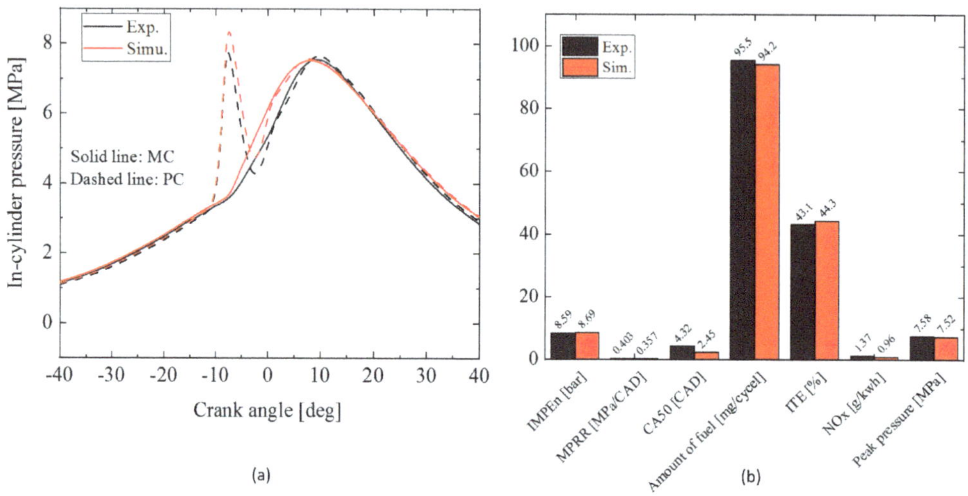

Figure 3. Validation of: (**a**) predicted pressure trace; and (**b**) predicted engine performance parameters for the reference case of Table 2.

3. Results and Discussion

The baseline case corresponded to a mid-load operation of a single-cylinder version of the Volvo D13C500 engine, which was extensively validated against experiments in previous studies [58,59]. The current investigation started with assessing the feasibility of converting this compression ignition (CI) diesel engine into operation with hydrogen in SI or PC modes. The practical goal was to implement a retrofit solution with minimal changes to an existing commercial engine architecture. The investigated controlling parameters included the compression ratio (CR), overall air–fuel ratio (λ), and spark timing (ST). In simulations, a maximum pressure rise rate (MPRR) of 1.5 MPa/deg, as suggested in [60], was used as the onset of abnormal combustion behavior. The calculated engine thermal efficiency, heat transfer loss, incomplete combustion loss, exhaust loss, and NO_x emissions, were collected and evaluated.

3.1. SI Combustion

The baseline case of H_2 spark ignition combustion is presented here as a reference case where the compression ratio was adjusted to 16.5:1 to account for the crevice volume which is not considered in the simulation as it would add unnecessary calculation cost and has negligible effect on the results. A lean mixture with $\lambda = 2.85$ was used to reduce the laminar flame speed and increase the minimum ignition energy to avoid auto-ignition and pre-ignition, respectively.

Figure 4 shows the pressure trace and heat release rate (HRR) of the baseline case where an abrupt pressure rise and HRR are observed. The MPRR, in this case, is 28.9 MPa/deg, which is considered too high for normal operation. This suggests that the mixture is highly reactive to yield nearly a HCCI mode of combustion. Based on the experimental work done by [27], it was not possible to reach the auto-ignition temperature of H_2 under lean conditions using a CR < 26:1 which raises some uncertainties about the chemical mechanism used in this study. The lack of the experimental data for H_2 combustion makes it a challenging task to validate this work. On the other hand, this study can provide general trends about the use of pure H_2 in CI diesel engines.

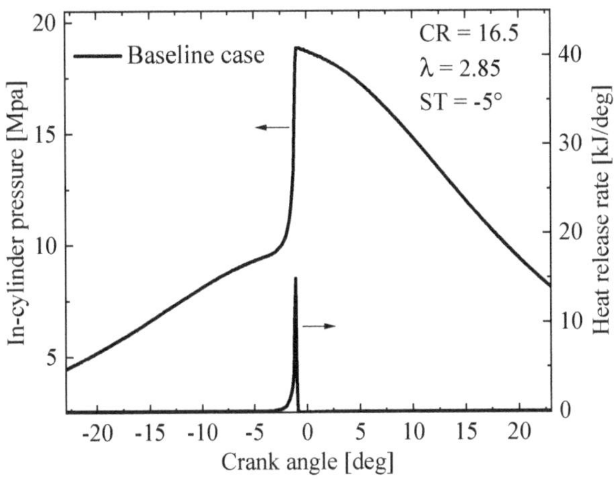

Figure 4. Predicted pressure and heat release rate traces for the baseline case.

Figure 5 shows the effect of using the spark plug as an ignition source. A slightly advanced combustion phasing is observed when the spark plug is used, while subsequent auto-ignition is evident from the steep MPRR in both cases (with and without spark ignition). This indicates that the spark plug assists in igniting the mixture leading to a spark-assisted compression ignition (SACI) combustion mode. Unlike in SI mode, knock in HCCI is mainly due to premature combustion in which ignition takes place before the desired time. It is challenging to control the ignition in such a combustion mode, which may limit the engine's operating range. It was observed that the HRR was slightly faster in the HCCI mode compared to the SACI mode.

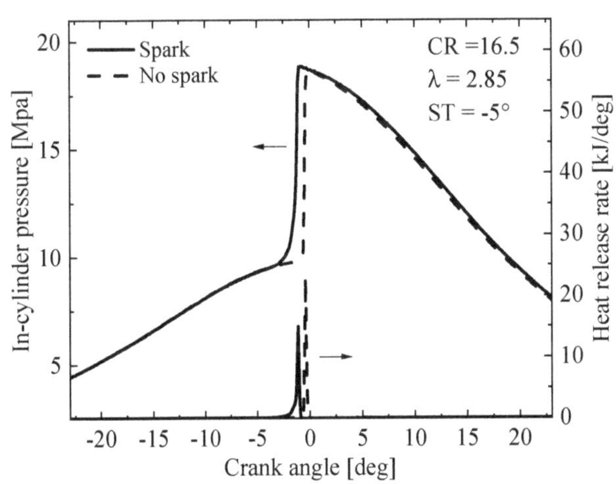

Figure 5. Effect of the spark plug on the predicted pressure and heat release rate traces for the baseline case.

Figure 6 shows the iso-surfaces of temperature at different times, where the occurrence of the auto-ignition is clearly seen. With SI, the auto-ignition was detected at 1.37 CAD bTDC (before the top dead center). As the flame propagated from the central position where the spark was located, a hot spot was observed near the cylinder walls. Without SI, the hot spot was observed at the same location at 1.47 CAD bTDC. It was thus concluded that in the baseline case with a high CR, the combustion is primarily dominated by auto-ignition rather than flame propagation.

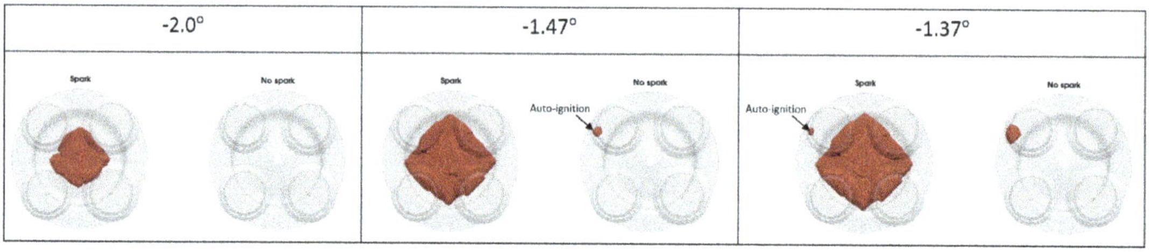

Figure 6. Comparison of temperature iso-surface (1600–3000 K) for spark vs. no-spark cases.

3.1.1. Effect of Compression Ratio

Engine CR is a critical factor directly affecting engine thermal efficiency and combustion. It is expected that the combustion duration decreases and the knocking tendency increases with an increased CR [61]. In the baseline case, a high MPRR was observed with an effective compression ratio of 16.5:1 even at very lean conditions (λ = 2.85). At high CRs, the compression temperature increased, which increased the reactivity of the fuel/oxidizer mixture. Thus, lowering the compression ratio may suppress the undesired auto-ignition/pre-ignition phenomena, making the combustion phasing easier to control with SI.

In this parametric study, at a fixed λ of 2.85, five CRs were selected ranging from 16.5 to 11.5 to investigate the effect on combustion performance and NO_x emissions. Figure 7 shows that as the CR is reduced, the combustion phasing becomes retarded, and the MPRR is mitigated. The heat transfer loss is also reduced due to the lower combustion temperature. Of these cases, a maximum indicated thermal efficiency (ITE) of 47.5% was obtained at the CR of 12.5.

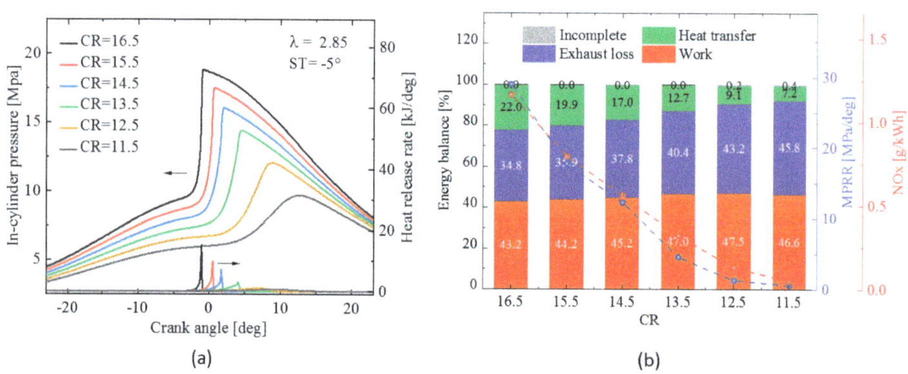

Figure 7. Comparison of: (**a**) pressure trace and heat release rate traces; and (**b**) efficiency for λ = 2.85 and ST = −5 CAD at different CRs.

Figure 8a shows the combustion duration, defined as the time interval between 5% (CA5) and 90% (CA90) of the total heat release, for different CRs at λ = 2.85. The results indicated that combustion duration is reduced almost linearly with an increased CR. As expected, the CR had a direct impact on the in-cylinder temperature, which explains the retarded combustion phasing and the lower NO_x emissions, as seen in Figure 8b. Note that the intake temperature and pressure remained constant as the CR was varied. With a CR below 12.5, the MPRR wwas kept under the threshold of 1.5 MPa/deg. For higher CRs, the rapid combustion over a short period indicated an auto-ignition event, as demonstrated by the iso-surfaces of temperature illustrated in Figure 9.

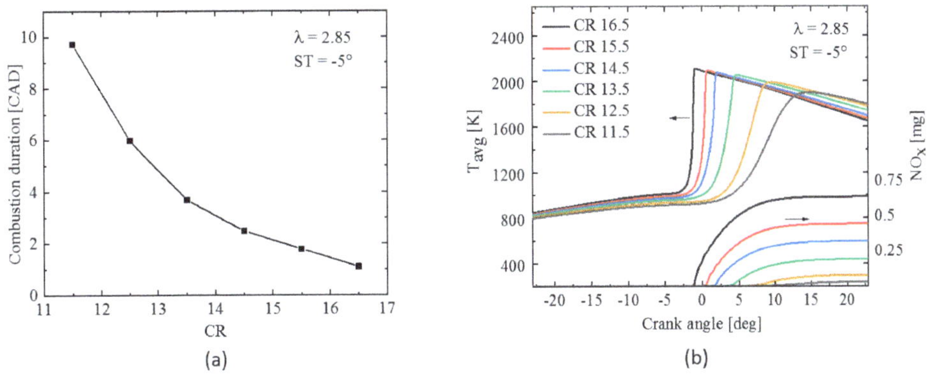

Figure 8. Comparison of: (**a**) combustion duration; and (**b**) in-cylinder average temperature for λ = 2.85 and ST = −5 CAD at different CRs.

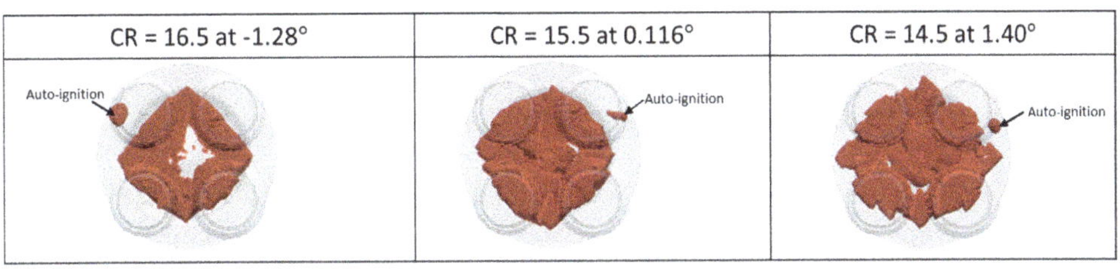

Figure 9. Comparison of iso-surface of temperature for λ = 2.85 and ST = −5 CAD at different CRs.

3.1.2. Effect of Air–Fuel Ratio (λ)

The effect of air–fuel ratio on auto-ignition was studied in this subsection. The previous case with a CR = 13.5 and λ = 2.85 was used as the baseline case. Two additional λ of 3.33 and 4 were also investigated. Note that the total injected fuel mass was kept constant by adjusting the intake pressure. For instance, the intake pressure was increased at a higher λ. It should be mentioned here that an intake pressure of 2.7 bar was used to reach a λ of 4. A further increment in λ would meet some challenges for practical limitations, such as a special boosting system, particularly at high engine loads.

The operation of the engine at leaner conditions would increase the minimum ignition energy, thus preventing pre-ignition events. In addition, it would mitigate the high reactivity of the hydrogen/air mixture resulting in lower laminar flame speed and longer auto-ignition delays. Moreover, the lean operation also leads to lowered adiabatic flame temperatures, reducing NOx emissions and heat transfer losses. An almost zero NOx level was achieved with leaner operation which was similarly reported by [62].

Figure 10 shows the pressure and HRRs traces at different λ with two CRs of 13.5 and 14.5. The results indicate that increasing λ beyond 2.85 effectively suppressed unexpected auto-ignition, as seen in Figure 10 (MPRR below 1.5 MPa/deg). However, increasing λ also resulted in a retarded combustion phasing and thus the higher incomplete combustion and exhaust losses. Of these cases, the case with a CR of 13.5 and λ= 3.33 achieved the highest ITE. The other cases suffered from misfire or auto-ignition issues. Therefore, further optimization by adjusting the spark timing should be performed to extend the high-performance operating range. The results obtained are consistent with what is reported by [63] where a CR = 10 and λ > 2.5 were used. The study showed that it was not feasible to operate the engine with a λ < 2.5 due to the high-pressure rise rate, which is generated because of the ultra-fast flame speed.

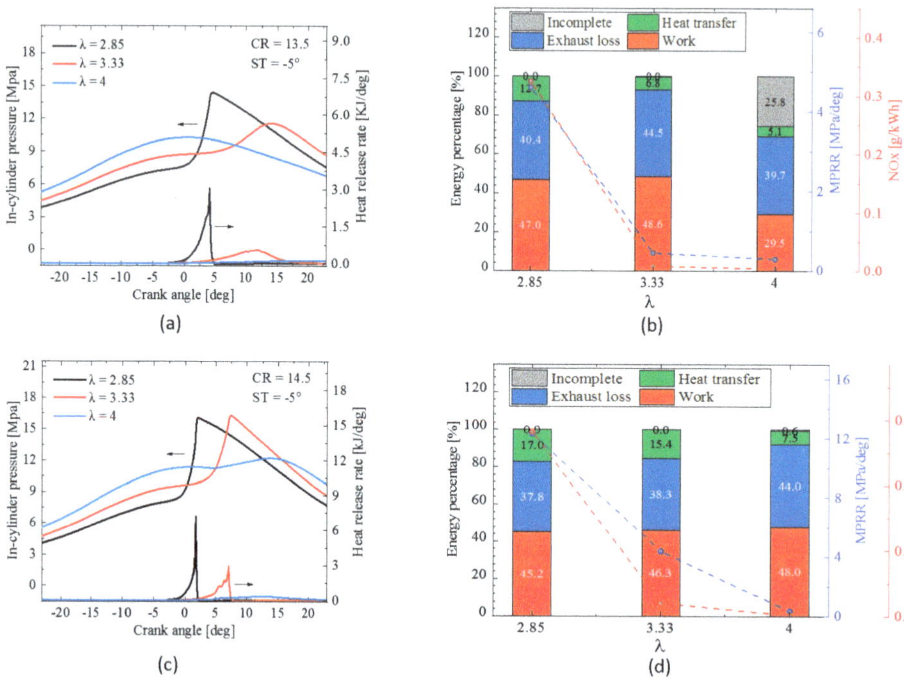

Figure 10. Comparison of: (**a**) pressure and heat release rate traces; (**b**) efficiency for CR = 13.5; (**c**) pressure trace and heat release rate; and (**d**) efficiency for CR = 14.5 using different λ at ST = −5 CAD.

3.1.3. Effect of Spark Timing

In this subsection, the effect of spark ignition timing was evaluated. The above cases with a CR of 13.5 and λ of 3.33 and 4 were used as the baseline cases. The spark timing (ST) sweep results are shown in Figures 11 and 12. By advancing the spark timing, the combustion and exhaust losses were reduced, while the heat transfer losses were increased due to the earlier combustion phasing. Within the limit of MPRR of 1.5 MPa/deg, the highest ITE of about 50% was obtained at a spark timing of 15 CAD bTDC with a CR of 13.5 and λ of 4.

In general, the advancement of the spark timing caused an increase in the MPRR due to near-TDC combustion heat release, which led to a shorter combustion duration. Comparing Figures 11 and 12 shows that it is preferable to go higher in CR as long as the mixture is lean enough to avoid auto-ignition. A higher achievable ITE was obtained at the higher CR (14.5) owing to the more isochoric combustion heat release near the TDC.

In addition, a high NO_x emission was generated when the spark time was advanced as the combustion temperature was increased. The overall level was always far below the Euro-6 limit (0.46 g/kW-h). In some cases, advancing the spark time may lead to lower NO_x emissions, but considering that NO_x emission has the unit of g/kWh, the total absolute NO_x in kg was higher when the spark time was advanced.

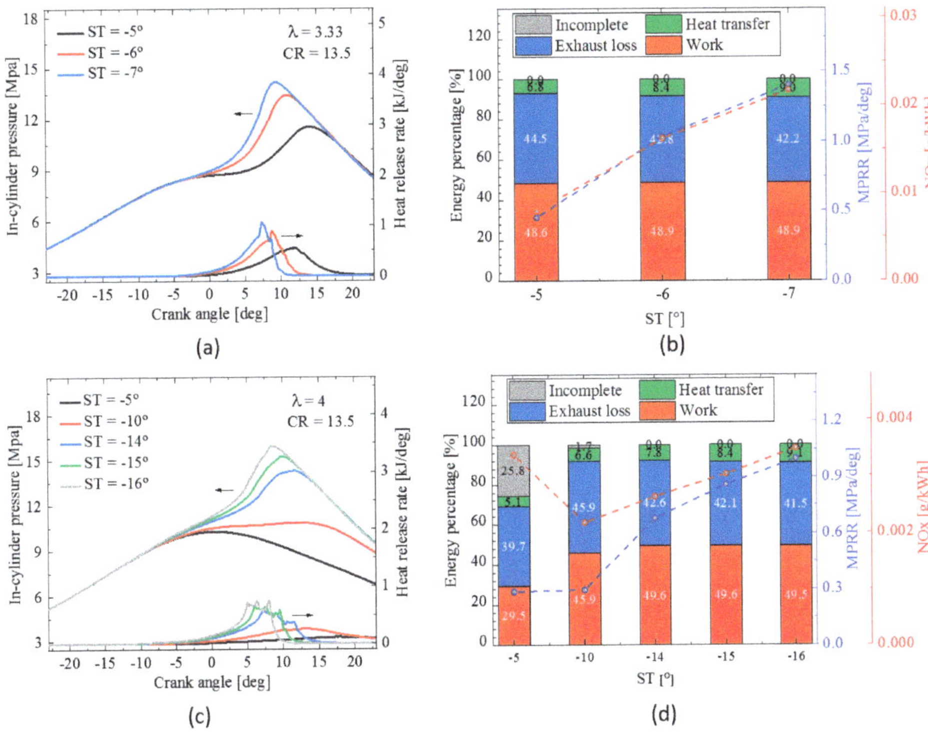

Figure 11. Comparison of: (**a**) pressure and heat release rate traces; (**b**) efficiency for and λ = 3.33; (**c**) pressure trace and heat release rate traces; and (**d**) efficiency for λ = 4 all set at CR =13.5 using different STs.

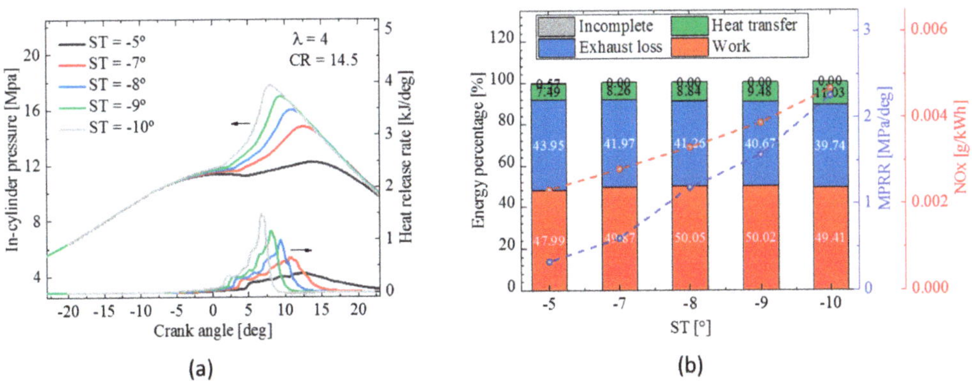

Figure 12. Comparison of: (**a**) pressure and heat release rate traces; and (**b**) efficiency for CR = 14.5 and λ = 4 using different STs.

The results obtained agree with [15] in terms of the optimized CR of 14.5 as they achieved a brake thermal efficiency of 38% for a 2 L four-cylinder engine. On the other hand, they used a λ = 1.81, while the optimized λ in this study was 4.

3.2. Pre-Chamber (PC) Combustion

The PC design was adopted based on the experiment done using methane as fuel [64]. The same engine configuration as in the SI mode was used, and the PC was mounted on the top of the cylinder head as indicated in Figure 1. Table 3 shows the PC specifications.

Table 3. PC specifications.

PC Throat Diameter (mm)	3.3
PC nozzle layer number	2
PC total nozzle number	12
PC nozzle diameter (mm)	1.5
PC nozzle angle (°)	134

3.2.1. Passive PC

First, a passive PC was used where all the fuel was supplied through the intake port, which is defined as PCFR of 0%. A baseline case with $\lambda = 4$ and CR = 14.5 was adopted as an optimum case in which the high MPRR was suppressed, and the highest efficiency was achieved in the SI mode. Note that as the PC was installed, the total chamber volume was increased, but the squish height remained the same, causing a slight drop in the compression ratio by about 0.5.

The ST was adjusted to reduce the MPRR and optimize the combustion performance. Figure 13 shows the predicted pressure trace, HRR, and efficiency analysis for using a passive PC with different spark timing. As seen in Figure 13a, when the spark time is advanced, the predicted peak pressure and heat release rate are elevated as the combustion duration is shortened, which in turn results in a higher MPRR, as seen in Figure 14. On the other hand, advancing the ST generally results in higher ITE due to the more advanced combustion phasing. However, an excessively early ST results in the over-high heat transfer loss and, thus, a declining ITE. The simulations show that with the current PC setup, it is easy to trigger an extreme MPRR within the PC, even at a near-TDC ST like 4 CA bTDC. Therefore, other solutions like optimizing the PC geometries, control strategies, and lower CRs should be used.

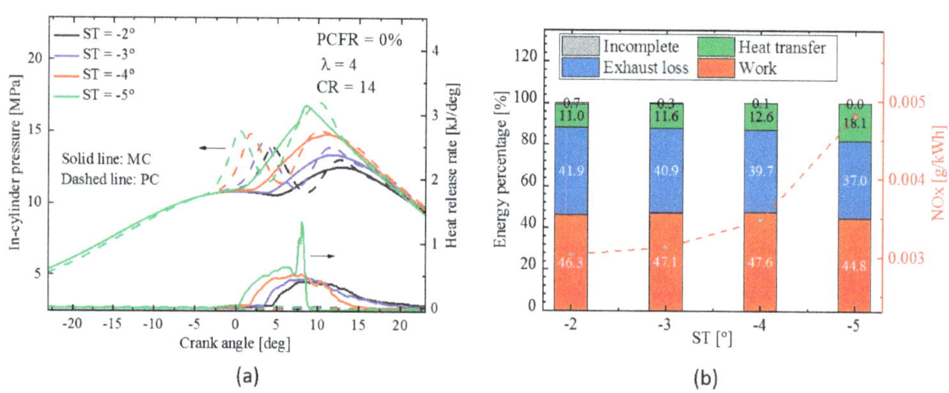

Figure 13. Predicted: (**a**) in-cylinder pressure and heat release rate traces; and (**b**) efficiency for PCFR = 0%, CR = 14, and $\lambda = 4$ using different STs.

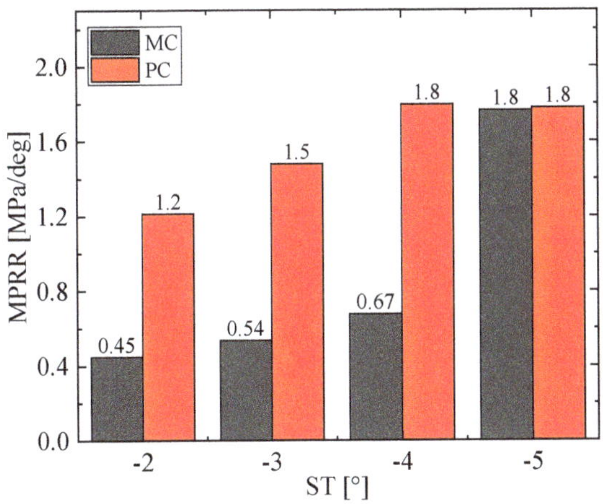

Figure 14. Predicted MPRR for PCFR = 0%, CR = 14, and λ = 4 using different STs.

As in the SI combustion mode, lowering the compression ratio from 14.5 to 13.5 can result in mitigating the high MPRR inside the PC and MC. Lowering the compression ratio also leads to lower combustion temperature and pressure, making both the MC and PC less prone to pre-ignition and knocking. As seen in Figure 15, at a lower compression ratio of 13, the ST needs to be advanced to increase efficiency. Note that as the ST is advanced, the combustion duration is shortened, resulting in faster and more abrupt combustion. This has a positive effect of lowering incomplete combustion losses. On the other hand, it has a negative effect of increasing heat transfer losses. It is also seen in Figure 15b that the reduction in the combustion losses resulted in an optimal ITE of about 47% at a ST of 5 CAD bTDC. Figure 16 shows the MPRR, and it indicates that the excessive pressure rise was significantly mitigated, as shown where the MPRR was still under the critical limit (1.5 MPa/deg).

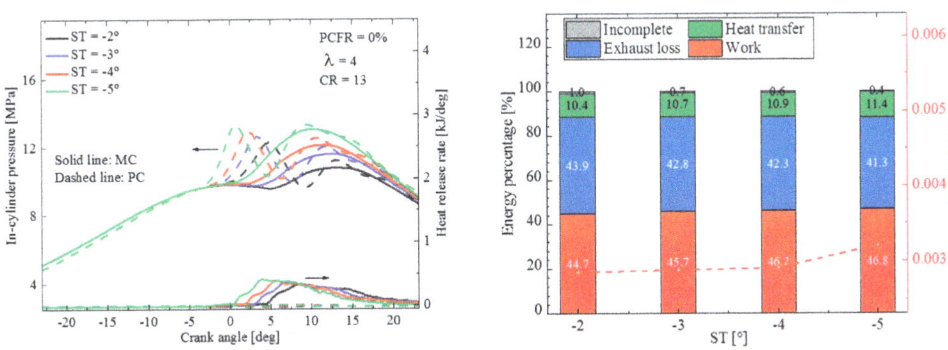

(a) (b)

Figure 15. Predicted: (**a**) in-cylinder pressure and heat release rate traces; and (**b**) efficiency for PCFR = 0%, CR = 13, and λ = 4 using different STs.

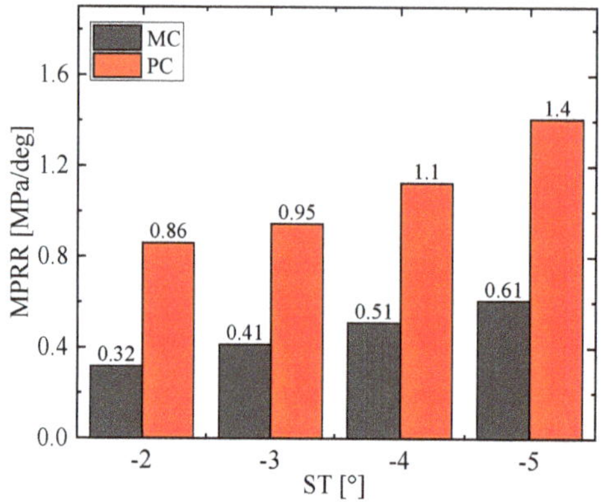

Figure 16. Predicted MPRR for PCFR = 0%, CR = 13, and λ = 4 using different STs.

3.2.2. Active PC

To assess the benefit of active PC, different values of PCFR were considered as the main parameter to control the combustion characteristics (including performance and emissions). The baseline case with a PCFR of 0% refers to a passive PC, while other PCFR values correspond to an active PC where a small percentage of the total fuel is provided through the PC while the majority is going through port injection. Figure 17 shows the predicted pressure and HRR traces with PCFR ranging from 0% up to 1.5% with an increment of 0.5%. As the PCFR is increased, so did the pressures in both MC and PC. Moreover, the combustion duration in the MC is also reduced, resulting in a more abrupt and higher heat release rate. However, the more advanced combustion phasing, and the resulting higher combustion temperature leads to a higher heat transfer loss and NOx emissions, as seen in Figure 17b. Due to the increased pressure difference between PC and MC, the faster and more intensely reacting jets are generated, as seen in Figure 18, which promotes the turbulent combustion within the MC and thus reduces the incomplete combustion loss, as indicated in Figure 17b. Nevertheless, as shown in Figure 19, the MPRR was overly high (>1.5 MPa/deg) with a non-zero PCFR.

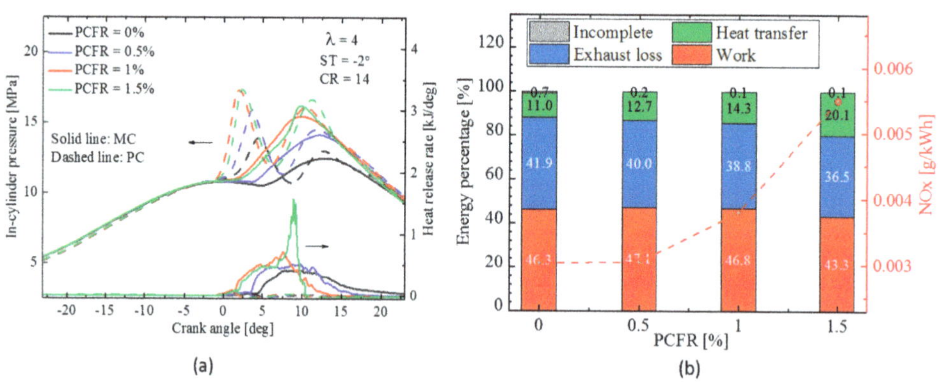

Figure 17. Predicted: (**a**) in-cylinder pressure and heat release rate traces; and (**b**) efficiency for CR = 14, and λ = 4, and ST = −2° using different PCFRs.

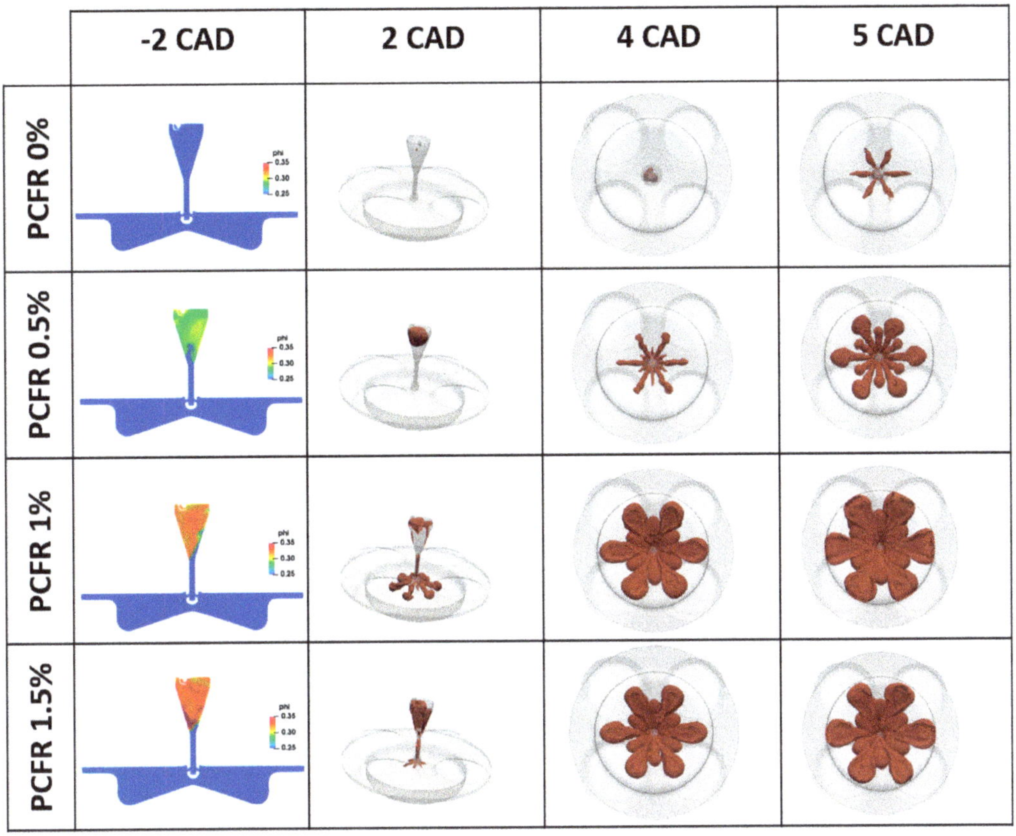

Figure 18. Comparison of φ and iso-surface (1600 K) of temperature using CR = 14, λ = 4, and ST = −2° using different PCFRs.

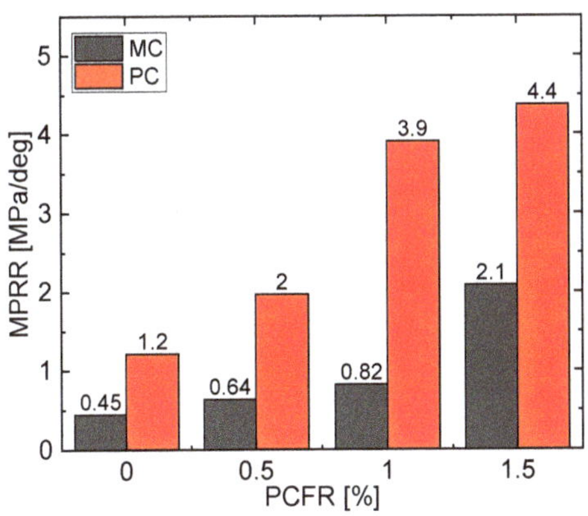

Figure 19. Predicted MPRR for ST = −2 CAD, CR = 14, and λ = 4 using different PCFRs.

3.3. Comparison between SI and PC Combustion Modes

A comparison between the SI and PC modes at similar operating conditions was summarized in this section. Figure 20 compares the pressure trace and HRR for λ = 4 in the

SI and PC modes. Compared to the SI mode, the PC mode resulted in a lower efficiency, which was mainly attributed to the significantly higher heat transfer losses. Figure 21 compares the rate of heat transfer for the optimized cases in the SI and PC. In both cases, the largest fraction of heat transfer losses was through the piston. In addition, the heat transfer losses through PC contributed to about 14.5% of the total heat transfer losses in the PC mode, while such a loss did not exist in the SI mode. As a result, the PC mode exhibited significantly higher heat transfer through the piston as the reactive turbulent jets from the PC enhanced the heat transfer rate through the piston, as seen in Figure 22. Despite the lower combustion temperature, the turbulent convection enhanced the heat transfer, yielding a higher total heat loss. The higher combustion temperature in the SI mode due to the higher CR does lead to higher NO_x emissions, although the NO_x emissions for both cases were low enough to raise practical concerns.

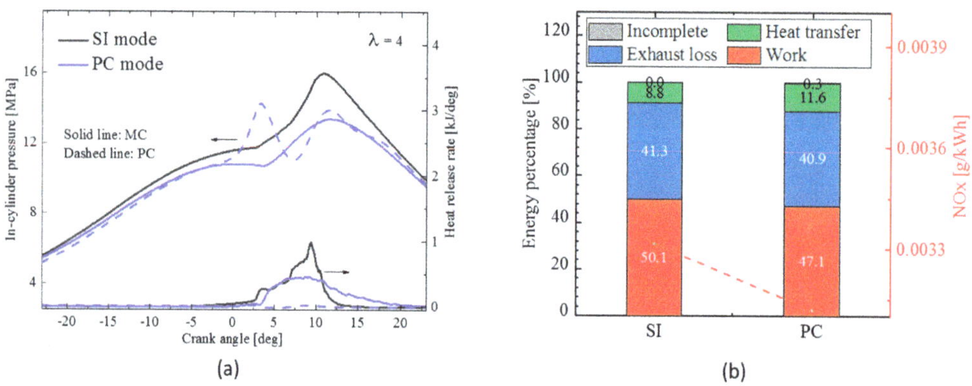

Figure 20. Predicted (**a**) in-cylinder pressure and heat release rate traces and (**b**) efficiency for SI mode (CR = 14.5 and ST = −8°) and passive PC mode (CR = 14 and ST = −3°) using λ = 4.

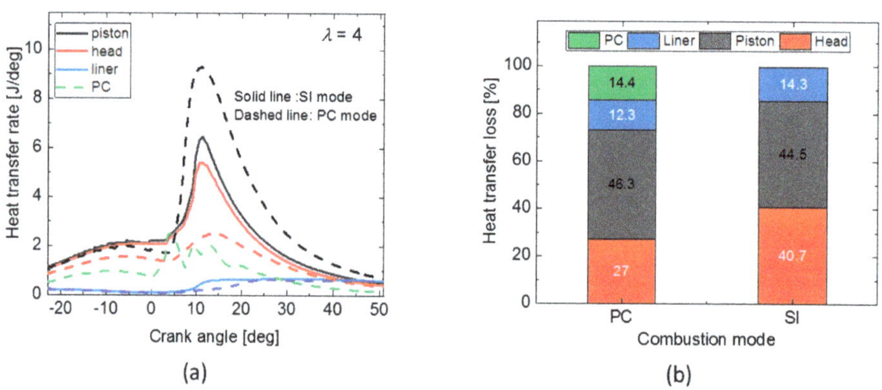

Figure 21. Comparison of the predicted: (**a**) heat transfer rates traces; and (**b**) % of heat transfer loss for SI mode (CR = 14.5 and ST = −8°) and passive PC mode (CR = 14 and ST = −3°) using λ = 4.

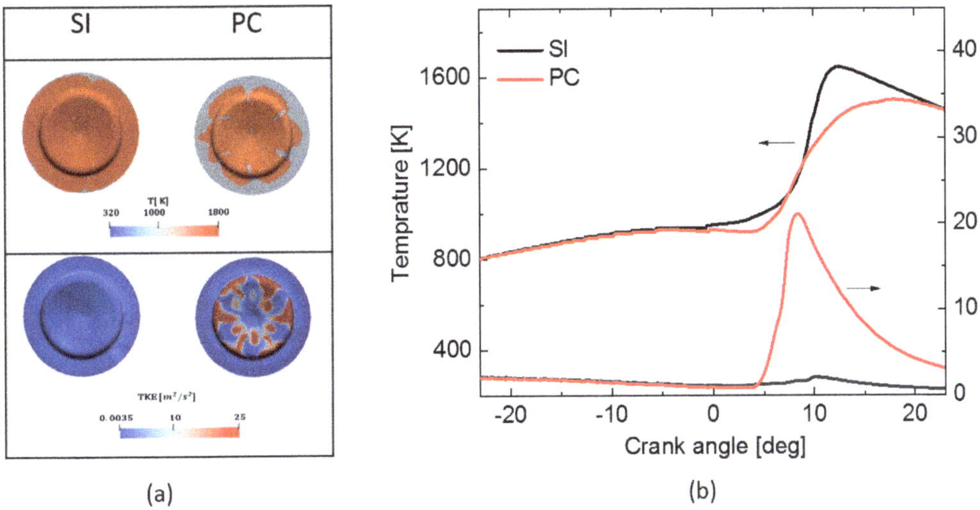

Figure 22. Comparison of the predicted piston (**a**) 3D distribution of temperature and TKE at 12 CAD, (**b**) near wall temperature and TKE for SI (CR = 14.5 and ST = −8°) and passive PC (CR = 14 and ST = −3°) modes using λ = 4.

4. Conclusions

In the current work, comprehensive CFD simulations of a heavy-duty engine converted to operate with pure H_2 under SI configuration were conducted. The effects of varying the air–fuel ratio, the compression ratio (CR), and the spark timing (ST) were investigated in order to identify the optimal range of operation at moderate load conditions. The results demonstrated that the optimal operation with SI mode would be at CRs of 13.5–14.5 and an air–fuel ratio of around 4. With such configurations, it was easier to adjust the ST to optimize the engine combustion performance by advancing the combustion phasing to improve the indicated efficiency while avoiding the abnormal auto-ignition that may cause high MPRR. The PC combustion mode was also investigated. The results indicated that the active PC would introduce higher MPRR and heat transfer losses inside both MC and PC compared to the passive PC. There was no additional means to optimize the passive PC at a high CR by adjusting ST due to the high MPRR. Thus, a reduction in CR combined with advanced ST was examined as a solution. In this case, the MPRR was within the normal range for engine operation, but the incomplete combustion losses increased.

As an overall comparison between SI and PC mode for the conditions under study, the PC mode generated a higher heat transfer loss owing to the significantly stronger jet flame-piston wall interaction and additional heat transfer through the PC assembly. Due to the high flame speed of H_2, even under ultra-lean conditions, a high ITE was achieved even at the SI mode by optimizing the ST and air–fuel mixture preparation. Moreover, combining moderate CRs with air dilution could be a potential solution to allow a larger range for engine controllability where PC can be more suitable compared to the normal SI mode. In the future, the numerical models used in this work will be validated against the experimental data under different loads.

Author Contributions: Conceptualization, H.A., M.S., M.B.H. and F.A.; Methodology, H.A.; Software, H.A.; Validation, H.A. and M.S.; Formal analysis, H.A.; Investigation, H.A., M.S., M.B.H., F.A., A.S.A. and B.M.; Resources, H.G.I.; Data curation, H.A.; Writing—original draft, H.A.; Writing—review & editing, M.S., M.B.H., X.L., M.A., F.A., A.S.A., B.M., E.C. and H.G.I.; Visualization, H.A., M.S., M.B.H. and F.A.; Supervision, M.B.H. and H.G.I.; Project administration, H.G.I.; Funding acquisition, A.S.A., B.M. and E.C. All authors have read and agreed to the published version of the manuscript."

Funding: This research was funded by Saudi Aramco Research and Development Center FUELCOM3 program under Master Research Agreement Number 6600024505/01.

Acknowledgments: The paper is based on work sponsored by Saudi Aramco Research and Development Center FUELCOM3 program under Master Research Agreement Number 6600024505/01. FUELCOM (Fuel Combustion for Advanced Engines) is a collaborative research undertaking between Saudi Aramco and KAUST intended to address the fundamental aspects of hydrocarbon fuel combustion in engines, and develop fuel/engine design tools suitable for advanced combustion modes. The computational simulations utilized the Shaheen supercomputer at KAUST Supercomputing Laboratory. The authors thank Convergent Science Inc. for providing the CONVERGE license.

Conflicts of Interest: The authors declare no conflict of interest.

Abbreviation

Abbreviation	Meaning
AMR	Adaptive mesh refinement
bTDC	Before top dead center
CA	Crank angle
CA5	Crank angle of 5% cumulative heat release
CA90	Crank angle of 90% cumulative heat release
CAD	Crank angle degree
CFD	Computational fluid dynamics
CI	Compression ignition
CO	Carbon monoxide
CR	Compression ratio
GHG	Greenhouse gas
HCCI	Homogenous charge compression ignition
HRR	Heat release rate
ICE	Internal combustion engine
IMEP	Indicated mean effective pressure
ITE	Indicated thermal efficiency
MC	Main chamber
MPRR	Maximum pressure rise rate
PC	Pre-chamber
PCFR	Pre-chamber fueling ratio
PFI	Port fuel injection
RNG	Renormalization group
RON	Research octane number
RPM	Revolution per minute
SACI	Spark assisted compression ignition
SI	Spark ignition
ST	Spark time
TDC	Top dead center
TKE	Turbulent kinetic energy
λ	Air–fuel ratio

References

1. McBain, S.; Teter, J. Tracking Transport 2021; IEA: International Energy Agency. Available online: https://policycommons.net/artifacts/1887399/tracking-transport-2021/2636759/ (accessed on 24 November 2022).
2. White, C.M.; Steeper, R.R.; Lutz, A.E. The hydrogen-fueled internal combustion engine: A technical review. *Int. J. Hydrog. Energy* **2006**, *31*, 1292–1305. [CrossRef]
3. Knop, V.; Benkenida, A.; Jay, S.; Colin, O. Modelling of combustion and nitrogen oxide formation in hydrogen-fuelled internal combustion engines within a 3D CFD code. *Int. J. Hydrog. Energy* **2008**, *33*, 5083–5097. [CrossRef]
4. Verhelst, S.; Maesschalck, P.; Rombaut, N.; Sierens, R. Increasing the power output of hydrogen internal combustion engines by means of supercharging and exhaust gas recirculation. *Int. J. Hydrog. Energy* **2009**, *34*, 4406–4412. [CrossRef]
5. Mohammadi, A.; Shioji, M.; Nakai, Y.; Ishikura, W.; Tabo, E. Performance and combustion characteristics of a direct injection SI hydrogen engine. *Int. J. Hydrog. Energy* **2007**, *32*, 296–304. [CrossRef]

6. Boretti, A.A.; Watson, H.C. Enhanced combustion by jet ignition in a turbocharged cryogenic port fuel injected hydrogen engine. *Int. J. Hydrog. Energy* **2009**, *34*, 2511–2516. [CrossRef]
7. Safari, H.; Jazayeri, S.A.; Ebrahimi, R. Potentials of NOX emission reduction methods in SI hydrogen engines: Simulation study. *Int. J. Hydrog. Energy* **2009**, *34*, 1015–1025. [CrossRef]
8. D'Errico, G.; Onorati, A.; Ellgas, S. 1D thermo-fluid dynamic modelling of an SI single-cylinder H_2 engine with cryogenic port injection. *Int. J. Hydrog. Energy* **2008**, *33*, 5829–5841. [CrossRef]
9. Kawahara, N.; Tomita, E. Visualization of auto-ignition and pressure wave during knocking in a hydrogen spark-ignition engine. *Int. J. Hydrog. Energy* **2009**, *34*, 3156–3163. [CrossRef]
10. Antunes, J.M.G.; Mikalsen, R.; Roskilly, A.P. An experimental study of a direct injection compression ignition hydrogen engine. *Int. J. Hydrog. Energy* **2009**, *34*, 6516–6522. [CrossRef]
11. Nieminen, J.; D'Souza, N.; Dincer, I. Comparative combustion characteristics of gasoline and hydrogen fuelled ICEs. *Int. J. Hydrog. Energy* **2010**, *35*, 5114–5123. [CrossRef]
12. Aleiferis, P.G.; Rosati, M.F. Controlled autoignition of hydrogen in a direct-injection optical engine. *Combust. Flame* **2012**, *159*, 2500–2515. [CrossRef]
13. Drell, I.L.; Belles, F.E. *Survey of Hydrogen Combustion Properties: Report 1383*; National Advisory Committee for Aeronautics: Cleveland, OH, USA, 1958.
14. Heywood, J.B. *Internal Combustion Engine Fundamentals*; McGraw-Hill Education: New York, NY, USA, 2018; ISBN 1260116107.
15. Tang, X.; Kabat, D.M.; Natkin, R.J.; Stockhausen, W.F.; Heffel, J. Ford P2000 hydrogen engine dynamometer development. *SAE Trans.* **2002**, *111*, 631–642.
16. Li, H.; Karim, G.A. Knock in spark ignition hydrogen engines. *Int. J. Hydrog. Energy* **2004**, *29*, 859–865. [CrossRef]
17. Al-Baghdadi, M.A.R.S. Effect of compression ratio, equivalence ratio and engine speed on the performance and emission characteristics of a spark ignition engine using hydrogen as a fuel. *Renew. Energy* **2004**, *29*, 2245–2260. [CrossRef]
18. Das, L.M. Hydrogen engines: A view of the past and a look into the future. *Int. J. Hydrog. Energy* **1990**, *15*, 425–443. [CrossRef]
19. Nagalingam, B.; Dubel, M.; Schmillen, K. *Performance of the Supercharged Spark Ignition Hydrogen Engine*; Indian Institute of Technology: Madras, India, 1983.
20. Natkin, R.J.; Tang, X.; Boyer, B.; Oltmans, B.; Denlinger, A.; Heffel, J.W. Hydrogen IC engine boosting performance and NOx study. *SAE Trans.* **2003**, *112*, 865–875.
21. Fischer, M.; Sterlepper, S.; Pischinger, S.; Seibel, J.; Kramer, U.; Lorenz, T. Operation principles for hydrogen spark ignited direct injection engines for passenger car applications. *Int. J. Hydrog. Energy* **2022**, *47*, 5638–5649. [CrossRef]
22. Oikawa, M.; Kojiya, Y.; Sato, R.; Goma, K.; Takagi, Y.; Mihara, Y. Effect of supercharging on improving thermal efficiency and modifying combustion characteristics in lean-burn direct-injection near-zero-emission hydrogen engines. *Int. J. Hydrog. Energy* **2022**, *47*, 1319–1327. [CrossRef]
23. Bao, L.; Sun, B.; Luo, Q. Experimental investigation of the achieving methods and the working characteristics of a near-zero NOx emission turbocharged direct-injection hydrogen engine. *Fuel* **2022**, *319*, 123746. [CrossRef]
24. Gürbüz, H.; Akçay, İ.H. Evaluating the effects of boosting intake-air pressure on the performance and environmental-economic indicators in a hydrogen-fueled SI engine. *Int. J. Hydrog. Energy* **2021**, *46*, 28801–28810. [CrossRef]
25. Niu, R.; Yu, X.; Du, Y.; Xie, H.; Wu, H.; Sun, Y. Effect of hydrogen proportion on lean burn performance of a dual fuel SI engine using hydrogen direct-injection. *Fuel* **2016**, *186*, 792–799. [CrossRef]
26. Molina, S.; Novella, R.; Gomez-Soriano, J.; Olcina-Girona, M. *Experimental Evaluation of Methane-Hydrogen Mixtures for Enabling Stable Lean Combustion in Spark-Ignition Engines for Automotive Applications*; No. 2022-01-0471. SAE Technical Paper; SAE: Warrendale, PA, USA, 2022.
27. Lee, K.J.; Kim, Y.R.; Byun, C.H.; Lee, J.T. Feasibility of compression ignition for hydrogen fueled engine with neat hydrogen-air pre-mixture by using high compression. *Int. J. Hydrog. Energy* **2013**, *38*, 255–264. [CrossRef]
28. Bunce, M.P.; Chinnathambi, P.; Blaxill, H.R. *The Effects of Pre-Chamber-Initiated Lean-Burn Combustion on Engine Efficiency and Emissions*; MAHLE Powertrain LLC: Plymouth, MI, USA, 2015.
29. *Control of Lean Mixtures*; SAE Technical Paper; SAE International: Warrendale, PA, USA, 1979.
30. Jamrozik, A. Lean combustion by a pre-chamber charge stratification in a stationary spark ignited engine. *J. Mech. Sci. Technol.* **2015**, *29*, 2269–2278. [CrossRef]
31. Liu, F.; Zhou, L.; Hua, J.; Liu, C.; Wei, H. Effects of pre-chamber jet ignition on knock and combustion characteristics in a spark ignition engine fueled with kerosene. *Fuel* **2021**, *293*, 120278. [CrossRef]
32. Novella, R.; Gómez-Soriano, J.; Martinez-Hernandiz, P.J.; Libert, C.; Rampanarivo, F. Improving the performance of the passive pre-chamber ignition concept for spark-ignition engines fueled with natural gas. *Fuel* **2021**, *290*, 119971. [CrossRef]
33. Benajes, J.; Novella, R.; Gómez-Soriano, J.; Martinez-Hernandiz, P.J.; Libert, C.; Dabiri, M. Evaluation of the passive pre-chamber ignition concept for future high compression ratio turbocharged spark-ignition engines. *Appl. Energy* **2019**, *248*, 576–588. [CrossRef]
34. Garcia-Oliver, J.M.; Niki, Y.; Rajasegar, R.; Novella, R.; Gomez-Soriano, J.; Martinez-Hernandiz, P.J.; Li, Z.; Musculus, M.P.B. An experimental and one-dimensional modeling analysis of turbulent gas ejection in pre-chamber engines. *Fuel* **2021**, *299*, 120861. [CrossRef]

35. Validi, A.; Schock, H.; Jaberi, F. Turbulent jet ignition assisted combustion in a rapid compression machine. *Combust. Flame* **2017**, *186*, 65–82. [CrossRef]
36. Ju, D.; Huang, Z.; Li, X.; Zhang, T.; Cai, W. Comparison of open chamber and pre-chamber ignition of methane/air mixtures in a large bore constant volume chamber: Effect of excess air ratio and pre-mixed pressure. *Appl. Energy* **2020**, *260*, 114319. [CrossRef]
37. Boretti, A.A. Modelling auto ignition of hydrogen in a jet ignition pre-chamber. *Int. J. Hydrog. Energy* **2010**, *35*, 3881–3890. [CrossRef]
38. Tanoue, K.; Kimura, T.; Jimoto, T.; Hashimoto, J.; Moriyoshi, Y. Study of prechamber combustion characteristics in a rapid compression and expansion machine. *Appl. Therm. Eng.* **2017**, *115*, 64–71. [CrossRef]
39. Biswas, S.; Tanvir, S.; Wang, H.; Qiao, L. On ignition mechanisms of premixed CH4/air and H2/air using a hot turbulent jet generated by pre-chamber combustion. *Appl. Therm. Eng.* **2016**, *106*, 925–937. [CrossRef]
40. Liu, Z.; Zhou, L.; Liu, B.; Zhao, W.; Wei, H. Effects of equivalence ratio and pilot fuel mass on ignition/extinction and pressure oscillation in a methane/diesel engine with pre-chamber. *Appl. Therm. Eng.* **2019**, *158*, 113777. [CrossRef]
41. da Costa, R.B.R.; Teixeira, A.F.; Rodrigues Filho, F.A.; Pujatti, F.J.P.; Coronado, C.J.R.; Hernández, J.J.; Lora, E.E.S. Development of a homogeneous charge pre-chamber torch ignition system for an SI engine fuelled with hydrous ethanol. *Appl. Therm. Eng.* **2019**, *152*, 261–274. [CrossRef]
42. Alvarez, C.E.C.; Couto, G.E.; Roso, V.R.; Thiriet, A.B.; Valle, R.M. A review of prechamber ignition systems as lean combustion technology for SI engines. *Appl. Therm. Eng.* **2018**, *128*, 107–120. [CrossRef]
43. Xu, G.; Kotzagianni, M.; Kyrtatos, P.; Wright, Y.M.; Boulouchos, K. Experimental and numerical investigations of the unscavenged prechamber combustion in a rapid compression and expansion machine under engine-like conditions. *Combust. Flame* **2019**, *204*, 68–84. [CrossRef]
44. Yousefi, A.; Birouk, M. Investigation of natural gas energy fraction and injection timing on the performance and emissions of a dual-fuel engine with pre-combustion chamber under low engine load. *Appl. Energy* **2017**, *189*, 492–505. [CrossRef]
45. Richards, K.J.; Senecal, P.K.; Pomraning, E. *CONVERGE 3.0*; Convergent Science: Madison, WI, USA, 2022.
46. Han, Z.; Reitz, R.D. Turbulence modeling of internal combustion engines using RNG κ-ε models. *Combust. Sci. Technol.* **1995**, *106*, 267–295. [CrossRef]
47. Amsden, A.A.; Findley, M. *KIVA-3V: A Block-Structured KIVA Program for Engines with Vertical or Canted Valves*; Lawrence Livermore National Lab.(LLNL): Livermore, CA, USA, 1997.
48. Nasrifar, K. Comparative study of eleven equations of state in predicting the thermodynamic properties of hydrogen. *Int. J. Hydrog. Energy* **2010**, *35*, 3802–3811. [CrossRef]
49. Park, J.-W.; Pei, Y.; Zhang, Y.; Zhang, A.; Som, S. Optimizing Hydrogen Kinetics for Zero-Carbon Emission Transport Technologies. In Proceedings of the International Petroleum Technology Conference, Riyadh, Saudi Arabia, 21–23 February 2022; OnePetro: Richardson, TX, USA, 2022.
50. Burke, M.P.; Chaos, M.; Ju, Y.; Dryer, F.L.; Klippenstein, S.J. Comprehensive H_2/O_2 kinetic model for high-pressure combustion. *Int. J. Chem. Kinet.* **2012**, *44*, 444–474. [CrossRef]
51. Bradley, D.; Lawes, M.; Liu, K.; Verhelst, S.; Woolley, R. Laminar burning velocities of lean hydrogen–air mixtures at pressures up to 1.0 MPa. *Combust. Flame* **2007**, *149*, 162–172. [CrossRef]
52. Silva, M.R.; Ben Houidi, M.; Hlaing, P.; Sanal, S.; Cenker, E.; AlRamadan, A.; Chang, J.; Turner, J.; Im, H. *The Effects of Piston Shape in a Narrow-Throat Pre-Chamber Engine*; SAE Technical Paper; SAE International: Warrendale, PA, USA, 2022.
53. Hlaing, P.; Silva, M.; Marquez, M.E.; Cenker, E.; Ben Houidi, M.; Im, H.G.; Turner, J.W.G.; Johansson, B. Estimates of the air-fuel ratio at the time of ignition in a pre-chamber using a narrow throat geometry. *Int. J. Engine Res.* **2021**. [CrossRef]
54. Silva, M.; Sanal, S.; Hlaing, P.; Cenker, E.; Johansson, B.; Im, H.G. *A Computational Investigation of Fuel Enrichment in the Pre-Chamber on the Ignition of the Main Chamber Charge*; In SAE Technical Paper Series; SAE International: Warrendale, PA, USA, 2021. [CrossRef]
55. Silva, M.; Sanal, S.; Hlaing, P.; Cenker, E.; Johansson, B.; Im, H.G. *Effects of Geometry on Passive Pre-Chamber Combustion Characteristics*; SAE Technical Paper; SAE International: Warrendale, PA, USA, 2020.
56. Sanal, S.; Silva, M.; Hlaing, P.; Cenker, E.; Johansson, B.; Im, H.G. *A Numerical Study on the Ignition of Lean CH 4/Air Mixture by a Pre-Chamber-Initiated Turbulent Jet*; SAE Technical Paper; SAE International: Warrendale, PA, USA, 2020.
57. Silva, M.; Liu, X.; Hlaing, P.; Sanal, S.; Cenker, E.; Chang, J.; Johansson, B.; Im, H.G. Computational assessment of effects of throat diameter on combustion and turbulence characteristics in a pre-chamber engine. *Appl. Therm. Eng.* **2022**, *212*, 118595. [CrossRef]
58. Aljabri, H.H.; Babayev, R.; Liu, X.; Badra, J.; Johansson, B.; Im, H.G. *Validation of Computational Models for Isobaric Combustion Engines*; SAE Technical Paper Series; SAE International: Warrendale, PA, USA, 2020. [CrossRef]
59. Aljabri, H.; Liu, X.; Allehaibi, M.; AlRamadan, A.S.; Badra, J.; Ben Houidi, M.; Johansson, B.; Im, H.G. *Computational Investigation of the Effects of Injection Strategy and Rail Pressure on Isobaric Combustion in an Optical Compression Ignition Engine*; SAE Technical Paper; SAE International: Warrendale, PA, USA, 2021.
60. Meininger, R.D.; Kweon, C.-B.M.; Szedlmayer, M.T.; Dang, K.Q.; Jackson, N.B.; Lindsey, C.A.; Gibson, J.A.; Armstrong, R.H. Knock criteria for aviation diesel engines. *Int. J. Engine Res.* **2017**, *18*, 752–762. [CrossRef]
61. Sahoo, S.; Srivastava, D.K. Effect of compression ratio on engine knock, performance, combustion and emission characteristics of a bi-fuel CNG engine. *Energy* **2021**, *233*, 121144. [CrossRef]
62. Sementa, P.; de Vargas Antolini, J.B.; Tornatore, C.; Catapano, F.; Vaglieco, B.M.; Sánchez, J.J.L. Exploring the potentials of lean-burn hydrogen SI engine compared to methane operation. *Int. J. Hydrog. Energy* **2022**, *47*, 25044–25056. [CrossRef]

63. Ge, H.; Lee, B.; Parameswaran, S.; Zhao, P. Combustion, knock and emissions of a port-fuled spark-ignition hydrogen engine: A CFD study. In Proceedings of the 12th U. S. National Combustion Meeting, Virtual, 14 May 2021.
64. Hlaing, P.; Echeverri Marquez, M.; Singh, E.; Almatrafi, F.; Cenker, E.; Ben Houidi, M.; Johansson, B. *Effect of Pre-Chamber Enrichment on Lean Burn Pre-Chamber Spark Ignition Combustion Concept with a Narrow-Throat Geometry*; SAE Technical Papers; SAE International: Warrendale, PA, USA, 2020; pp. 1–21. [CrossRef]

energies

MDPI

Article

Numerical Study of the Influence of the Thermal Gas Expansion on the Boundary Layer Flame Flashback in Channels with Different Wall Thermal Conditions

Kai Huang [1,2], Damir M. Valiev [1,2,3,*], Hongtao Zhong [4] and Wenhu Han [5]

1 Center for Combustion Energy, Tsinghua University, Beijing 100084, China
2 Key Laboratory for Thermal Science and Power Engineering of Ministry of Education, Department of Energy and Power Engineering, Tsinghua University, Beijing 100084, China
3 Department of Applied Physics and Electronics, Umeå University, 90187 Umeå, Sweden
4 Department of Mechanical Engineering, Stanford University, Stanford, CA 94305, USA
5 State Key Laboratory of Explosion Science and Technology, Beijing Institute of Technology, Beijing 100081, China
* Correspondence: dvaliev@mail.tsinghua.edu.cn

Abstract: In recent years, boundary layer flame flashback (BLF) has re-emerged as a technological and operational issue due to the more widespread use of alternative fuels as a part of a global effort to promote carbon neutrality. While much understanding has been achieved in experiments and simulations of BLF in the past decades, the theoretical modeling of BLF still largely relies on the progress made as early as the 1940s, when the critical gradient model (CGM) for the laminar flame flashback was proposed by Lewis and von Elbe. The CGM does not account for the modification of the upstream flow by the flame, which has been recently shown to play a role in BLF. The aim of the present work is to gain additional insight into the effects of thermal gas expansion and confinement on the flame-flow interaction in laminar BLF. Two-dimensional simulations of the confined laminar BLF in a channel are performed in this work. The parametric study focuses on the channel width, the thermal gas expansion coefficient, and the heat losses to the wall. This study evaluates the influence of these factors on the critical condition for the flame flashback. By varying the channel width, it is demonstrated that at the critical condition, the incoming flow in narrow channels is modified globally by the thermal gas expansion, while in wider channels, the flow modification by the flame tends to be more local. In narrow channels, a non-monotonic dependence of the critical-condition centerline velocity on the channel width has been identified. The variation of the heat loss to the wall confirms that the wall's thermal conditions can significantly alter the flashback limit, with the flashback propensity being larger when the thermal resistance of the wall is high. To assess the general applicability of the CGM, the flame consumption speed and the flow velocity near the wall are quantified. The results confirm that the assumption of flame having no influence on the upstream flow, employed in the CGM, is not fulfilled under confinement for a realistic thermal gas expansion. This results in a general disagreement between the simulations and the CGM, which implies that the thermal expansion effects should be accounted for when considering the confined boundary layer flashback limits. It is shown that the critical velocity gradient increases with the gas expansion coefficient for the given channel width and wall thermal condition.

Keywords: boundary layer flashback; laminar flame; thermal gas expansion; numerical simulation

1. Introduction

A wider global utilization of alternative fuels, in particular hydrogen-enriched fuels, has the potential to minimize carbon emissions and accelerate the transition to a carbon-neutral society. At the same time, their introduction may be met with operational challenges, such as flashback, that can lead to combustion device failure [1–4]. For that reason, the phenomenon of flame flashback has gained renewed interest from the combustion community in recent years.

Flame flashback mechanisms can be generally divided into (a) the flashback in the core flow, (b) the flashback due to pressure-heat release coupling, (c) the combustion-induced vortex breakdown (CIVB), and (d) the flashback in the wall boundary layer [1,5]. The flame flashback in the boundary layer can be further categorized into the unconfined flashback of the flame that is initially stabilized at the exterior of the channel, and the confined flame flashback in the interior of the channel [6]. The present study focuses on the latter.

One of the early systematic studies of the critical condition for the onset of the boundary layer flame flashback was done by Lewis and von Elbe [6]. In their work, a critical gradient model (CGM) was developed that characterizes the relationship between the burning velocity and the incoming flow velocity under the critical condition for a flame flashback. Within the CGM framework, it is assumed that the flame will be blown out if the incoming velocity exceeds the flame-burning velocity at some specific distance from the wall. Additional details on the CGM are provided in Section 2.2.

Subsequently, boundary layer flashback was extensively studied experimentally over the decades [1,7–9]. As the computational methods and resources evolved, numerical studies of flame flashback became feasible. In Ref. [10], a 2D cylindrical geometry and a one-step chemical reaction were employed to demonstrate that the critical velocity gradient depends non-monotonically on the tube radius. The range of the tube radii considered was limited due to the computational limitations at the time. In a subsequent study [11], experimental and numerical studies of the unconfined flame flashback were performed for a wide range of slit and pyrex tube burners. In Ref. [11], a modified method of calculating the critical velocity gradient was suggested to address the non-linearity of the velocity profile near the wall in narrow tubes. More recently, Kurdyumov et al. [12–15] conducted a series of studies on the flame flashback for different Lewis numbers and thermal wall conditions. In particular, in Ref. [14], a parametric study for the unburnt-to-burnt density ratio (thermal gas expansion) θ was performed, which was taken in the range of 5–8. It was concluded that for the range of θ considered, the critical condition for the flame flashback weakly depends on θ.

The present study addresses several outstanding questions. First, it is aimed at further clarifying the dependence of the critical condition for the flame flashback on the channel width. In Refs. [6,7] it was concluded that the critical velocity gradient is mostly independent of the tube diameter (except for a very narrow channel close to the quenching limit). At the same time, in Refs. [10,11], a fairly strong dependence on the tube diameter was identified. Moreover, it has been shown experimentally [16] that the confinement leads to a shift of the flashback limits towards higher critical velocity gradients. In the present study, we further address this discrepancy, using the 2D planar geometry.

Second, we further investigate the dependence of the critical velocity gradient on the thermal gas expansion for a large yet realistic range of the unburnt-to-burnt density ratios θ, beyond that considered in Ref. [14]. While for most laboratory flames, the unburnt-to-burnt density ratio does not typically exceed 10, it can reach a value of about 14 for the ethylene/oxygen mixtures [17] and attain similar values during cryogenic combustion [18]. In light of these findings, consideration of an extended range of the unburnt-to-burnt density ratios is of fundamental importance.

Third, it has been demonstrated experimentally [19,20] and numerically [21] that the approaching flow is modified significantly during the boundary layer flashback. Thus, the intrinsic assumption of the flame having a negligible effect on the upstream flow field, employed in CGM, is compromised. In the present study, an attempt to comprehensively assess the applicability of the CGM for evaluating the laminar BLF at realistically large thermal expansion coefficients is done. A combined effect of the wall heat loss and the thermal gas expansion on the critical condition for the flame flashback is analyzed, with a focus on the critical velocity gradient and the flow structure around the flame tip.

The present paper is organized as follows: in Section 2, we present the numerical approach and further describe the background of the present work. In Section 3, we present and discuss the results of the numerical simulations. The paper is concluded with a short summary. Nomenclature is provided at the end of the paper.

2. Numerical Approach and Theoretical Background

2.1. Basic Equations and Numerical Setup

In this section, a description of the numerical simulations and setup is provided. The flame dynamics are simulated using a 2-D Cartesian Navier–Stokes solver [22] that is based on the cell-centered finite-volume scheme. The solver was initially developed for applications in aerodynamics [23] and later was extensively utilized in various studies of of laminar flame dynamics [24–26]. The equations in tensor form read

$$\frac{\partial \rho}{\partial t} + \frac{\partial}{\partial x_i}(\rho u_i) = 0, \quad (1)$$

$$\frac{\partial}{\partial t}(\rho u_i) + \frac{\partial}{\partial x_j}(\rho u_i u_j + \delta_{ij} P - \eta_{ij}) = 0, \quad (2)$$

$$\frac{\partial}{\partial t}\left(\rho\varepsilon + \frac{1}{2}\rho u_i u_i\right) + \frac{\partial}{\partial x_j}\left(\rho u_j h + \frac{1}{2}\rho u_i u_i u_j + q_j - u_i \eta_{ij}\right) = 0, \quad (3)$$

$$\frac{\partial}{\partial t}(\rho Y) + \frac{\partial}{\partial x_i}\left(\rho u_i Y - \frac{\mu}{Sc}\frac{\partial Y}{\partial x_i}\right) = -\frac{\rho Y}{\tau_R}\exp(-E_a/R_p T), \quad (4)$$

where Y is the mass fraction of the fuel mixture, $\varepsilon = QY + C_V T$ is the internal energy, $h = QY + C_p T$ is the enthalpy, Q is the energy released in the reaction, and C_V and C_p are the heat capacities at the constant volume and pressure, respectively. The stress tensor η_{ij} and the energy diffusion vector q_j are given as

$$\eta_{ij} = \mu\left(\frac{\partial u_i}{\partial x_j} + \frac{\partial u_j}{\partial x_i} - \frac{2}{3}\frac{\partial u_k}{\partial x_k}\delta_{ij}\right), \quad (5)$$

$$q_j = -\mu\left(\frac{C_p}{Pr}\frac{\partial T}{\partial x_j} + \frac{Q}{Sc}\frac{\partial Y}{\partial x_j}\right), \quad (6)$$

where μ is the dynamic viscosity and Pr and Sc are the Prandtl and Schmidt numbers, respectively. To avoid the thermal-diffusive instability, we take the unity Lewis number $Le \equiv Pr/Sc = 1$, with $Pr = Sc = 0.75$. The thermal and chemical parameters of the fuel mixture are chosen to reproduce the most important properties of typical laboratory flames and, thus, are not specific to a particular fuel. The dynamic viscosity is $\mu = 1.7 \times 10^{-5}$ Ns/m^2. The fuel mixture and burnt matter are assumed to be perfect gases with the equation of state $P = \rho R_p T/m$ and with constant molar mass $m = 2.9 \times 10^{-2}$ kg/mol. $C_V = 5R_p/2m$, $C_p = 7R_p/2m$, where $R_p \approx 8.31$ J/(mol · K) is the universal gas constant. The adiabatic index is $\gamma \equiv C_P/C_V = 1.4$. The initial temperature of the fuel mixture is $T_0 = 300$ K, initial pressure is $P_0 = 10^5 Pa$, and the thermal expansion ratio (coefficient) is $\theta = T_b/T_0 = \rho_0/\rho_b = 4$–$12$, with subscript "$b$" standing for the burnt gas. Thus, the energy released in the reaction is $Q = (\theta - 1)C_P T_0$. The initial flame Mach number is $Ma_0 = S_L/c_{s0} = 0.004$. The right-hand side of Equation (4) describes a single-step irreversible reaction of the first order, with the reaction rate obeying the Arrhenius law with the activation energy E_a, and the factor of the time dimension τ_R. In the present simulations, the scaled activation energy (Arrhenius number) is $\epsilon = E_a/R_pT_b = 5$. The factor τ_R was adjusted to obtain a particular value of the planar flame velocity S_L by solving the associated eigenvalue problem for a standard one-dimensional flame similar to that addressed in Ref. [27]. The flame thickness is conventionally defined [28,29] as $\delta_L \equiv \mu/(Pr\rho_0 S_L)$, where ρ_0= 1.16 kg/m^3 is the initial mixture density.

Subsonic outflow is applied at the outlet boundary. The non-reflecting inlet boundary condition [28] is used at the inflow boundary, similar to previous studies [30–32]. A parabolic velocity profile [33] $u_x(y) = U_{max}(1 - ((R-y)/R)^2)$ is implemented at the inlet, where U_{max} is the maximum velocity at the centerline, and R and $u_x(y)$ are the channel half-width and the velocity component in the x direction, respectively. The symmetry boundary condition is used at the channel centerline, and a non-slip wall boundary condition [26] is used at the bottom wall, see Figure 1. The wall thermal boundary condition is formulated as $\partial\tilde{T}/\partial(y/\delta_L) = -B(\tilde{T} - 1/\theta)$, where the temperature is scaled as $\tilde{T} = T/T_b$, and the heat transfer parameter B is a dimensionless number that characterizes the thermal resistance of the wall. The approach is similar to that used in Refs. [34–36], although the variable scaling is different. $B = 0$ corresponds to adiabatic walls, and $B = \infty$ corresponds to isothermal walls. The initial internal structure of the flame was imitated by the analytical solution similar to that of Zeldovich and Frank–Kamenetskii [37,38], see Refs. [26,39] for details on the initial solution setting. For the presentation of the results, it is convenient to use non-dimensional variables: length is scaled by δ_L, velocity is scaled by S_L, and time is scaled by δ_L/S_L.

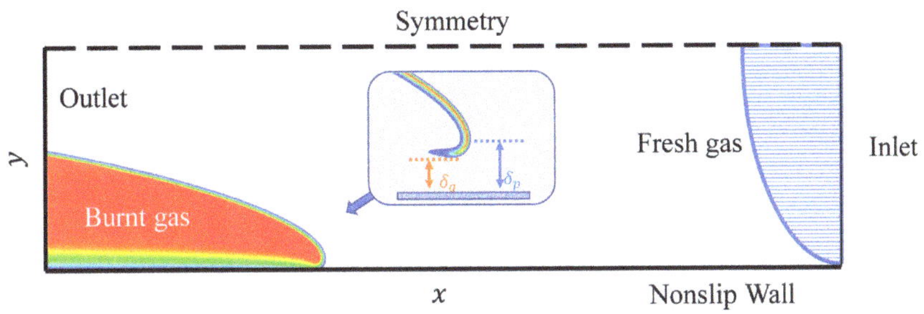

Figure 1. Sketch of the numerical setup, δ_p is the penetration distance, and δ_q is the quenching distance (see the inset).

In the present simulations, a variable-resolution mesh is employed [26,39], offering considerable savings in computing resources. In such a mesh, a subdomain with a higher resolution contains the flame front and the other parts of the calculation domain use a lower resolution, see, for example, Ref. [39]. Square cells of the size 0.1–0.2δ_L are used in the flame-containing subdomain to ensure a proper resolution of the reaction zone and isotropic flame propagation properties. The grid resolution varies gradually from one subdomain to another along the channel. The total length of the channel L is typically 100 times that of half-width ($L = 100R$) for eliminating the influence of the inlet boundary condition on the flame dynamics and ensuring that the critical condition can be attained far enough from the inlet. The flame is ignited at the bottom left corner as a sector of burnt gas with radius 5δ_L for each case, similar to Ref. [31]. The critical condition, at which the flame becomes stagnant within the channel, is attained by gradually adjusting the total flow rate at the inlet.

Resolution tests are performed for the cell sizes l/δ_L = 0.05–0.4 in the flame subdomain. In the test case, a flame propagation velocity U_{tip} is quantified for the channel of half-width 60δ_L, with $\theta = 8$, $U_{max}/S_L = 3$, and the heat transfer parameter $B = 10$. The resolution tests are summarized in Table 1, demonstrating the convergence of U_{tip}. The constant refinement ratio $r = 2$ is applied. The observed order p of the truncation rate decay [40,41] is $p = 2.58$ for $l = 0.4, 0.2, 0.1$, and $p = 2.0$ for $l = 0.2, 0.1, 0.05$, in a reasonable agreement with the overall order of approximation. Based on the resolution tests, the cell size in the flame subdomain is taken as $l = 0.2\delta_L$ for $R/\delta_L = 20$–200. For the very small channel half-widths $R/\delta_L = 5$–20, an even smaller cell size $l = 0.1\delta_L$ is used.

Table 1. Resolution tests for the flame flashback speed U_{tip} ($\theta = 8$, $B = 10$, $R = 60\delta_L$, $U_{max} = 3S_L$).

l/δ_L	U_{tip}/S_L	$\Delta U_{tip}/U_{tip}$
0.40	0.402	N/A
0.20	0.474	15.15%
0.10	0.486	2.46%
0.05	0.489	0.74%

2.2. Critical Condition for the Flame Flashback

In the context of the boundary layer flame flashback, two length scales were previously discussed, the quenching distance δ_q and the penetration distance δ_p [1,42,43]. The two distances, δ_q and δ_q, are shown in Figure 2, with a more detailed description of the illustration provided below. Quenching distance δ_q is the gap between the wall surface and the location above the wall where the burning rate becomes non-zero [1,44]. Various ways were suggested to measure it experimentally, such as relating it to the minimum distance between the two parallel plates, for which the flame propagation is possible, or estimating it from the minimum flame diameter in a free space [38,43,45].

However, δ_q is of limited use for the description of the boundary layer flashback [8,10]. To tackle the problem of the boundary layer flashback propensity, another length scale, called penetration distance δ_p, is introduced. Historically, it has been defined in different ways. According to a simpler definition, δ_p is the distance from the wall at which the flame consumption speed becomes equal to the laminar burning velocity. Another, more rigorous definition, equates δ_p to the distance from the wall to the point where the profile of the flame consumption speed $s_c(y)$ becomes tangential to the profile of the velocity of the plane-parallel flow [7], as illustrated in Figure 2. Here, the profile of the flame consumption speed $s_c(y)$ can be calculated as $s_c(y) = (\rho_0 Y_0 S_L)^{-1} \int_{-\infty}^{\infty} w(x,y)dx$, where $w(x,y)$ is the local reaction rate, see Equation (4).

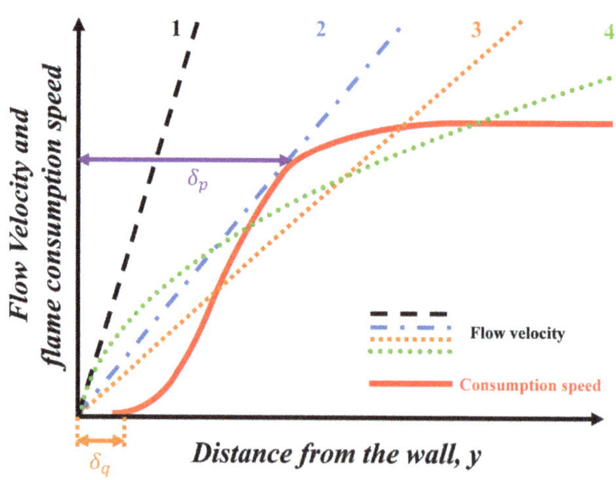

Figure 2. Illustration of the critical condition for the boundary layer flashback. Solid line depicts the profile of the flame consumption speed $s_c(y)$, lines 1 to 4 stand for various profiles of the incoming flow velocity.

The main idea behind the critical gradient model [6,7] (abbreviated as CGM in the text) is summarized as follows. In Figure 2, curve 1 corresponds to the case when the velocity of the incoming flow is greater than the flame consumption speed for all y; the flame will be thus blown downstream by the incoming flow. Curve 2 corresponds to the critical condition, where there is a balance of the flame consumption speed and the incoming flow at the distance δ_p from the wall. Curves 3 and 4 correspond to a flashback situation when the flame propagates upstream. For the very narrow channels, the velocity profile near the wall may deviate from a linear one substantially, with curve 4 illustrating this situation.

The near-wall velocity gradient g for the case of a wide channel can be calculated as

$$g = \lim_{y \to 0} \left(-\frac{du_x}{dy} \right). \tag{7}$$

Based on the assumptions of the CGM, the penetration distance can be obtained as

$$\delta_p = \frac{S_L}{g_c}, \tag{8}$$

where g_c is the near-wall velocity gradient at the critical condition, which can be calculated from Equation (7).

Early on, it has been reported by Harris et al. [43] that Equation (7) becomes inapplicable at small channel widths, as the velocity profile near the wall is not sufficiently linear in narrow channels. Despite that, Equation (7) has been widely used in experimental studies [6,7,43]. In the numerical and experimental study by Mallens et al. [11], the issue of the non-linearity of the velocity profile at the penetration distance in narrow channels has been addressed. An alternative expression for the critical velocity gradient has been suggested [11]:

$$g_c = \frac{2U_{max}}{R} \left(1 - \frac{\delta_p}{R} \right). \tag{9}$$

The velocity distribution in the fully developed laminar channel flow is [33] $u_x = U_{max}(1 - ((R-y)/R)^2)$, and the velocity gradient for any y can be expressed as $g = 2U_{max}(1 - y/R)/R$. For the larger channel widths, $\delta_p \ll R$, and, as the result, $\frac{\delta_p}{R} \to 0$, therefore the velocity gradient near the wall can be taken as $g = \frac{2U_{max}}{R}$.

2.3. *Influence of the Thermal Expansion on the Planar Flame Structure*

To gain the initial understanding of the influence of the thermal gas expansion on the planar adiabatic flame structure for $\theta = 4, 8, 12$, we plot the calculated normalized profiles of temperature T^+, fuel mass fraction Y^+, and reaction rate w^+ in Figure 3. It can be observed that the difference in the temperature distribution is rather small, as well as the difference in the mass fraction profiles, while the difference in reaction rate profiles is more evident. The reaction zone is visibly wider for the lower thermal gas expansion coefficients (see the one for $\theta = 4$) and is thinner for the higher values of θ. It is noted that the scaled activation energy $\epsilon = E_a/R_P T_b$ is kept constant. The variation of the reaction zone thickness can be attributed to the variation of the Zeldovich number $Ze = \epsilon(1 - 1/\theta)$, which defines the relative thickness of the reaction zone within the standard flame structure [38]. Here we would like to point out that δ_L, as defined in Section 2.1, is a mathematical parameter of length dimension rather than a realistic flame thickness, with the latter typically several times larger than δ_L [39,46].

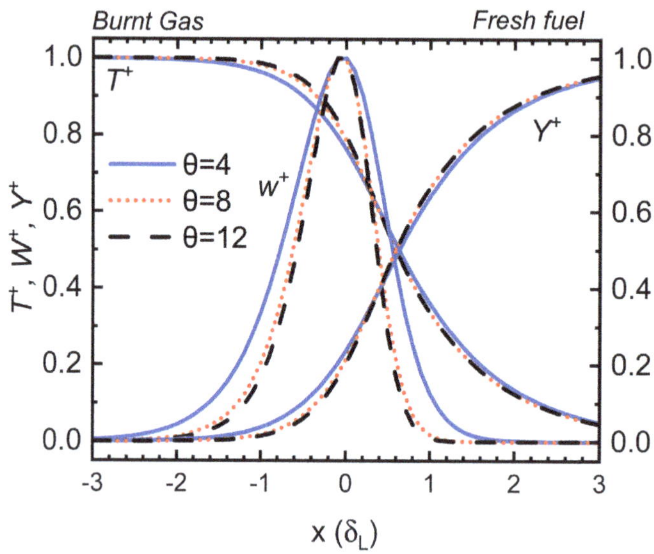

Figure 3. Influence of the thermal gas expansion on the flame structure. Shown is the variation of the normalized temperature T^+, fuel mass fraction Y^+, and reaction rate w^+ with the vertical distance y for different thermal expansion coefficients $\theta = 4, 8$, and 12, for the freely propagating planar flame.

It should be noted that in the presence of heat loss near the cold wall, it is expected that the profile of the flame consumption speed in the wall-normal direction may differ for different θ due to different spatial distributions of the reaction rate. We examine the effects of thermal gas expansion on the flame flashback limit and flame structure in more detail in Sections 3.1.3 and 3.2.

3. Results and Discussion

In this section, we examine the influence of the thermal gas expansion coefficient θ on the flame flashback for different channel widths R and wall thermal conditions defined by heat transfer coefficient B. Different values of θ can be representative of different characteristics of gaseous fuels, such as level of preheating, diluent mass fraction, etc. One of the advantages of the present numerical simulation that employs the chemical kinetics in the form of a one-step reaction is that we can keep the majority of the physical parameters unchanged and only modify one of them independently, for example, the thermal expansion coefficient θ. It is noted that for different $\theta = 4$–12, the laminar flame speed S_L is therefore constant. For the higher θ, the burnt gas temperature T_b will be higher. As we will see, the flame flashback can be affected by the wall's thermal conditions, because heat loss to the wall can significantly change the shape of the reaction zone, as well as the flow structure in the vicinity of the flame.

3.1. The Effect of Channel Width

In this subsection, we investigate the effects of the channel width on the flame flashback, in particular focusing on the critical condition for a flame flashback. We perform the numerical study for different channel half-widths ($R/\delta_L = 6$–200) and for different thermal expansion coefficients ($\theta = 4, 8, 12$).

The flame response to the change of the channel width is found to be qualitatively different for more narrow and more wide channels. There exists a lower limit of the channel width, for which the flame can exist, with the flame being quenched in narrower channels. In narrow channels with a width up to about $20\delta_L$, the flame shape at the critical condition is visibly different as compared to that in more wide channels. The response of the critical velocity gradient to the channel width is also qualitatively different for narrower and wider channels, as summarized further. Based on these qualitative differences, we distinguish the two ranges of channel widths, dubbed as "small channel width" ($R \lessapprox 20\delta_L$) and "medium channel width" ($R > 20\delta_L$). Channels with the width $R > 200\delta_L$ are presently not considered due to the computational limitations of the multi-parameter study.

3.1.1. Small Channel Width

In Figure 4, reaction zone shapes are shown for different channel widths at the critical condition for $\theta = 4$. The shape of the reaction zone is illustrated using the isolines of the normalized reaction rate ($w^+ = 0.01$). In the following, we refer to the reaction zone shape as the "flame shape." For the small channel widths, the flame shape is noticeably affected by both the flow profile and the heat loss to the wall. The lower limit for the channel half-width, below which the flame is quenched, is found to be around $R = 6\delta_L$. For $R/\delta_L = 6.5$–10, the flame shape is convex. With the further increase in the channel width, the flame near the channel's centerline gets impacted by the flow more strongly and starts to bend. When the channel half-width reaches $R = 12\delta_L$, the flame near the centerline becomes nearly flat. A similar transition from a "mushroom"-shaped flame to a tulip-shaped flame was observed by Hackert et al. [47] for stationary flames in ducts. A flame shape at the critical condition, similar to that obtained at $R = 12\delta_L$, was experimentally observed by Kim et al. [48].

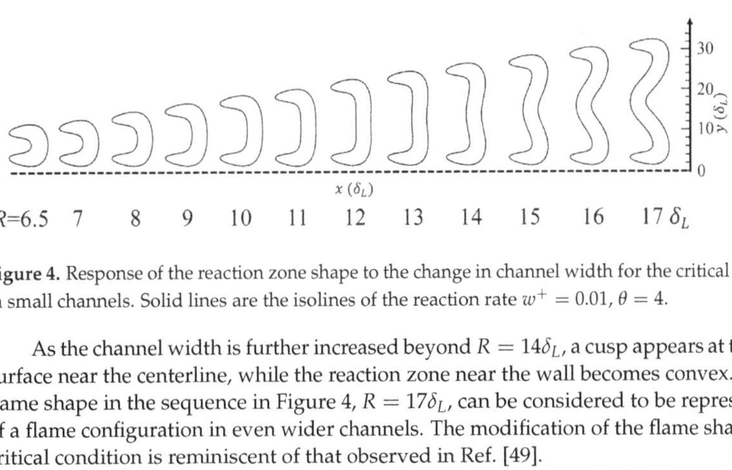

Figure 4. Response of the reaction zone shape to the change in channel width for the critical condition in small channels. Solid lines are the isolines of the reaction rate $w^+ = 0.01$, $\theta = 4$.

As the channel width is further increased beyond $R = 14\delta_L$, a cusp appears at the flame surface near the centerline, while the reaction zone near the wall becomes convex. The last flame shape in the sequence in Figure 4, $R = 17\delta_L$, can be considered to be representative of a flame configuration in even wider channels. The modification of the flame shape at the critical condition is reminiscent of that observed in Ref. [49].

Figure 5 illustrates the dependence of the critical centerline velocity $U_{max,c}$ (centerline velocity at the critical condition) on the channel half-width R. A non-monotonic dependence of U_{max} on R is observed within the range of $R/\delta_L = 6.5$–14. In this range of R, the critical centerline velocity U_{max} first increases and then decreases with the channel width. Comparing Figures 4 and 5, it can be noted that the flame shape change can be correlated with the changes in the trend for the critical centerline velocity $U_{max,c}$. A local minimum in $U_{max,c}(R)$ corresponds to the formation of the flame cusp at the centerline. It is worth noting that the higher the gas expansion coefficient θ, the more prominent is the decreasing trend in the critical centerline velocity $U_{max,c}$ in the range $R = 10$–14.

Once the flame front attains a concave shape at the channel centerline, the critical $U_{max,c}$ starts to increase again for $R > 14\delta_L$. The critical centerline velocity increases with the channel width, preventing the flame from propagating upstream near the channel centerline. This leaves the only option for the flame to propagate upstream closer to the wall, where the upcoming flow velocity is lower. Lee and Tsai [50] provide a similar explanation of the transition from "mushroom"-shaped flames to tulip-shaped flames for steadily propagating flames in tubes.

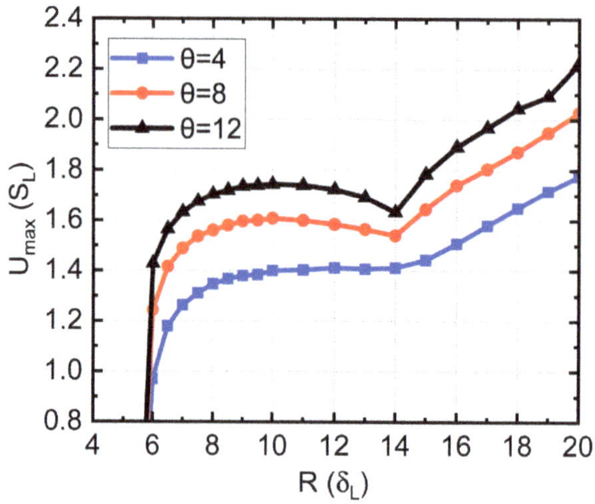

Figure 5. Centerline velocity $U_{max,c}$ versus channel half-width R at the critical condition. $\theta = 4, 8, 12$.

In Figure 6, the flame shape at the critical condition is represented by the normalized reaction rate and temperature isolines for different channel widths. Streamlines are plotted in white. For the smallest channel half-width $R = 6.5\delta_L$, the flame is convex at the centerline, and the streamlines are deflected from the centerline towards the walls. For the channel half-width $R = 12\delta_L$ the flame front at the centerline is flat; thus, the flame "tip" is still at the centerline, so qualitatively, it is similar to that of $R = 6.5\delta_L$. For $R = 20\delta_L$, the leading point shifts to the vicinity of the wall, where the flame propagation is able to balance the lower velocity. It can be seen that the streamlines are affected by the flame only within a limited distance ahead of the flame tip at the critical condition.

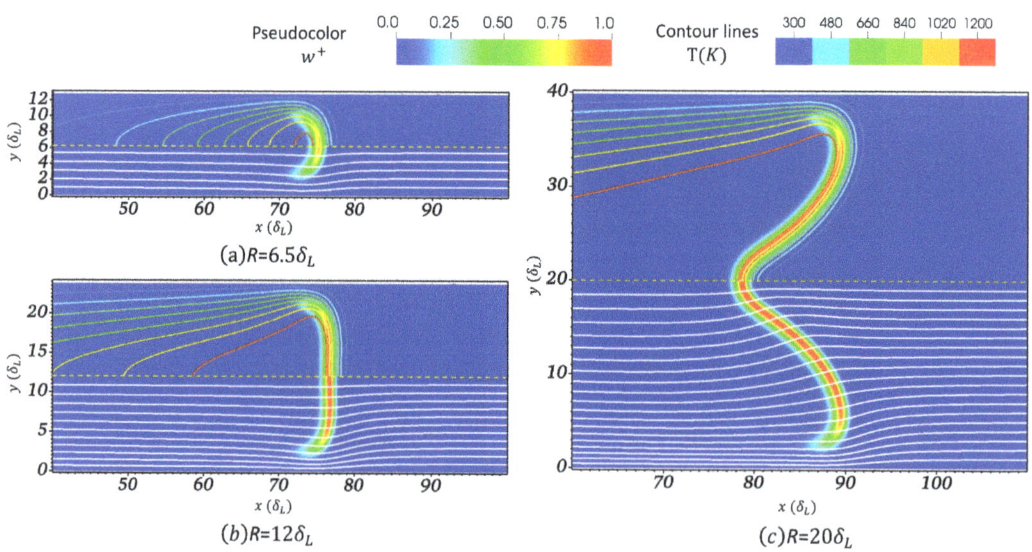

Figure 6. Comparison of the flame shapes for different channel widths at the critical condition. Flame shapes are illustrated by the reaction rate field and the temperature contour lines. Streamlines are shown in white. Channel half-width $R = 6.5, 12, 20\delta_L$, $\theta = 4$, isothermal wall.

3.1.2. Critical Velocity Gradients for Small to Medium Channel Widths

In this subsection, we quantify the critical velocity gradients for the channel half-widths within the range $R/\delta_L = 6$–200 and discuss the influence of the definition of the critical velocity gradient on its trends for the small channel widths. First, we define the critical velocity gradient, which is commonly used as a simplified local characteristic of the flow ahead of a flame tip during the flashback. In accordance with the discussion of Section 2, we consider and compare the two definitions of the critical velocity gradient, $g_1 = \frac{2U_{max}}{R}$ and $g_2 = \frac{2U_{max}}{R}(1 - \frac{\delta_p}{R})$. Figure 7 illustrates the dependence of the critical velocity gradients g_1 and g_2 on the channel half-width R. It is evident that Figure 7a,b show different trends for smaller channel widths.

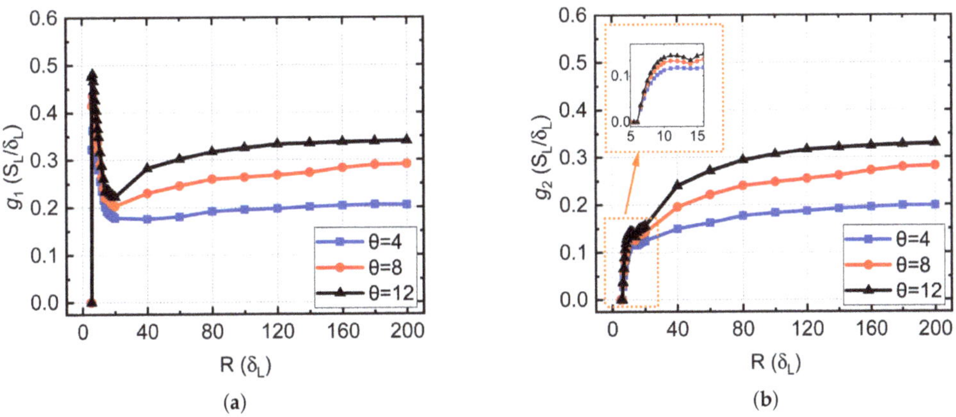

Figure 7. Dependence of $g_1 = \frac{U_{max}}{R}$ (**a**) and $g_2 = \frac{U_{max}}{R}(1 - \frac{\delta_p}{R})$ (**b**) on the channel half-width R, isothermal walls.

It is noted that Lewis and colleagues [6,7,43] used $g_1 = \frac{2U_{max}}{R}$, and U_{max} was calculated based on the mass flow rate. The term δ_p/R in g_2 is expected to be small for the experimental studies in wider channels and tubes. For the case of small channel widths, $R = 6 - 20\delta_L$, Figure 7a shows a decreasing trend for g_1, while Figure 7b shows an increasing trend for g_2, indicating an inconsistency in the two definitions for this range of channel widths. Harris et al. [43] pointed out that the CGM model may not be suitable for burners with a very small size. If the channel width is similar to the quenching distance in magnitude, the parabolic shape of the velocity profile at the critical condition would have an impact on the difference between g_1 and g_2. It is noted that in Figure 7, as the channel width increases beyond $20\delta_L$, g_1 and g_2 converge for all θ. The reason is that, as the channel width increases, δ_p/R becomes smaller, and the velocity profile in the vicinity of δ_p becomes almost linear, see Figure 2 for reference. Figure 7 also indicates that for the larger expansion coefficients θ, the critical gradient is larger, i.e. the flashback propensity is higher for the larger thermal expansion.

In Figure 8, the effect of the channel width on the flow structure at the critical condition is further illustrated. The normalized reaction rate w^+ is shown with contour lines. The negative values of the vertical component of the velocity $u_y \leq 0$ (downward direction of the flow) are shown in uniform gray. It is seen that, as the channel width increases, the gray zone ahead of the flame is shrinking for $R = 20, 40, 60\delta_L$. However, the difference between the gray zones ahead of flame for $R = 60$ and $120\delta_L$ is small. This is in agreement with the convergent trend for the critical velocity gradient for the larger channel widths in Figure 7. It can be attributed to the diminishing effect of the channel confinement on the flow around the flame tip. It is seen that the wall-normal component of the velocity just above the flame tip becomes larger as the channel width increases from $R = 20\delta_L$ to $60\delta_L$.

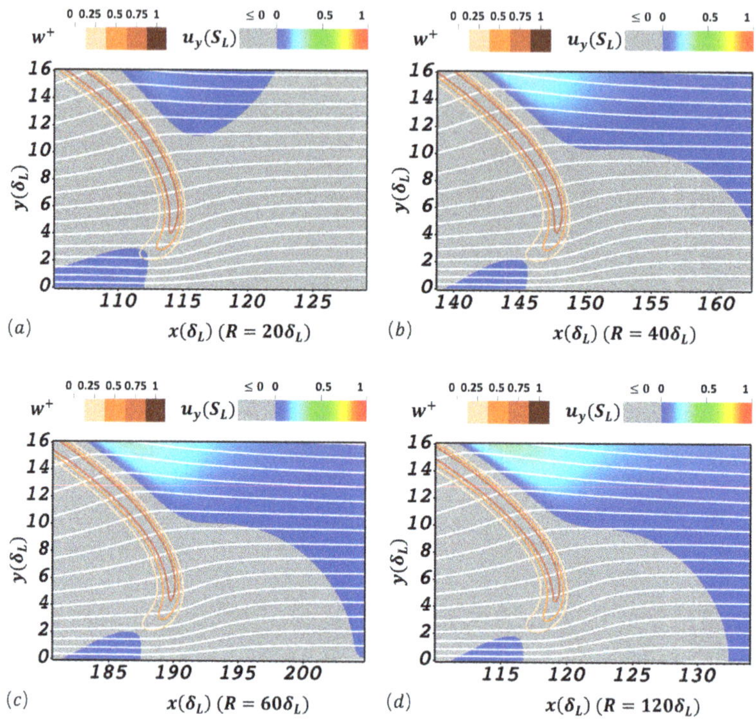

Figure 8. Flame shapes (based on the reaction rate isolines) and streamlines for different channel widths at the critical condition, $\theta = 4$. Subplots (**a–d**) are for $R = 20, 40, 60, 120\delta_L$ respectively. The normalized reaction rate is illustrated by the contour lines. Pseudo-color field shows the vertical component of the velocity u_y. $u_y \leq 0$ is highlighted in gray.

The dependence of the critical velocity gradient on the channel width was discussed by Lewis et al. [6], Lee and Tien [10], and Mallens et al. [11]. The general conclusion is similar: when the channel width is large enough, the critical velocity gradient tends to be independent of the channel width.

The comparison of the consumption speed $s_c(y)$ and the horizontal component of the upstream velocity $u_x(y)$ at the critical condition is shown in Figure 9 for different channel widths and $\theta = 4$. It is seen from Figure 9 that the CGM doesn't appear to be directly applicable, irrespective of the channel width considered, as generally $u_x(y)$ is not tangent to $s_c(y)$ at the critical condition, thus contradicting the basic assumptions of the CGM, see Section 2.2. For a smaller channel half-width $R = 20$, s_c is generally larger than u_x in the vicinity of the wall ($y < \delta_p$) at the critical condition. For the larger channel half-widths, $R = 60$ and $120\delta_L$, the agreement with CGM is somewhat better, the latter showing the best agreement with CGM. Based on Figures 8 and 9, it can be suggested that a stronger influence of thermal gas expansion in narrower channels due to a stronger confinement effect leads to a stronger modification of the flow upstream by the flame. As the result, the flashback can happen at a lower critical gradient as compared to that predicted using the CGM, as corroborated by Figure 9.

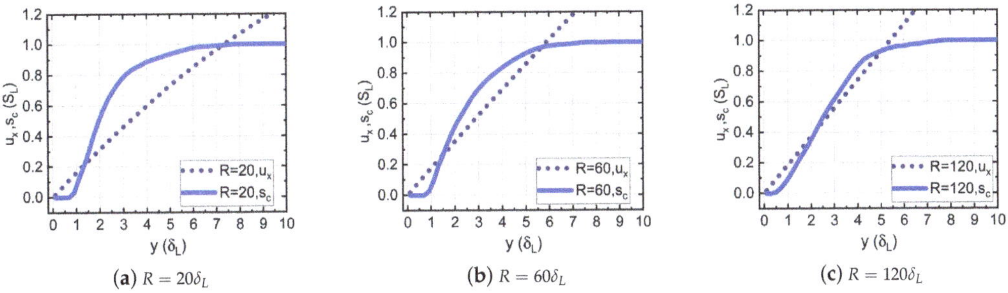

Figure 9. Comparison of the local consumption speed $s_c(y)$ and the incoming flow velocity $u_x(y)$ for $R = 20, 60, 120\delta_L$ (subplots (**a–c**), respectively) at the critical condition ($\theta = 4$).

3.1.3. Effect of Thermal Expansion in Channels of Different Width

In this subsection, the effects of the thermal expansion coefficient on the flame characteristics at the critical condition are studied for a range of channel half-widths, and for $\theta = 4, 8, 12$. The dependence of the critical velocity gradient g_2 on the thermal expansion coefficient θ in channels with isothermal walls is shown in Figure 10 for different channel half-widths R. It is seen that the larger the thermal expansion coefficient θ, the larger the critical velocity gradient for a given channel width. This indicates that the gas expansion at the flame front has an impact on the flow upstream of the flame, promoting the upstream flame propagation and affecting the boundary layer flashback limits.

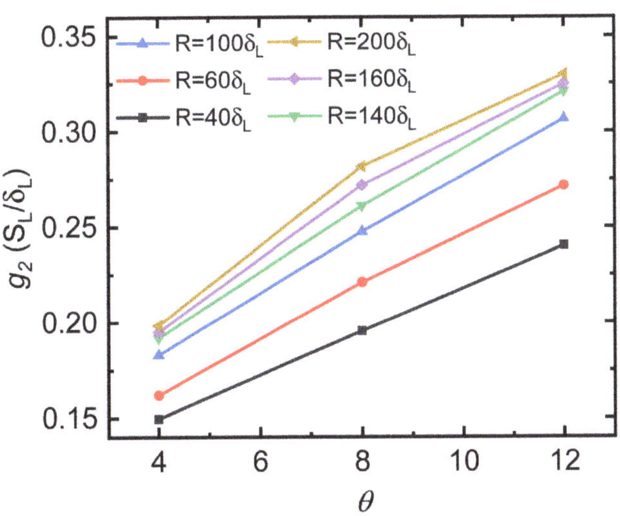

Figure 10. Critical inlet velocity gradient g_2 versus thermal expansion coefficient θ for different channel half-widths R, isothermal wall.

In Figure 11, the flame shapes near the wall are shown for different θ and two different channel widths under the critical condition. From Figure 11, we can see that δ_q at the critical condition (as illustrated in Figure 2) does not significantly change with θ in the chosen range of R. For the higher θ, the reaction zone given by w^+ looks thinner, which can be explained by the same reasoning as presented in the discussion of Figure 3: the smaller the θ, the smaller the Zeldovich number $Ze = \epsilon(1 - 1/\theta)$, which leads to a thicker reaction zone relative to the total flame thickness. As seen from Figure 11, when flame shapes for different channel widths are overlapped, the flame portion close to the wall remains almost unchanged when R is varied, but the change in θ results in a more significant elongation of the flame tip.

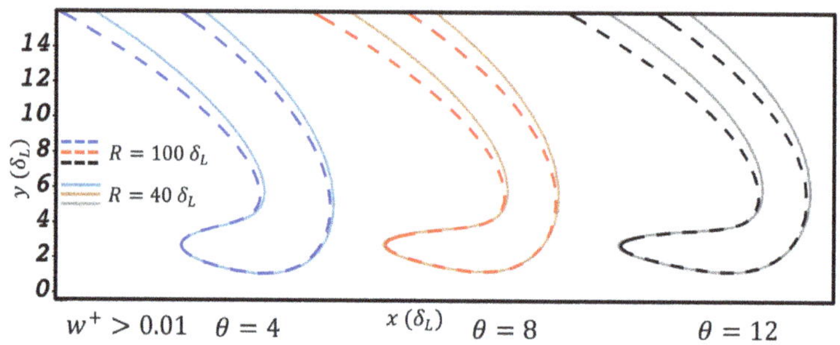

Figure 11. Flame shape under critical velocity condition for different channel half-widths $R = 40, 100\delta_L$ and gas expansion coefficients $\theta = 4, 8, 12$ (isothermal walls). Lines show the zone of normalized reaction rate $w^+ > 0.01$.

To evaluate the validity of CGM at different θ, the flame consumption speed $s_c(y)$ and the incoming flow velocity $u_x(y)$ under the critical condition are plotted in Figure 12 for two values of channel half-width, $R = 40\delta_L$ (a) and $R = 100\delta_L$(b). It is seen that for most cases, there is a disagreement with CGM. It is assumed that if the upstream flow remains unperturbed by the flame, there will be a tangency point for $s_c(y)$ and $u_x(y)$, as illustrated in Figure 2. From Figure 12a, we notice that for all θ, s_c exceeds u_x in a certain range of y at the critical condition, which contradicts the CGM. In Figure 12b, there is a visible agreement with CGM for $\theta = 8$, however, this appears to be a coincidence, as for the other two values of θ there is no agreement with CGM. From both Figure 12a,b, it can be noted that with an increasing θ, the slope of u_x will increase at the critical condition, in agreement with the results of Figure 10, while s_c profile will be modified with θ only slightly for a given R.

Overall, from Figure 12, we can conclude that a straightforward application of CGM to $s_c(y)$ and $u_x(y)$, similar to Figure 2, is not very meaningful for realistic gas expansion, as the assumption of no upstream flow modification by the flame is not fulfilled for the range of θ and R considered. The $s_c(y)$ and $u_x(y)$ profiles at the critical condition are found to be strongly dependent on the thermal expansion coefficient θ and the channel half-width R.

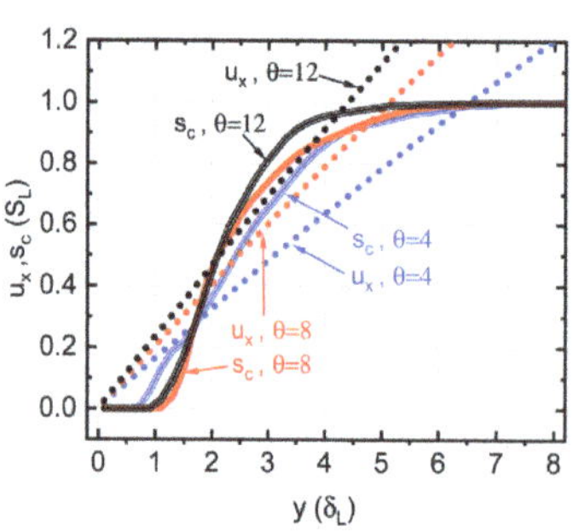

(a) $R = 40\delta_L$, s_c and u_x, $\theta = 4, 8, 12$

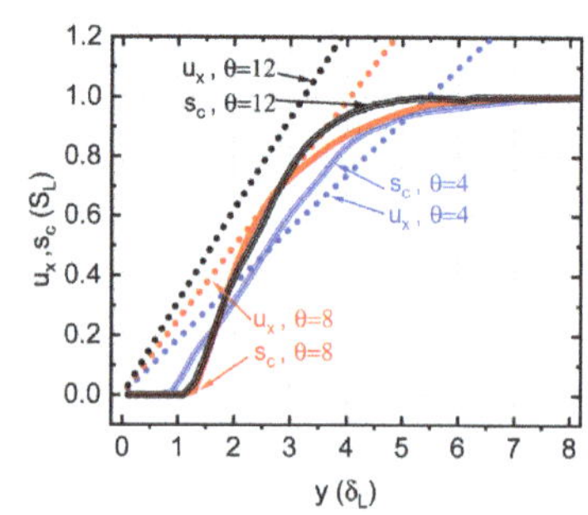

(b) $R = 100\delta_L$, s_c and u_x, $\theta = 4, 8, 12$

Figure 12. Comparison of the consumption speed $s_c(y)$ and the incoming flow velocity $u_x(y)$ under the critical condition for the channel half-widths $R = 40, 100\delta_L$ and $\theta = 4, 8, 12$ (isothermal walls).

3.2. Effect of Wall Thermal Conditions for Different Thermal Gas Expansion Coefficients

Wall thermal condition is another key factor that affects the flame flashback propensity. In the present numerical study, we use the wall thermal boundary condition described in Section 2.1, that employs the heat transfer parameter B, to regulate the heat flux at the wall. In this section, the channel half-width is set to a constant $R = 60\delta_L$. In the parametric study, three values of θ are used, $\theta = 4, 8, 12$, and the heat transfer coefficient B ranges from 0 to $+\infty$.

Figure 13 shows the critical velocity gradient g_2 for different thermal gas expansion coefficients $\theta = 4$–12 at the different heat transfer parameters $B = 0, 0.01, \cdots, 10^4$ for the same channel half-width $R = 60\delta_L$. It is seen that with the increase of the thermal expansion coefficient θ for a given heat transfer parameter B, the critical velocity gradient becomes larger. Comparing the critical velocity gradients g_2 for different wall thermal conditions in Figure 13b, we can see that for the adiabatic wall, the critical velocity gradient g_2 is higher than that for an isothermal wall and all values of B in between, in qualitative agreement with the results of Ref. [14]. The difference in g_2 for adiabatic and isothermal walls is 5–10 times.

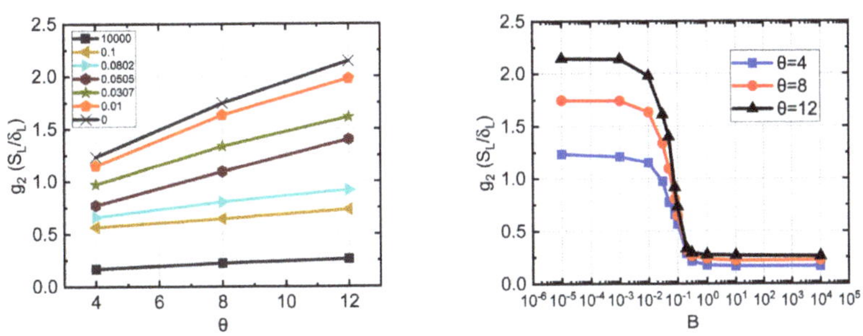

(**a**) Critical velocity gradient g_2 versus θ for given B (**b**) Critical velocity gradient g_2 versus B for given θ

Figure 13. The influence of wall thermal condition ($B = 0, 0.01, \cdots, 10^4$) and thermal gas expansion ($\theta = 4, 8, 12$) on the critical velocity gradient g_2 ($R = 60\delta_L$).

In order to better understand the flame and flow configuration at the critical condition and different B and θ, the reaction rate fields and streamlines are visualized in Figure 14. It is seen that, for a given θ, as the heat transfer parameter B decreases from the very large values (at which B is effectively near-infinity, i.e., the wall is isothermal) to near-zero, the penetration distance δ_p decreases significantly, disappearing at lower B. Thus, for a given θ, as B is decreased, the flame moves closer to the wall and becomes harder to be blown downstream.

It is also seen that as the thermal expansion coefficient θ is increased, the divergence of the streamlines increases, indicating the increased propensity of the flame to modify the flow even in a critical condition. This tendency is most noticeable for the larger B (Figure 14a–c).

Finally, we compare the consumption speed $s_c(y)$ and the incoming flow velocity profiles $u_x(y)$ at the critical condition in Figure 15 for different θ and B.

It is seen that with the increase of the heat transfer coefficient B, the height (y-coordinate) of the cross-point between $s_c(y)$ and $u_x(y)$ is increasing as well. It is noted that for $B = 0.1$, flame consumption speed s_c is below S_L wall for all θ considered. The results indicate that the concept of the critical gradient model (CGM) [6] may, in principle, become inapplicable when the thermal resistance of the walls is high. Based on Figures 14 and 15, it can be suggested that the influence of the thermal gas expansion on the flame-flow interaction is more significant at larger heat transfer coefficients B.

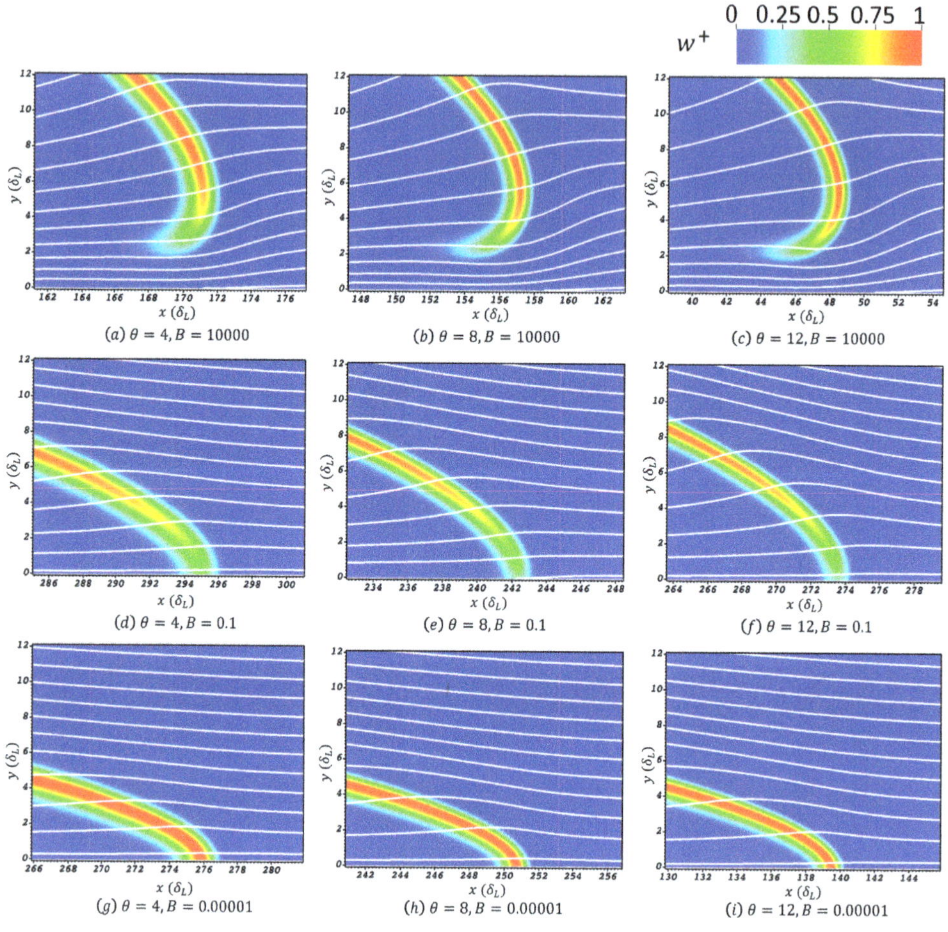

Figure 14. Flame shapes (given by the normalized reaction rate w^+ and streamlines at the critical condition for $\theta = 4, 8, 12$ and $B = 10^5, 0.1, 10^{-5}$ ($R = 60\delta_L$).

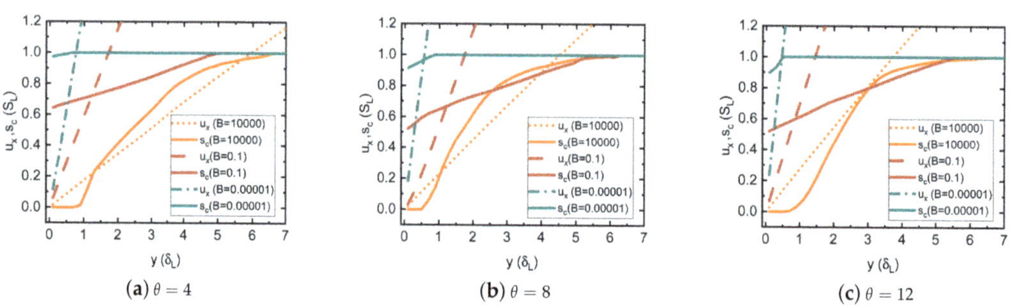

Figure 15. Comparison of the consumption speed $s_c(y)$ and the incoming flow velocity $u_x(y)$ for heat transfer coefficients $B = 10^4, 0.1, 10^{-5}$ and thermal gas expansion coefficients $\theta = 4, 8, 12$ at the critical condition ($R = 60\delta_L$).

4. Summary

This work is aimed at providing additional insights into tackling the flame flashback problem. The combined effects of the thermal gas expansion, wall thermal conditions, and channel width on the limits of the confined laminar boundary layer flashback in channels are investigated. It is shown that for the small channel widths, the flow upstream of the flame is globally modified at the critical condition. For the larger channel widths, the critical velocity gradient is nearly independent of the channel width. In order to assess the applicability of the critical gradient model for the confined boundary layer flashback, the flame consumption speed and the incoming flow velocity profile near the wall are quantified. Two-dimensional fields of the reaction rate and the streamlines are analyzed for a range of thermal expansion coefficients and wall heat transfer parameters. It is demonstrated that the thermal gas expansion affects the flame shape and the flow ahead of the flame significantly in the critical condition. Generally, this results in a lack of agreement of the simulation results with the critical gradient model (CGM). It is concluded that a straightforward application of the CGM may not be meaningful, as the upstream flow can be modified significantly by the thermal gas expansion, thus invalidating the key assumption of the CGM.

The critical velocity gradient is shown to considerably depend on the thermal expansion coefficient in both isothermal and adiabatic limits for the wall thermal condition. The effects of the channel width, the thermal gas expansion, and the wall thermal conditions, on the critical condition for the confined laminar BLF, are summarized as follows:

- In narrow channels, with the channel half-width $R \lesssim 20\delta_L$, the centerline velocity at the critical condition $U_{max,c}$ depends on the channel width non-monotonically;
- In wider channels, $R > 20\delta_L$, the critical velocity gradient g_c grows gradually with the channel width, reaching a nearly constant value for a given gas expansion coefficient;
- The critical velocity gradient increases with the gas expansion coefficient for given channel width and wall thermal condition;
- It is confirmed that the critical velocity gradient for the adiabatic wall can be an order of magnitude higher than that for the isothermal wall.

From a practical standpoint, it can be concluded that the flashback propensity is larger when the thermal resistance of the wall is high, and when the unburnt-to-burnt density ratio is large.

Author Contributions: Conceptualization, D.M.V.; methodology, K.H., D.M.V. and H.Z.; software, H.Z., K.H. and D.M.V.; validation, K.H.; resources, D.M.V.; writing—original draft preparation, K.H. and D.M.V.; writing—review and editing, H.Z. and W.H.; investigation, K.H.; visualization, K.H.; supervision, D.M.V.; funding acquisition, D.M.V. and W.H. All authors have read and agreed to the published version of the manuscript.

Funding: This research was funded by the National Natural Science Foundation of China (NSFC) grant numbers 52176118 and 51750110503, and by the State Key Laboratory of Explosion Science and Technology (Beijing Institute of Technology) through the opening of Project No. KFJJ21–10M. W.H. is supported by Shaanxi Institute of Applied Physical Chemistry grant number 6142602210408. Computations and data handling were enabled by resources provided by the Swedish National Infrastructure for Computing (SNIC) at High Performance Computing Center North (HPC2N), partially funded by the Swedish Research Council through Grant Agreement No. 2018-05973. The APC was funded by MDPI.

Institutional Review Board Statement: Not applicable.

Data Availability Statement: The data presented in this study are available on request from the corresponding author. The data are not publicly available due to large disk usage.

Acknowledgments: The authors are grateful to Kun Zhang for implementing the thermal boundary conditions. D.V. is grateful to Andrea Gruber for useful discussions.

Conflicts of Interest: The authors declare no conflict of interest. The funders had no role in the design of the study; in the collection, analyses, or interpretation of data; in the writing of the manuscript; or in the decision to publish the results.

Abbreviations and Nomenclature

BLF	boundary layer flashback
CGM	critical gradient model
CIVB	combustion-induced vortex breakdown
B	wall heat transfer parameter
c_s	sound speed
C_P	heat capacity at constant pressure
C_V	heat capacity at constant volume
E_a	activation energy
g	velocity gradient
l	cell size
Le	Lewis number
Ma	flame propagation Mach number
P	pressure
p	order of truncation error decay
Pr	Prandtl number
q_j	energy diffusion vector
Q	energy release from the reaction
R	channel half-width
R_p	universal gas constant
Sc	Schmidt number
s_c	consumption speed
S_L	unstretched laminar burning velocity
t	time
T	temperature
U, u	velocity
w	reaction rate
x	axial coordinate
y	radial coordinate
Y	mass fraction of fuel
Ze	Zeldovich number
δ_{ij}	Kronecker delta
δ_L	laminar flame thickness
δ_q	quenching distance
δ_p	penetration distance
ε	total energy per unit volume
ϵ	Arrhenius number
η_{ij}	stress tensor
γ	adiabatic index
θ	gas expansion ratio
μ	dynamic viscosity
ρ	density
τ_R	factor of time dimension in the Arrhenius law
Subscripts	
0	initial value
b	burnt gas
c	critical
max	maximum (centerline)
tip	flame tip
Superscripts	
$+$	normalized value
Other designations	
$\sim$	scaled value

References

1. Kalantari, A.; McDonell, V. Boundary layer flashback of non-swirling premixed flames: Mechanisms, fundamental research, and recent advances. *Prog. Energy Combust. Sci.* **2017**, *61*, 249–292. [CrossRef]
2. Leicher, J.; Schaffert, J.; Cigarida, H.; Tali, E.; Burmeister, F.; Giese, A.; Albus, R.; Görner, K.; Carpentier, S.; Milin, P.; et al. The Impact of Hydrogen Admixture into Natural Gas on Residential and Commercial Gas Appliances. *Energies* **2022**, *15*, 777. [CrossRef]
3. Jiang, X.; Tang, Y.; Liu, Z.; Raman, V. Computational Modeling of Boundary Layer Flashback in a Swirling Stratified Flame Using a LES-Based Non-Adiabatic Tabulated Chemistry Approach. *Entropy* **2021**, *23*, 567. [CrossRef] [PubMed]
4. Vigueras-Zuniga, M.O.; Tejeda-del Cueto, M.E.; Vasquez-Santacruz, J.A.; Herrera-May, A.L.; Valera-Medina, A. Numerical Predictions of a Swirl Combustor Using Complex Chemistry Fueled with Ammonia/Hydrogen Blends. *Energies* **2020**, *13*, 288. [CrossRef]
5. Eichler, C.T. Flame Flashback in Wall Boundary Layers of Premixed Combustion Systems. Ph. D. Thesis, Technical University of Munich, Munich, Germany, 2011.
6. Lewis, B.; von Elbe, G. Stability and structure of burner flames. *J. Chem. Phys.* **1943**, *11*, 75–97. [CrossRef]
7. Von Elbe, G.; Mentser, M. Further studies of the structure and stability of burner flames. *J. Chem. Phys.* **1945**, *13*, 89–100. [CrossRef]
8. Berlad, A.L.; Potter, A.E. Relation of boundary velocity gradient for flash-back to burning velocity and quenching distance. *Combust. Flame* **1957**, *1*, 127–128. [CrossRef]
9. Potter, A.E. 5-Flame Quenching. In *Progress in Combustion Science and Technology*; Ducarme, J., Gerstein, M., Lefebvre, A.H., Eds.; Pergamon Press: New York, USA, 1960; Volume 1, pp. 145–181. [CrossRef]
10. Lee, S.T.; T'ien, J.S. A numerical analysis of flame flashback in a premixed laminar system. *Combust. Flame* **1982**, *48*, 273–285. [CrossRef]
11. Mallens, R.M.M.; De Goey, L.P.H. Flash-back of Laminar Premixed Methane/Air Flames on Slitand Tube Burners. *Combust. Sci. Technol.* **1998**, *136*, 41–54. [CrossRef]
12. Kurdyumov, V.; Fernández, E.; Liñán, A. Flame flashback and propagation of premixed flames near a wall. *Proc. Combust. Inst.* **2000**, *28*, 1883–1889. [CrossRef]
13. Kurdyumov, V.N.; Fernandez-Tarrazo, E. Lewis number effect on the propagation of premixed laminar flames in narrow open ducts. *Combust. Flame* **2002**, *128*, 382–394. [CrossRef]
14. Kurdyumov, V.; Fernández-Tarrazo, E.; Truffaut, J.M.; Quinard, J.; Wangher, A.; Searby, G. Experimental and numerical study of premixed flame flashback. *Proc. Combust. Inst.* **2007**, *31*, 1275–1282. [CrossRef]
15. Jiménez, C.; Fernández-Galisteo, D.; Kurdyumov, V.N. DNS study of the propagation and flashback conditions of lean hydrogen-air flames in narrow channels: Symmetric and non-symmetric solutions. *Int. J. Hydrogen Energy* **2015**, *40*, 12541–12549. [CrossRef]
16. Eichler, C.; Baumgartner, G.; Sattelmayer, T. Experimental investigation of turbulent boundary layer flashback limits for premixed hydrogen-air flames confined in ducts. *J. Eng. Gas Turbines Power* **2012**, *134*, 011502. [CrossRef]
17. Wu, M.H.; Wang, C.Y. Reaction propagation modes in millimeter-scale tubes for ethylene/oxygen mixtures. *Proc. Combust. Inst.* **2011**, *33*, 2287–2293. [CrossRef]
18. Kuznetsov, M.; Denkevits, A.; Veser, A.; Friedrich, A. Flame propagation regimes and critical conditions for flame acceleration and detonation transition for hydrogen-air mixtures at cryogenic temperatures. *Int. J. Hydrogen Energy* **2022**, *47*, 30743–30756. [CrossRef]
19. Ebi, D.; Clemens, N.T. Experimental investigation of upstream flame propagation during boundary layer flashback of swirl flames. *Combust. Flame* **2016**, *168*, 39–52. [CrossRef]
20. Ranjan, R.; Clemens, N.T. Insights into flashback-to-flameholding transition of hydrogen-rich stratified swirl flames. *Proc. Combust. Inst.* **2021**, *38*, 6289–6297. [CrossRef]
21. Gruber, A.; Chen, J.H.; Valiev, D.; Law, C.K. Direct numerical simulation of premixed flame boundary layer flashback in turbulent channel flow. *J. Fluid Mech.* **2012**, *709*, 516–542. [CrossRef]
22. Eriksson, L.E. *A Third Order Accurate Upwind-Biased Finite-Volume Scheme For Unsteady Compressible Viscous Flow*; Volvo Aero Corporation: Trollhättan, Sweden, 1993.
23. Eriksson, L.E. Flow solution on a dual-block grid around an airplane. *Comput. Methods Appl. Mech. Eng.* **1987**, *64*, 79–93. [CrossRef]
24. Valiev, D.; Bychkov, V.; Akkerman, V.; Law, C.K.; Eriksson, L.E. Flame acceleration in channels with obstacles in the deflagration-to-detonation transition. *Combust. Flame* **2010**, *157*, 1012–1021. [CrossRef]
25. Demirgok, B.; Ugarte, O.; Valiev, D.; Akkerman, V. Effect of thermal expansion on flame propagation in channels with nonslip walls. *Proc. Combust. Inst.* **2015**, *35*, 929–936. [CrossRef]
26. Valiev, D.; Bychkov, V.; Akkerman, V.; Eriksson, L.E.; Law, C.K. Quasi-steady stages in the process of premixed flame acceleration in narrow channels. *Phys. Fluids* **2013**, *25*, 096101. [CrossRef]
27. Han, Y.; Modestov, M.; Valiev, D.M. Effect of momentum and heat losses on the hydrodynamic instability of a premixed equidiffusive flame in a Hele–Shaw cell. *Phys. Fluids* **2021**, *33*, 103608. [CrossRef]
28. Poinsot, T.; Veynante, D. *Theoretical and Numerical Combustion*; RT Edwards, Inc.: Morningside, QLD, Australia, 2005.

29. Adebiyi, A.; Abidakun, O.; Akkerman, V. Acceleration of Premixed Flames in Obstructed Pipes with Both Extremes Open. *Energies* **2020**, *13*, 4094. [CrossRef]
30. Feng, R.; Gruber, A.; Chen, J.H.; Valiev, D.M. Influence of gas expansion on the propagation of a premixed flame in a spatially periodic shear flow. *Combust. Flame* **2021**, *227*, 421–427. [CrossRef]
31. Ding, S.; Huang, K.; Han, Y.; Valiev, D. Numerical study of the influence of wall roughness on laminar boundary layer flashback. *Phys. Rev. Fluids* **2021**, *6*, 023201. [CrossRef]
32. Feng, R.; Valiev, D. Influence of gas expansion on the velocity and stability limits of stationary curved flames in channels. *Combust. Sci. Technol.* **2022**. [CrossRef]
33. Schlichting H, G.K. *Boundary-Layer Theory*, 9th ed.; Springer: Berlin/Heidelberg, Germany, 2016.
34. Kagan, L. On the transition from deflagration to detonation in narrow channels. *Math. Model. Nat. Phenom.* **2007**, *2*, 16. [CrossRef]
35. Kagan, L.; Sivashinsky, G. On the Transition from Deflagration to Detonation in Narrow Tubes. *Flow Turbul. Combust.* **2010**, *84*, 423–437. [CrossRef]
36. Huang, J.; Han, W.; Gao, X.; Wang, C. Effects of heat loss and viscosity friction at walls on flame acceleration and deflagration to detonation transition. *Chin. Phys. B* **2019**, *28*, 074704. [CrossRef]
37. Zeldovich, I.B.; Barenblatt, G.I.; Librovich, V.B.; Makhviladze, G.M. *Mathematical Theory of Combustion and Explosions*; Consultants Bureau: New York, NY, USA, 1985.
38. Law, C.K. *Combustion Physics*; Cambridge University Press: Cambridge, UK, 2006.
39. Valiev, D.; Bychkov, V.; Akkerman, V.; Eriksson, L.E.; Marklund, M. Heating of the fuel mixture due to viscous stress ahead of accelerating flames in deflagration-to-detonation transition. *Phys. Lett. A* **2008**, *372*, 4850–4857. [CrossRef]
40. Versteeg, H.; Malalasekera, W. *An Introduction to Computational Fluid Dynamics: The Finite Volume Method*; Pearson Education Limited: London, UK, 2007.
41. AlKhabbaz, M.; Kodakoglu, F.; Valiev, D.; Akkerman, V. Impacts of wall conditions on flame acceleration at the early stages of burning in channels. *Phys. Rev. Fluids* **2022**, *7*, 013201. [CrossRef]
42. Dugger, G.L. Flame stability of preheated propane-air mixtures. *Ind. Eng. Chem.* **1955**, *47*, 109–114. [CrossRef]
43. Harris, M.; Grumer, J.; Elbe, G.V.; Lewis, B. Burning velocities, quenching, and stability data on nonturbulent flames of methane and propane with oxygen and nitrogen 1: Application of theory of ignition, quenching, and stabilizationto flames of propane and air. *Symp. Combust. Flame Explos. Phenom.* **1948**, *3*, 80–89. [CrossRef]
44. Alliche, M.; Haldenwang, P.; Chikh, S. Extinction conditions of a premixed flame in a channel. *Combust. Flame* **2010**, *157*, 1060–1070. [CrossRef]
45. Takizawa, K.; Igarashi, N.; Takagi, S.; Tokuhashi, K.; Kondo, S. Quenching distance measurement of highly to mildly flammable compounds. *Fire Saf. J.* **2015**, *71*, 58–68. [CrossRef]
46. Akkerman, V.; Bychkov, V.; Petchenko, A.; Eriksson, L.E. Flame oscillations in tubes with nonslip at the walls. *Combust. Flame* **2006**, *145*, 675–687. [CrossRef]
47. Hackert, C.; Ellzey, J.; Ezekoye, O. Effects of thermal boundary conditions on flame shape and quenching in ducts. *Combust. Flame* **1998**, *112*, 73–84. [CrossRef]
48. Kim, N.I.; Kataoka, T.; Maruyama, S.; Maruta, K. Flammability limits of stationary flames in tubes at low pressure. *Combust. Flame* **2005**, *141*, 78–88. [CrossRef]
49. Bioche, K.; Vervisch, L.; Ribert, G. Premixed flame-wall interaction in a narrow channel: Impact of wall thermal conductivity and heat losses. *J. Fluid Mech.* **2018**, *856*, 5–35. [CrossRef]
50. Lee, S.T.; Tsai, C.H. Numerical investigation of steady laminar flame propagation in a circular tube. *Combust. Flame* **1994**, *99*, 484–490. [CrossRef]

applied sciences

MDPI

Article

Effect of Zr Modification on NH_3-SCR Reaction Performance of Cu-Ce/SAPO-34 Catalysts

Chongfei Liu [1,2], Xuetao Wang [1,2,*], Lili Xing [1,2], Xingxing Cheng [3], Xingyu Zhang [4], Haojie Li [1,2] and Mengjie Liu [1,2]

1 Department of Energy and Power Engineering, Henan University of Science and Technology, Luoyang 471003, China
2 Longmen Laboratory, Luoyang 471000, China
3 National Engineering Lab of Coal-fired Pollution Emission Reduction, Shandong University, Jinan 250061, China
4 Department of Energy and Power Engineering, Qilu University of Technology, Jinan 250353, China
* Correspondence: wxt7682@163.com

Abstract: Molecular sieve catalysts containing transition metals have been attracting attention for their potential applications in various fields, including environmental and industrial catalysis. A Cu-Ce-Zr/SAPO-34 series of molecular sieve catalysts were prepared by the impregnation method, and the effect of Zr introduction on the selective catalytic reduction of NO by Cu-Ce/SAPO-34 molecular sieve catalysts was explored. Through various characterization methods, the physical and chemical properties of the catalysts were analyzed, and the denitration mechanism of the molecular sieve catalyst was discussed. This study found that the total acid content of the acid sites on the catalyst surface decreased with the introduction of Zr, leading to a decrease in the denitration efficiency of the catalyst. At 350–400 °C, the denitration efficiency of the 4Cu-4Ce-4Zr/SAPO-34 catalyst was over 80%, and at 400–500 °C, it was over 99%. Moreover, excessive metal Zr could destroy its CHA structure and decrease the denitration efficiency of the catalyst. This study analyzed the reaction mechanism of NH_3-SCR of Zr-modified polymetallic zeolites and the effect of Zr modification on the NH3-SCR reaction results. This study contributes to the understanding of the performance of molecular sieve catalysts containing transition metals. Reliable conclusions were obtained, which offer data support for future research in the field of NH_3-SCR.

Keywords: Cu-Ce-Zr/SAPO-34; molecular sieve catalyst; selective catalytic reduction; modification

1. Introduction

Nitrogen oxides (NO_x) are currently the main atmospheric pollutants in the world. They not only endanger the natural environment but also affect human social production and living activities, and they even endanger human health. Selective catalytic reduction (SCR) technology is currently a mature denitration technology, which is widely used in thermal power plants, waste incineration furnaces, boiler facilities, diesel engines, gas turbines, and other scenarios. The core of SCR technology is the catalyst, which is the main factor of catalyst performance. The main factors affecting catalyst performance include the catalyst carrier, auxiliary pharmacy, active components, precursors, and the preparation method [1–4].

In terms of the selection of active components, researchers have conducted deep explorations of metal oxides such as Cu, Fe, and Mn and have found that they all show advantages under various working conditions. Researchers [5] found that, under the same support and preparation method, the denitration efficiency of catalysts followed the following order: Cu–Fe-based > Cu-based > Fe-based, that is, bimetallic oxide catalysts are superior to single-metal-oxide ones, and Cu-based catalysts are superior to Fe-based catalysts. According to Yan's research [6], under the same conditions, the denitration efficiency of catalysts followed the following order: Cu–Mn > Cu > Mn; when the reaction conditions were higher than 200 °C, the order was: Cu > Cu–Mn > Mn, indicating that Cu catalysts are superior to Mn catalysts, and bimetallic oxide catalysts are superior to single-metal-oxide catalysts at a temperature lower than 200 °C. In conclusion, it was proposed to select Cu as the active component of the catalyst. In terms of the carrier, most mainstream researchers choose zeolite molecular sieves, active carbon, TiO_2, etc., as the carrier of catalysts. Zeolite molecular sieves are widely used in various fields because of their porosity, selective adsorption of molecules through channels, large specific surface area, and the fact that they are excellent catalyst carriers. As reported in the literature in [7], the denitrification performance of Cu Mn/SAPO-34 is generally better than that of Cu Mn/ZSM-5; Kouadio Brou Albert pointed out that [8] Cu/SAPO-34 catalysts have a wide active temperature window, excellent hydrothermal stability, and resistance to alkali metal and alkaline earth metal poisoning. According to the actual conditions of the laboratory, an SAPO-34 molecular sieve was proposed to be used as the catalyst carrier.

To obtain a superior denitration performance and a wider range of applications, SCR denitration catalysts are mostly bimetallic or even polymetallic in order to obtain the physical and chemical properties of various metals. Many scholars [9,10] have found that doping with metal Ce has a positive effect on the denitration ability of Cu-based molecular sieve catalysts. Ce can improve the thermal stability of the catalyst, improve the distribution of active substances on the catalyst surface, and enhance the redox performance of the catalyst. Therefore, Ce was proposed to be used as a metal in a multi-metal-oxide catalyst. According to Zhang's research [11], Zr can increase the oxygen vacancy concentration of Cu–Ce metal oxide catalysts, enhance their oxidation–reduction capacity, and improve the catalysts' adsorption capacity for NO_x. However, it will slightly reduce their denitration performance. Some scholars [12] found that doping with Zr can inhibit the aggregation of Cu on the catalyst surface, promote the dispersion of Cu, increase the concentration of Cu species on the catalyst surface, and improve the resistance to Ca poisoning. However, there is still no consensus on the reasons as to why Zr doping leads to a decline in the catalyst denitration performance, nor is there a consensus on the reasons behind the advantages and disadvantages of Zr participating in denitration reactions and their mechanisms. Therefore, Zr was selected as the last metal of the polymetallic oxide catalyst.

In previous research, it was found that a 4Cu-4Ce/SAPO-34 (the mass fraction of Cu and Ce being 4%) bimetallic oxide catalyst had the best performance in all aspects. On this basis, Zr was added to the catalystand the influence of different Zr loading amounts on the denitration performance of the catalyst was explored to obtain the best Zr loading. The physical and chemical properties of the catalyst were studied through XRD, SEM, NH_3-TPD, in situ DRIFTS, XPS, and other characterization techniques. The possible NH_3-SCR reaction path of Cu-Ce-Zr/SAPO-34 was proposed.

2. Materials and Methods

2.1. Preparation of Catalysts

The 4Cu-4Ce-xZr/SAPO-34 catalyst (x being the mass fraction of metal Zr in the catalyst) with different Zr loading amounts was prepared by the immersion method. Copper nitrate, cerium nitrate, and zirconium nitrate were mixed together in correct proportions. Then, the SAPO-34 molecular sieve was placed in a muffle furnace at 400 °C for 2 h to remove impurities and water. The mass fraction of fixed Cu and Ce was 4%, while the mass fraction of Zr was changed between 0%, 2%, 4%, 6%, and 8%, respectively. Five portions of impregnating solution were prepared with the corresponding proportion. the SAPO-34 carrier was fully dissolved in the impregnating solution according to the corresponding mass. It was heated and stirred with a magnetic stirrer for 5 h to form a paste. An ultrasonic oscillator was used to vibrate the mixture for 40 min so that the active substance could fully adhere to the carrier surface. Then, it was dried at a constant temperature of 110 °C in a constant-temperature blast-drying oven overnight. The five obtained samples were fully ground to a powder and then put into a muffle furnace for roasting at 450 °C for 2 h. After being ground again, they were bagged for standby and were recorded as 4Cu-4Ce-xZr/SAPO-34 (where x was the mass fraction of metal Zr in the catalyst).

2.2. Experimental Conditions and Equipment

2.2.1. Catalytic Activity Test

The denitration performance of the catalyst samples was tested using a VDRT-200SCR catalyst evaluation device (Quzhou Ward Instrument Co., Ltd., Quzhou, China). The test required 200 mg of the catalyst. The simulated flue gas consisted of 0.1% NH_3, 0.1% NO, and 1% O_2. N_2 was used as a balance gas. The evaluation device was equipped with a mass flow meter, which could accurately control the gas flow. The simulated flue gas flow was 200 mL/min, the space velocity was 22,000 h^{-1}, and the temperature range was 100–500 °C. The temperature was increased by 50 °C within 5 min, kept constant for 30 min, then increased by 50 °C within 5 min, and this cycle was repeated. The volume concentration of NO_x after the reaction was recorded. The NO_x concentration in the reaction tail gas was detected using a German Testo 350 Pro flue gas analyzer. The following formula was used to calculate the denitration efficiency η:

$$\eta = (NO_{x,in} - NO_{x,out})/NO_{x,in} \times 100\%$$

where $NO_{x,in}$ represents the volume concentration of NO_x before the reaction and $NO_{x,out}$ is the volume concentration of NO_x after the reaction.

2.2.2. Characterization of the Catalyst

The crystal phase composition of the catalyst was characterized by an X-ray diffractometer (DX-2700B, Dandong Haoyuan Instrument Co., Ltd., Dandong, China). The scanning angle was 6–70°, and the scanning rate was 2°/min. Field emission scanning electron microscopy (JSM-7800F, Japan Electron Optics Laboratory Co., Ltd., Tokyo, Japan) was used to test the surface micromorphology and element distribution of the catalyst samples. A TP-5076TPD/TPR dynamic adsorption instrument from Tianjin Xianquan was used to conduct the temperature-programmed adsorption (NH_3-TPD) experiment on the catalyst to test the acidity of the catalyst. Fourier transform infrared absorption spectroscopy (IRT racer-100, Shimadzu Corporation, Kyoto, Japan) was used to determine the surface functional groups of the catalyst samples and the types of acidic sites of the catalyst samples.

3. Results and Discussion

3.1. NH_3-SCR Catalytic Activity

Figure 1 shows the test results of the catalyst denitration performance. In the temperature range of 150–500 °C, the NO_x conversion increased with increasing the temperature. In the temperature range of 200–350 °C, the increase in the denitration performance of the catalyst was the most obvious. 4Cu-4Ce/SAPO-34 showed the most excellent denitration performance in the temperature range of 200–500 °C, which was consistent with the conclusion that metal Zr reduced the overall denitration performance of the catalyst to a certain extent [11]. At temperatures below 200 °C, the denitration performance of the 4Cu-4Ce-4Zr/SAPO-34 catalyst was better than that of 4Cu-4Ce/SAPO-34, and we inferred that an appropriate Zr doping could improve the low-temperature activity of the denitration catalyst. Among all the trimetallic catalysts, the 4Cu-4Ce-4Zr/SAPO-34 catalyst had the best denitration performance. In the range of 350~500 °C, the denitration efficiency was above 80%. In the range of 400~500 °C, the denitration efficiency was more than 99%. The reason why the Zr reduced the denitration efficiency of the catalyst and the reason for the better denitration performance of the 4Cu-4Ce-4Zr/SAPO-34 catalyst need to be further explored.

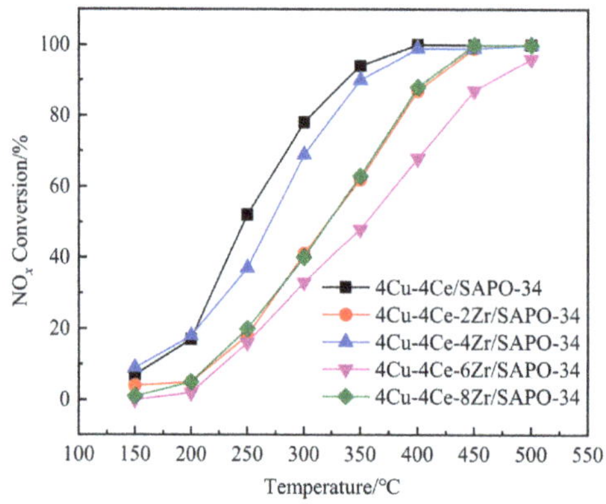

Figure 1. NO_x conversion of catalysts with different Zr loadings.

3.2. XRD

Figure 2 shows the XRD spectrum of the prepared catalyst and SAPO-34. In the figure, it can be seen that 9.47°, 12.84°, 16.04°, 17.72°, 20.52°, 25.03°, 25.88°, and 30.58° were the characteristic peaks of the SAPO-34 molecular sieve support [13], which was a typical CHA structure with a high crystallinity. The corresponding characteristic peaks of the metal oxides were also detected in the experiment. The characteristic peaks detected at 35.3° and 38.5° were associated with CuO [14], 28.2° and 33.1° were associated with CeO_2 [15], and 33.8°, 50.7°, and 60.1° were associated with ZrO_2 [16]. Five catalysts showed the same characteristic peaks of SAPO-34, while their peaks were slightly different. Comparing the characteristic peaks of 4Cu-4Ce/SAPO-34, SAPO-34, and 4Cu-4Ce-4Zr/SAPO-34, it was found that the loading of Cu and Ce reduced the characteristic peaks of SAPO-34. After the Zr modification, the characteristic peaks of SAPO-34 were enhanced, which proved that a small amount of Zr could promote the crystallization growth of the SAPO-34 zeolite molecular sieve to a certain extent. It also enhanced the CHA skeleton structure of SAPO-34 [14]. However, the characteristic peaks of SAPO-34 of 4Cu-4Ce-6Zr/SAPO-34 and 4Cu-4Ce-8Zr/SAPO-34 almost disappeared. It was speculated that a large number of Zr oxides appeared in the channel and outer surface of the crystal, resulting in a decrease in its crystallinity, the disappearance of the characteristic peaks, and the destruction of the crystal structure [17,18]. The characteristic peaks of Cu, Ce, and Zr were not detected in the five catalysts, indicating that the above metal species were dispersed on the surface of the catalyst in an amorphous form, which was conducive to the NH_3-SCR reaction [19].

3.3. NH_3-TPD

Figure 3 shows the NH_3-TPD spectrum of the prepared catalyst. The figure shows the temperature of the acid center of ammonia desorption. Three desorption peaks were found at about 77 °C, 161.1 °C, and 371.1 °C, representing the different types of acid sites on the catalyst surface. In general, the intensity of the acid sites was proportional to the temperature of the desorption peak, and the number of acid sites was proportional to the area of the desorption peak. The desorption peaks at about 77 °C and 161.1 °C were attributed to weak acidic sites bound to NH_3, while the desorption peaks at 371.1 °C were attributed to strong acidic sites bound to NH_3 [20]. Compared with the desorption peaks of 4Cu-4Ce/SAPO-34, 4Cu-4Ce-4Zr/SAPO-34, and 4Cu-4Ce-8Zr/SAPO-34, the number of weak acid sites at 77 °C decreased, and the intensity increased with the increase in the Zr mass fraction. In addition, the number of acidic sites of 4Cu-4Ce/SAPO-34 was the largest. The number and intensity of weak acid sites at 161.1 °C increased first and then decreased, and the number and intensity of acidic sites of 4Cu-4Ce-4Zr/SAPO-34 were the largest. The catalysts with the strongest acidic sites at about 371.1 °C were 4Cu-4Ce/SAPO-34 and 4Cu-4Ce-6Zr/SAPO-34. In conclusion, the total number of acidic sites of 4Cu-4Ce/SAPO-

34 was the largest, and thus the denitration efficiency was the highest. This shows that the Zr loading reduced the total number of acidic sites of 4Cu-4Ce/SAPO-34, thus reducing the denitration efficiency. The denitration efficiency of 4Cu-4Ce-4Zr/SAPO-34 was only lower than that of 4Cu-4Ce/SAPO-34, which could be attributed to the fact that the number and intensity of weak acid sites at 161.1 °C increased due to the introduction of the appropriate amount of Zr. Similarly, the poor denitration efficiency of 4Cu-4Ce-6Zr/SAPO-34 could be attributed to the excessive metal Zr, which led to a reduction in the number of acidic sites. The type of specific acid sites at each temperature needs to be further concluded through in situ DRIFTS tests.

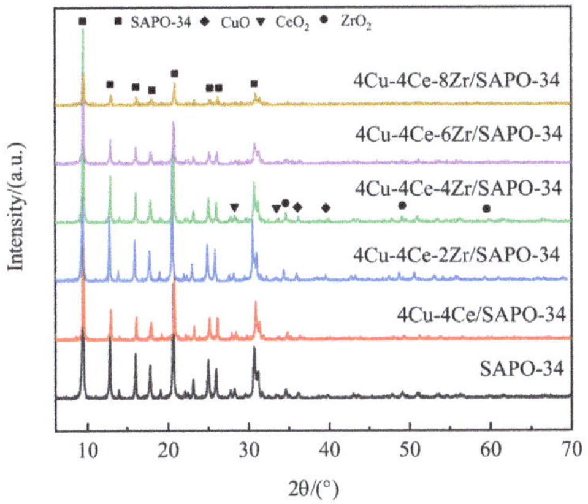

Figure 2. XRD spectra of catalysts with different Zr loadings.

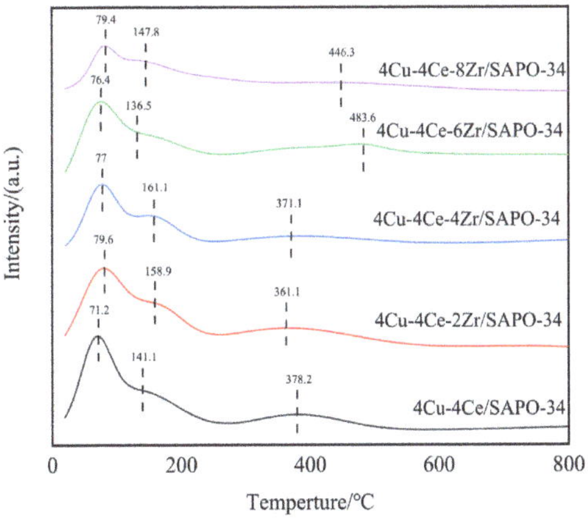

Figure 3. NH_3-TPD spectra of catalysts with different Zr loadings.

3.4. SEM

Figure 4 shows SEM photos of some of the prepared catalysts. SEM characterization is mainly used to explore the crystal state, microstructure, and adhesion state of active components on catalyst surfaces. As shown in Figure 4, all the catalysts presented a cube state related to the CHA structure of SAPO-34. The grain size was about 4 µm. The surface of the 4Cu-4Ce/SAPO-34(a) catalyst was covered with randomly shaped materials, and it was speculated that Cu and Ce oxides were attached to the crystal surface after interaction [21]. The XRD results showed that the characteristic peaks of the 4Cu-4Ce-2Zr/SAPO-34(b) and 4Cu-4Ce-4Zr/SAPO-34(c) samples were high, that is, both the crystallinity and strength were high. Compared with the SEM photos, it was found that the

4Cu-4Ce-2Zr/SAPO-34(b) and 4Cu-4Ce-4Zr/SAPO-34(c) cubes had complete structures and flat surfaces, and each particle was in an independent state, which confirmed the conclusion of the XRD analysis. For the 4Cu-4Ce-6Zr/SAPO-34(d) catalyst, it was found that most of cubic structures were damaged, the surface was rough, and most of the particles were bonded together after being damaged, which was consistent with the conclusion that the XRD results showed that the characteristic peaks disappeared, the crystallization decreased, and the grain structure was damaged.

Figure 4. SEM images of 4Cu-4Ce/SAPO-34 (**a**), 4Cu-4Ce-2Zr/SAPO-34 (**b**), 4Cu-4Ce-4Zr/SAPO-34 (**c**), and 4Cu-4Ce-6Zr/SAPO-34 (**d**) (the magnifications is 10,000 and 1000 times).

3.5. *In Situ DRIFTS Measurements*

To ascertain the types of acidic sites on the surface of the 4Cu-4Ce-4Zr/SAPO-34 catalyst and the paths of the NH_3-SCR denitration reaction, samples of the 4Cu-4Ce-4Zr/SAPO-34 catalyst were measured and analyzed by in situ DRIFTS measurements.

3.5.1. NH_3 Adsorption Experiment

Figure 5 shows the in situ DRIFTS spectra of NH_3 adsorbed onto the 4Cu-4Ce-4Zr/SAPO-34 catalyst. The test temperature was 100 °C, and the characteristic curves of the samples at 3, 6, 10, 20, 30, and 40 min were recorded. In Figure 5, it can be seen that the characteristic peaks appeared at 1097, 1128, 1400, 1458, 1508, 1651, 2360, 3034, and 3117 cm^{-1}. The adsorption peak at 3117 cm^{-1} in the N-H stretching vibration area was attributed to the symmetric vibrations of NH_3 in the Lewis acid coordination state [22], the peak at 3034 cm^{-1} was attributed to the Brønsted acid site with the deformation vibrations of NH_4^+ in an ionic state adsorbed on the peak [23], the peak at 2360 cm^{-1} was attributed to the N-N stretching vibrations of surface N_2 molecules [24], and the peak at 1651 cm^{-1} was attributed to the symmetric bending vibrations of N-H bonds in covalently adsorbed NH_3 in the Lewis acid sites. The peaks at 1458 and 1400 cm^{-1} were classified as the deformed vibrations of NH_4^+ in the ionic state on the Brønsted acid site [25,26], and the peaks at 1508, 1128, and 1097 cm^{-1} were regarded as the symmetric vibrations of NH_3 coordinated by the Lewis acid site [27]. According to the conclusions of the NH_3-TPD analysis, the majority of the weak acid sites were considered to be weak Lewis acid sites, and the quantity of strong acid sites was less than that of Brønsted acid sites. In general, there were both Lewis acid sites and Brønsted acid sites on the surface of the 4Cu-4Ce-4Zr/SAPO-34 catalyst. Lewis acid sites were dominant due to their large amount, which played an important role in the NH_3-SCR activity of 4Cu-4Ce-4Zr/SAPO-34.

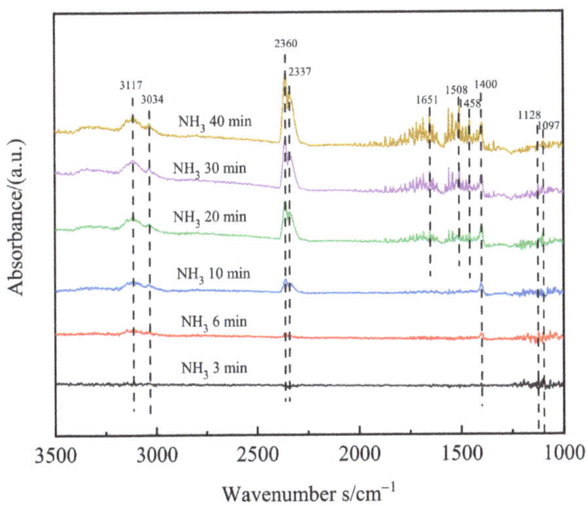

Figure 5. In situ DRIFTS of NH_3 adsorbed on the 4Cu-4Ce-4Zr/SAPO-34 catalyst.

3.5.2. Transient Reaction of NO+O_2 after NH_3 Adsorption Saturation

After NH_3 was saturated for 40 min, NO+O_2 was introduced for transient reaction, and the samples were scanned at 3, 6, 10, 20, and 30 min to analyze the change in the curve. As shown in Figure 6, with the introduction of NO+O_2, the characteristic peaks of Lewis acid at 3117 and 3034 cm^{-1} disappeared, and the N-N extension vibration at 2360 cm^{-1} was enhanced, which may be due to the reaction between adsorbed NH_3 and gaseous NO+O_2, resulting in the formation of N_2, which enhanced the characteristic peak at 2360 cm^{-1} attributed to N_2. the rest of the characteristic peaks did not change significantly, but new characteristic peaks were formed at 1383 cm^{-1} and 1265 cm^{-1}, among which the adsorption species at 1265 cm^{-1} belonged to the bridged nitrate species [26,28], and the adsorption peak at 1383 cm^{-1} was the strongest and belonged to the CIS-$N_2O_2^{2-}$ species [29,30], indicating that new nitrate species were generated on the catalyst surface. In general, the characteristic peaks of Lewis acid disappear, which proved that the adsorbed NH_3 reacted with gaseous NO+O_2. This process followed the E-R mechanism, while the characteristic peaks of Brønsted acid did not change significantly. Therefore, the weak Lewis acid site played a leading role in the reaction process.

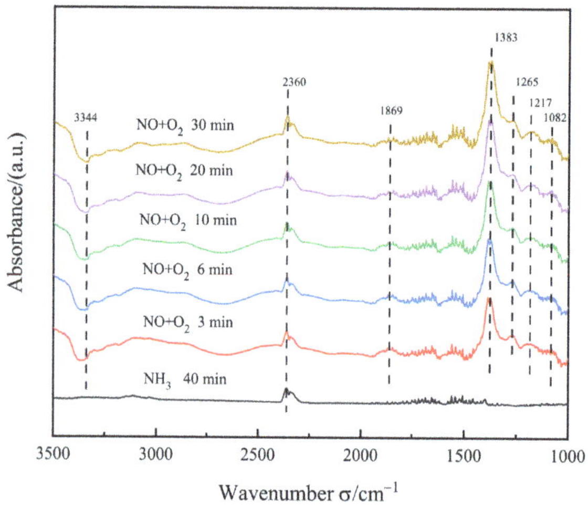

Figure 6. In situ DRIFTS of NO+O_2 reacting with pre-adsorbed NH_3 over the 4Cu-4Ce-4Zr/SAPO-34 catalyst.

3.5.3. NO+O_2 Adsorption Experiment

Figure 7 shows the in situ DRIFTS spectra of NO+O_2 adsorption on the 4Cu-4Ce-4Zr/SAPO-34 catalyst. The experimental temperature was kept at 150 °C, the characteristic curves of the samples were recorded at 3, 6, 10, 20, 30, and 40 min, and characteristic peaks appeared at 1244, 1384, 2268, 2397, and 3109 cm^{-1}. The characteristic peak at 1384 cm^{-1} was the strongest, which was classified as belonging to a CIS-$N_2O_2^{2-}$ species, the peak at 1244 cm^{-1} was considered to be due to a bridging nitrate species [31,32], and the characteristic peaks at 2268 and 2397 cm^{-1} were mainly composed of NOH^+ and HNO_2^+ and were formed by NO adsorption on the catalyst surface [33]. The peak at 3109 cm^{-1} was the hydroxyl vibration peak formed by the oxygen in the mixed gas and the H bond in the molecular sieve framework. The in situ DRIFTS spectrum of NO+O_2 adsorption on the 4Cu-4Ce-4Zr/SAPO-34 catalyst hardly changed with time, indicating that the adsorption of NO onto the 4Cu-4Ce-4Zr/SAPO-34 catalyst reached a saturated state at the beginning, and this may be related to metallic Zr.

Figure 8 shows the in situ DRIFTS spectra of 4Cu-4Ce/SAPO-34 and 4Cu-4Ce-4Zr/SAPO-34 after adsorbing NO+O_2 for 40 min up to saturation. Compared with 4Cu-4Ce/SAPO-34, 4Cu-4Ce-4Zr/SAPO-34 had more adsorption peaks and higher peaks, indicating that there were more nitrate species adsorbed onto the surface of 4Cu-4Ce-4Zr/SAPO-34, which proved that metal Zr could enhance the adsorption capacity of the catalyst for NO_x [11,34], which facilitated the reaction with NH_3-SCR.

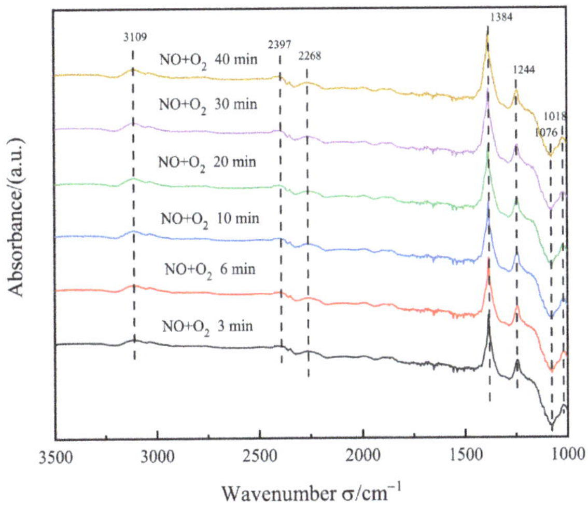

Figure 7. In situ DRIFTS of $NO+O_2$ adsorbed over the 4Cu-4Ce-4Zr/SAPO-34 catalyst.

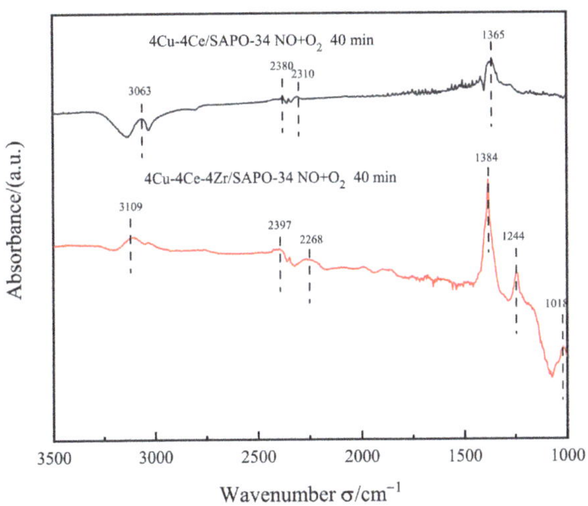

Figure 8. In situ DRIFTS of $NO+O_2$ adsorbed over the 4Cu-4Ce-4Zr/SAPO-34 and 4Cu-4Ce/SAPO-34 catalyst.

3.5.4. Transient Reaction of NH_3 after $NO+O_2$ Adsorption Saturation

$NO+O_2$ was adsorbed on the surface of the 4Cu-4Ce-4Zr/SAPO-34 catalyst for 40 min. After the adsorption was saturated, a transient reaction experiment of NH_3 was carried out. The sample was scanned at 3, 6, 10, 20, and 30 min to analyze its changes. Figure 9 shows the characteristic reaction curve of the above experiment. It can be seen that after NH_3 was introduced, the bridging nitrate characteristic peak at 1244 cm^{-1} and the monodentate nitrate characteristic peak at 1384 cm^{-1} did not change significantly. However, new characteristic peaks were formed at 1624 and 3334 cm^{-1}, and the characteristic peak at 1624 cm^{-1} was attributed to weakly adsorbed NO_2 [35]. It was speculated that the introduction of NH_3 might enhance the conversion and adsorption of NO. The characteristic peak at 3334 cm^{-1} was considered to be due to the NH_3 species adsorbed on the Lewis acid site [36,37]. During the reaction process, the main characteristic peaks of the nitrate species formed on the catalyst surface did not change significantly, and new NH_3 adsorption species were formed, but both of them did not change significantly after they became stable, indicating that no obvious reaction occurred between the adsorbed NO and NH_3, that is, the process did not follow the L-H mechanism.

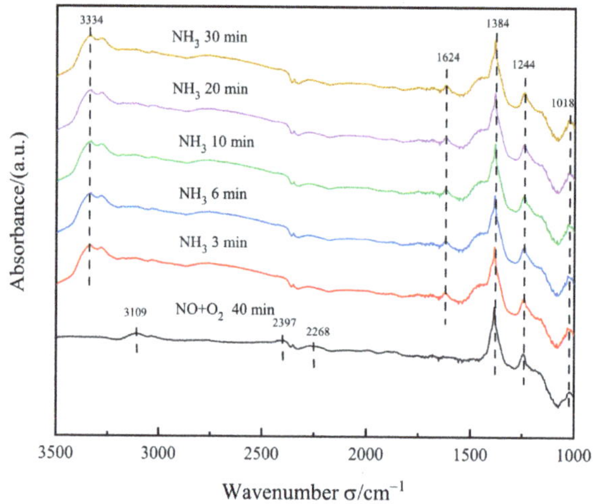

Figure 9. In situ DRIFTS of NH_3 reacting with pre-adsorbed $NO+O_2$ over the 4Cu-4Ce-4Zr/SAPO-34 catalyst.

3.6. XPS

Figures 10–12 show the XPS spectra of Cu 2p, Ce 3d, and Zr 3d of 4Cu-4Ce/SAPO-34 and 4Cu-4Ce-4Zr/SAPO-34 catalysts, respectively, in which the surface valence structure of the catalyst and the influence of Zr doping on the catalyst were analyzed.

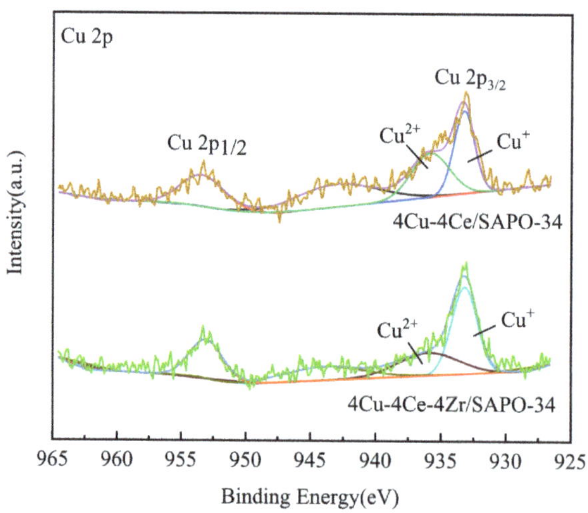

Figure 10. Cu 2p XPS spectra of 4Cu-4Ce/SAPO-34 and 4Cu-4Ce-4Zr/SAPO-34.

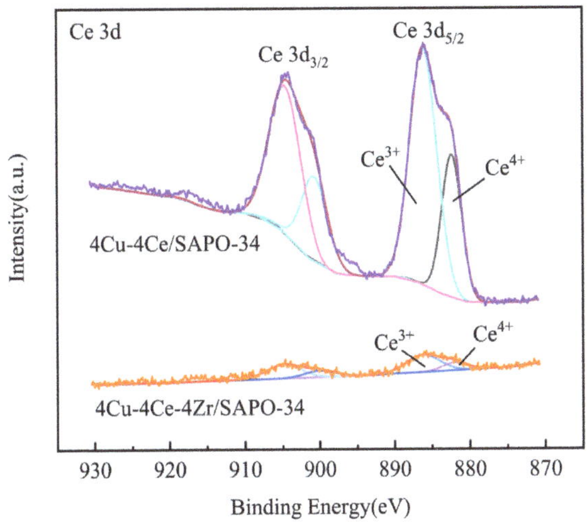

Figure 11. Ce 3d XPS spectra of 4Cu-4Ce-4Zr/SAPO-34 and 4Cu-4Ce/SAPO-34.

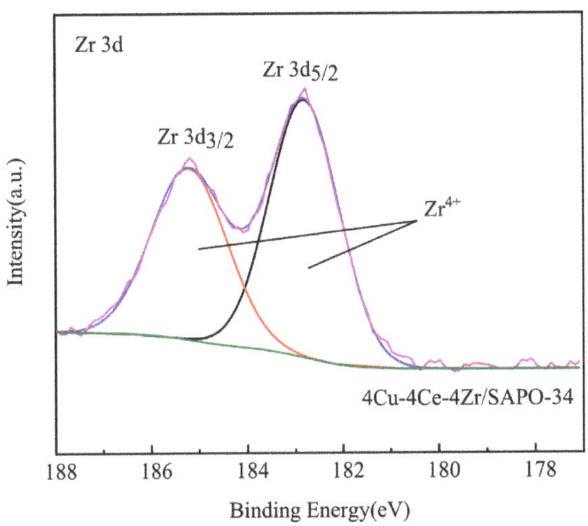

Figure 12. Zr 3d XPS spectrum of 4Cu-4Ce-4Zr/SAPO-34.

There were two characteristic peaks of Cu $2p_{3/2}$ and Cu $2p_{1/2}$ orbits in the Cu 2p spectrum of 4Cu-4Ce/SAPO-34, as shown in Figure 10. According to the reference [38,39], Cu $2p_{3/2}$ can be resolved into two characteristic peaks. The characteristic peak with a binding energy of 933.3 eV was due to Cu^+, and the characteristic peak with a binding energy of 935.7 eV was due to Cu^{2+}. Similarly, the characteristic peak with a binding energy of 953.6 eV at Cu $2p_{1/2}$ was classified as Cu^+, and the satellite peak with a binding energy of 943.1 eV was classified as Cu^{2+}. After doping with Zr, the binding energies of the Cu^+ characteristic peak were 933.2 eV and 953 eV, and the binding energies of the Cu^{2+} characteristic peak were 936 eV were 943.9 eV, which was similar with the characteristic peak of the binding energy of 4Cu-4Ce/SAPO-34 copper ions. As for the area of the characteristic peak, after doping with Zr, the area of the characteristic peak of Cu^+ increased by about 15%. Cu^+ can promote the catalytic performance of a catalyst in the low-temperature range [40,41], so the doping with Zr improved the low-temperature performance of the catalyst. The characteristic peak area of Cu^{2+} decreased by about 40%, and the total characteristic peak area of Cu^{+}+Cu^{2+} decreased by about 8.3%. Cu^{+}+Cu^{2+} played a key role in the performance of the catalyst, which might be the reason for the decline in the catalyst performance after Zr doping.

In the Ce 3d spectrum of 4Cu-4Ce/SAPO-34, as shown in Figure 11, there were characteristic peaks of Ce $3d_{5/2}$ and Ce $3d_{3/2}$ orbits. According to reference [42,43], both Ce $3d_{5/2}$ and Ce $3d_{3/2}$ can be resolved into two characteristic peaks. The characteristic peaks with binding energies of 886.1 eV and 904.5 eV were classified as Ce^{3+}, which were caused by its photoelectron emission, while the characteristic peaks with binding energies of 882.4 eV and 900.6 eV were caused by photoelectron emission of Ce^{4+}. After doping with Zr, the binding energies of the characteristic peaks of Ce^{3+} were 886.2 eV and 904.9 eV, and those of the characteristic peaks of Ce^{4+} were 882.5 eV and 900.9 eV. The binding energies of the four characteristic peaks were slightly increased, which confirmed that Zr could increase the stability of the catalyst. The area of the characteristic peak of Ce^{3+} decreased by 90.6%, the area of the characteristic peak of Ce^{4+} decreased by about 92.3%, and the total area of the characteristic peak of Ce^{3+}+Ce^{4+} decreased by about 91.1%. The transition from Ce^{4+} to Ce^{3+} enabled the Ce to have a strong redox performance and oxygen storage performance [43,44]. The addition of Zr caused a significant decrease in the total amount of Ce^{3+}+Ce^{4+}, which was one of the reasons for the decline in the performance of the catalyst after Zr doping.

The Zr 3d spectrum of 4Cu-4Ce-4Zr/SAPO-34 shown in Figure 12 contains the characteristic peaks of Zr $3d_{5/2}$ and Zr $3d_{3/2}$ orbits. According to the reference [45,46], the characteristic peak binding energy of the Zr $3d_{3/2}$ orbit was 185.18 eV, and that of the Zr $3d_{5/2}$ orbit was 182.78 eV. Both characteristic peaks belonged to Zr^{4+}. Therefore, it was concluded that Zr on the 4Cu-4Ce-4Zr/SAPO-34 surface only existed in the form of Zr^{4+}.

4. Conclusions

The overall denitration performance of 4Cu-4Ce-4Zr/SAPO-34 was only inferior to that of 4Cu-4Ce/SAPO-34. At 350~400 °C, the denitration efficiency was more than 80%, and at 400~500 °C, the denitration efficiency was more than 99%. Zr could improve the low-temperature performance of NH_3-SCR below 200 °C but reduced its overall denitration performance. The main reason was that Zr could reduce the amount of Cu and Ce metal ions on the catalyst surface, thus decreasing the total acid content of the acidic sites on the catalyst surface.

An appropriate amount of Zr could increase the relative crystallinity of the molecular sieve. There was an interaction between Zr and the CHA structure of SAPO-34, which could enhance its skeleton structure. In addition, Cu, Ce, and Zr were well dispersed, but excessive Zr damage the CHA structure, leading to a decline in the catalyst's denitration performance.

There were Lewis acid sites and Brønsted acid sites on the surface of the 4Cu-4Ce-4Zr/SAPO-34 catalyst, which was dominated by Lewis acid sites. Zr could increase the amount of weak Lewis acid sites and enhance the catalyst's adsorption capacity for NO_x, which is of great significance for its NH_3-SCR reaction performance. The reaction process only followed the E-R mechanism, not the L-H mechanism.

Author Contributions: Conceptualization, X.W.; software, X.Z.; formal analysis, C.L. and M.L.; investigation, X.W., L.X., X.C., X.Z. and H.L.; resources, C.L., X.Z. and H.L.; data curation, C.L., L.X., H.L. and M.L.; writing—original draft, C.L.; writing—review & editing, L.X.; supervision, X.W. and X.C.; project administration, X.C., H.L. and M.L.; funding acquisition, X.W. All authors have read and agreed to the published version of the manuscript.

Funding: This research was funded by the National Natural Science Foundation of China (No. 50806020), Excellent Youth Foundation of Henan Scientific Committee (No. 114100510010), Henan Province Science and Technology Research Project (No. 152102210280), and The Natural Science Foundation of Shandong Province (No. ZR2021QE066).

Conflicts of Interest: The authors declare no conflict of interest.

References

1. Wang, D.; Hui, S.; Liu, C. Effect of the different types of titanium dioxide carrier on the DeNO$_X$ performance of V-Ti catalysts. *Environ. Chem.* **2014**, *33*, 5.
2. Meng, L.; Fang, J.; Guan, X. Effect of Active Component Doping on the Denitration Properties of Manganese Based Catalyst at Low Temperature. *Mater. Rev.* **2017**, *31*, 35–39, 56.
3. Wu, Y.; Liang, H.; Chen, X.; Chen, C.; Wang, X.; Dai, C.; Hu, L.; Chen, Y. Effect of preparation methods on denitration performance of V-Mo/TiO_2 catalyst. *J. Fuel Chem. Technol.* **2020**, *48*, 189–196. [CrossRef]
4. Chen, G.; Fang, J.; Ma, T.; Wang, L. Effect of Different Precursor Solution on the Denitrification Performance of Mn/TiO_2 Catalyst. *Bull. Chin. Ceram. Soc.* **2018**, *37*, 2274–2279.
5. Wang, A.; Wang, Y.; Walter, E.D.; Washton, N.M.; Guo, Y.; Lu, G.; Peden, C.H.F.; Gao, F. NH_3-SCR on Cu, Fe and Cu+Fe exchanged beta and SSZ-13 catalysts: Hydrothermal aging and propylene poisoning effects. *Catal. Today* **2019**, *320*, 91–99. [CrossRef]
6. Yan, Q.; Chen, S.; Qiu, L.; Gao, Y.; O'Hare, D.; Wang, Q. The synthesis of CuyMnzAl1-zOx mixed oxide as low-temperature NH_3-SCR catalyst with enhanced catalytic performance. *Dalton Trans.* **2017**, *47*, 2992–3004. [CrossRef] [PubMed]
7. Zhao, S.; Huang, L.; Jiang, B.; Chen, M.; Zhang, J.; Hu, Y. Stability of Cu–Mn bimetal catalysts based on different zeolites for NO$_X$ removal from diesel engine exhaust. *Chin. J. Catal.* **2018**, *39*, 800–809. [CrossRef]
8. Albert, K.B.; Fan, C.; Pang, L.; Chen, Z.; Ming, S.; Albert, T.; Li, T. The influence of chemical poisoning, hydrothermal aging and their co-effects on Cu-SAPO-34 catalyst for NO$_X$ reduction by NH_3-SCR. *Appl. Surf. Sci.* **2019**, *479*, 1200–1211. [CrossRef]
9. Cao, Y.; Zou, S.; Lan, L.; Yang, Z.; Xu, H.; Lin, T.; Gong, M.; Chen, Y. Promotional effect of Ce on Cu-SAPO-34 monolith catalyst for selective catalytic reduction of NO$_X$ with ammonia. *J. Mol. Catal. A Chem.* **2015**, *398*, 304–311. [CrossRef]
10. Zhou, J.; Zhao, C.; Lin, J.; Yang, H.; Zhou, R. Promotional effects of cerium modification of Cu-USY catalysts on the low-temperature activity of NH_3-SCR. *Catal. Commun.* **2018**, *114*, 60–64. [CrossRef]
11. Zhang, X.; Cheng, X.; Ma, C.; Wang, X.; Wang, Z. Effect of a ZrO_2 support on Cu/Fe_2O_3–CeO_2/ZrO_2 catalysts for NO removal by CO using a rotary reactor. *Catal. Sci. Technol.* **2018**, *8*, 5623–5631. [CrossRef]
12. Xue, H.; Meng, T.; Liu, F.; Guo, X.; Wang, S.; Mao, D. Enhanced resistance to calcium poisoning on Zr-modified Cu/ZSM-5 catalysts for the selective catalytic reduction of NO with NH_3. *RSC Adv.* **2019**, *9*, 38477–38485. [CrossRef] [PubMed]
13. Wang, C.; Wang, J.; Wang, J.; Wang, J.; Shen, M. Promotional effect of ion-exchanged K on the low-temperature hydrothermal stability of Cu/SAPO-34 and its synergic application with Fe/Beta catalysts. *Front. Environ. Sci. Eng.* **2021**, *15*, 30. [CrossRef]
14. Gao, Z.; Zhao, D.; Yang, Y.; Jiang, X.; Tian, Y.; Ding, T.; Li, X. Influence of copper locations on catalytic properties and activities of Cu/SAPO-34 in C_3H_6-SCR. *Ind. Eng. Chem. Res.* **2021**, *60*, 6940–6949. [CrossRef]
15. Salah Aldeen, O.D.A.; Mahmoud, M.Z.; Majdi, H.S.; Mutlak, D.A.; Uktamov, K.F.; Kianfar, E. Investigation of Effective Parameters Ce and Zr in the Synthesis of H-ZSM-5 and SAPO-34 on the Production of Light Olefins from Naphtha. *Adv. Mater. Sci. Eng.* **2022**, *2022*, 6165180. [CrossRef]
16. Huang, F.; Cao, J.; Wang, L.; Wang, X.; Liu, F. Enhanced catalytic behavior for methanol to lower olefins over SAPO-34 composited with ZrO_2. *Chem. Eng. J.* **2020**, *380*, 122626. [CrossRef]
17. Aghaei, E.; Haghighi, M.; Pazhohniya, Z.; Aghamohammadi, S. One-pot hydrothermal synthesis of nanostructured ZrAPSO-34 powder: Effect of Zr-loading on physicochemical properties and catalytic performance in conversion of methanol to ethylene and propylene. *Microporous Mesoporous Mater.* **2016**, *226*, 331–343. [CrossRef]
18. Varzaneh, A.Z.; Towfighi, J.; Mohamadalizadeh, A. Comparative study of naphtha cracking over SAPO-34 and HZSM-5: Effects of cerium and zirconium on the catalytic performance. *J. Anal. Appl. Pyrolysis* **2014**, *107*, 165–173. [CrossRef]
19. Tian, S.; Ji, S.; Lü, D.; Bai, B.; Sun, Q. Preparation of modified Ce-SAPO-34 catalysts and their catalytic performances of methanol to olefins. *J. Energy Chem.* **2013**, *22*, 605–609. [CrossRef]
20. Bin, F.; Song, C.; Lv, G.; Song, J.; Wu, S.; Li, X. Selective catalytic reduction of nitric oxide with ammonia over zirconium-doped copper/ZSM-5 catalysts. *Appl. Catal. B Environ.* **2014**, *150*, 532–543. [CrossRef]
21. Chen, Q.; Yang, Y.; Luo, H.; Liu, Z.; Tong, Z.; Tao, C.; Du, J. Ce regulated surface properties of Mn/SAPO-34 for improved NH_3-SCR at low temperature. *RSC Adv.* **2020**, *10*, 40047–40054. [CrossRef] [PubMed]
22. Wang, S.; Fan, C.; Zhao, Z.; Liu, Q.; Xu, G.; Wu, M.; Chen, J.; Li, J. A facile and controllable in situ sulfation strategy for CuCeZr catalyst for NH_3-SCR. *Appl. Catal. A Gen.* **2020**, *597*, 117554. [CrossRef]
23. Zhao, H.; Xie, G.; Liu, Z.; Liu, Y. A Combined in-situ Diffuse Reflectance FTIR and On-line Mass Spectroscopy Study of Surface Acidity and Reactivity over a CuO/Al_2O_3 Catalyst. *Acta Chim. Sin.* **2008**, *66*, 1021–1027.
24. Zhang, Q.; Wang, X. Preparation and properties of Ce-Mn/ZSM-5 catalysts modified with different metals. *J. Fuel Chem. Technol.* **2019**, *47*, 1265–1272.
25. Guo, J.; Yang, W.; Zhang, Y.; Gan, L.; Fan, C.; Chen, J.; Peng, Y.; Li, J. A multiple-active-site Cu/SSZ-13 for NH_3-SCO: Influence of Si/Al ratio on the catalytic performance. *Catal. Commun.* **2020**, *135*, 105751. [CrossRef]
26. Zhang, Q.; Fan, J.; Ning, P.; Song, Z.; Liu, X.; Wang, L.; Wang, J.; Wang, H.; Long, K. *In situ* DRIFTS investigation of NH_3-SCR reaction over CeO_2/zirconium phosphate catalyst. *Appl. Surf. Sci.* **2018**, *435*, 1037–1045. [CrossRef]
27. Wang, S.; Chen, Z.; He, B.; Yan, Z.; Wang, H.; Liu, L.; Wang, X. In situ DRIFTS investigation on CeOx catalyst supported by fly-ash-made porous cordierite ceramics for low-temperature NH_3-SCR of NO$_X$. *Catalysts* **2019**, *9*, 496. [CrossRef]

28. Xie, S.; Li, L.; Jin, L.; Wu, Y.; Liu, H.; Qin, Q.; Wei, X.; Liu, J.; Dong, L.; Li, B. Low temperature high activity of M (M = Ce, Fe, Co, Ni) doped M-Mn/TiO_2 catalysts for NH_3-SCR and in situ DRIFTS for investigating the reaction mechanism. *Appl. Surf. Sci.* **2020**, *515*, 146014. [CrossRef]
29. Zhang, L.; Pierce, J.; Leung, V.L.; Wang, D.; Epling, W.S. Characterization of Ceria's Interaction with NO_X and NH_3. *J. Phys. Chem. C* **2013**, *117*, 8282–8289. [CrossRef]
30. Gao, C.; Yang, G.; Huang, X.; Yang, Q.; Li, B.; Wang, D.; Peng, Y.; Li, J.; Lu, C.; Crittenden, J. Key intermediates from simultaneous removal of NOx and chlorobenzene over a V_2O_5–WO_3/TiO_2 catalyst: A combined experimental and DFT study. *Catal. Sci. Technol.* **2021**, *11*, 7260–7267. [CrossRef]
31. Liang, H.; Gui, K.; Zha, X. DRIFTS study of γFe_2O_3 nano-catalyst for low-temperature selective catalytic reduction of NO_X with NH_3. *Can. J. Chem. Eng.* **2016**, *94*, 1668–1675. [CrossRef]
32. Mihaylov, M.Y.; Zdravkova, V.R.; Ivanova, E.Z.; Aleksandrov, H.A.; Petkov, P.S.; Vayssilov, G.N.; Hadjiivanov, K.I. Infrared spectra of surface nitrates: Revision of the current opinions based on the case study of ceria. *J. Catal.* **2021**, *394*, 245–258. [CrossRef]
33. Weng, X.; Dai, X.; Zeng, Q.; Liu, Y.; Wu, Z. DRIFT studies on promotion mechanism of $H_3PW_{12}O_{40}$ in selective catalytic reduction of NO with NH_3. *J. Colloid Interface Sci.* **2016**, *461*, 9–14. [CrossRef] [PubMed]
34. Morandi, S.; Prinetto, F.; Ghiotti, G.; Castoldi, L.; Lietti, L.; Forzatti, P.; Daturi, M.; Blasin-Aubé, V. The influence of CO_2 and H_2O on the storage properties of Pt-Ba/Al_2O_3 LNT catalyst studied by FT-IR spectroscopy and transient microreactor experiments. *Catal. Today* **2014**, *231*, 116–124. [CrossRef]
35. Jia, Y.; Jiang, J.; Zheng, R.; Guo, L.; Yuan, J.; Zhang, S.; Gu, M. Insight into the reaction mechanism over PMoA for low temperature NH_3-SCR: A combined In-situ DRIFTs and DFT transition state calculations. *J. Hazard. Mater.* **2021**, *412*, 125258. [CrossRef]
36. Chen, L.; Si, Z.; Wu, X.; Weng, D. DRIFT study of CuO–CeO_2–TiO_2 mixed oxides for NO_X reduction with NH_3 at low temperatures. *ACS Appl. Mater. Interfaces* **2014**, *6*, 8134–8145. [CrossRef]
37. Wei, L.; Wang, Z.; Liu, Y.; Guo, G.; Dai, H.; Cui, S.; Deng, J. Support promotion effect on the SO_2 and K^+ co-poisoning resistance of MnO_2/TiO_2 for NH_3-SCR of NO. *J. Hazard. Mater.* **2021**, *416*, 126117. [CrossRef] [PubMed]
38. Wang, J.; Peng, Z.; Qiao, H.; Yu, H.; Hu, Y.; Chang, L.; Bao, W. Cerium stabilized Cu-SSZ-13 Catalyst for the Catalytic Removal of NO_X by NH_3. *Ind. Eng. Chem. Res.* **2016**, *55*, 1174–1182. [CrossRef]
39. Wilken, N.; Nedyalkova, R.; Kamasamudram, K.; Li, J.; Currier, N.W.; Vedaiyan, R.; Yezerets, A.; Olsson, L. Investigation of the Effect of Accelerated Hydrothermal Aging on the Cu Sites in a Cu-BEA Catalyst for NH_3-SCR Applications. *Top. Catal.* **2013**, *56*, 317–322. [CrossRef]
40. He, P.; Shen, D.; Liu, G. NH_3-SCR performance of modified SAPO-34 molecular sieve. *J. Southeast Univ. (Nat. Sci. Ed.)* **2017**, *47*, 513–520.
41. Chen, D.; Yan, Y.; Guo, A.; Rizzotto, V.; Lei, H.; Qiao, Z.; Liang, H.; Jabłońska, M.; Jiang, X.; Jiang, J.; et al. Mechanistic insights into the promotion of low-temperature NH_3-SCR catalysis by copper auto-reduction in Cu-zeolites. *Appl. Catal. B Environ.* **2023**, *322*, 122118. [CrossRef]
42. Yan, Q.; Gao, Y.; Li, Y.; Vasiliades, M.A.; Chen, S.; Zhang, C.; Gui, R.; Wang, Q.; Zhu, T.; Efstathiou, A.M. Promotional effect of Ce doping in $Cu_4Al_1O_X$–LDO catalyst for low-T practical NH_3-SCR: Steady-state and transient kinetics studies. *Appl. Catal. B Environ.* **2019**, *255*, 117749. [CrossRef]
43. Gao, X.; Jiang, Y.; Fu, Y.; Zhong, Y.; Luo, Z.; Cen, K. Preparation and characterization of CeO_2/TiO_2 catalysts for selective catalytic reduction of NO with NH_3. *Catal. Commun.* **2010**, *11*, 465–469. [CrossRef]
44. Xiao, M.; Zhang, X.; Yang, Y.; Cui, X.; Chen, T.; Wang, Y. M (M = Mn, Co, Cu)-CeO_2 catalysts to enhance their CO catalytic oxidation at a low temperature: Synergistic effects of the interaction between Ce^{3+}-M^{x+}-Ce^{4+} and the oxygen vacancy defects. *Fuel* **2022**, *323*, 124379. [CrossRef]
45. Raveendra, G.; Li, C.; Liu, B.; Cheng, Y.; Meng, F.; Zhong, L. Synthesis of lower olefins from syngas over Zn/Al_2O_3–SAPO-34 hybrid catalysts: Role of doped Zr and influence of the Zn/Al_2O_3 ratio. *Catal. Sci. Technol.* **2018**, *8*, 3527–3538. [CrossRef]
46. Li, W.; Li, M.; Liu, H.; Jia, W.; Yu, X.; Wang, S.; Zeng, X.; Sun, Y.; Wei, J.; Tang, X.; et al. Domino transformation of furfural to γ-valerolactone over SAPO-34 zeolite supported zirconium phosphate catalysts with tunable Lewis and Brønsted acid sites. *Mol. Catal.* **2021**, *506*, 111538. [CrossRef]

aerospace

MDPI

Article

Performance and Weight Parameters Calculation for Hydrogen- and Battery-Powered Aircraft Concepts

Michal Janovec [1,*], Viktor Babčan [1], Branislav Kandera [1], Kristína Šajbanová [1], Filip Škultéty [1] and Ľuboš Halvoň [2]

[1] Faculty of Operation and Economics of Transport and Communications, University of Žilina, SK-010-26 Žilina, Slovakia; babcan@stud.uniza.sk (V.B.); branislav.kandera@uniza.sk (B.K.); sajbanova@stud.uniza.sk (K.Š.); filip.skultety@uniza.sk (F.Š.)

[2] Department of Remote Sensing, National Forest Centre, SK-960-01 Zvolen, Slovakia; lubos.halvon@nlcsk.org

* Correspondence: michal.janovec@uniza.sk

Abstract: This article describes the creation of a program that would be useful for calculating mathematical models in order to estimate the weight of aircraft components. Using several parameters, it can calculate other parameters of civil transport aircraft powered by batteries or fuel cells. The main goals of this research were to add the missing dimensions and parameters to the aircraft database, create a simple but effective program for creating mathematical models, and use this program to find technological barriers to battery or hydrogen fuel-cell-powered aircraft concepts. The article introduces the reader to the problem of calculating OEW (operating empty weight) using Breguet–Leduc equations. A calculation model was created for OEW calculation. The result of this work is the verification of a mathematical model for battery-powered electric aircraft of the CS-23 (European Aviation Safety Agency Certification Specification for Normal, Utility, Aerobatic, and Commuter Category Aeroplanes) category by comparing the program's outputs with real aircraft. Subsequently, the results of mathematical models are shown in graphs that specify the space of possible concepts of aircraft powered by batteries or fuel cells, sorted by the number of passengers and the range of the aircraft, delimited by two or three criteria, respectively.

Keywords: battery-powered aircraft; Breguet–Leduc equations; energy efficiency factor; hydrogen propulsion; fuel cell; energy density; gravimetric efficiency; volumetric mass density; mass efficiency factor

1. Introduction

During the latest years of air transport aircraft development, the Achilles' heel of many aircraft manufacturers was the changing nature of the market demand. Two examples of this phenomenon that stand out are the Aérospatiale-BAC (British Aircraft Corporation) Concorde and the Airbus A380. The Concorde was designed as a response to the ever-increasing need for speed in air travel in the late 1960s. One of the main reasons for its retirement in 2003, however, was the fact that first-class passengers preferred the comfort of a subsonic aircraft cabin over the possibility of shaving a few hours off their travel time [1]. The market research for the A380 development dates to the early 1990s when air travel was booming, and the market showed an unanswered demand for a high-capacity subsonic aircraft [2]. However, during its development and production, the demand shifted towards relatively smaller aircraft, which can achieve a higher load factor on a point-to-point, rather than a hub-and-spoke, system. This fact sealed the fate of the A380, and its potential successors, with the end of production in 2021 [3].

In comparison with the challenges stated above, today's development of air transport aircraft has one big advantage—the undoubted need for a carbon-neutral airliner. Climate change is slowly moving from being only a threat into a full-blown emergency, taking human lives daily and showing a clear link with possible widespread famine, the worldwide migration crisis, biodiversity collapse, and much more in the future [4]. According to carbon emissions data provided by Emission Database for Global Atmospheric Research (EDGAR), international aviation will account for approximately 1% of total global emissions in 2020. The IEA states that between 2013 and 2019, prior to the COVID-19 epidemic, global carbon emissions from air transport already exceeded 70% of the ICAO's projected values. By 2050, aviation will be responsible for 25% of global carbon emissions if current trends continue [5]. One can easily conclude that the demand for carbon-neutral air travel will only grow. The research and development of carbon emissions in aviation for the period from 1992 to 2021 is outlined in the article A Bibliometric Analysis and Visualization of Aviation Carbon Emissions Studies [5]. The use of sustainable aviation fuels (SAFs) can also play a key role in achieving future emission targets; bio and synthetic fuels are currently the most viable option, but hydrogen is also being considered as a long-term solution [6]. It is hydrogen that proves to be a suitable fuel for industries that pollute the environment, such as aviation. It is abundant, clean, and produces no carbon emissions, only water after use, which has the potential to cool the environment. However, other issues such as storage, costs, and the entire production chain require significant efforts to implement hydrogen [7].

Propulsion systems powered by proton exchange membrane fuel cells (PEMFCs) emit only water vapor and would reduce the climate impact of aviation. Research into the feasibility of hydrogen-powered PEMFC propulsion systems for applications in various aircraft categories was carried out by Schmelcher and Häßy [8]. The authors Trainelli et al. [9] focused on the sizing methodology of the hybrid electric drive of the aircraft based on fuel cells. The method is based on an accurate physical model of the fuel cell module, integrated within a procedure that, given aircraft and mission parameters, estimates the corresponding sizing and performance.

The possibility of using liquid hydrogen and fuel cells was also addressed by Nicolay et al. [10]. The research focused on finding an efficient configuration of the aircraft and suitable components related to propulsion, such as a hydrogen tank, fuel cells, and electric motors.

A clear goal was set during the 2021 United Nations Climate Change Conference (COP26): securing global net-zero emissions by midcentury while keeping 1.5 °C of temperature rise within reach. Hydrogen (H_2) propulsion technology provides one of the most promising opportunities to meet this target and decarbonize global aviation. Challenges and recommendations for future research for hydrogen-powered aircraft and fuel cells are provided by research by Gao et al. [11].

In order to design fuel cell systems for passenger aircraft, it is necessary to specify the requirements that the system has to fulfill. A statistical approach to the analysis of such requirements was elaborated in their research by Kadyk et al. [12]. By using Monte Carlo-based sampling of flight mission profiles, the range of necessary component sizes, as well as optimal degrees of hybridization with a battery, is explored, and design options are evaluated.

Electric and hybrid aircraft drives may initially find use in light-sport aircraft and regional aircraft. The feasibility of an electrified drive for light-sport aircraft is confirmed by the authors McQueen et al. [13]. The simulations performed in this research theoretically confirm that hybrid-electric aircraft could achieve 44% of the range of conventionally powered aircraft. Therefore, it is still necessary to improve the power-to-weight ratio of the existing battery technology before electrified propulsion becomes a contender in the light-sport aircraft segment.

The authors Palaia and Salem [14] deal with the possible application of hybrid-electric drives in regional aircraft, whose research results show that the hybrid-electric drive integrated into regional aircraft provides direct fuel savings, which have a positive impact on environmental protection.

Over recent years, many prototypes of hybrid aircraft have been proposed. In 2017, Frosina et al. [15] developed a simple model to evaluate the performance of a light aircraft equipped with a hybrid-electric propulsion system. Marinaro et al. [16], in their article, present a numerical study of a light electric aircraft whose architecture based on batteries is converted to an architecture based on fuel cells with the contribution of a battery pack that essentially intervenes in the take-off phase and, in addition, of a pack of ultracapacitors to prevent a high discharge rate of the battery. The final intent is to explore the range extension without an increase in maximum take-off weight.

The authors Staack et al. [17] deal with the analysis of system components and possible construction concepts of a fully electric aircraft. These limitations are illustrated with a conceptual study of a full-electric FAR/CS-23 commuter aircraft and realistic estimations of its performance.

This article focuses on better understanding the relationships between different dimensions and parameters of modern commercial aircraft. It uses a large database of nearly all commercial aircraft in service today. This database is enriched and corrected to create a unique collection of information that has not yet been accessible. The development of a code that draws links and trends from the database is then described, and the initial results serve to picture a technological barrier of different new propulsion technologies. As we focused on battery and hydrogen technologies, we dedicated the theoretical description to these two, focusing on the key numerical values that may serve as their comparison.

2. Theoretical Description of Aircraft Propulsion Technologies

Of all the technologies proposed to supply and produce a thrust of a future air transport aircraft, two of them show the best potential to be used in the future: battery and hydrogen technology. Batteries are widely used nowadays in road vehicles, and their capacities are growing with leap improvements linked to discoveries of new chemistries every couple of years. This improvement makes the overall long-term curve of their energy capacity show an exponential improvement. Hydrogen, which is mostly used as a fuel cell, is still finding its way into road vehicles and aircraft alike [18]. However, its recent improvements and experimentations show some advantages of this technology.

2.1. Liquified Hydrogen

The liquified hydrogen technology seems to be the best choice out of all hydrogen storage technologies due to its high natural energy density, 120 MJ/kg, and its potential to reach relatively high gravimetric and volumetric efficiency [19].

Gravimetric efficiency, also called gravimetric capacity, refers to the ratio of the mass of the propellant alone versus the mass of the entire storage system, including the propellant [20], and is commonly presented as the percentage of hydrogen by weight. As we concluded from many articles, the current values of gravimetric efficiencies of this technology vary greatly. The state-of-the-art value is in the range of 8–25% [21], but values up to 30% can be expected by 2035 [22,23]. In the work of Marciello et al. [24], it was assumed that by 2050, hydrogen fuel cells would be advanced enough to completely replace kerosene-based engines.

Volumetric efficiency is typically represented in absolute values of grams per liter, and we will use a numerically equivalent SI unit of kilograms per cubic meter. The volumetric density of liquid hydrogen is 71 kg/m^3 (at −252.87 °C and 1.013 bar) [25]. As for the volumetric density of the entire system, the same as gravimetric efficiency, this measure also shows large variations in different studies. The state-of-the-art value is around 40–50 kg/m^3 (56–70%) [22,26], and we may anticipate an increase up to 60 kg/m^3 (85%) in the upcoming decade, which is the value used in this calculation.

2.2. Gaseous Hydrogen

Nowadays, gaseous hydrogen is widely used in the automotive industry. Its tanks are usually compressed to 700 bar, which gives a gravimetric efficiency of 5.7% for the Toyota Mirai tank [26]. This value is rounded up to 6% for future calculations and expects an increase to 10% in the upcoming decade. The volumetric density of gaseous hydrogen is 42 kg/m^3 (at 25 °C and 700 bar). In the same case of Toyota Mirai, the volumetric energy density is up to 4.9 MJ/dm^3, which gives us a very high volumetric density of the entire tank system of 40 kg/m^3 (95%). Therefore, this value is used to represent the current practice. As for the values in the upcoming decade, Rivard et al. [27] set the ultimate value to 6.1 MJ/dm^3 (considering tanks with higher pressure), so the rough mean value between the present and the ultimate one is taken, namely 5.5 MJ/dm^3. An illustration of a hydrogen tank can be seen in the following Figure 1, where total fuel is represented by the blue area and the mass of the rest counts as the delta system mass.

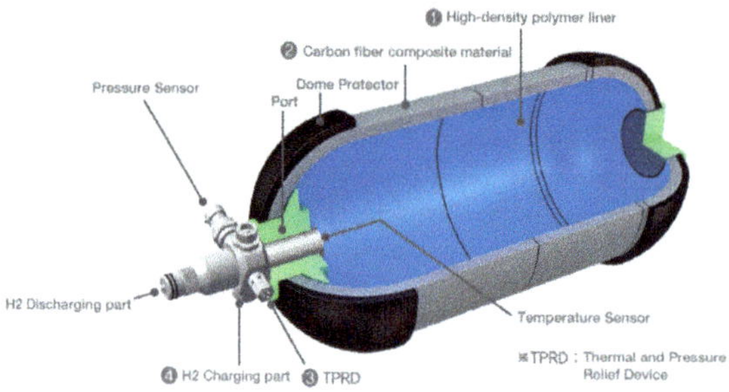

Figure 1. Hydrogen tank. Adapted with permission from Ref. [28]. 2023, Doosanmobility.

2.3. *Battery*

We know about two types of battery technology that show an improvement strong enough to be considered potential energy vectors of a future commercial electric aircraft:

- Lithium–sulfur

This type promises substantial improvements when compared to lithium-ion batteries. It was used, for example, in the Airbus Zephyr solar aircraft [29]. Lithium–sulfur (Li–S) battery is identified as one of the most promising next-generation energy storage systems due to its ultra-high theoretical energy density up to 2600 Wh/kg (which is approx. 9 MJ/kg) [30]. A disadvantage of this type is that the batteries can experience rapid loss in capacity over multiple cycles due to changes in electrode volume. This phenomenon is currently being tackled by the use of binders within the cathode of the battery cell [31]. The current applications of lithium–sulfur batteries have relatively low volumetric mass density. However, there is a technology of compressed graphene/carbon nanotubes (G/CNT) aerogels, which allows for a high volumetric mass capacity, namely 1640 kg/m^3 [32].

- Metal–air

This type typically uses a sodium/potassium hydroxide or sodium chloride solution as the electrolyte with a metal anode and an ambient air cathode. It is being considered for the Alice electric business aircraft of the Eviation company [33]. These battery cells' theoretical maximum energy density is 46 MJ/kg. The proposed technologies for putting this theory into practice are met with many drawbacks, such as the need for flammable organic electrolytes or noble metal nanomaterials to use as electrocatalysts. It is also important to mention the fact that recharging these batteries is still very problematic. Therefore, the solution to this could be a concept of battery packs, which would be quickly and easily removable [34]. For the purpose of further calculations, a hypothesis of achieving an energy density of these battery cells of 20 MJ/kg is proposed. As for the volumetric mass density, a reference value of 1500 kg/m^3 is used, similar to the one achievable today.

2.4. *Conventional Kerosene (Jet A-1)*

Given the fact that virtually all modern airliners use the concept of a wet wing, the added weight of the tank system is very low. A 10% decrease in energy density is used to account for fuel pumps, ventilation parts, sealant, and other parts needed in such a structure. The values of the volumetric mass density of Jet A-1 were computed on the whole scale of its possible values, varying with temperature.

All the hypotheses above led to a comparison of the resulting values in Table 1.

Table 1. Comparison of alternative energy vectors.

	Jet A-1	Near Future			2035		
		Gaseous H_2	Liquid H_2	Battery	Gaseous H_2	Liquid H_2	Battery
Energy Density (fuel alone) (MJ/kg)	42.8	120	120	9	120	120	20
Gravimetric efficiency	90%	6%	10%	60%	10%	30%	75%
Energy Density (fuel + tank system) (MJ/kg)	38.52	7.2	12	5.4	12	35.7	15
Vol. Mass Density (fuel alone) (kg/m^3)	775–840	42	71	N/A	N/A	71	
Vol. Mass Density (fuel and tank system) (kg/m^3)	~775–840	40	45	1640	N/A	60	1500
Vol. Energy Density (fuel and tank system) (MJ/dm^3)	~33.17–35.95	4.9	5.4	14.76	5.5	7.2	30

In conclusion, can be said that the only technology that has a chance of matching the energy density of kerosene is liquid H_2. Regarding volumetric energy density, the only technology matching kerosene values is a battery. Both of those technologies have their specific characteristics, and they should be seen as partners rather than competitors.

3. Presentation of the Stakes, Tools, and Objectives

When planning a new concept of an aircraft, the first task we must do is to define the design space. The design space of conception can be defined through three directions, also known as the 3 main axes:

1. Technologies: hydrogen, electric, etc.
2. Service: payload vs. range, altitude, speed, cost, etc.
3. Aircraft category: VTOL/STOL (vertical/short take-off and landing), Cargo/PAX (passenger), etc.

The question is, therefore: how do we proceed when exploring this 3-dimensional space? Our proposition is to start by defining tools that will enable us to rank the solutions in the design space.

3.1. Evaluation Tools

The first step of a database analysis is to define the evaluation criteria. We start by stating 3 typical evaluation criteria from the service quality point of view of a certain transport service:

- PK: Npax × distance;
- MK: payload × distance;
- LT: lifetime of the device (expressed in amount of using cycles).

Next, we have criteria from the point of view of the physical cost of a transport service. We can divide those according to two domains, with the first being the domain of energies:

- DE: direct energy consumption (e.g., the amount of fuel burnt);
- PE: primary energy consumed (e.g., if the fuel is Hydrogen that has been produced by electrolysis;
- RE: recycling energy (spread over lifetime).

The second is the domain of materials:

- MTOW: the maximum weight of the device;
- OEW: the empty weight of the device;
- MSIJ: the mass of the various materials used to build the device times the "scarcity index" of the materials;
- EE: embodied energy (spread over a lifetime).

We can then combine these criteria to evaluate the efficiency of an aircraft. These combined criteria can be, for example:

- Energetic transport criteria: PK/MTOW, PK/DE, PK/PE, PK.LT/(PE + EE), PK.LT/(PE + EE + RE), etc.
- Mass transport criteria: MK/MTOW, MK/OEW, MK.LT/Σ MSI, etc.

With these evaluations, we can define limits for new aircraft (battery or hydrogen-powered) to find out if the aircraft concept is viable or not. Among all the criteria we determined, we chose to focus on the PK/MTOW.

3.2. *Completion of the Database*

At the beginning of our research, we worked on an incomplete database of 324 civil aircraft. The given database contained errors and blank spaces for some data. It was also divided into two separate files (commercial aircraft and business plus general aviation), where sometimes one or both of the files entirely lacked the required parameters. Our first task was, therefore, to fill in the missing data and check the existing ones.

To find the data, we used both the literature and the Internet. From the literature, we used the following two encyclopedias:

- Jane's: All the World's Aircraft—In Service [35];
- Élodie Roux: Avions civils à réaction [36].

On the Internet, the most complex data we found came from the following websites (both accessed on 13 March 2023):

- www.jetav.com
- www.skybrary.aero

However, certain data were untraceable in any of these sources. It was mostly the case of more detailed dimensions, such as the HTP (horizontal taiplane), VTP (vertical tailplane) area, or propeller diameter. Therefore, we had to measure these dimensions directly from the aircraft blueprints. The blueprints we used usually came from websites the-blueprints.com or drawingdatabase.com.

To measure the needed dimensions, we used ImageJ software, in which we always set the scale first on a horizontal dimension (for example, fuselage length) and then checked it on a vertical one (for example, wingspan) or vice versa in order to check for the correctness of the blueprint or potential distortion due to non-linear zooming of the image. Finally, to find the area of the HTP, we used a simple method of extending the leading and the trailing edges. In the case of the VTP of smaller GA aircraft, the area was approached by connecting the root of the fin with the lower inboard corner of the rudder, as shown in Figure 2.

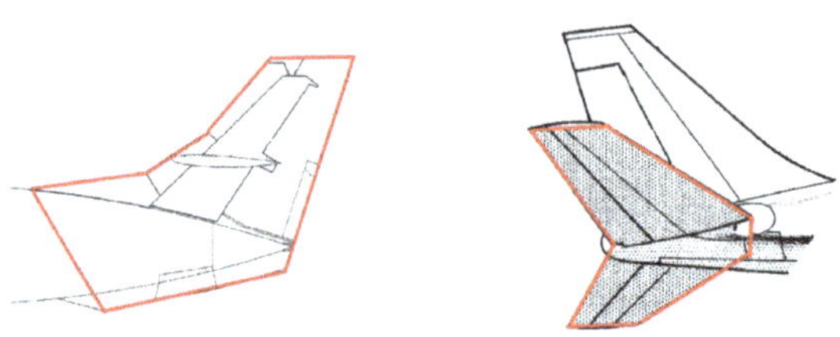

Figure 2. Methods of measuring HTP and VTP surfaces.

4. Method for the Identification of the Masses

Having established our main criterion and introduced the database, let us now analyze the database on the performance aspects to set up hard and soft limits for our future calculations. First of all, we have to specify that we are interested in "small" aircraft of the CS-23 category—aircraft whose mass is less than 5700 kg.

4.1. *PK/MTOW vs. Range*

As we can see in Figure 3, if we plot the PK/MTOW (marked vs. nominal range of the aircraft in our database), we see that the aircraft is well dispersed around the regression line and that their relative errors are increasing with higher nominal range and higher

PK/MTOW. We can therefore define a lower limit of a minimal PK/MTOW regarding range, which will accept 100% of the database's aircraft. In other words, we set up a maximum weight of an aircraft carrying a given number of passengers over a given distance. The formula for this limit is very simple:

$$(\text{PK}/\text{MTOW})_{\text{mini}} = \frac{\text{range}_{(\text{km})}}{670} \tag{1}$$

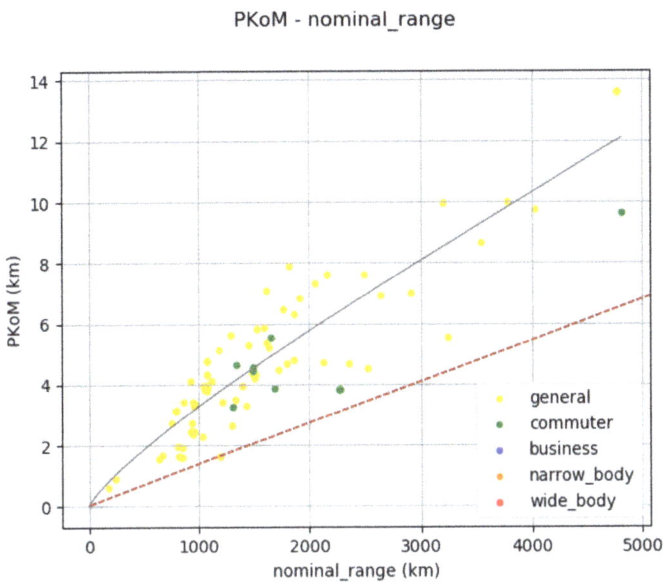

Figure 3. Minimal PK/MTOW.

In Figure 3, we marked this function by a red hatched line. Let us stress the fact that we will consider this a limit for any new aircraft concepts, i.e., we will not build an aircraft below this line because it would be simply too heavy for its mission.

4.2. Npax vs. Range

Having defined a lower limit in terms of the range of our conceptual aircraft, let us now set up an upper limit. For this, we chose the number of passengers as the defing factor. If we plot the number of passengers on the range (Figure 4), we can see that an upper constraint can almost exclusively limit the existing aircraft. This constraint gives us the minimum distance that can be accepted for a given passenger capacity. Note that we are not discussing a limit here, but only a constraint that shows us the contemporary practice among airplane manufacturers. Once again, its formula is very simple, and one can see that it corresponds to 19 passengers over 1300 km:

$$\text{p}/\text{k}_{\text{max}} = \frac{19}{1300 \text{ km}} = 0.0146 \tag{2}$$

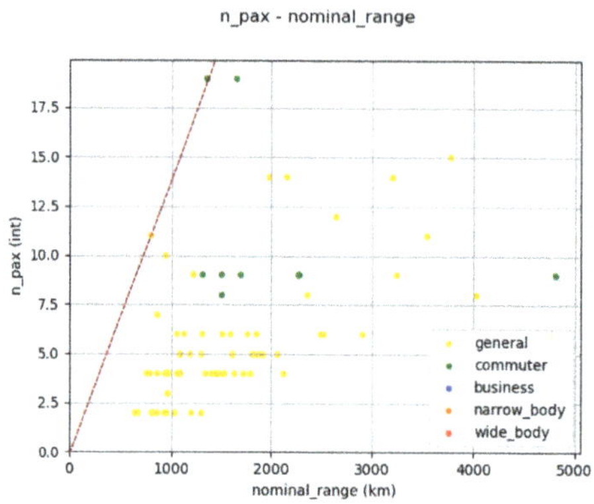

Figure 4. Maximal number of PAX per range.

4.3. OEW Estimation

If we plot the OEW vs. MTOW of our database (Figure 5), we can see a fairly linear regression.

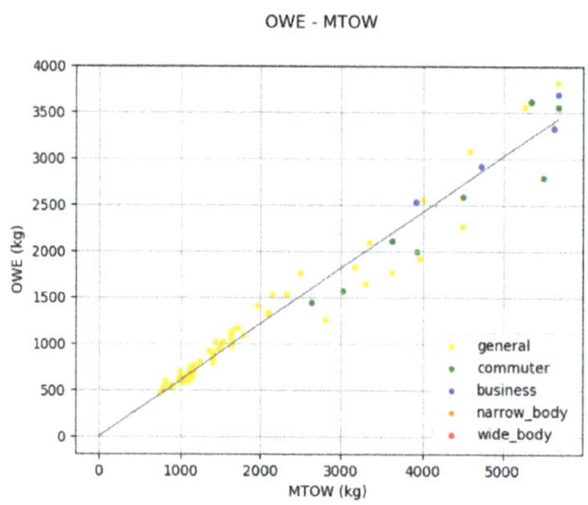

Figure 5. OEW vs. MTOW.

This regression can be expressed mathematically as:

$$\text{owe}_{\text{basic}} = -9.63\text{e} - 7 \cdot \text{mtow}^2 + 0.61 \cdot \text{mtow} \pm 10\% \quad (3)$$

We notice 2 points about the regression:

1. That the regression is quite robust, i.e., it remains valid whether the aircraft has a turboprop or a piston engine. We know that the power density of an electric engine is similar to the density of a turbine of a turboprop (considering that the propeller remains the same).
2. That the regression is independent of an aircraft's design capacity and range. In other words, the regression is valid, whatever the mass we add to the OEW.

We will therefore develop a model of calculation of the OEW. This model is explained by the two schemes in Figure 6. We assume that we have this relationship between the OEW and the MTOW of a conventional aircraft, and we will keep these relations, but we will adjust the masses of the components inside (add the new engine, the mass of the new system, etc.). We assume that the regression is still valid to link MTOW with the basic part of the new OEW.

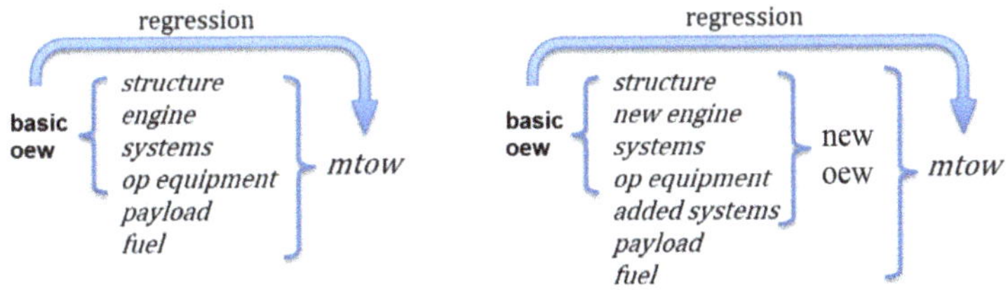

Figure 6. Comparison of OEW/MTOW regression for conventional and hydrogen/battery-powered aircraft.

To put into practice the principle in Figure 6, we remove the propulsion of a piston engine and add the mass of an electric motor and the mass of the batteries. This process is shown step by step in Figure 7 below:

$$\text{oew}_{\text{new}} = -9.63e-7\cdot mtow^2 + 0.61\cdot mtow$$

$$-\left(\frac{pw_{max}}{piston_{power\,density}}\right) \quad \text{Remove basic engine mass}$$

$$+\left(\frac{pw_{max}}{motor_{power\,density}}\right)+\left(\frac{pw_{max}}{power\_elec_{power\,density}}\right) \quad \text{Add new engine mass}$$

$$+\left(\frac{pw_{cruise}\cdot distance/tas}{battery_{energy\,density}}\right) \quad \text{Add battery mass}$$

Figure 7. Example of OEW estimation in case of replacement of a piston engine by an electroprop with batteries.

4.4. Installed Power Estimation

In order to scale the new engines, we need to approximate the necessary power of an aircraft. We find that the total power (power multiplied by the number of engines) of our database's aircraft is strongly associated with their MTOW. Therefore, we can proceed with a similar approach to the OEW estimation. In Figure 8 below, we see that the regression is quadratic.

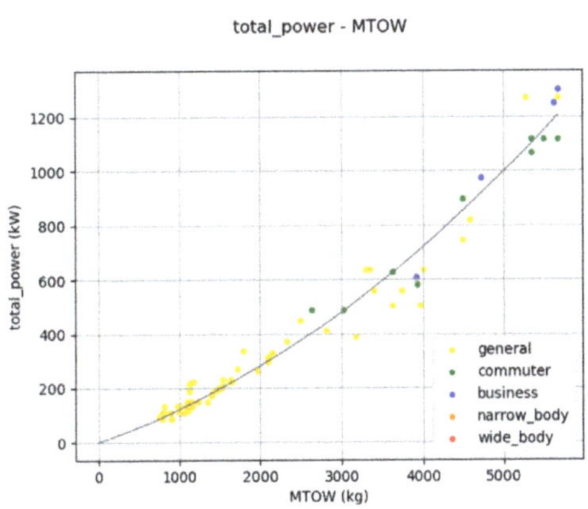

Figure 8. Total power vs. MTOW.

The following equation can be used to express this regression mathematically:

$$pw_{max} = 0.0197\cdot mtow^2 + 100.6\cdot mtow \pm 20\% \tag{4}$$

Our model will use this equation to scale the selected power plant and its required systems.

4.5. Breguet–Leduc Demonstration

In order to calculate OEW, we first need to calculate the fuel required for a certain mission. We perform this by using a few aircraft parameters, such as speed, SFC (specific fuel consumption), lift-over-drag ratio, or engine efficiencies; we are going to derive the Breguet–Leduc formula. We will first introduce some general equations needed to build the fundamental principle of these formulas. Then, we will divide the calculations into six groups, according to the required propulsion system, and express the range of an exemplary aircraft of each group. Those expressions will then help us to calculate the total fuel, including all the different types of reserve fuel.

4.5.1. General Equations

Firstly, we will take the basic fuel flow formula:

$$\mathrm{ff} = \frac{\mathrm{dm}}{\mathrm{dt}} \tag{5}$$

Then we will divide it by speed expressed in a similar manner:

$$\mathrm{v} = \frac{\mathrm{dx}}{\mathrm{dt}} \tag{6}$$

It will give us an expression that will become useful in the next calculations:

$$\frac{\mathrm{dm}}{\mathrm{dx}} = \frac{\mathrm{ff}}{\mathrm{v}} \tag{7}$$

Now, let us take the two basic flight formulas. The lift and the drag formulas, respectively, are:

$$\mathrm{mg} = \frac{1}{2}\rho \mathrm{S} \mathrm{v}^2 \mathrm{C_z} \tag{8}$$

$$F = \frac{1}{2}\rho S v^2 C_x \tag{9}$$

By dividing the lift formula by the drag formula, we obtain an expression of lift to drag ratio:

$$\frac{mg}{F} = \frac{C_z}{C_x} = \frac{L}{D} \tag{10}$$

For the purpose of the following calculations, we will express the thrust from the last equation:

$$\mathrm{F} = \frac{\mathrm{mg}}{\frac{\mathrm{L}}{\mathrm{D}}} \tag{11}$$

4.5.2. Range Calculation

Knowing that each combination of engine type and energy source type requires its own equation, we divided them in the following way:

1. Jet/fan engine
    - Conventional (turbojet/turbofan);
    - Electric (electrofan);
    - Battery;
    - Fuel cell (H_2).
2. Propeller engine
    - Conventional (piston or turboprop);
    - Electric (electroprop);
    - Battery;
    - Fuel cell (H_2).

Conventional Turbojet/Turbofan

From the definition of thrust-specific fuel consumption (TSFC), we know that it corresponds to the fuel consumption (fuel flow) of an engine per unit of thrust. Therefore, if we take Equation (7) and replace the fuel flow with the TSFC multiplied by thrust from Equation (11), we will obtain the following equation:

$$\frac{dm}{dx} = TSFC\frac{mg}{\frac{L}{D}}\frac{1}{v} \tag{12}$$

The development of this equation will lead us to the following expression of dx:

$$dx = \frac{\frac{L}{D}v}{TSFC*g}\frac{dm}{m} \tag{13}$$

As we will use this formula only for turbofan aircraft, we can replace the speed by a product of the Mach number and a speed of sound at 35,000 ft ISA (International Standard Atmosphere):

$$dx = \frac{\frac{L}{D}*M*a}{TSFC*g}\frac{dm}{m} \tag{14}$$

The integration of this equation will give us a final Breguet-Leduc formula for turbofan aircraft:

$$Range = \frac{\frac{L}{D}*M*a}{TSFC*g}\log\left(\frac{m_i}{m_f}\right) \tag{15}$$

where m_i stands for the initial mass of the aircraft (take-off mass) and m_f stands for its final mass (landing mass).

Conventional Propeller Engine

The final Breguet-Leduc formula for propeller-powered combustion engine aircraft is:

$$Range = \frac{\eta_p\frac{L}{D}}{PSFC*g}\log\left(\frac{m_i}{m_f}\right) \tag{16}$$

Battery-Powered Electrofan

In the case of electric motors, we need to split the power variable into two separate variables, the electric power W_e and the motor power W_m. The electric power represents the power coming from the batteries, and the motor power is the one on the shaft of the engine and thus equivalent to the shaft power of a turboprop engine. The ratio between W_e and W_m is expressed simply by the efficiency of the electric power train including power electronics and the motor η_m:

$$\eta_m = \frac{W_m}{W_e} \tag{17}$$

The final formula for expressing the battery-powered electrofan range is:

$$Range = \frac{\eta_f\eta_m\frac{L}{D}}{mg}E_{battery} \tag{18}$$

Fuel-Cell-Powered Electrofan

In the case of a hydrogen fuel cell system, we will use a method similar to the one for conventional fuel since, in both cases, we are reducing the amount—and therefore the weight—of the remaining fuel.

The required expression of the range is:

$$Range = \frac{\eta_f*\eta_m*\eta_{fc}*HHV*\frac{L}{D}}{g}\log\left(\frac{m_i}{m_f}\right) \tag{19}$$

Battery-Powered Electroprop

This system is, by its principle, identical to a battery-powered electrofan, with the only difference of using a propeller instead of a fan. Its range formula will therefore be the same as the one for an electrofan, with the only difference being the use of propeller efficiency η_p instead of fan efficiency η_f:

$$\text{Range} = \frac{\eta_p \eta_m \frac{L}{D}}{mg} E_{battery} \tag{20}$$

Fuel-Cell-Powered Electroprop

This system is once again identical to a fuel-cell-powered electrofan, with the only difference of using a propeller instead of a fan. To obtain its range formula will once again use the one for an electrofan and substitute fan efficiency η_f instead of propeller efficiency η_p:

$$\text{Range} = \frac{\eta_p * \eta_m * \eta_{fc} * HHV * \frac{L}{D}}{g} \log\left(\frac{m_i}{m_f}\right) \tag{21}$$

4.6. OEW Calculation

We are going to try to find a formula that would give an aircraft's operating empty weight (OEW) as a function of the aircraft's maximum take-off weight (MTOW) and its type on the basis of its mission. First, we need to identify the MTOW as follows:

$$\text{MTOW} = \text{OEW} + \text{payload} + \text{fuel} + \text{fuel\_reserve} \tag{22}$$

As the MTOW will be considered known and the OEW will be the result of the final formula, we will now concentrate on the other members of the sum.

4.6.1. Payload

The payload will be simply the number of passengers N_{pax} multiplied by a selected average mass of a passenger, including their baggage M_{pax}:

$$\text{Payload} = N_{pax} * M_{pax} \tag{23}$$

4.6.2. Leg Fuel

Leg fuel equals the fuel used during a flight leg. It refers to the distance between the two waypoints divided by groundspeed and multiplied by the fuel burn rate for the leg type.

Conventional Turbojet/Turbofan

In this part, we will calculate the trip fuel, that is, the fuel needed for a particular route, excluding all reserves. First, we will calculate it for the case of turbofan aircraft. Let us start where we left off—with the Breguet–Leduc formula:

$$\text{Range} = \frac{\frac{L}{D} * M * a}{TSFC * g} \log\left(\frac{m_i}{m_f}\right) \tag{24}$$

If we simplify the speed into v and approximate the initial and final masses as MTOW and (MTOW-leg fuel) respectively, we obtain:

$$\text{Range} = \frac{\frac{L}{D} * v}{TSFC * g} \log\left(\frac{\text{MTOW}}{\text{MTOW} - \text{leg fuel}}\right) \tag{25}$$

The development of this equation will lead us to:

$$\frac{TSFC*g*Range}{\frac{L}{D}*v} = \log\left(\frac{1}{1-\frac{\text{leg fuel}}{MTOW}}\right) \tag{26}$$

thus:

$$\left(1-\frac{\text{leg fuel}}{MTOW}\right)e^{\left(\frac{TSFC*g*Range}{\frac{L}{D}*v}\right)} = 1 \tag{27}$$

and finally:

$$\text{leg fuel} = MTOW\left[1-e^{-\left(\frac{TSFC*g*Range}{\frac{L}{D}*v}\right)}\right] \tag{28}$$

Battery-Powered Electrofan

The leg fuel is represented by the electric energy in the battery, $E_{battery}$, which is easily expressed from the range Formula (18) as:

$$E_{battery} = \frac{MTOW*g*Range}{\eta_f*\eta_m*\frac{L}{D}} \tag{29}$$

Fuel-Cell-Powered Electrofan

If we take the range expression (19) and express the weights in the same manner as before, we obtain:

$$Range = \frac{\eta_f*\eta_m*\eta_{fc}*HHV*\frac{L}{D}}{g}\log\left(\frac{MTOW}{MTOW-\text{leg fuel}}\right) \tag{30}$$

A similar development as the one used above leads us to:

$$\frac{g*Range}{\eta_f*\eta_m*\eta_{fc}*HHV*\frac{L}{D}} = \log\left(\frac{1}{1-\frac{\text{leg fuel}}{MTOW}}\right) \tag{31}$$

thus:

$$\left(1-\frac{\text{leg fuel}}{MTOW}\right)e^{\left(\frac{g*Range}{\eta_f*\eta_m*\eta_{fc}*HHV*\frac{L}{D}}\right)} = 1 \tag{32}$$

and finally:

$$\text{leg fuel} = MTOW\left[1-e^{-\left(\frac{g*Range}{\eta_f*\eta_m*\eta_{fc}*HHV*\frac{L}{D}}\right)}\right] \tag{33}$$

Conventional Propeller Engine

By developing (16) in a similar manner as for the conventional turbofan, we obtain:

$$\text{leg fuel} = MTOW\left[1-e^{-\left(\frac{PSFC*g*Range}{\eta_p*\frac{L}{D}}\right)}\right] \tag{34}$$

Battery-Powered Electroprop

Same as for the range formula, the electroprop leg fuel formula is equivalent to the electrofan one, only with the propeller efficiency:

$$E_{battery} = \frac{MTOW*g*Range}{\eta_p*\eta_m*\frac{L}{D}} \tag{35}$$

Fuel-Cell-Powered Electroprop

Once again, we obtain the electroprop formula by using the same method:

$$\text{leg fuel} = \text{MTOW}\left[1 - e^{-\left(\frac{g*\text{Range}}{\eta_p*\eta_m*\eta_{fc}*\text{HHV}*\frac{L}{D}}\right)}\right] \tag{36}$$

4.6.3. Reserve Fuel

The reserve fuel of most commercial flights of turbofan aircraft consists of three main parts: diversion fuel, contingency fuel, and holding fuel.

Diversion Fuel

We will count with an average diversion distance R_{div} (e.g., a route from the destination to the alternate airport) of 200 NM. To approximate the necessary fuel, we will use the same reasoning as for the trip fuel. The only difference is the range substituted by R_{div}:

$$\text{diversion fuel} = \text{MTOW}\left[1 - e^{-\left(\frac{\text{TSFC}*g*R_{div}}{\frac{L}{D}*v}\right)}\right] \tag{37}$$

Contingency Fuel

Contingency fuel is calculated simply by adding 3% or 5% to the trip fuel; therefore:

$$\text{contingency fuel} = k_r*\text{leg fuel} \tag{38}$$

where k_r is a factor of 0.03 or 0.05.

Holding Fuel

The final reserve is calculated in terms of time and usually equals to 30 min of flight time in a holding configuration. We will, therefore, derive it by using a combination of Equations (5) and (11) and the definition of the TSFC, which will give us the following:

$$\frac{dm}{dt} = \frac{\text{SFC}*mg}{\frac{L}{D}} \tag{39}$$

Therefore:

$$\frac{dm}{m} = \frac{\text{SFC}*g}{\frac{L}{D}}dt \tag{40}$$

This, after integration, equals to:

$$\log\frac{m_i}{m_f} = \frac{\text{SFC}*g}{\frac{L}{D}}\Delta \tag{41}$$

At this point, we can write the final mass as the initial mass minus Δm, where Δm corresponds to the fuel burned in the final holding phase. After developing the logarithm, we find the following:

$$\text{holding fuel} = (\text{MTOW} - \text{leg fuel})\left[1 - e^{-\left(\frac{\text{SFC}*g}{\frac{L}{D}}\Delta t\right)}\right] \tag{42}$$

We realize that the resulting formula is, in fact, a variation of the leg fuel formula, with two differences:

- Updating $MTOW$ to $MTOW - leg\ fuel$;
- Expressing $\frac{Range}{v}$ as Δt.

We can therefore obtain the formulas for all the other propulsion systems by following only these two steps. For the fuel-cell-powered electrofan, battery-powered electrofan, conventional propeller engine, fuel-cell-powered electroprop, and battery-powered electroprop, respectively, we obtain the following formulas:

$$E_{battery} = \frac{(MTOW - leg\ fuel) * g * \Delta t}{\eta_f * \eta_m * \frac{L}{D}} \quad (43)$$

$$holding\ fuel = (MTOW - leg\ fuel)\left[1 - e^{-\left(\frac{g*TAS*\Delta t}{\eta_f*\eta_m*\eta_{fc}*HHV*\frac{L}{D}}\right)}\right] \quad (44)$$

$$holding\ fuel = (MTOW - leg\ fuel)\left[1 - e^{-\left(\frac{SFC*g*TAS}{\frac{L}{D}*\eta_p}\Delta t\right)}\right] \quad (45)$$

$$E_{battery} = \frac{(MTOW - leg\ fuel) * g * TAS * \Delta t}{\eta_p * \eta_m * \frac{L}{D}} \quad (46)$$

$$holding\ fuel = (MTOW - leg\ fuel)\left[1 - e^{-\left(\frac{g*TAS*\Delta t}{\eta_p*\eta_m*\eta_{fc}*HHV*\frac{L}{D}}\right)}\right] \quad (47)$$

With the newly acquired relations, we can update Formula (22) for a conventional fuel turbofan as follows:

$$\begin{aligned} MTOW = OWE + N_{pax}M_{pax} + MTOW\left[1 - e^{-\left(\frac{TSFC*g*Range}{\frac{L}{D}*v}\right)}\right] \\ +MTOW\left[1 - e^{-\left(\frac{TSFC*g*R_{div}}{\frac{L}{D}*v}\right)}\right] + k_r * leg\ fuel \\ +(MTOW - leg\ fuel)\left[1 - e^{-\left(\frac{SFC*g}{\frac{L}{D}}\Delta t\right)}\right] \end{aligned} \quad (48)$$

5. Construction of a Model for Estimating the Masses of Aircraft Components

Once the criteria are defined, we need to calculate the parameters of the new concepts. We will simplify the calculation with the following assumptions:

- We focus only on the cruise leg, which we extend from the departure airport all the way to the destination airport.
- We consider climb energy as the potential energy acquired by the aircraft.
- As we use the Breguet–Leduc formula, we consider the trajectory to be static for the entire duration of the flight.
- As stated before, we will consider only small aircraft of the CS-23 category, meaning aircraft with MTOW < 5700 kg.
- The aircraft will therefore have a range of approximately 2000 km and lift to drag ratio of around 10.
- Regarding the mass, we make a small hypothesis that the hydrogen-powered aircraft is so light that even on long missions, we will not have more than 5% mass variation. This hypothesis is explained mathematically by the formula below:

$$\frac{m_{fuel}}{mass} = \frac{\left(\frac{mass \cdot g}{\frac{L}{D}} \cdot dist\right) \cdot \left(\frac{1}{\eta_{thpr}}\right) \cdot \left(\frac{1}{121e6}\right)}{mass} \approx 0.05 \quad (49)$$

where η_{thpr} is the efficiency in transforming fuel heat flow into propulsive power, and we approximate it as 0.3. Finally, if we consider a battery-powered aircraft, the mass is rigorously constant. Therefore, we can consider that the mass of our concept aircraft is constant.

5.1. Software Implementation

The python software created to build models prepares the raw data for and then executes the calculation of the abovementioned formulas. We are going to describe its calculation process step by step.

5.1.1. Coefficient Setup

In this part, we will set up all the coefficients required for future calculations. First, the software allows us to set a temperature deviation from the International Standard Atmosphere.

Next, we set the heating value of the combustion of kerosene to 43 MJ/kg and the one of hydrogen to 121 MJ/kg.

We can now set all the efficiencies of propulsion system components (Table 2). We set the propeller efficiency η_p to 0.8 and the fan efficiency η_f to 0.82. Regarding the electric motor, we set its motor efficiency to 0.93, as it is the minimal value of the efficiency of electric aviation motors of the Magnix company. This company is one of the most successful in developing electric motors for aviation, as their motor Magni500 powered the first-ever electrified commercial airplane, the De Havilland Beaver seaplane of Harbour Air, in 2019 [37]. The last required efficiency is one of the fuel cells. The maximum theoretical conversion efficiency of fuel cells is 82.7% [38]. One of the leading companies in fuel cell solutions, Horizon, proposes fuel cells with efficiencies of 40% and has shown a peak efficiency of 59% during tests of its newer products [39].

Table 2. Comparison of Fuel Cell Technologies. Adapted with permission from Ref. [40]. 2023, Energy.gov.

Fuel Cell Type	Common Electrolyte	Operating Temperature	Typical Stack Size	Electrical Efficiency (LHV)
Polymer Electrolyte Membrane (PEM)	Perfluorosulfonic acid	<120 °C	<1 kW–100 kW	60% direct H_2; 40% reformed fuel
Alkaline (AFC)	Aqueous potassium hydroxide soaked in a porous matrix or alkaline polymer membrane	<100 °C	1–100 kW	60%
Phosphoric Acid (PAFC)	Phosphoric acid soaked in a porous matrix or imbibed in a polymer membrane	150–200 °C	5–400 kW, 100 kW module (liquid PAFC); <10 kW (polymer membrane)	40%
Molten Carbonate (MCFC)	Molten lithium, sodium, and/or potassium carbonates, soaked in a porous matrix	600–700 °C	300 kW–3 MW, 300 kW module	50%
Solid Oxide (SOFC)	Yttria stabilized zirconia	500–1000 °C	1 kW–2 MW	60%

Let us now move on to setting the power densities of different propulsion solutions. The power density, also known as the power-to-weight ratio, expresses the nominal power of a certain propulsion system divided by its weight. We will use these densities as a simple way of calculating a transformation from one propulsion system to another. We set the turbofan power density to 7 kW/kg [41], the turboprop to 5 kW/kg [42,43], and the piston to 1 kW/kg, which is the order of magnitude for a typical Lycoming used in a Cessna 172 [44]. As for electric motors, for the purpose of further calculations, we are going to split the power densities into one value for the electric motor itself, one for a propeller installed on an electric motor (electroprop), and one for a fan installed in the same way (electrofan). For the electric motor itself, we will use the Magni500 once more as an example, with its power density of 4.2 kW/kg [45]. For the propeller, we take an example of the Bombardier Q400 with its 6-bladed all-composite Dowty R408, which weighs around 251.7 kg [46]. Given the fact that the engine's nominal power is 3781 kW, the ratio is 15 kW/kg. As for the fan, we took the approximation performed by Periklis Lolis [47], which gives the mass of the fan of a CFM56-7B27 engine, including its frames, as 1226.6 kg. This engine is used on 737–900 W, for which we calculated a power of 10,146 kW. The ratio will therefore be 8.3 kW/kg.

To be able to compare the use of battery-powered motors with conventional engines, we will also need the battery energy density and volumetric density. We will use projected values for two commonly used chemistries, Lithium-ion and Lithium–Sulfur [48]. The best-performing Lithium-ion chemistry, Li-NMC441, has a gravimetric density of around 250 Wh/kg and a volumetric density of around 400 Wh/dm^3, which translates to 1600 kg/m^3. The Lithium–Sulfur also shows values of around 400 Wh/kg for the gravimetric density, but the volumetric density is projected to be as high as 400 Wh/dm^3, which translates to 1000 kg/m^3. We will pre-set the coefficient to the values of the Lithium-ion, but they can be easily rewritten to Lithium–Sulfur or any other chemistry as required.

Finally, to be able to compare the use of fuel-cell-powered motors with conventional engines, we will add gravimetric and volumetric densities for each of the fuel cell components, which are the fuel cell itself, its cooling unit, and, finally, the liquid or gas hydrogen tank. For the gravimetric density of a fuel cell, we will use, once again, the Horizon fuel cell, which shows a density of 5 kW/kg [49]. For its cooling unit, we will make an approximation of 6 kW/kg, which corresponds to the data mentioned by Canders et al. [50]. To represent the gravimetric and volumetric densities of hydrogen tanks, we will use the gravimetric and volumetric index. As stated in Section 1, for liquified hydrogen, the state-of-the-art value for the gravimetric index is around 10% and around 45 kg/m^3 for the volumetric index. In the case of compressed gas hydrogen, the values are about half of those for liquified, that is, 6% for the gravimetric and 40 kg/m^3 for the volumetric index.

5.1.2. PAX Allowance

For the purpose of further calculations, we need to set the average weight of a passenger, including their luggage for different types of flights. We label this variable as the PAX allowance and pre-set it with the following values in Table 3.

Table 3. Cruise, diversion, and holding altitudes as per aircraft type.

Range (km)	**<1000**	**1000–8000**	**>8000**
PAX allowance (kg)	90	110	150

5.1.3. L/D

To calculate the lift-over-drag ratio, the software at first sets L/D limit values. These values are also coupled with MTOW limit values. The aircraft's MTOW is then compared with these limits. If the MTOW is found to be outside of the limits, its L/D is automatically assigned as the limit L/D value. If it is found to be between the limits, the L/D is calculated as:

$$\frac{L}{D} = \left(\frac{L}{D}\right)_{min} + \frac{\left[\left(\frac{L}{D}\right)_{max} - \left(\frac{L}{D}\right)_{min}\right] * (MTOW - MTOW_{min})}{MTOW_{max} - MTOW_{min}} \tag{50}$$

5.1.4. PSFC and TSFC

In the case of PSFC (propeller-specific fuel consumption) and TSFC, the calculation is the same, with the only difference of using the shaft power as an input for the PSFC and the equivalent shaft power developed during take-off for TSFC, as defined in the database.

We will describe the case of PSFC: The pre-set limit values of PSFC are coupled with limit values of shaft power. In the case of hydrogen-powered aircraft, the PSFC limit values are multiplied by the ratio of the fuel heating value of kerosene divided by the one of hydrogen. The reason is the formula for SFC itself:

$$SFC = \frac{V_0}{\eta_{th} * \eta_{pr} * FHV} \tag{51}$$

where η_{th} and η_{pr} stand for the thermal and propulsive efficiency, respectively, and V_0 is the air speed upstream of the engine.

Once the limit values are established, the aircraft's max shaft power (P) is then compared with the limits (P_{max} and P_{min}). If it is found to be between the limits, then it is calculated as:

$$PSFC = PSFC_{max} - \frac{(PSFC_{max} - PSFC_{min}) * (P - P_{min})}{P_{max} - P_{min}} \tag{52}$$

5.1.5. ref_power and ref_OEW

In this part, we set up the regression curves for the required total power for an airplane with a given MTOW and the averaged OEW for an airplane with a given MTOW, respectively. It is, in both cases, a least square linear regression, which forms the following:

$$y = (a * MTOW + b) * MTOW + c \tag{53}$$

where a, b, and c are the coefficients that are calculated by the software but have to be reset manually into the code whenever the database changes. For our calculations, we chose a polynomial equation of a second degree, but it is possible to change the degree should we require the function to follow any database trend more closely.

5.1.6. Cruise, Diversion, and Holding Altitudes

We assign cruise, diversion, and holding altitudes to each aircraft according to their type in the following order in Table 4:

Table 4. Cruise, diversion, and holding altitudes as per aircraft type.

Aeroplane Type	Cruise Altitude (ft)	Diversion Altitude (ft)	Holding Altitude (ft)
General aviation	5000	5000	1500
Commuter	20,000	10,000	1500
Business, Narrow body, Widebody	35,000	25,000	1500

5.1.7. Reserve Data

This part specifies the reserve fuel parameters for each airplane type. It assigns contingency fuel percentage, diversion leg length, and holding time in the following manner in Table 5:

Table 5. Percentage of contingency fuel, length of diversion, and holding time as per aircraft type.

Airplane Type	Contingency Fuel (%)	Diversion Length (NM)	Holding Time (min)
General aviation	0	0	30
Commuter	0	0	30
Business, Narrow body	5	200	30
Widebody	3	200	30

5.1.8. Leg Fuel

We now finally calculate the leg fuel (also referred to as trip fuel) required for a given distance.

Conventional Turbojet/Turbofan

We take the aforementioned Equation (28) and adjust it to suit the input data. We replace MTOW with a variable called start_mass, which corresponds to the mass at the beginning of a flight segment, so in this case, it would be the actual TOW. We also replace speed with the TAS (true airspeed) and range by the given distance.

$$\text{fuel} = \text{start\_mass}\left[1 - e^{-\left(\frac{\text{TSFC}*g*\text{distance}}{\text{TAS}*\frac{L}{D}}\right)}\right] \quad (54)$$

Fuel-Cell-Powered Electrofan

$$\text{fuel} = \text{start\_mass}\left[1 - e^{-\left(\frac{g*\text{distance}}{\eta_f*\eta_m*\eta_{fc}*\text{HHV}*\frac{L}{D}}\right)}\right] \quad (55)$$

where η_f stands for fan efficiency, η_m for motor efficiency, and η_{fc} for fuel cell efficiency.

Battery-Powered Electrofan

With this type of technology, the calculated value will be energy expressed in Joules as opposed to fuel mass expressed in kilograms.

$$E_{\text{battery}} = \frac{\text{start\_mass}*g*\text{distance}}{\eta_f*\eta_m*\frac{L}{D}} \quad (56)$$

Conventional Propeller

$$\text{fuel} = \text{start\_mass}\left[1 - e^{-\left(\frac{\text{PSFC}*g*\text{distance}}{\eta_p*\frac{L}{D}}\right)}\right] \quad (57)$$

Fuel-Cell-Powered Electroprop

$$\text{fuel} = \text{start\_mass}\left[1 - e^{-\left(\frac{g*\text{distance}}{\eta_p*\eta_m*\eta_{fc}*\text{HHV}*\frac{L}{D}}\right)}\right] \quad (58)$$

Battery-Powered Electroprop

$$E_{\text{battery}} = \frac{\text{start\_mass}*g*\text{distance}}{\eta_p*\eta_m*\frac{L}{D}} \quad (59)$$

Finally, we add 5% of the calculated leg fuel (or battery energy required for a leg) on top in order to take into account the climb phase of each flight.

5.1.9. Holding Fuel

Conventional Turbojet/Turbofan

Same as for the leg fuel, we take the appropriate Equation (42) and replace (MTOW—leg fuel) by start_mass, which in this case corresponds to the mass at the beginning of the holding phase.

$$\text{holding fuel} = \text{start\_mass}\left[1 - e^{-\left(\frac{\text{TSFC}*g*\Delta t}{\frac{L}{D}}\right)}\right] \quad (60)$$

We follow the same logic for all the remaining propulsion setups, mostly updating the leg fuel formulas with the relevant variables:
Fuel-Cell-Powered Electrofan

$$\text{holding fuel} = \text{start\_mass}\left[1 - e^{-\left(\frac{g*TAS*\Delta t}{\eta_f*\eta_m*\eta_{fc}*HHV*\frac{L}{D}}\right)}\right] \tag{61}$$

Battery-Powered Electrofan

$$E_{battery} = \frac{start_{mass}*g*TAS*\Delta t}{\eta_f*\eta_m*\frac{L}{D}} \tag{62}$$

Conventional Propeller Engine

$$\text{holding fuel} = \text{start\_mass}\left[1 - e^{-\left(\frac{PSFC*g*TAS*\Delta t}{\eta_p*\frac{L}{D}}\right)}\right] \tag{63}$$

Fuel-Cell-Powered Elecroprop

$$\text{holding fuel} = \text{start\_mass}\left[1 - e^{-\left(\frac{g*TAS*\Delta t}{\eta_p*\eta_m*\eta_{fc}*HHV*\frac{L}{D}}\right)}\right] \tag{64}$$

Battery-Powered Electroprop

$$E_{battery} = \frac{start\_mass*g*TAS*\Delta t}{\eta_p*\eta_m*\frac{L}{D}} \tag{65}$$

5.1.10. Total Fuel

Having performed all the calculations above, we can now calculate the total fuel required for a certain mission. The software adds all the different parts of reserve fuel to the leg fuel according to the formula below:

$$\text{OWE} = \text{MTOW} - \text{payload} - \text{total fuel} \tag{66}$$

At this stage, we also calculate the weight of the fuel tanks. In the case of gas and liquid hydrogen storage, we use the corresponding gravimetric index as follows:

$$\text{fuel tanks} = \text{total fuel}*\left(\frac{1}{\text{gravimetric index}} - 1\right) \tag{67}$$

In the case of battery storage, we simply divide total fuel by the battery energy density. Finally, in the case of conventional fuel, we do not take the weight of the fuel tanks into account, as their weight tends to be negligible compared to OEW.

5.1.11. OEW Structure

Here, we calculate the change in aircraft's mass in case of a theoretical propulsion system change. We use the predefined power densities each time to calculate the mass of the engine using its nominal power. Then, we save this initial mass as a variable. We follow the same process for the target propulsion system, making sure to add all required weights. This means the electric motor plus fan or propeller mass in the case of a battery-powered aircraft. In the case of a hydrogen fuel-cell-powered aircraft, we add the calculated mass of the fuel cell and its cooling system on top. Once again, we save the newly calculated target mass as a variable. We now set up a new variable called delta_engine_mass, which will be simply the difference between the target engine mass and the initial engine mass, and its result is finally added or subtracted from the OEW.

5.1.12. Design

Along with basic information about an aircraft's design, this function allows for the calculation of four following evaluation criteria:

Mass Efficiency Factor

This criterion is calculated simply as:

$$\text{mass efficiency factor} = \frac{N_{pax} * \text{distance}}{\text{MTOW}} \tag{68}$$

It represents the transport service as compared to the mass to be invested. In other words, it tells us how much mass we need to invest in order to provide a certain service. We can, therefore, think of it as of a dimensionless EROI ratio (energy return on investment).

Energy Efficiency Factor

This criterion is very similar to the previous one but uses the total energy as a comparing factor.

$$\text{energy efficiency factor} = \frac{N_{pax} * \text{distance}}{\text{total energy}} \tag{69}$$

The total energy represents the energy of the total fuel required for a certain mission (including reserves), which is either calculated by multiplying the fuel mass by its heating value or taken directly as the necessary energy in the case of battery-powered aircraft.

System Energy Density

The energy density of a certain energy stockage system is calculated as the total energy potentially exploitable from the fuel or battery divided by the mass of the latter.

$$\text{system energy density} = \frac{\text{total energy}}{\text{delta system mass} + \text{total fuel}} \tag{70}$$

In the case of kerosene, we take the mass of the fuel solely and neglect the relatively small mass of the tanks. In the case of all other systems, we calculate the added mass and then add it as delta system mass to the total fuel mass. The delta system mass can be the mass of the tank holding the liquid or gas hydrogen or the mass of the so-called battery pack, which surrounds the battery itself.

6. Application to the Concept of Battery-Powered Aircraft and Its Limits (Research of the Technological Frontier according to the Energy Density)

To put our model in the trial, we decided to focus on the general aviation domain, as there are already multiple electric aircraft in service or in development that we can use as a clear indicator of whether our model is precise. Therefore, we wanted to see whether we can use it for further exploration of the domain of battery-powered aircraft.

6.1. Chosen Aircraft Examples

We chose to compare our model with two real aircraft, the Socata TB20 Trinidad and the H55-Bristell Energic. Let us now introduce the two with their specifications.

6.1.1. Socata TB20

The Socata TB20 is one of the most common light single-engine piston aircraft developed and manufactured in France. The TB within the designation stands for "Tarbes", which is the city where the aircraft is manufactured. It was introduced in 1975, and until 2007, there were 2150 units produced [51].

6.1.2. H55-Bristell Energic

The H55-Bristell Energic is one of the first light single-engine electric aircraft proposed for serial production. It is developed by a consortium of a Czech aircraft manufacturer, BRM Aero, based in Kunovice, and a Swiss company, H55. BRM Aero supplies the aircraft itself, as it is a derivative of their own Bristell B23. H55 supplies the propulsion system, including its batteries, as it is the company behind the Solar Impulse project. The aircraft is currently undergoing testing and it should be certified under CS-23 by 2024.

Specifications of H55-Bristell Energic are an engine power of 100 kW [52] and the range, which was calculated by taking its endurance of 90 min and making a hypothesis of a cruise speed of 130 km/h. To be compliant with the regulations, we need to have a 30 min reserve and, therefore, only 60 min of trip time, which gives us a range of 130 km.

6.2. Model Validation

Let us now assemble the key aircraft specifications, which we will use as inputs for our software. As one can see in Table 6 below, for the TB20, we interpolated the number of passengers as four and a half, as the aircraft max occupancy is four to five passengers.

Table 6. Model validation software inputs.

Airplane Type	Nb PAX	Cruise Speed (km/h)	Range (km)
Socata TB20	4.5	280	1300
H55-Bristell Energic	2	130	130

If we run the calculation with this input, we obtain the output of a conventional propeller engine-powered aircraft and a battery-powered electroprop, respectively, in Tables 7 and 8:

Table 7. Model validation software outputs for a conventional propeller engine-powered aircraft with Socata TB20 specifications.

Parameter	Value
Max. power	192 kW
Mission fuel	161 kg
Reserve fuel	16 kg
Total fuel	177 kg
Mission energy	1924 kWh
Reserve energy	194 kWh
MTOW	1478 kg
OWE	896 kg
Payload	405 kg
PK/MTOW minimum	2.6 pk/kg
PK/MTOW	3.96 pk/kg
PK/Energy	2.76 pk/kWh
Max. distance vs. PK/M	2573 km

Table 8. Model validation software outputs for a battery-powered electroprop with H55-Bristell Energic specifications.

Parameter	Value
Max. power	102 kW
Cruise power	29 kW
Battery mass	230 kg
Mission energy	32 kWh
Reserve energy	14 kWh
MTOW	866 kg
OEW	686 kg
Payload	180 kg
PK/MTOW minimum	0.26 pk/kg
PK/MTOW	0.30 pk/kg
PK/Energy	5.64 pk/kWh
Max. distance vs. PK/M	153 km

We can now compare the highlighted data with the real ones from the manufacturer, as stated in Table 9.

Table 9. Specifications comparison of selected airplanes with manufacturer data.

Airplane Type	Max. Power (kW)	MTOW (kg)	OEW (kg)
Real: Socata TB20	186	1400	900
Model: Socata TB20 (deviation from real)	192 (+3%)	1478 (+6%)	896 (−1%)
Real: H55-Bristell Energic	100	850	660
Model: H55-Bristell Energic (deviation from real)	102 (+2%)	866 (+2%)	686 (+4%)

We can see that for these specific aircraft and in those specifications, our model overestimates the real values slightly, but its deviations stay lower than 10%.

6.3. Technological Frontier Exploration

Using the limits defined in Section 4 and our battery aircraft model, we can depict the space of possible designs on a Npax vs. Range in Figure 9 below:

In Figure 9, the color field shows $\frac{PK/M}{PK/M_{min}}$, and the yellow line highlights the case when PK/M = PK/M_{min}. As established in Section 4, this line will be considered a lower limit of a possible design, meaning that we will not build an aircraft below this line.

The grey line shows the previously developed P/k_{max} = 0.0146. This line represents an upper limit of a possible design, and the design space that does not satisfy the constraint is once again covered by hashes. As stated before, this limit is not a hard one. It just shows the current practice among aircraft manufacturers.

The designs producing the highest level of service (PK maximum) should be in the white ellipse of Figure 10. As we can see, this area is mostly considered unusable due to the upper limit. We will, therefore, divide it into two points, A and B, which will represent the maximal level of service possible with the limits chosen and the maximal range with 19 passengers if we disregard the soft limit, respectively.

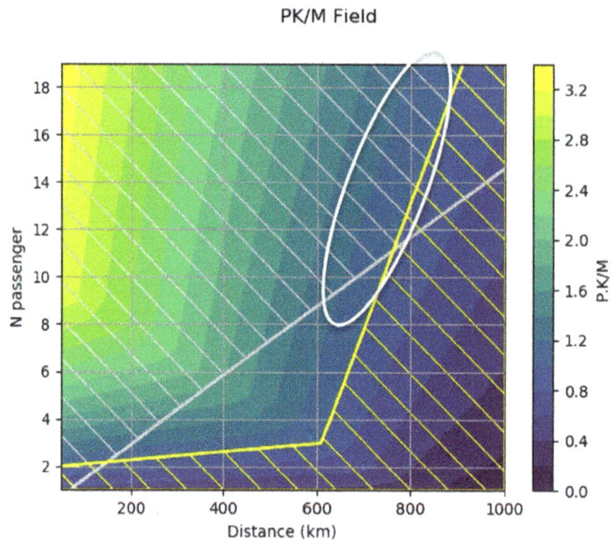

Figure 9. Design space, Npax vs. Range.

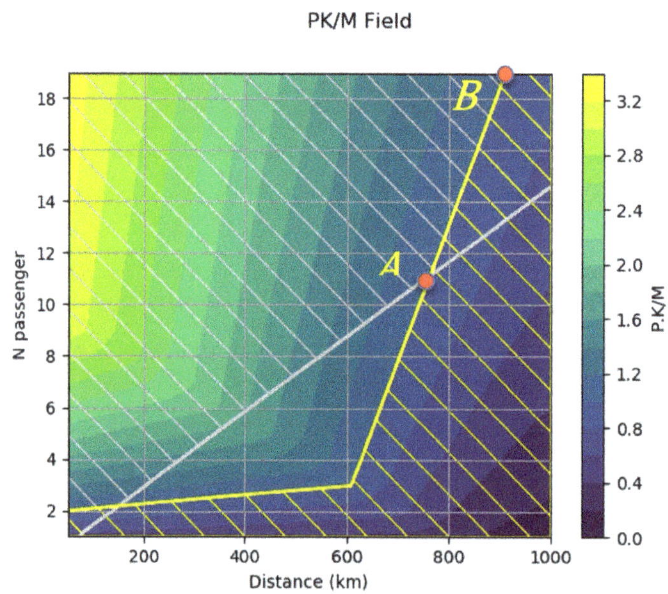

Figure 10. Points of the maximum level of service in the design space.

As we established earlier, the most important factor in battery-powered aircraft development is the battery power density. It will, therefore, also change the field and the $\frac{PK/M}{PK/M_{min}}$ position of the PK/M = PK/M_{min} line. In the following, we will follow the trajectories of points A and B when the battery energy density varies from 200 Wh/kg to 600 Wh/kg (Figure 11).

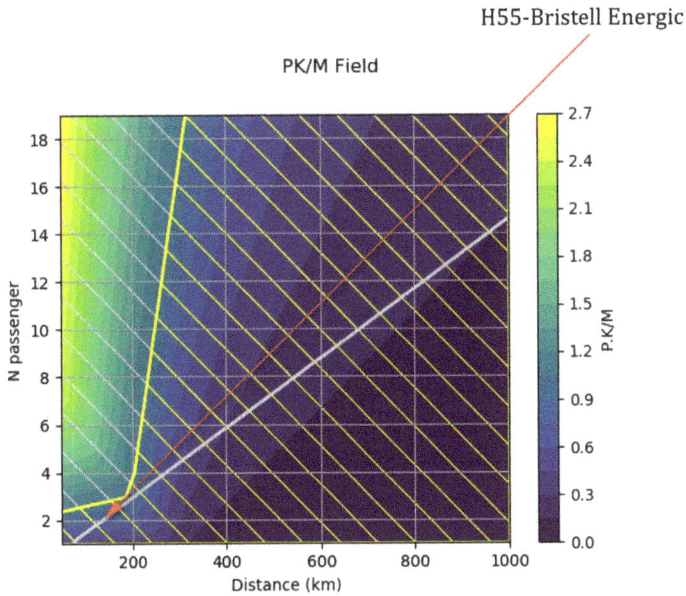

Figure 11. Design space with a 200 Wh/kg technology—case of H55-Bristell Energic.

In Figure 12, we can see a comparison of points A, which is the maximal constrained level of service, and point B, the maximal absolute level of service for 19 passengers, for battery energy densities varying from 200 to 600 Wh/kg.

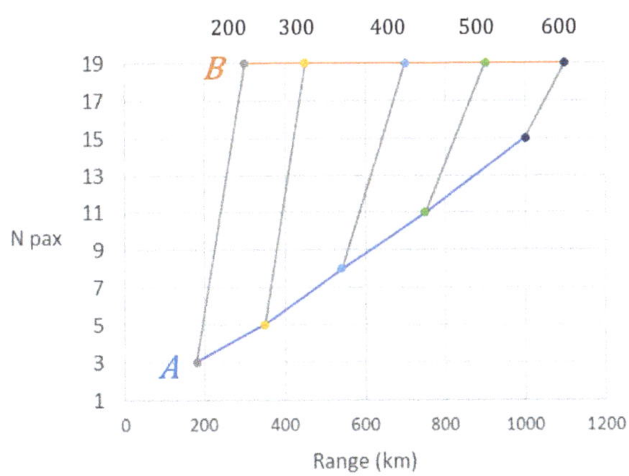

Figure 12. Comparison of points A and B for battery energy densities of 200–600 Wh/kg.

It is important to mention that these results focus on the small aircraft domain, the CS23. We could continue to have a higher number of passengers and a higher range with a commuter aircraft category, for example, but we can conclude that, according to our model, this technology is interesting mostly for shorter distances than the usual practice among aircraft manufacturers.

7. Application to the Concept of Hydrogen-Powered Aircraft and Its Limits (Research of the Technological Frontier according to the Energy Density)

Due to a lack of complete information about flying hydrogen-powered aircraft (the Boeing fuel cell demonstrator, for example, was using batteries along with a fuel cell and therefore was not powered solely by hydrogen [53]), we had to skip the model validation part and concentrate on the technological frontier research, which is, in this case, more complex. As we are concentrating on the small aircraft domain of the CS-23 category, we are going to consider fuel cell technology powered by hydrogen in the form of gas. The reason for this hypothesis is that the fuel cell technology with gas hydrogen seems to be the best one, as it was proven, for example, by the Boeing demonstrator mentioned above or the Zero Avia's modified Piper Malibu [54].

Technological Frontier Exploration

We are going to use the same Npax vs. Range figure as in the previous section and we add one more limitation, which is simply the weight limitation to keep the MTOW of the aircraft to less than 5700 kg in order to stay in the CS-23 category. This limitation is marked by the purple line in the figures below.

Unlike battery technology, hydrogen has not one but three defining parameters. It is the fuel cell system power density, the cooling power density, and the fuel cell efficiency. The first two of the three are given in kW/kg, and the efficiency is a unitless variable. The power densities of a fuel cell and its cooling are in the order of 5 and 6 kW/kg, respectively. However, in this case, we are considering a fully integrated system designed for affordable production and safe service. We are, therefore, going to downgrade the aforementioned values to 2 and 3 kW/kg, respectively, in the first instance. As for the fuel cell efficiency, we are going to start with the value of 0.4 declared by the Horizon fuel cell company. We can see the results for this first instance with the lowest values in Figure 13 below:

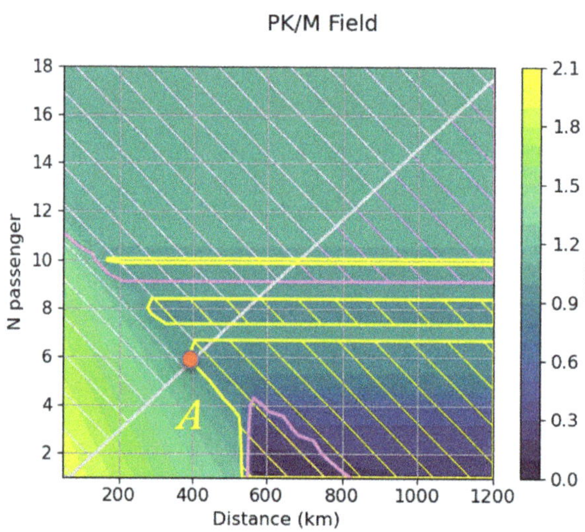

Figure 13. Design space with the lowest fuel cell technology.

If we ignore the holes in both the PK/M and maximal MTOW limitations, which are caused by the software not being adapted for these low values, we can see that the point with the highest level of service is situated roughly at 6 passengers and 400 km of range.

We will now increase each of the parameters (the fuel cell system power density, the cooling power density, and the fuel cell efficiency) one by one to demonstrate their effect on the overall concept (Figure 14). Let us start by increasing the fuel cell efficiency to 0.7 while keeping the other parameters constant.

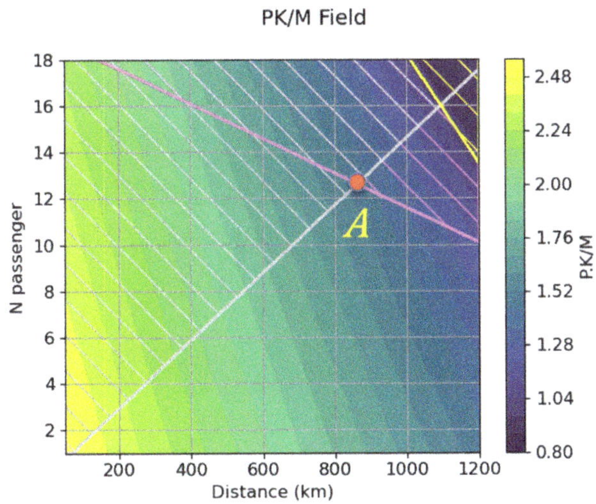

Figure 14. Design space with fuel cell efficiency of 0.7.

We can see that we are no longer limited by the PK/M, and the mass limitation puts us at 12 passengers and a little bit over 800 km of range.

If we bring the fuel cell efficiency back to its default value of 0.4 and increase the fuel cell system power density to 5 kW/kg while keeping the cooling system power density at 3 kW/kg, we obtain the following Figure 15.

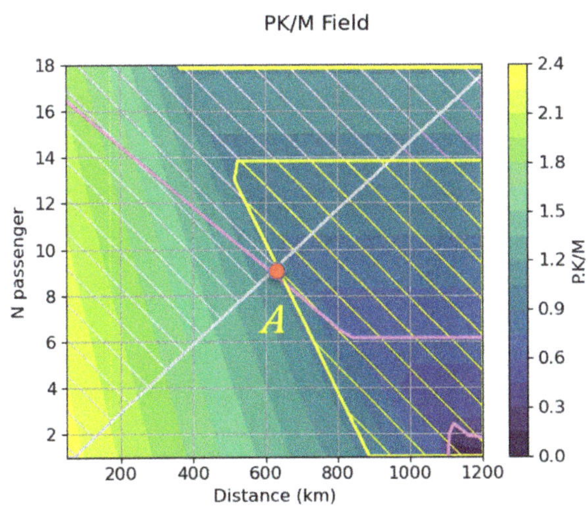

Figure 15. Design space with the fuel cell system power density of 5 kW/kg.

In this case, all three limitations intersect roughly at the same point of 9 passengers and a bit over 600 km of range.

Let us now increase the cooling system power density to 5 kW/kg while holding the remaining parameters at their minimum values (Figure 16).

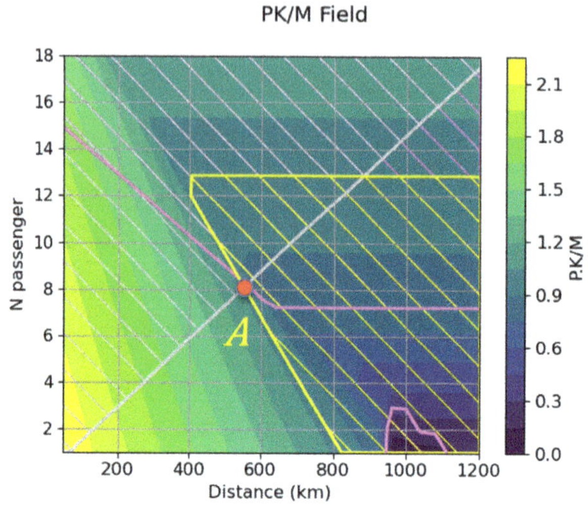

Figure 16. Design space with cooling system power density of 5 kW/kg.

We notice that we obtain a very similar, albeit slightly lower, level of service than when solely increasing the fuel cell system power density.

Finally, we increase all the parameters at once, that is, 5 kW/kg for both the fuel cell and the cooling system power density and 0.7 for the fuel cell efficiency. Notice in Figure 17 below that we increased the maximal range to 2400 km.

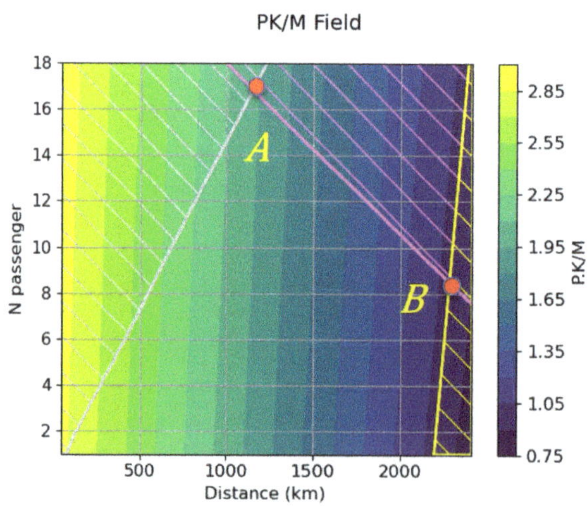

Figure 17. Design space with the highest fuel cell technology.

We marked the point of highest service (17 passengers over around 1200 km) as point A and the point of highest range, which is around 2250 km, with 8 passengers on board, as point B. Let us now look at a comparison of points A in all the abovementioned cases (Figure 18):

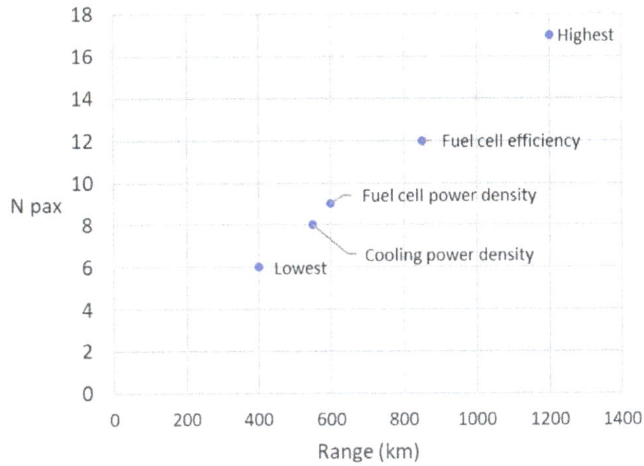

Figure 18. Comparison of points A for different hydrogen technologies.

As we can see, the fuel cell efficiency increase from 0.4 to 0.7 clearly outperforms both the fuel cell and the cooling system power density increases from 2 to 5 and 3 to 5 kW/kg, respectively. Out of these two, the fuel cell power density shows better improvement. However, this is mostly because we kept the initial cooling power density higher than one of the fuel cells. If they were kept at the same values, these two variables would have a very similar effect.

In conclusion, we would like to stress the hypothesis that we proposed. The study focuses on the mass aspect of commercial aircraft. Nothing says a priori that the fuel cell aircraft is feasible. The problems with feasibility may arise mainly because of thermal evacuation problems. Moreover, we did not take into account this technology's need for volume. Large tanks of compressed hydrogen could radically change the fuselage of such an aircraft concept. However, we assumed that we could always obtain a lift to drag ratio of 14, whatever the embarked technology.

8. Discussion

The scope of this research is to provide a comprehensive analysis of the potential of battery- and hydrogen-fuel-powered aircraft as alternatives to traditional fossil-fuel-powered aviation systems. As a part of the research work, a program to calculate mathematical models that estimate the weight of aircraft components has been developed, with a specific focus on batteries and fuel cells. The study has identified the technological barriers to the development of these technologies and has created a database and software for exploring and comparing them.

The results of the study have shown that both battery- and hydrogen fuel-cell-powered aircraft have the potential to replace traditional engines, with power density and efficiency improvements making them viable alternatives. However, there are limitations to the adoption of these technologies, particularly in terms of the difficulty of putting hydrogen propulsion technology into place.

One of the main limitations of this research is that it primarily focuses on the power density and efficiency trade-offs of various advanced engines, electrolytic batteries, fuel cells, and cooling technologies. While this is an essential aspect of understanding the potential of these technologies, there are other factors to consider, such as the economics of implementation and the regulatory framework. For example, the cost of hydrogen fuel cells is currently high, and this could be a significant barrier to their adoption. Additionally, the regulatory framework will need to be adapted to accommodate these new technologies, which may take time and require significant investment.

Another limitation is that the research focuses on general aviation, and there may be differences in the application of these technologies in other areas, such as commercial aviation. The study does not explore the potential for hybrid systems, which could combine the benefits of both battery and fuel cell technologies.

Ideas for further work could include an exploration of the economics of implementing these technologies, including the cost of production, maintenance, and infrastructure. Additionally, further research could explore the regulatory framework and policy changes required to support the adoption of these technologies. Another area of research could be the development of hybrid systems that combine the benefits of both battery and fuel cell technologies.

In conclusion, this research provides a valuable contribution to the understanding of the potential of battery and hydrogen-fuel-powered aircraft. It highlights the need for interdisciplinary research and collaboration to find sustainable solutions to contemporary challenges. While there are limitations to this study, the results demonstrate the potential of these technologies and provide a foundation for further research. Ultimately, the adoption of these technologies will require significant investment, innovation, and collaboration between industry, policymakers, and researchers.

9. Conclusions

This article presents a program for calculating mathematical models to estimate the weight of aircraft components, focusing on battery- and hydrogen fuel-cell-powered aircraft concepts. The research aimed to add missing dimensions and parameters to the aircraft database, create an effective program for creating mathematical models, and identify technological barriers. This article showed a new possibility of comparing new technologies by using data from conventional ones. It introduced the two most important propulsion technologies, showed their past and current development, and compared their core values critically. It then described the creation of the database and its exploiting software. Finally, it presented a model validation for battery technology and a search of the technological frontier of both technologies. The results showed overall better values of the level of service of the hydrogen technology, defining the fuel cell efficiency as its most important parameter. However, this lead over battery technology is the difficulty of putting the hydrogen propulsion technology in place. It is best demonstrated by the fact that small battery-powered aircraft have already reached serial production (such as the Alpha Electro of the Pipistrel company [55]), whereas purely hydrogen-powered aircraft are still waiting for their first pioneering model. Nevertheless, batteries and hydrogen should be regarded as partners, completing rather than competing with each other.

The research primarily focuses on calculating power density and efficiency trade-offs for various advanced engines and turbines, electrolytic batteries, fuel cells, and cooling technologies. One of the critical aspects of this work is to evaluate the potential of hydrogen- and battery-powered aircraft to replace traditional fossil-fuel-powered aviation systems, which contribute significantly to the carbon footprint and air pollution. The analysis results indicate that the power density and efficiency of electric motors and fuel cells have been improving with advancements in technology, making them viable alternatives to traditional engines. The authors also highlight the importance of cooling technologies for high-power density electrical machines in aviation applications.

Apart from the technical insights, the article demonstrates the importance of collaborative and interdisciplinary research in finding sustainable solutions to contemporary challenges, such as reducing carbon emissions in aviation. The authors have integrated findings from various sources to provide a comprehensive overview of the current and potential technology for hydrogen- and battery-powered aircraft, which can benefit designers, engineers, policymakers, and regulators.

Overall, this work presents valuable contributions to the ongoing discourse about sustainable aviation and provides a starting point for further research and development of hydrogen- and battery-powered aircraft concepts.

Author Contributions: Conceptualization, M.J. and V.B.; Methodology, M.J., V.B. and B.K.; Software, V.B.; Validation, V.B. and M.J.; Formal Analysis, F.Š. and K.Š.; Investigation, M.J. and K.Š.; Resources, M.J. and Ľ.H.; Data Curation, V.B.; Writing—Original Draft Preparation, M.J., V.B. and B.K.; Writing—Review and Editing, M.J., K.Š., F.Š. and Ľ.H.; Visualization, M.J. and V.B.; Supervision, K.Š. and B.K.; Project Administration, M.J., F.Š. and Ľ.H. All authors have read and agreed to the published version of the manuscript.

Funding: This work was supported under the project of Operational Programme Integrated Infrastructure: Research and development of contactless methods for obtaining geospatial data for forest monitoring to improve forest management and enhance forest protection, ITMS code 313011V465. The project is co-funded by the European Regional Development Fund.

Data Availability Statement: The article includes the original contributions presented in the study. Further inquiries can be directed to the corresponding author.

Conflicts of Interest: The authors declare no conflict of interest.

Nomenclature

Abbreviations	
BAC	British Aircraft Corporation
CS-23	European Aviation Safety Agency Certification Specification for Normal, Utility, Aerobatic, and Commuter Category Aeroplanes
DE	Direct energy consumption
EE	Embodied energy
EROI	Energy return on investment
G/CNT	Graphene/carbon nanotubes
HHV	Hydrogen heating value
HTP	Horizontal tailplane
ISA	International Standard Atmosphere
LT	Lifetime
MK	"Mass-kilometer" (Payload times Distance)
MSIJ	Mass of the various materials used to build the device times the "scarcity index" of the materials
MTOW	Maximum take-off weight
Npax	Number of passengers
OP equipment	Operational equipment
OEW	Operating empty weight
PAX	Passenger
PE	Primary energy consumed
PK	"Passenger-kilometer" (Npax times Distance)
PK/M	PK/MTOW = (Passenger capacity times Range) divided by Maximum take-off weight
PSFC	Propeller-Specific Fuel Consumption
RE	Recycling energy
SFC	Specific Fuel Consumption
SI	International System of Units
TAS	True airspeed
TOW	Take-off weight
TSFC	Thrust-Specific Fuel Consumption
VTOL/STOL	Vertical/Short take-off and landing
VTP	Vertical tailplane

List of Symbols

a	Speed of sound
CO_2	Carbon dioxide
C_x	Drag coefficient
C_z	Lift coefficient
D	Drag
E	Energy
F	Force; Engine thrust
ff	Fuel flow
g	Gravitational constant
H_2	Hydrogen
k_r	Contingency fuel factor
L	Lift
m	Mass
M	Mach number
M_{pax}	Average mass of a passenger and their baggage
ρ	Density
pw	Power
R_{div}	Average diversion distance
S	Wing area
t	Time
v	Speed
W_e	Electric power; Energy flow
W_{fuel}	Fuel energy
W_m	Motor power
W_{shaft}; P	Shaft power
x	Distance
η_f	Fan efficiency
η_{fc}	Efficiency of a fuel cell
η_m	Motor efficiency
η_p	Propeller efficiency
η_{thpr}	Thermopropulsive efficiency

References

1. Supersonic Flight: Will It Ever Rise Out of the Ashes of Concorde? Available online: http://www.theguardian.com/technology/2018/jun/05/supersonic-flight-concorde-crash (accessed on 25 June 2022).
2. Airbus Hits the Road with A3xx. (Airbus Industrie). Available online: https://web.archive.org/web/20121106060040/http://www.highbeam.com/doc/1G1-16444324.html (accessed on 25 June 2022).
3. Zhang, B. An American Airlines VP Reveals Why the Airbus A380 doesn't Work for the World's Biggest Airline. Available online: https://www.businessinsider.com/american-airlines-exec-reveals-why-company-doesnt-use-airbus-a380-2019-4 (accessed on 18 February 2022).
4. Vermeulen, S.; Zougmoré, R.; Wollenberg, E.; Thornton, P.; Nelson, G.; Kristjanson, P.; Kinyangi, J.; Jarvis, A.; Hansen, J.; Challinor, A.; et al. Climate change, agriculture and Food Security: A global partnership to link research and action for low-income agricultural producers and consumers. *Curr. Opin. Environ. Sustain.* **2012**, *4*, 128–133. [CrossRef]
5. Li, X.; Tang, J.; Li, W.; Si, Q.; Guo, X.; Niu, L. A bibliometric analysis and visualization of aviation carbon emissions studies. *Sustainability* **2023**, *15*, 4644. [CrossRef]
6. Cabrera, E.; de Sousa, J.M. Use of sustainable fuels in aviation—A Review. *Energies* **2022**, *15*, 2440. [CrossRef]
7. Yusaf, T.; Fernandes, L.; Abu Talib, A.R.; Altarazi, Y.S.; Alrefae, W.; Kadirgama, K.; Ramasamy, D.; Jayasuriya, A.; Brown, G.; Mamat, R.; et al. Sustainable aviation—Hydrogen is the future. *Sustainability* **2022**, *14*, 548. [CrossRef]
8. Schmelcher, M.; Häßy, J. Hydrogen Fuel Cells for Aviation? Available online: https://elib.dlr.de/188835/1/ISABE_2022_291_full.pdf (accessed on 13 March 2023).
9. Trainelli, L.; Saluci, F.; Comincini, D.; Riboldi, C.E.D.; Rolando, A. Sizing and Performance of Hydrogen-Driven Airplanes. Available online: https://mahepa.eu/wp-content/uploads/2020/06/Sizing-and-Performance-of-Hydrogen-Driven-Airplanes.pdf (accessed on 13 March 2023).
10. Nicolay, S.; Karpuk, S.; Liu, Y.; Elham, A. Conceptual Design and optimization of a general aviation aircraft with fuel cells and hydrogen. *Int. J. Hydrog. Energy* **2021**, *46*, 32676–32694. [CrossRef]
11. Gao, Y.; Jausseme, C.; Huang, Z.; Yang, T. Hydrogen-powered aircraft: Hydrogen–electric hybrid propulsion for aviation. *IEEE Electrif. Mag.* **2022**, *10*, 17–26. [CrossRef]

12. Kadyk, T.; Schenkendorf, R.; Hawner, S.; Yildiz, B.; Römer, U. Design of Fuel Cell Systems for Aviation: Representative Mission Profiles and Sensitivity Analyses. *Front. Energy Res.* **2019**, 7. [CrossRef]
13. McQueen, M.; Karataş, A.E.; Bramesfeld, G.; Demir, E.; Arenas, O. Feasibility Study of electrified light-sport aircraft powertrains. *Aerospace* **2022**, *9*, 224. [CrossRef]
14. Palaia, G.; Abu Salem, K. Mission Performance Analysis of hybrid-electric regional aircraft. *Aerospace* **2023**, *10*, 246. [CrossRef]
15. Frosina, E.; Caputo, C.; Marinaro, G.; Senatore, A.; Pascarella, C.; Di Lorenzo, G. Modelling of a hybrid-electric light aircraft. *Energy Procedia* **2017**, *126*, 1155–1162. [CrossRef]
16. Marinaro, G.; Di Lorenzo, G.; Pagano, A. From a battery-based to a PEM fuel cell-based propulsion architecture on a lightweight full electric aircraft: A Comparative Numerical Study. *Aerospace* **2022**, *9*, 408. [CrossRef]
17. Staack, I.; Sobron, A.; Krus, P. The potential of full-electric aircraft for civil transportation: From the breguet range equation to operational aspects. *CEAS Aeronaut. J.* **2021**, *12*, 803–819. [CrossRef]
18. Hoff, T.; Becker, F.; Dadashi, A.; Wicke, K.; Wende, G. Implementation of fuel cells in aviation from a maintenance, repair and overhaul perspective. *Aerospace* **2022**, *10*, 23. [CrossRef]
19. Ferretto, D.; Viola, N. Preliminary design and simulation of a thermal management system with integrated secondary power generation capability for a Mach 8 aircraft concept exploiting liquid hydrogen. *Aerospace* **2023**, *10*, 180. [CrossRef]
20. Broom, D.P.; Webb, C.J.; Fanourgakis, G.S.; Froudakis, G.E.; Trikalitis, P.N.; Hirscher, M. Concepts for improving hydrogen storage in nanoporous materials. *Int. J. Hydrog. Energy* **2019**, *44*, 7768–7779. [CrossRef]
21. Ni, M. An overview of hydrogen storage technologies. *Energy Explor. Exploit.* **2006**, *24*, 197–209. [CrossRef]
22. Preuster, P.; Alekseev, A.; Wasserscheid, P. Hydrogen Storage Technologies for Future Energy Systems. *Annu. Rev. Chem. Biomol. Eng.* **2017**, *8*, 445–471. [CrossRef]
23. Tzimas, E.; Filiou, C.; Peteves, S.; Veyret, J. Hydrogen Storage: State-of-the-Art and Future Perspective. EUR 20995 EN. cat. No. LD-NA-20995-EN-C; 2003. JRC26493. Available online: https://publications.jrc.ec.europa.eu/repository/handle/JRC26493 (accessed on 20 February 2022).
24. Marciello, V.; Di Stasio, M.; Ruocco, M.; Trifari, V.; Nicolosi, F.; Meindl, M.; Lemoine, B.; Caliandro, P. Design exploration for sustainable regional hybrid-electric aircraft: A study based on technology forecasts. *Aerospace* **2023**, *10*, 165. [CrossRef]
25. NIST Chemistry Webbook, SRD 69. Available online: https://webbook.nist.gov/cgi/fluid.cgi?Action=Load&ID=C1333740&Type=SatT&Digits=5&PLow=.5&PHigh=1.5&PInc=.1&RefState=DEF&TUnit=K&PUnit=atm&DUnit=kg%2Fm3&HUnit=kJ%2Fmol&WUnit=m%2Fs&VisUnit=uPa%2As&STUnit=N%2Fm (accessed on 15 May 2022).
26. Hwang, H.T.; Varma, A. Hydrogen storage for fuel cell vehicles. *Curr. Opin. Chem. Eng.* **2014**, *5*, 42–48. [CrossRef]
27. Rivard, E.; Trudeau, M.; Zaghib, K. Hydrogen storage for mobility: A Review. *Materials* **2019**, *12*, 1973. [CrossRef]
28. Hydrogen Tank: Doosan Mobility Innovation. Available online: http://www.doosanmobility.com/en/products/hydrogen-tank/ (accessed on 16 January 2023).
29. Airbus. Airbus Partners with Amprius, Leader in High Energy Density Battery Technology. Available online: https://www.airbus.com/en/newsroom/press-releases/2019-10-airbus-partners-with-amprius-leader-in-high-energy-density-battery (accessed on 27 February 2023).
30. Bi, C.X.; Zhao, M.; Hou, L.P.; Chen, Z.X.; Zhang, X.Q.; Li, B.Q.; Yuan, H.; Huang, J.Q. Anode material options toward 500 wh kg−1lithium–sulfur batteries. *Adv. Sci.* **2021**, *9*, 2103910. [CrossRef]
31. Kim, S.; Oguchi, H.; Toyama, N.; Sato, T.; Takagi, S.; Otomo, T.; Arunkumar, D.; Kuwata, N.; Kawamura, J.; Orimo, S.-I. A complex hydride lithium superionic conductor for high-energy-density all-solid-state lithium metal batteries. *Nat. Commun.* **2019**, *10*, 1081. [CrossRef]
32. Shi, H.; Zhao, X.; Wu, Z.-S.; Dong, Y.; Lu, P.; Chen, J.; Ren, W.; Cheng, H.-M.; Bao, X. Free-standing integrated cathode derived from 3D graphene/carbon nanotube aerogels serving as binder-free sulfur host and interlayer for ultrahigh volumetric-energy-density lithium sulfur batteries. *Nano Energy* **2019**, *60*, 743–751. [CrossRef]
33. Randall, A.C.; Moeini, H. Eviation Aircraft Sets Sights on Kokam Batteries. Available online: https://www.electrive.com/2018/02/15/eviation-aircraft-sets-sights-kokam-batteries/ (accessed on 11 May 2022).
34. Liu, X.; Iocozzia, J.; Wang, Y.; Cui, X.; Chen, Y.; Zhao, S.; Li, Z.; Lin, Z. Noble metal–metal oxide nanohybrids with tailored nanostructures for efficient solar energy conversion, photocatalysis and environmental remediation. *Energy Environ. Sci.* **2017**, *10*, 402–434. [CrossRef]
35. All the World's Aircraft: In Service Yearbook 20/21. Available online: https://shop.janes.com/all-the-world-s-aircraft-in-service-yearbook-20-21-6541-3000008244 (accessed on 11 April 2022).
36. Les Editions Elodie Roux-Avions Civils Raction: Plan 3 Vues et Donnes Caractristiques. Available online: http://elodieroux.com/EditionsElodieRouxAvions.html (accessed on 25 July 2022).
37. *Roux Élodie Avions Civils À Réaction: Plan 3 Vues et Données Caractéristiques*; Editions Elodie Roux: Blagnac, France, 2007.
38. Haseli, Y. Maximum conversion efficiency of hydrogen fuelcells. *Int. J. Hydrog. Energy* **2018**, *43*, 9015–9021. [CrossRef]
39. Horizon 5000W PEM Fuel Cell SKU: FCS-C5000. Category: 1000W to <5000W Fuel Cell Stacks. Brands: Horizon. Available online: https://www.fuelcellearth.com/fuel-cell-products/horizon-5000w-pem-fuel-cell/ (accessed on 15 December 2022).
40. Comparison of Fuel Cell Technologies. Available online: https://www.energy.gov/eere/fuelcells/comparison-fuel-cell-technologies (accessed on 13 March 2023).

41. Luongo, C.A.; Masson, P.J.; Nam, T.; Mavris, D.; Kim, H.D.; Brown, G.V.; Waters, M.; Hall, D. Next generation more-electric aircraft: A potential application for HTS superconductors. *IEEE Trans. Appl. Supercond.* **2009**, *19*, 1055–1068. [CrossRef]
42. Picard, B.; L.-Blais, A.; Picard, M.; Rancourt, D. *Power-Density vs Efficiency Trade-Off for a Recuperated Inside-Out Ceramic Turbine (ICT)*; Microturbines, Turbochargers, and Small Turbomachines; Steam Turbines: Phoenix, AZ, USA, 2019; Volume 8. [CrossRef]
43. GE Turbines and Small Engines Overview. deBock-Arpa-e INTEGRATE Meeting. Available online: https://arpa-e.energy.gov/sites/default/files/14_deBock_GE%20Turbines%20and%20small%20engines%20overview%20-%20ARPA-e%20INTEGRATE%20V2.pdf (accessed on 17 April 2022).
44. Lycoming IO-390. Available online: http://all-aero.com/index.php/contactus/64-engines-power/13224-lycoming-io-390 (accessed on 6 January 2022).
45. H3X. H3X Claims It's Tripled the Power Density of Electric Aircraft Motors, New Atlas. Available online: https://newatlas.com/aircraft/h3x-electric-aircraft-motor-power-density/ (accessed on 19 May 2022).
46. FlyRadius. Available online: https://www.flyradius.com/bombardier-q400/propeller-dowty-r408 (accessed on 27 September 2022).
47. Lolis, P. Development of a Preliminary Weight Estimation Method for Advanced Turbofan Engines, CERES Home. Cranfield University. Available online: http://dspace.lib.cranfield.ac.uk/handle/1826/9244 (accessed on 24 February 2022).
48. Cheng, L.; Curtiss, L.A.; Zavadil, K.R.; Gewirth, A.A.; Shao, Y.; Gallagher, K.G. Sparingly solvating electrolytes for high energy density lithium–sulfur batteries. *ACS Energy Lett.* **2016**, *1*, 503–509. [CrossRef]
49. Horizon to deliver 370 kW PEMFC stacks for heavy-duty vehicles. *Fuel Cells Bull.* **2019**, *2019*, 3. [CrossRef]
50. Canders, W.-R.; Hoffmann, J.; Henke, M. Cooling Technologies for high power density electrical machines for aviation applications. *Energies* **2019**, *12*, 4579. [CrossRef]
51. SOCATA TB20, SKYbrary Aviation Safety. Available online: https://skybrary.aero/aircraft/tb20 (accessed on 27 February 2023).
52. H55. Available online: https://www.h55.ch/ (accessed on 15 July 2022).
53. Lapeña-Rey, N.; Mosquera, J.; Bataller, E.; Ortí, F. *The Boeing Fuel Cell Demonstrator Airplane*; SAE Technical Paper Series; SAE International: Warrendale, PA, USA, 2007. [CrossRef]
54. Redactoramexico. Powercell Sweden Receives New Follow-on Order for Aircraft Fuel Cell Systems from Zeroavia Worth Msek 7.4, Hydrogen Central. Available online: https://hydrogen-central.com/powercell-sweden-aircraft-fuel-cell-systems-zeroavia/ (accessed on 15 July 2022).
55. Pipistrel Aircraft. Why Alpha Electro? Available online: https://www.pipistrel-aircraft.com/aircraft/electric-flight/alpha-electro/ (accessed on 25 July 2022).

aerospace

MDPI

Article

Combustion Regimes in Turbulent Non-Premixed Flames for Space Propulsion

Daniel Martinez-Sanchis [1,*], Andrej Sternin [1,2], Oskar Haidn [1] and Martin Tajmar [2]

[1] Department of Space Propulsion, Technical University of Munich, 85748 Garching, Germany; andrej.sternin@tum.de (A.S.); oskar.haidn@tum.de (O.H.)
[2] Chair of Space Systems, Technical University of Dresden, 01062 Dresden, Germany; martin.tajmar@tu-dresden.de
* Correspondence: daniel.martinez@tum.de

Abstract: Direct numerical simulations of non-premixed fuel-rich methane–oxygen flames at 20 bar are conducted to investigate the turbulent mixing burning of gaseous propellants in rocket engines. The reacting flow is simulated by using an EBI-DNS solver within an OpenFOAM frame. The transport of species is resolved with finite-rate chemistry by using a complex skeletal mechanism that entails 21 species. Two different flames at low and high Reynolds numbers are considered to study the sensitivity of the flame dynamics to turbulence. Regime markers are used to measure the probability of the flow to burn in premixed and non-premixed conditions at different regions. The local heat release statistics are studied in order to understand the drivers in the development of the turbulent diffusion flame. Despite the eminent non-premixed configuration, a significant amount of combustion takes place in premixed conditions. Premixed combustion is viable in both lean and fuel-rich regions, relatively far from the stoichiometric line. It has been found that a growing turbulent kinetic energy is detrimental to combustion in fuel-rich premixed conditions. This is motivated by the disruption of the local premixed flame front, which promotes fuel transport into the diffusion flame. In addition, at downstream positions, higher turbulence enables the advection of methane into the lean core of the flame, enhancing the burning rates in these regions. Therefore, the primary effect of turbulence is to increase the fraction of propellants burnt in oxygen-rich and near-stoichiometric conditions. Consequently, the mixture fraction of the products shifts towards lean conditions, influencing combustion completion at downstream positions.

Keywords: turbulence; combustion; diffusion flames; space propulsion; CFD; quasi DNS

1. Introduction

Methane is a promising propellant for the new generation of space propulsion systems [1,2]. Compared to kerosene, this fuel has superior cooling capabilities and a specific impulse [3]. In addition, it produces a negligible amount of soot [4] and displays low toxicity [2]. These characteristics confer a competitive edge to methane over the established propellants in the space industry, such as hydrazine or the previously mentioned kerosene [2,5]. Consequently, substantial research activities have been fostered over the last decade to investigate the usage of green propellants [6–8], particularly methane for space propulsion applications. Nevertheless, critical aspects related to turbulent ignition, mixing, and combustion challenge the development of new thrust chambers propelled by methane [9,10]. Computational fluid dynamics (CFD) software provides essential support in this process. The conventional methods applied therein, such as Reynolds-averaged Navier–Stokes (RANS) solvers, require turbulent combustion models to predict unclosed turbulent terms. This question has challenged the combustion community during the last century. Reacting flows escalate the inherent complexity of turbulence with the introduction of additional mechanisms of vorticity generation and dissipation [11,12]. Turbulence and combustion can promote or inhibit each other, depending on the overlap between their characteristic time and length scales [13–15]. Combustion in rocket engines is a particularly challenging case. Due to the high pressures and the absence of nitrogen, the flow's reactivity is strongly accentuated, dwindling the chemical length and timescales [16]. As a consequence, the Damköhler number is very high, enabling the occurrence of exotic phenomena such as flame-generated turbulence [17,18] and intermittency between combustion in lean and fuel-rich conditions [19]. Knowledge of these phenomena is essential in the formulation of effective turbulent combustion models. Therefore, it is necessary to investigate the conditions in which the combustion of injected propellants develops in a rocket combustor. The present text addresses this question by using direct numerical simulations, representing turbulent mixing and combustion near the injection region of a methane–oxygen engine at high pressure. The resulting databases are post-processed to investigate the occurrence of combustion in different mixing configurations within this environment.

The rest of the paper is organized as follows: The next section provides a theoretical background, which provides necessary concepts concerning turbulence and non-premixed combustion. The third section describes the numerical simulation strategy. The numerical results are analyzed in the fourth section. Finally, the fifth section summarizes the concluding remarks of this research.

2. Theoretical Background

The present section introduces the basic notions that are required to discuss the results of the conducted numerical simulations. The text is structured in three parts. First, the basic concepts of turbulence for variable-density flows are presented. Second, the basic parameters for characterizing turbulent diffusion flames are presented. Third, the classification techniques of local combustion regimes are described, along with their phenomenological interpretation.

2.1. Turbulence

Turbulent flows are characterized by an unstable nature, which precludes reaching steady-state conditions. Due to their chaotic features [20,21], a statistical approach is deemed convenient. For a turbulent quantity q, the most immediate indicator is its Reynolds number, i.e., time average, which is defined as follows:

$$\overline{q} = \frac{1}{\tau}\int_0^{\tau} q(t)dt \tag{1}$$

In variable-density flows, such as turbulent flames, it is convenient to introduce density weighting to account for the specific volume variations. The Favre average [22] arises from the following consideration:

$$\widetilde{q} = \frac{1}{\tau\overline{\rho}}\int_0^{\tau} q(t)\rho(t)dt \tag{2}$$

The averaged values can be taken as a reference to address the instantaneous values as fluctuations. Therefore, the Reynolds fluctuations are defined as $q'(t) = q(t) - \overline{q}$ and the Favre fluctuations as $q'' = q(t) - \widetilde{q}$. Higher-order statistics are used to quantify the statistics of these fluctuations. The most relevant indicator is the turbulent kinetic energy, which is essentially the variance of the velocity fluctuations, as follows:

$$\widetilde{k} = \frac{\overline{\rho u_i'' u_i''}}{\overline{\rho}} \tag{3}$$

The turbulent kinetic energy is a fundamental manifestation of turbulence, as it indicates the magnitude of the flow unsteadiness and its influence on the momentum conservation equations. Therefore, it denotes the degree of forced convection, which is essential in turbulent combustion and mixing processes. Indeed, turbulent combustion models, such as the eddy dissipation concept (EDC) [23,24] and the flamelet model [25], contain implicit or explicit dependencies on the turbulent kinetic energy [26].

In turbulent premixed flames, turbulent kinetic energy enhances the burning rate, although the flame speed sensitivity to $\tilde{k}$ decreases with growing turbulence following the so-called bending effect [27]. This is caused by a change in the overlap between chemical and turbulent scales [28–30]. If the vortexes are significantly larger than the flame front, their effect on flame development is mainly an increase in the effective surface, due to wrinkling. If the turbulent structures are smaller, the enhancing effects of turbulence primarily occur in terms of diffusion increase. In both cases, it is expected that greater turbulence yields higher flame speeds. Nonetheless, previous research [31] has suggested the possibility of quenching in premixed flames with high turbulent intensity; however, there is no scientific consensus concerning this claim [32]. In diffusion flames, if the chemistry is very fast compared to the turbulence, an enhancement of the burning rate with respect to chemistry takes place [33]. This is primarily motivated by the fact that the mixing intensity grows with turbulent kinetic energy. However, if turbulence becomes significantly faster than the mixing process, it can disrupt the local diffusion flame structure producing local quenching [15].

2.2. Turbulent Diffusion Flames

In a diffusion flame, the oxidizer and the fuel are simultaneously mixing and reacting. The development of these processes is measured with the mixture fraction and the progress variable, which are briefly described in this sub-section.

The mixture fraction Z is used to assess the local mixing conditions, regardless of the combustion progress. For a chemical mixture consisting of N species, this parameter is determined as follows:

$$Z = \sum_{k=1}^{N} Y_k Z_k \tag{4}$$

where Y_k and Z_k stand for the mass fraction and the mixture fraction of the kth chemical species, respectively. The mixture fraction of a molecule is its hydrocarbon content mass fraction. For a generic chemical species with the formulation $C_{n_C}H_{n_H}O_{n_O}$, this parameter is determined as follows:

$$Z_k = \frac{A_C n_C + A_H n_H}{A_C n_C + A_H n_H + A_O n_O} \approx \frac{12n_C + n_H}{12n_C + n_H + 16n_O} \tag{5}$$

where A_C, A_H, and A_O stand for the atomic mass weights of carbon, hydrogen, and oxygen, respectivley. As can be seen, the mixture fraction is zero when only the oxidizer is present, and it tends to unify when the mixture consists only of the fuel. Near the injection site, the mixture fraction profile presents sharp gradients, indicating the mixing of clusters with $Z \approx 1$ and $Z \approx 0$. At downstream positions, the Z field is smoothed, tending to the global mixture fraction $Z_{flame} = \dot{m}_{CH4} / \left(\dot{m}_{CH_4} + \dot{m}_{O_2}\right)$.

Besides the mixing of propellants, it is also necessary to evaluate the local development of chemical reactions. The progress variable is used for this purpose. For non-premixed combustion, the progress variable can be determined as follows, as suggested by Bray et al. [34]:

$$c = \frac{Y_c}{Y_{c,eq}(Z)} \tag{6}$$

where Y_c is the reaction progress variable and $Y_{c,eq}(Z)$ is the equilibrium composition for the present mixture fraction. The reaction progress variable is a scalar that indicates the development of the combustion process. For methane combustion, the definitions devised by Pierce and Moin [35] are often used. However, in space propulsion applications, due to the absence of inert gases, high temperatures take place, with a significant proliferation of free radicals and hydroxyl molecules. For this reason, the mixture fraction definition used is the same in recent works [17,36], as follows:

$$Y_c = \frac{Y_{CO_2}}{M_{CO_2}} + \frac{Y_{CO}}{M_{CO}} + \frac{Y_{H_2O}}{M_{H_2O}} + \frac{Y_O}{M_O} + \frac{Y_H}{M_H} + \frac{Y_{OH}}{M_{OH}} \tag{7}$$

where M_l denotes the molecular weight of the lth chemical species. If the chemical reactions are very fast compared to the turbulent processes, then the progress variable and the mixture fraction are algebraically related [37]. Therefore, the progress variable dynamics are driven by the mixing intensity. This physical phenomenon is evaluated with the scalar dissipation of the mixture fraction, which is defined as follows:

$$\chi = 2D_m \left(\frac{\partial Z}{\partial x_i} \right)^2 \tag{8}$$

where D_m is the mass diffusivity of the reacting mixture. The scalar dissipation rate is analogous to the dissipation rate of the turbulent kinetic energy. Indeed, closure models [25,38,39] relate them as follows:

$$\widetilde{\chi} \propto \frac{\widetilde{\varepsilon}}{\widetilde{k}} \widetilde{Z''^2} = \frac{\sqrt{\widetilde{k}}}{\widetilde{L}} \widetilde{Z''^2} \tag{9}$$

where $\widetilde{L} = \widetilde{k}^{\frac{3}{2}} / \widetilde{\varepsilon}$ denotes the length scale of the large eddies. In stable non-premixed flamelets with fast chemistry, the heat release rate scales with the scalar dissipation rate at stoichiometric conditions [33] are $\dot{q} \propto \chi_{st}$. Following (9), it is clear that the local mixing intensity and the burning rate are primarily driven by the turbulent kinetic energy.

In anchored flames with a stable recirculation zone, a standing flame can be maintained, regardless of the injection velocity, since continuous mixing is enabled near the injection posts. Therefore, as the velocity increases, the heat release rate grows as well, due to the greater rate of incoming fresh reactants. This fact can hinder the comparison between flames burning with different injection velocities. Such is the case for the numerical simulations performed in the present work. For this reason, the heat release rate should be normalized with the mean flow rate at the inlet to account for this fact. One possible normalization is as follows:

$$\frac{1}{\delta_q} = \frac{\dot{q}}{\dot{V} Z_{flame} E} \tag{10}$$

where E is the specific energy content of methane ($E \approx 55.5$ MJ/kg) and $\dot{V} = \rho u_{inj}$ is the volumetric flow rate at injection. This value has dimensions of 1/length, indicating the longitude required for the mixture to obtain all of the available energy. Therefore, one suitable reference value to normalize it is the laminar flame thickness at stoichiometric conditions, as follows:

$$\frac{\delta_{L0}}{\delta_q} = \frac{\dot{q}\delta_{L0}}{\dot{V} Z_{flame} E} \tag{11}$$

The most important point of this normalization is the correction with respect to the incoming velocity, which allows the comparison of flows with different injection mass flow rates. More sophisticated normalizations might be possible, but the one presented in this paper suffices within the present work's scope.

2.3. Flame Classification

For stability and performance purposes, liquid rocket engines typically operate in non-premixed conditions with fuel excess. Despite this global configuration, at local and instantaneous levels, combustion can take place in premixed and lean conditions as well. The occurrence of these processes is dependent on the turbulent mixing of the propellants.

Direct numerical simulations allow the detailed observation of these local phenomena. The present sub-section discusses the objective criteria to quantify and classify the local burning regime based on the instantaneous physical fields. The presented methods will be applied to the output of the performed numerical simulations to investigate the occurrence of the different burning conditions.

2.3.1. Fuel-Rich vs. Lean

The classical procedure to calculate the stoichiometric ratio of a hydrocarbon is based on the assumption that the reactants are oxidized into a mixture that is entirely composed of water and carbon dioxide. For the case of methane, this corresponds to the following global reaction scheme:

$$CH_4 + 2O_2 \rightarrow 2H_2O + CO_2 \tag{12}$$

which yields a stoichiometric mass O/F ratio of approximately 4, corresponding to $Z_{st} \approx 0.2$. In atmospheric combustion applications that operate in lean conditions, this approximation is quite accurate, since the carbon monoxide content is very low compared to that of carbon dioxide. However, for combustion in space propulsion applications, fuel-rich mixtures are desired in order to maximize the specific impulse. In addition, the high temperatures significantly promote carbon monoxide generation. As a consequence, the equilibrium mass fraction of carbon monoxide significantly outweighs that of carbon dioxide. In an extreme case, the corresponding adjusted global reaction can be written as follows:

$$CH_4 + \frac{3}{2}O_2 \rightarrow 2H_2O + CO \tag{13}$$

which implies a stoichiometric O/F of roughly 3, corresponding to $Z_{st} \approx 0.25$. Besides the significant portion of carbon monoxide, additional particularities of combustion in the frame of space propulsion further entangle the equilibrium composition. Due to the high pressures and the absence of nitrogen, very high temperatures are reached in the flame ($T_{ad} > 3400$ K). Consequently, significant mass fractions of free radicals and hydroxyl are present at equilibrium. All of these substantial deviations from the assumption in (13) demand the redefinition of the stoichiometric condition. A reasonable criterion is to choose the O/F that maximizes the volumetric heat release. The integrated heat release rate is calculated as a function of the mixture fraction using Cantera [40] and a complex chemical mechanism [41]. The result is displayed in Figure 1. As can be seen, the mixture fraction that maximizes heat release takes place at values higher than the classical assumption of $Z_{st} \approx 0.2$. More specifically, the maximum value is found at $Z \approx 0.25$. This is in good agreement with the previous reasoning, since $Y_{CO,eq}/Y_{CO_2,eq} \approx 4$. Therefore, it can be expected that the redefined stoichiometric value is closer to the one in the assumption in Figure 1. Besides the graph in Figure 1, additional evidence for the better suitability of an $O/F \approx 3$ as a stoichiometric value is provided in the discussion of the results.

The shift in the maximum heat release is relevant, since most models for non-premixed combustion require knowledge of the properties at stoichiometric conditions. Within the present work, it is assumed that the stoichiometric mixture fraction in high-pressure methane–oxygen combustion corresponds to $Z_{st} \approx 0.25$. It is important to note that this specific definition is assumed to address the decoupling between the local maximum heat release and the standard stoichiometric oxidizer-to-fuel ratio. From a global standpoint, the classical definition should be considered.

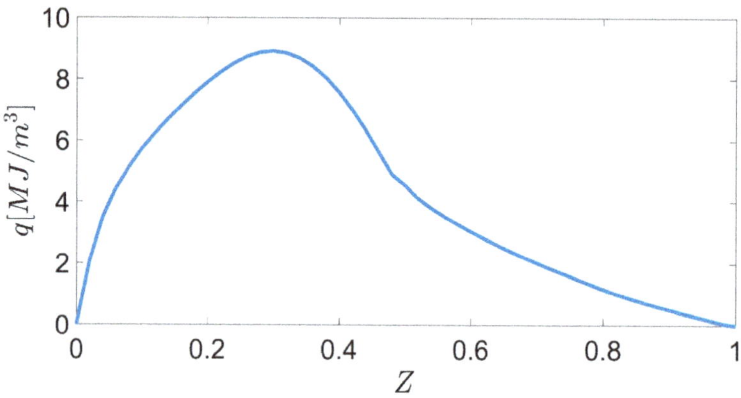

Figure 1. Integral volumetric heat release as a function of the mixture fraction for methane–oxygen combustion at 20 bar.

2.3.2. Flame Index

In burning devices that operate in non-premixed conditions, the reacting flow simultaneously displays both premixed and non-premixed configurations at a local level. These variations are driven by the local mixing conditions, which can enable premixed combustion despite the predominant diffusive nature of the flame. Regime markers can be used to identify the local burning regime. The most commonly used marker is the flame index FI, as defined by Yamashita et al. [42], as follows:

$$\mathrm{FI} = \frac{1}{2} + \frac{\nabla Y_{CH_4} \cdot \nabla Y_{O_2}}{2|\nabla Y_{CH_4}||\nabla Y_{O_2}|} \tag{14}$$

The original formulation has been re-normalized so that zero corresponds to non-premixed conditions and unity corresponds to premixed conditions. The relationship between this numerical value and the local combustion regime can be easily deduced by the physical meaning of the mathematical expression. If the gradients of the fuel and the oxidizer are aligned, then the local mixing conditions can be deemed premixed. Contrarily, the opposed gradients of the reactants correspond to a counterflow non-premixed configuration. There are several available definitions of a flame index that have been proposed in recent works [43–48]. These markers have been extensively reviewed in a recent study [49], where it was found that most definitions yield similar classifications overall. In the frame of the present work, the definition of Yamashita et al. was the one considered, due to its simplicity and extended usage.

An example of the different classifiers presented in this section can be visualized in Figure 2. In this figure, the instantaneous detail of the shear layer for one of the simulated flames is displayed. This graph shows the flame index, mixture fraction, and temperature in order to illustrate the different regimes in which combustion can take place in space propulsion applications. As can be seen, the central region of the shear layer presents the classical counterflow structure of a diffusion flame. Heat release is concentrated in the nearby stoichiometric conditions, where the local O/F maximizes the reacting capabilities of the flow. However, in certain positions far away from the stoichiometric line, the gradients of methane and oxygen are aligned, denoting local premixed combustion. This circumstance is very specific to space propulsion applications. Due to the absence of nitrogen, the flammability limits are significantly widened, enabling combustion in extremely fuel-rich and/or oxygen-rich conditions. Direct numerical simulations enable accessibility to every relevant physical field. Therefore, they offer the possibility to study the occurrence of these events and how they influence the combustion development process.

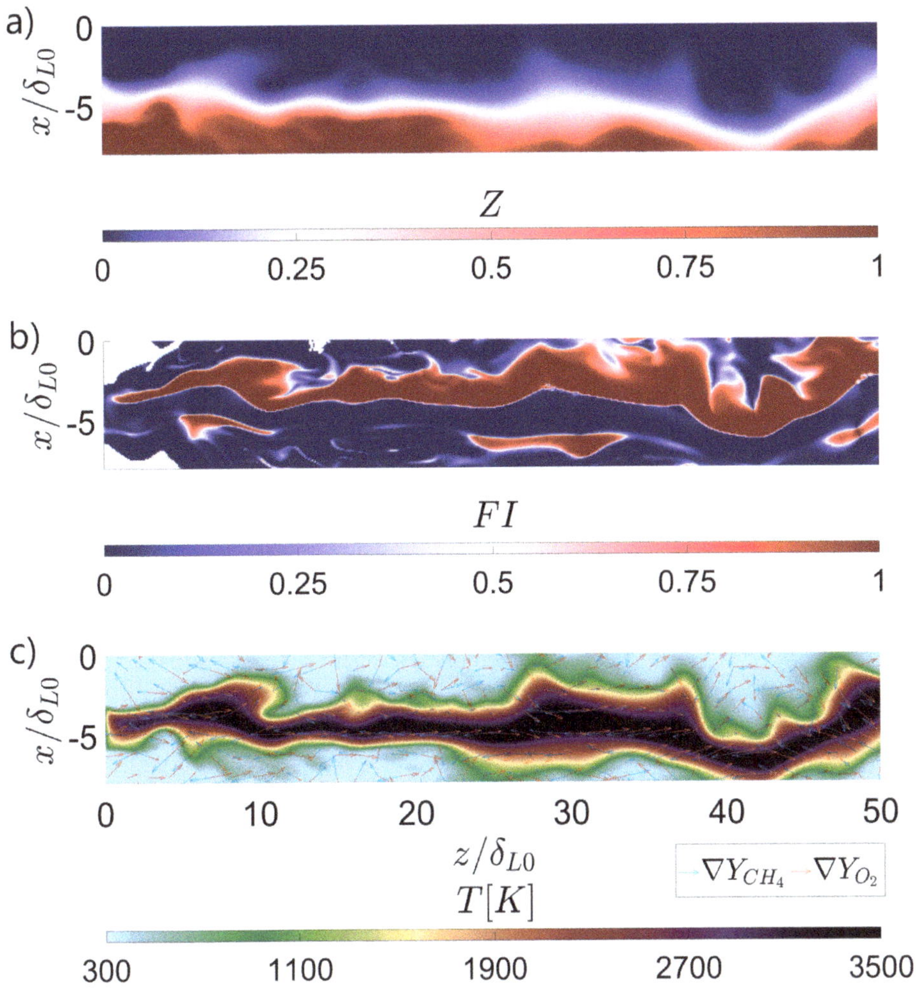

Figure 2. Example of instantaneous flame classification parameters in a reacting shear layer: mixture fraction (**a**), flame index (**b**), and temperature, unity fuel, and oxidizer gradients (**c**). Only the x and z components of the normalized gradients are displayed.

3. Numerical Simulation Case Setup

The numerical simulations aim to represent the turbulent combustion process that occurs in modern rocket engines in locations near to injection and far away from walls. The turbulent reacting flow is resolved with the compressible solver EBI-DNS [50–53], which is embedded in the open-source software OpenFOAM [54,55]. This code uses an implicit scheme to resolve the transient conservation equations for mass, momentum, energy, and species with the finite volume method (FVM) [56,57]. EBI-DNS has been applied and validated in several combustion-related problems in recent years [58–62]. A recent study by Zirwes et al. [63] comprehensively assessed the numerical accuracy of this solver for a wide variety of setups, with different reactive and compressibility features. This research shows that EBI-DNS is capable of providing results that are comparable with high-fidelity explicit solvers under a wide range of scenarios. Direct validation of DNS for space propulsion applications is unfeasible, due to the prohibitive computational cost that would be required for a complete system simulation. This difficulty can be addressed by applying convenient normalizations over the results of a reduced engine instance. Following such an approach, a good agreement between experimental and simulated combustion completion points was reported in a previous research paper [64].

The Cantera [40] routines are used for the computation of detailed chemistry and transport properties, using the mixture-averaged transport model described by Kee et al. [65]. The reaction mechanism developed by Slavinskaya et al. [41] is used to determine the chemical source terms of methane combustion using the finite-rate method. This skeletal mechanism consists of 21 species and 97 reactions, and it was devised for combustion in high-pressure space propulsion applications. This mechanism has been applied in several numerical studies of combustion for scale rocket combustors [66,67]. Despite the relatively high complexity of the used reaction scheme, the formation of large aromatic compounds and soot has not been considered. Such a simplification is appropriate for methane combustion, where soot formation is negligible. Nonetheless, it is important to keep in mind the possibility of soot-related phenomena being ignored in the fuel-rich side of the flame. The relevant data concerning combustion are provided in Table 1.

Table 1. Main turbulence and combustion parameters of the simulated flame.

Symbol	Description	Value
δ_{L0}	Laminar flame thickness at stoichiometric conditions	5.505 μm
s_{L0}	Laminar flame speed at stoichiometric conditions	2.574 m/s
Z_{flame}	Global mixture fraction	0.33
T_i	Injection temperature	300 K
p	Combustion pressure	20 bar

The inlet boundary conditions at the injection site in the main simulation are patched using the results of a precursor simulation. In this prior simulation, periodic synthetic turbulence is generated at the inlet via the procedure described in [68]. This scheme is derived from the original formulation proposed by Kraichnan and the subsequent improvements suggested by Shur et al. [69] and Morsbach et al. [70]. These methods are based on discretizing a reference turbulent kinetic energy spectrum in harmonics with random amplitude and phases. These waves are thereafter superposed to provide a synthetic velocity field that complies with the prescribed 2nd order turbulent statistics. The statistics were selected to resemble the output of RANS simulations for a scale methane rocket combustor discussed in a previous work [71]. The described simulation strategy is summarized in Figure 3.

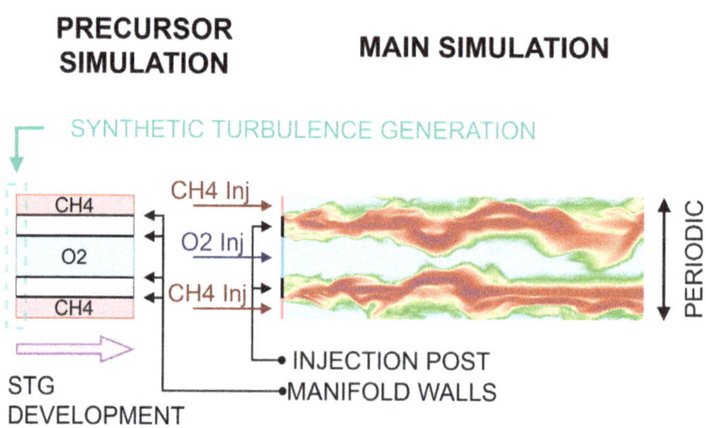

Figure 3. Scheme of the conducted simulation strategy.

Due to the high computational cost associated with DNS, the hydraulic diameter of the injector ϕ_H is one order of magnitude smaller than the actual setup. As a consequence, the global Reynolds number $Re = U\phi_H/\nu$ and the turbulent Reynolds number $Re_T = k^2/\nu\varepsilon$ are smaller than the reference device. Despite this shift, most of the relevant turbulent and chemical features are preserved. Therefore, the simulations can be assumed to capture the relevant aspects of turbulent mixing and combustion. Relevant turbulent flow parameters at the injector plane are provided in Table 2.

Table 2. Main turbulence and combustion parameters of the simulated flame.

Symbol	Description	Sα	Sδ
$\tilde{\eta}_{CH_4} - \tilde{\eta}_{O_2}$	Average Kolmogorov scale	2.33–2.51 μm	0.517–0.558 μm
$Re_{T,CH_4} - Re_{T,O_2}$	Average turbulent Reynolds number	100–115	452–516
$Re_{CH_4} - Re_{O_2}$	Global Reynolds number	1291–1425	5810–6411
U_i	Injection velocity	8.3 m/s	37.5 m/s

The main simulation comprises a volume of 0.3 × 0.2 × 3 mm, resolved with 144 × 216 × 2160 cubic cells with identical geometry. Periodicity is imposed in the directions perpendicular to the injection. The grid size is small enough to resolve the smallest Kolmogorov eddies in both of the simulations, following the standard requirements [72]. The flame front is resolved with varying resolutions, since the flame front thickness varies. The chosen grid size ensures a resolution of the instantaneous flame front realizations in the order of 10 cells. This is generally enough, although a higher resolution would be desirable. The convergence of the turbulent statistics for the used setup has been addressed in a prior work [36]. The effect of increasing the resolution is marginal. More specifically, a grid size 2.5 times larger yields maximum errors in the order of 5% and axially averaged properties below 1%. A finer grid (halving the cell volume) leads to discrepancies in the axially averaged properties in the order of 0.5%. Therefore, it can be concluded that the resolution of the used mesh is high enough to allow the usage of the simulation results for turbulence research purposes. The timestep size is chosen in an adaptive fashion to ensure that the maximum acoustic CFL number is below 0.9. This corresponds to maximum convective CFL numbers in the order of 0.05 for Sα and 0.25 for Sδ. Moreover, due to the implicit nature of the used solver, the fulfillment of the general condition $CFL < 1$ is not required to achieve numerical stability. Such low CFL numbers, coupled with the cubic spatial differentiation scheme, ensure the minimization of numerical dissipation, as discussed elsewhere [62].

4. Results Analysis

This section is devoted to the discussion and analysis of the results obtained in the performed numerical simulations. The discussion is structured in three parts. The first and second sub-sections i.e., Sections 4.1 and 4.2, discuss the overall results of simulations Sα and Sδ, respectively. The flame development is discussed from a phenomenological standpoint to identify the main events and the driving physical mechanisms. In the third sub-section (i.e., Section 4.3), the different interactions between turbulent mixing and combustion are discussed in order to investigate the sensitivity of the burning process, with respect to the turbulent kinetic energy.

4.1. Simulation Sα

The Simulation Sα is characterized by low inlet turbulence, with a flow behavior close to laminarity. An example of instantaneous flame development can be observed in Figure 4. In this illustration, the flame classification is displayed at the top, and both a positive and a negative heat release rate can be visualized at the center and bottom, respectively.

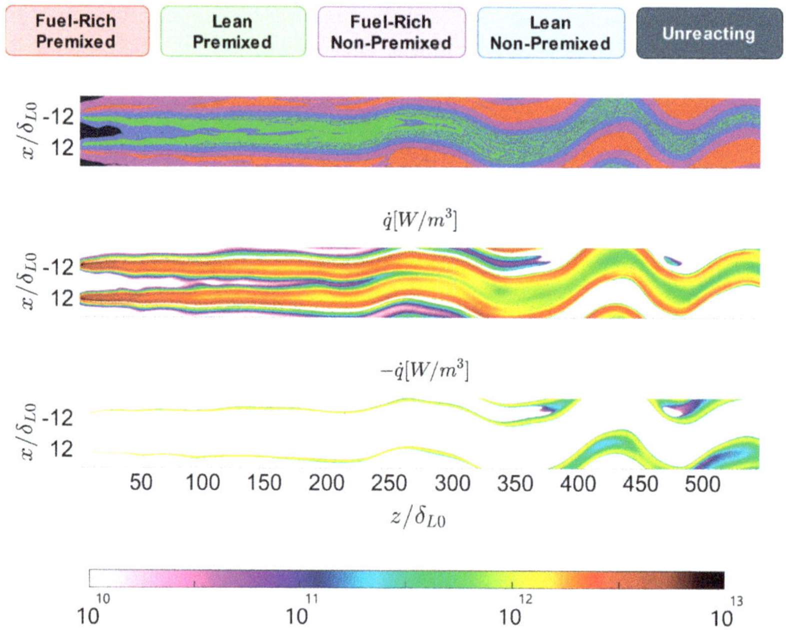

Figure 4. Example of instantaneous flame classification and heat release rate for simulation $S\alpha$.

The flow structure near injection resembles a triple flame, with a non-premixed branch surrounded by lean and fuel-rich premixed branches. Negative heat release can be observed in the transition from fuel-rich premixed to fuel-rich non-premixed conditions. This process is on account of the cracking reactions, which can lead to endothermic results due to the formation of larger hydrocarbons (especially ethylene). The fuel-rich premixed branch displays a discontinuous axial development, with local breakups. This is caused by the forced advection of methane into the flame core. Due to its higher diffusivity, methane benefits more than oxygen from the moderate turbulence. This different sensitivity to turbulence shifts the local O/F in the thick premixed regions outside of the flammability limits. Contrarily, the lean premixed branch displays a more stable configuration. As expected, most of the heat release is obtained from the non-premixed flamelets. Nevertheless, a significant amount of combustion occurs in the lean premixed side, while both of the shear layers are still present. This trend is interrupted by $z \approx 250\ \delta_{L0}$, when both of the shear layers merge into a single structure. From this point, the combustion at the lean core significantly decreases, and the majority of chemical reactions develop on the non-premixed side.

Figure 5 was elaborated to statistically summarize the observed processes. The right manifold can be used to investigate the probability of occurrence of the different classifications. Most of the observations take place in non-premixed conditions, ranging from moderately lean to significantly fuel rich. This is in good agreement with the example in Figure 4, where it can be observed that the non-premixed flamelet is predominant. There is a significant portion of data in the premixed conditions with extremely lean and fuel-rich mixtures. However, only the lean premixed regions deliver heat release rates in the same order of magnitude as the non-premixed flamelets. Intermediate flame indexes are rather unusual. This is due to the quasi-bimodal distribution of the flame index, which can be observed in the examples in Figure 2 and by other researchers [49]. These graphs provide valuable insights concerning the overall combustion process. However, it is important to remark on their strong dependency on geometry. For example, if the simulation domain were significantly longer, the statistics would shift toward equilibrium conditions. This would imply a greater concentration of points towards $Z = Z_{global}$, with a lower averaged heat release rate.

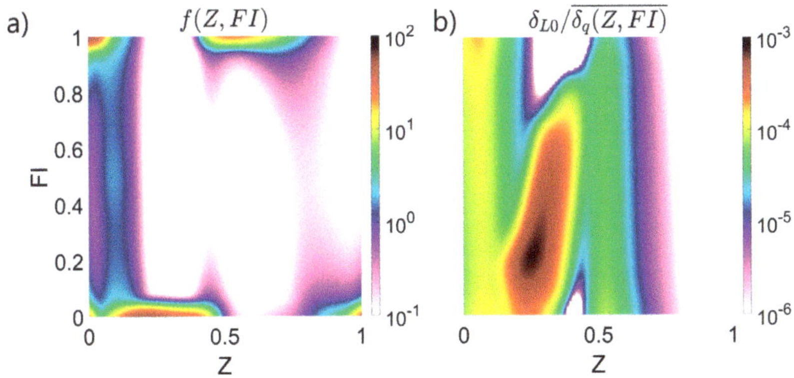

Figure 5. Statistical distribution of the normalized heat release rate, with respect to the local combustion regime for Simulation Sα: Joint probability density as a function of the flame index FI and the mixture fraction Z (**a**) and the average heat release rate as a function of the flame index FI and the mixture fraction Z (**b**).

4.2. Simulation Sβ

An analogous analysis was conducted with the database of Simulation Sδ. Unlike Sα, this simulation presents a Reynolds number high enough to display fully turbulent features. An example of an instantaneous flame classification and heat release rate is shown in Figure 6.

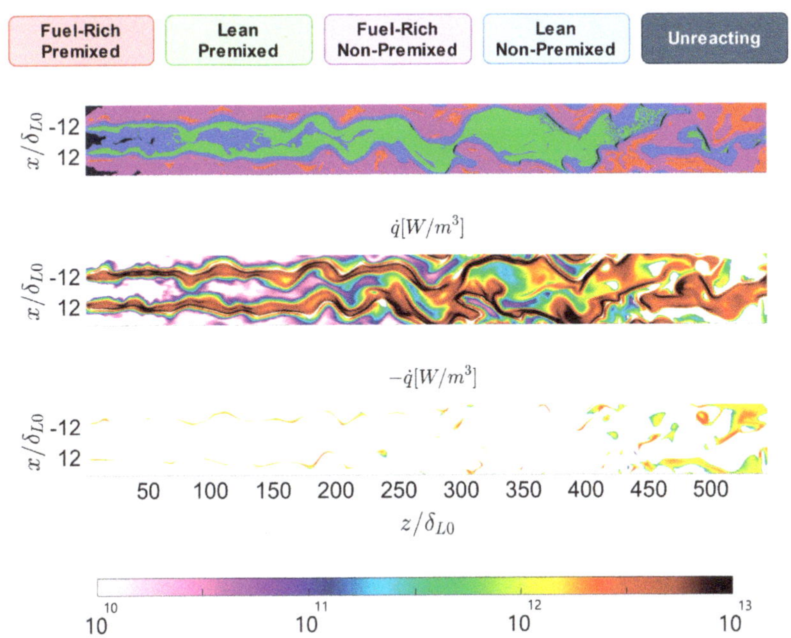

Figure 6. Example of instantaneous flame classification and heat release rate for simulation Sδ.

As can be seen, there is a greater disruption in the fuel-rich premixed region, as this combustion regime only occurs in isolated regions. This is caused by the greater turbulent kinetic energy, which, furthermore, promotes the transport of methane into the flame core. Another remarkable fact is that the negative heat release caused by cracking reactions near to the injection site takes place intermittently. These variations are caused by the flame stretching, which can locally promote or inhibit the heat transfer from the warm products to the cold methane. If heat transfer is promoted, the initial transitory ethylene formation is dampened, due to the local enhancement of the burning rate. The opposite occurs if the flame stretching defocuses the local heat fluxes from the products to the reactants. This effect is discussed in greater detail in the following sub-section. Negative heat release can be observed at downstream positions as well; however, this result is mainly motivated by recombination reactions of the combustion products.

Overall, the absolute heat release rate is substantially larger in Sδ compared to Sα. This is mainly a consequence of the greater mass flow rate of the propellants, which enhances the incoming rate of fresh reactants and the turbulent kinetic energy $\widetilde{k}$, which is proportional to the scalar dissipation rate $\widetilde{\chi}$. Nevertheless, the flow in simulation Sδ is able to keep up a significant burning rate in the flame core, with lean premixed conditions following the merging of the shear layers. To complete this section, the 2D joint functions are displayed in Figure 7. Few differences can be observed in the 2D joint probability density function. The most remarkable one is the greater prevalence of lean premixed configurations. This is caused by the greater presence of this regime at downstream positions, as can be seen by comparing Figures 4 and 6. Concerning the normalized heat release, it can be seen that it is higher overall, regardless of the considered configuration. However, the enhancements are more remarkable in the lean regions. The differences in the heat release trends are discussed in greater detail in the following sub-section.

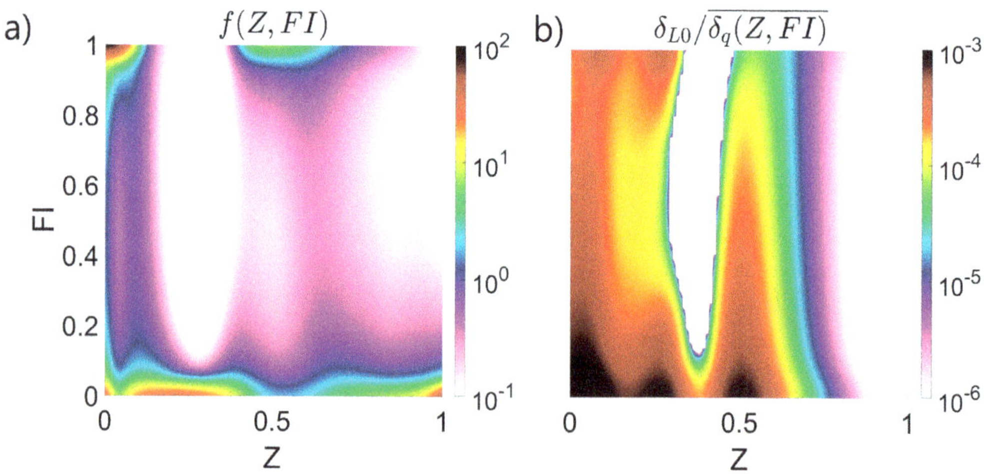

Figure 7. Statistical distribution of the heat release rate, with respect to the local combustion regime for simulation $S\delta$: Joint probability density as a function of the flame index FI and the mixture fraction Z (**a**) and the average heat release rate as a function of the flame index FI and the mixture fraction Z (**b**).

4.3. Comparison

Despite the remarkably different flow characteristics, for both of the flames, the chemistry remains very fast compared to the turbulence. To prove this claim, the progress variable, as a function of the mixture fraction, is represented in Figure 8. As can be seen, both of the flames follow the trend expected for fast chemistry [33,37]. Flame $S\delta$ displays a slight deviation, due to its higher turbulent kinetic energy, but the differences are negligible overall. The maximum value of the progress variable is obtained at $Z \approx 0.25$, which further supports the discussed necessity to consider a shift in the stoichiometric value. The maximum variance of the progress variable conditional to Z is in the order of 2×10^{-3}, corresponding to $Z \approx Z_{st}$, while, for the rest of the domain, $\widetilde{c''^2} < 10^{-4}$ holds. Therefore, it can be assumed that combustion is limited by mixing, since, for a given mixture fraction, the corresponding progress variable is reached almost immediately. The turbulent timescales near injection can be used to examine the validity of the flamelet assumption as well. The mean turbulent timescale in this region is in the order of 50 µs and 10 µs for simulations $S\alpha$ and S_δ, respectively. Since the chemical timescale is in the order of 2 µs, it is evident that chemistry is significantly faster in both cases. These points are crucial for the interpretation of the results discussed in this section. Since $c = f(Z)$, the local heat release rate $\dot{q}$, which scales with $\dot{c}$, is essentially dependent on the mixture fraction dynamics. Therefore, the influence that turbulence exerts on the flame development occurs in terms of heat and mass transfer enhancements caused by forced convection.

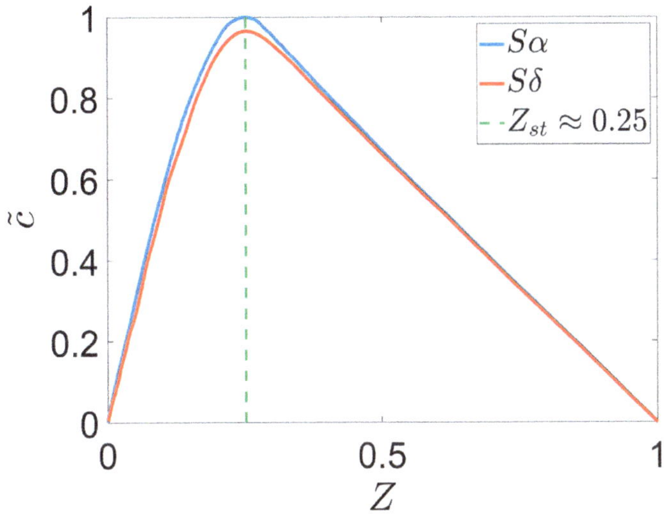

Figure 8. Progress variable $\tilde{c}$ as a function of the instantaneous mixture fraction.

To further study the influence of turbulence in the mixing process, the probability distributions in Figures 5 and 7 are compared. To assess the sensitivity of these joint distributions, with respect to inlet turbulence, the relative increase is displayed in Figure 9. This graph provides a rough indication of the effects that a turbulence enhancement has with respect to mixing. As can be seen, there is a greater amount of data toward the horizontal center. This is due to improved mixing, which promotes the concentration of data, with respect to the global mixture fraction, i.e., $Z_{flame} \approx 0.33$. However, the probability of presenting extreme conditions, i.e., $(Z = 0, Z = 1)$, is enhanced as well. The motivation for this difference is the smaller residence time of the molecules in simulation $S\delta$. Since turbulence and mean velocity increase simultaneously, the enhanced mixing is accompanied by a shorter stay time, which increases the possibility of an unmixed cluster reaching the outlet. These effects illustrate that, although these comparative analyses provide valuable information, they should be approached carefully and critically.

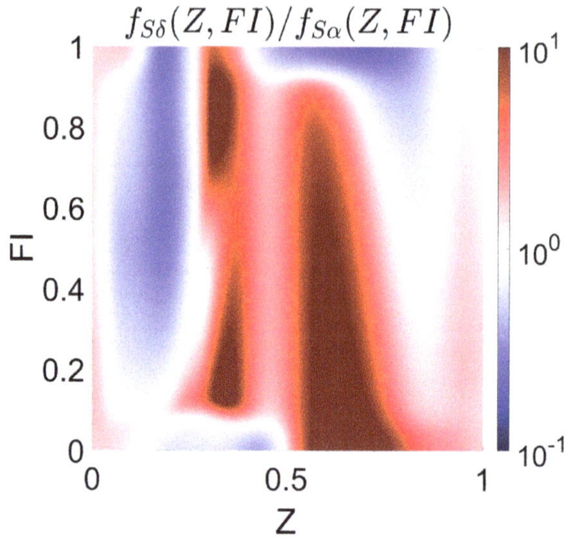

Figure 9. Relative probability density increase in Simulation Sδ, with respect Simulation Sα as a function of the mixture fraction Z and the flame index FI.

To study the variations in heat release, this value was averaged conditionally to the mixture fraction for both of the simulations. The result is displayed in Figure 10. As can be seen, Sδ provides a higher heat release overall, with the only exception being the extremely lean regions.

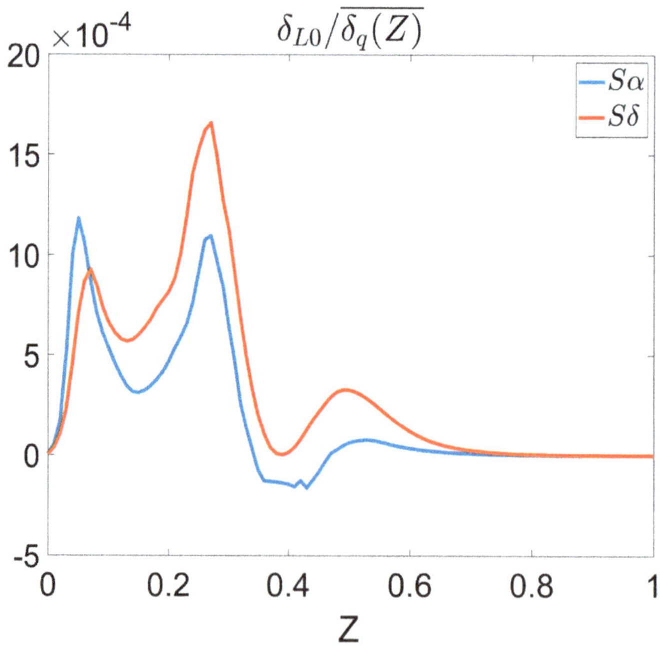

Figure 10. Average normalized heat release rate conditional to the mixture fraction or the simulated flames.

Thanks to the fact that c and Z are algebraically related, most of the dependencies with the geometry and residence time are eliminated. Therefore, the differences in Figure 10 are linked to the mixing enhancement driven by turbulence. These effects are strongly dependent on which part of the flame is considered. To simplify the phenomenological description, the different trends are commented on in two different positions: near injection and downstream (where both of the shear layers have merged). The physical driving mechanisms in both of these cases are commented on in the text below.

4.3.1. Near Injection Mixing

To approach the turbulent mixing dynamics near injection, details of the simulated flame are presented in Figure 11. In this figure, the instantaneous stoichiometric line is presented along with the line marking the instantaneous border between the premixed and non-premixed classifications. Therefore, the region between these two isolines corresponds to the lean side of the local diffusion flame. Overall, flame $S\alpha$ exhibits a greater amount of normalized heat release. $S\delta$ presents similar values in the stoichiometric line, but it only keeps up in the stretched regions due to the enhanced convection. Moreover, in these positions, the negative heat release caused by cracking reactions can be suppressed thanks to the high local curvature, which fosters the heat transfer from the stoichiometric line to the fuel-rich zone. Indeed, it can be seen that the regions with non-negative heat release only appear where there is significant stretching.

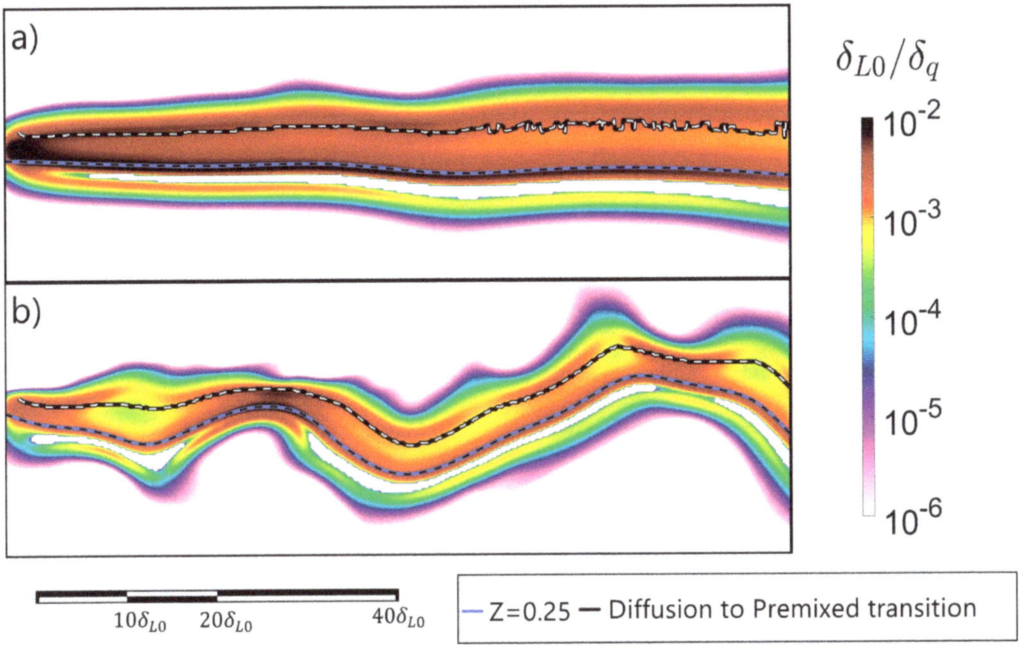

Figure 11. Detail of heat release near injection: Simulation $S\alpha$ (**a**) and simulation $S\delta$ (**b**).

4.3.2. Downstream Mixing

The flame development dynamics are dramatically altered once the merging of the shear layers takes place. To study the driving mechanisms, details of the local heat release for both simulations are presented in Figure 12. In this illustration, the transparency is varied in order to identify the local burning regime. As can be seen, the lean side of the diffusion flame generates similar heat release in both of the simulations, although, in Simulation Sδ, the observed values are slightly larger.

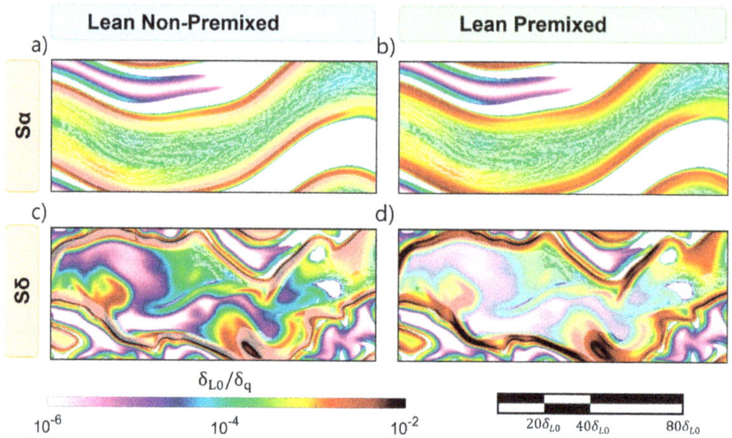

Figure 12. Details of normalized heat release $1/\delta_q$ right after the reactive shear layers merge: Simulation Sα with a mask indicating lean non-premixed regions (**a**), Simulation Sα with a mask indicating lean premixed regions (**b**), Simulation Sδ with a mask indicating lean non-premixed regions (**c**), and Simulation Sδ with a mask indicating lean premixed regions (**d**).

The main differences between the simulations take place at the flame core, where the mixture fraction is significantly below its stoichiometric value. In Simulation Sα, intermittency between premixed and non-premixed combustion can be observed. This result points to unstable burnability, due to insufficient mixing intensity. This behavior contrasts with what is observed in Sδ, where the flame core burns mostly in the premixed conditions. There, the heat release greatly varies between the two flames, since it depends on the local strain and availability of reactants. In both cases, a heat release rate drop can be observed from the lean edge of the diffusion flame to the oxygen core. However, heat release near the diffusive region is significantly larger in Sα, and the values decrease faster. This decreasing trend is motivated by the lower availability of burnable reactants, as the mixture fraction tends to Z_{flame}. Although heat release at the core is smaller in Simulation Sα, the significantly higher burning rate close to the diffusive side enables a greater heat release in lean conditions overall. This process is behind the trend observed in Figure 10 towards the lean side.

The differences in the burning rate between the simulations are primarily driven by the different evolution of the scalar dissipation. To discuss the influence of this variable, instantaneous 2D cuts of its value are displayed in Figure 13.

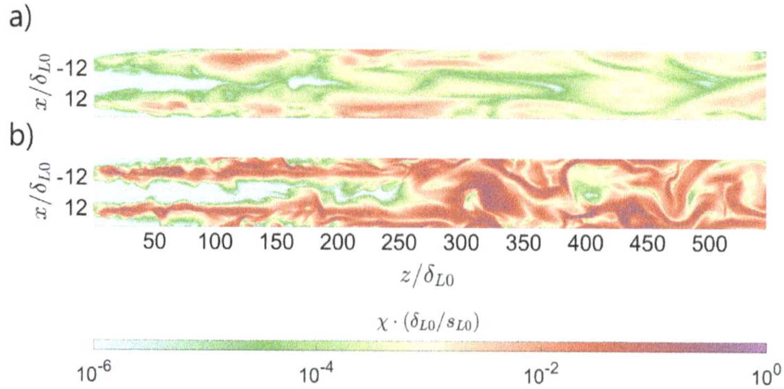

Figure 13. Instantaneous scalar dissipation rate: Simulation Sα (**a**), Simulation Sδ (**b**).

As expected, the scalar dissipation rate is greater in Sδ overall. This is caused by the larger turbulent kinetic energy at the inlet. The most remarkable difference between the simulations corresponds to the behavior following the merging of the shear layers. In Simulation Sα, the mixing intensity decreases after the shear layers collapse. Contrarily, in Sδ, an increase in scalar dissipation follows the merging of the shear layers. This is motivated by the larger wrinkling of the shear layer, due to the integrated effects of the turbulence departing from the injection site. In Sα, the shear layers are mostly aligned with the injection direction and, consequently, between themselves. Therefore, their merging produces little additional shear, and global dissipation dominates. Nevertheless, in Sδ, turbulence has enough energy to create substantial oscillations of the diffusive flame structures prior to their merging. In such a scenario, the collapse becomes more abrupt, enhancing the local mixing intensity. This phenomenon enables larger heat release rates, as premixed combustion becomes viable at the core, as displayed in Figure 12 for Sδ. If the scalar dissipation rate becomes too small, the possibility of combusting in a premixed condition in a stable way is disabled. This is evident in $S\alpha$, but it can occur in localized regions in Sδ as well, as can be observed in Figures 6 and 12.

5. Conclusions

Non-premixed methane–oxygen turbulent diffusion flames were studied using direct numerical simulations with finite-rate chemistry. The main purpose of this research was to study the flame development in the frame of modern space propulsion systems. A novel definition of the stoichiometric oxidizer-to-fuel ratio was proposed to capture the effects of significant carbon monoxide formation. The proposed definition is better suited to reaction engines, since the amount of carbon dioxide in equilibrium is small compared to carbon dioxide, altering the main assumption in which the classical definition of stoichiometric conditions is grounded.

The redefined stoichiometric value was used in conjunction with flame regime markers to classify the local combustion regime. Therefore, two main degrees of freedom were considered: fuel-rich vs. oxygen-rich and premixed vs. non-premixed conditions. The statistics of the heat release rate were analyzed conditional to these conditions, with varying Reynolds numbers, to investigate the influence of turbulence in the flame development process. This investigation was performed by a comparative analysis of one mildly turbulent flame (Sα) and one fully turbulent one (Sδ). For both of the flames, the chemistry was very fast compared to the turbulence. Therefore, combustion progress is limited by mixing, and differences in the heat release dynamics can be primarily attributed to the forced convection variations. It has been found that, despite the eminent non-premixed configuration of the burning device, a significant amount of heat release occurs in lean premixed conditions. This process occurs primarily at the flame core, following the merging of the shear layers. With growing turbulence, this merging becomes more abrupt, due to the increased decorrelation between the two interacting diffusive flame structures.

The results presented here point out several challenges concerning combustion in systems that resort to using oxygen as an oxidizer. Due to the wider flammability limits, combustion in lean premixed conditions can become significant. This effect indicates the necessity to consider a combination of premixed and non-premixed combustion models in turbulent combustion simulations for space propulsion systems. In addition, the definition of the stoichiometric conditions should be redefined to account for the large amount of carbon monoxide in equilibrium.

Author Contributions: Conceptualization, D.M.-S. and A.S.; methodology, D.M.-S. and O.H.; software, A.S.; validation, D.M.-S. and A.S.; formal analysis, D.M.-S.; investigation, D.M.-S.; resources, A.S.; data curation, D.M.-S.; writing—original draft preparation, D.M.-S.; writing—review and editing, D.M.-S.; visualization, D.M.-S.; supervision, O.H. and M.T.; project administration, O.H.; funding acquisition, A.S. and O.H. All authors have read and agreed to the published version of the manuscript.

Funding: This research was funded by the Bayerische Forschungsstiftung (BFS), "Weiterentwicklung der CFD-Simulationsmethoden für H_2-O_2-Hochduck-Dampfgeneratoren zur Stromnetzstabilisierung". Project number: AZ-1517-21.

Data Availability Statement: The authors will share the used datasets upon reasonable request.

Acknowledgments: The authors thank the Leibniz-Rechenzentrum (LRZ) for providing the computational resources for performing the numerical simulations and post-processing. Special thanks go to Martin Ohlerich for his decisive aid and support that allowed this project to make the most of SuperMUC NGen's capabilities. This research was founded by the Bavarian Research Foundation (Bayerische Forschungsstiftung), within the project "$H_2 - O_2$-Brennkammer im Kraftwerk", with reference Grant No. AZ-1517-21.

Conflicts of Interest: The authors declare no conflict of interest.

References

1. Haeseler, D.; Bombelli, V.; Vuillermoz, P.; Lo, R.; Marée, T.; Caramelli, F. Green Propellant Propulsion Concepts for Space Transportation and Technology Development Needs. In Proceedings of the 2nd International Conference on Green Propellants for Space Propulsion, Sardinia, Italy, 7–8 June 2004.
2. Sackheim, R.L.; Masse, R.K. Green Propulsion Advancement: Challenging the Maturity of Monopropellant Hydrazine. *J. Propuls. Power* **2014**, *30*, 265–276. [CrossRef]
3. Turner, M.J.L. Liquid Propellant Rocket Engines. In *Rocket and Spacecraft Propulsion*; Springer: Berlin/Heidelberg, Germany, 2009; pp. 67–108.
4. Nickerson, G.R.; Johnson, C.W. A Soot Prediction Model for the TDK Computer Program. In Proceedings of the AIAA/SAE/ASME/ASEE 28th Joint Propulsion Conference and Exhibit, Nashville, TN, USA, 6–8 July 1992.
5. Ciezki, H.K.; Zhukov, V.; Werling, L.; Kirchberger, C.; Naumann, C.; Friess, M.; Riedel, U. Advanced Propellants for Space Propulsion—A Task within the DLR Interdisciplinary Project "Future Fuels". In Proceedings of the 8th European Conference for Aeronautics and Space Sciences (EUCASS), Madrid, Spain, 1–4 July 2019.
6. Kajon, D.; Liuzzi, D.; Boffa, D.; Rudnykh, M.; Drigo, D.; Arione, L.; Ierardo, N.; Sirbi, A. Oxygen-methane combustion studies in the In Space Propulsion Programme. In Proceedings of the 4th European Conference for Aerospace Sciences (EUCASS), Madrid, Spain, July 2019.
7. Ciezki, H.N.M.; Werling, L. Trends in Research and Development on Green Chemical Propulsion for Orbital Systems. In Proceedings of the 7th International Conference on Recent Advances in Space Technologies, RAST 2015, Istanbul, Turkey, 16–19 June 2015.
8. Werling, L.; Freudenmann, D.; Ricker, S.C.; Wilhelm, M.; Lauck, F.; Strauss, F.; Manassis, K.; Kurilov, M.; Petrarolo, A.; Hörger, T.; et al. Research and Test Activities on Advanced Rocket Propellants at DLR's Institute of Space Propulsion in Lampoldshausen. In Proceedings of the 9th European Conference for Aeronautics and Space Sciences, Lille, France, 27 June–1 July 2022.
9. Haidn, O.; Ordonneau, G.; Soller, S.; Onofri, M. Development of the liquid oxygen and methane M10 rocket engine for the Vega-E upper stage. In Proceedings of the 8th European Conference for Aerospace Sciences, Saint Petersburg, Russia, 4–8 July 2011.
10. Haidn, O.J.; Celano, M.P.; Luo, M.; Roth, C.; Silvestri, S.; Slavinskaya, N.A. On Methane/ Oxygen Combustion for Rocket Applications. In Proceedings of the International Symposium on Innovation and Prospects of Liquid Propulsion, Xi'an, China, 4–6 September 2016.
11. Friedrich, R. Modelling of Turbulence in Compressible Flows. In *Transition, Turbulence and Combustion Modelling: Lecture Notes from the 2nd ERCOFTAC Summerschool held in Stockholm, 10–16 June 1998*; Springer: Dordrecht, The Netherlands, 1999; pp. 243–348. [CrossRef]
12. Echekki, T.; Mastorakos, E. Concepts, Governing Equations and Modelling Strategies. In *Turbulent Combustion*; Springer: Dordrecht, The Netherlands, 2011; pp. 19–39.
13. Chakraborty, N.; Katragadda, M.; Cant, R.S. Statistics and Modelling of Turbulent Kinetic Energy Transport in Different Regimes of Premixed Combustion. *Flow Turbul. Combust.* **2010**, *87*, 205–235. [CrossRef]
14. Chakraborty, N. Influence of Thermal Expansion on Fluid Dynamics of Turbulent Premixed Combustion and Its Modelling Implications. *Flow Turbul. Combust.* **2021**, *106*, 753–848. [CrossRef]
15. Cuenot, B.; Poinsot, T. Effects of curvature and unsteadiness in diffusion flames. Implications for turbulent diffusion combustion. In *Symposium (International) on Combustion*; Elsevier: Amsterdam, The Netherlands, 1994; Volume 25, pp. 1383–1390. [CrossRef]
16. Martinez-Sanchis, D. A Flame Control Method for Direct Numerical Simulations of Reacting Flows in Rocket Engines. Master's Thesis, Institute of Space Propulsion, Technical University of Munich, Munich, Germany, 2021.
17. Martinez-Sanchis, D.; Banik, S.; Sternin, A.; Sternin, D.; Haidn, O.; Tajmar, M. Analysis of turbulence generation and dissipation in shear layers of methane–oxygen diffusion flames using direct numerical simulations. *Phys. Fluids* **2022**, *34*, 045121. [CrossRef]
18. Martínez-Sanchis, D.; Sternin, A.; Tagscherer, K.; Sternin, D.; Haidn, O.; Tajmar, M. Interactions Between Flame Topology and Turbulent Transport in High-Pressure Premixed Combustion. *Flow Turbul. Combust.* **2022**, *109*, 813–838. [CrossRef]

19. Martinez-Sanchis, D.; Sternin, A.; Shvab, J.; Haidn, O.; Hu, X. An Eddy Dissipation Concept Performance Study for Space Propulsion Applications. *Aerospace* **2022**, *9*, 476. [CrossRef]
20. Ruelle, D. Strange attractors as a mathematical explanation of turbulence. In *Statistical Models and Turbulence: Proceedings of a Symposium Held at the University of California, San Diego (La Jolla) 15–21 July 1971*; Springer: Berlin/Heidelberg, Germany, 1972.
21. Ruelle, D.; Takens, F. On the nature of turbulence. *Commun. Math. Phys.* **1971**, *20*, 167–192. [CrossRef]
22. Favre, A. *Problems of Hydrodynamics and Continuum Mechanics*; SIAM: Philadelphia, PA, USA, 1969.
23. Magnussen, B.F. On the Structure of Turbulence and a Generalized Eddy Dissipation Concept for Chemical Reaction in Turbulent Flow. In Proceedings of the Nineteenth AIAA Meeting, St. Louis, MO, USA, 12–15 January 1981. [CrossRef]
24. Magnussen, B.F. The Eddy Dissipation Concept—A Bridge Between Between Science and Technology. In Proceedings of the ECCOMAS Thematic Conference on Computational Combustion, Lisbon, Portugal, 21–24 June 2005.
25. Peters, N. Laminar diffusion flamelet models in non-premixed turbulent combustion. *Prog. Energy Combust. Sci.* **1984**, *10*, 319–339. [CrossRef]
26. Veynante, D.; Vervisch, L. Turbulent Combustion Modeling. *Prog. Energy Combust. Sci.* **2002**, *28*, 193–266. [CrossRef]
27. Nivarti, G. The Bending Effect in Turbulent Flame Propagation. Doctoral Dissertation, University of Cambridge, Cambridge, UK, 2017.
28. Damköhler, G. Der einfluss der turbulenz auf die flammengeschwindigkeit in gasgemischen. *Z. Elektrochem. Angew. Phys. Chem.* **1940**, *46*, 601–652.
29. Nivarti, G.; Cant, S. Direct Numerical Simulation of the bending effect in turbulent premixed flames. *Proc. Combust. Inst.* **2017**, *36*, 1903–1910. [CrossRef]
30. Nivarti, G.V.; Cant, S. Scalar transport and the validity of Damköhler's hypotheses for flame propagation in intense turbulence. *Phys. Fluids* **2017**, *29*, 085107. [CrossRef]
31. Bradley, D. How fast can we burn? In *Symposium (International) on Combustion*; Elsevier: Amsterdam, The Netherlands, 1992; Volume 24, pp. 247–262. [CrossRef]
32. Driscoll, J.F. Turbulent premixed combustion: Flamelet structure and its effect on turbulent burning velocities. *Prog. Energy Combust. Sci.* **2008**, *34*, 91–134. [CrossRef]
33. Bilger, R.W. The Structure of Diffusion Flames. *Combust. Sci. Technol.* **1976**, *13*, 155–170. [CrossRef]
34. Bray, K.; Domingo, P.; Vervisch, L. Role of the progress variable in models for partially premixed turbulent combustion. *Combust. Flame* **2005**, *141*, 431–437. [CrossRef]
35. Pierce, C.D.; Moin, P. Progress-variable approach for large-eddy simulation of non-premixed turbulent combustion. *J. Fluid Mech.* **2004**, *504*, 73–97. [CrossRef]
36. Martinez-Sanchis, D.; Sternin, A.; Haidn, O.; Tajmar, M. Turbulent combustion statistics in a diffusion flame for space propulsion applications. *Phys. Fluids* **2022**, *34*, 125115. [CrossRef]
37. Burke, S.P.; Schumann, T.E.W. Diffusion Flames. *Ind. Eng. Chem.* **1928**, *20*, 998–1005. [CrossRef]
38. Peters, N. *Turbulent Combustion*; Cambridge University Press: Cambridge, UK, 2000.
39. Poinsot, T.; Veynante, D. Turbulent non premixed flames. In *Theoretical and Numerical Combustion*; Aquaprint: Bordeaux, France, 2012; pp. 287–348.
40. Goodwin, D.; Moffat, H.; Speth, R. Cantera: An Object-Oriented Software Toolkit for Chemical Kinetics, Thermodynamics and Transport Processes Version 2.3.0b. 2017. Available online: https://zenodo.org/record/48735 (accessed on 1 January 2023).
41. Salavinskaya, N.A.; Haidn, O.J. Reduced Chemical Model for High Pressure Methane Combustion with PAH Formation. In Proceedings of the 46th AIAA Aerospace Sciences Meeting and Exhibit, Reno, NV, USA, 7–10 January 2008.
42. Yamashita, H.; Shimada, M.; Takeno, T. A numerical study on flame stability at the transition point of jet diffusion flames. In *Symposium (International) on Combustion*; Elsevier: Amsterdam, The Netherlands, 1996; Volume 26, pp. 27–34. [CrossRef]
43. Fiorina, B.; Gicquel, O.; Vervisch, L.; Carpentier, S.; Darabiha, N. Approximating the chemical structure of partially premixed and diffusion counterflow flames using FPI flamelet tabulation. *Combust. Flame* **2005**, *140*, 147–160. [CrossRef]
44. Nguyen, P.-D.; Vervisch, L.; Subramanian, V.; Domingo, P. Multidimensional flamelet-generated manifolds for partially premixed combustion. *Combust. Flame* **2010**, *157*, 43–61. [CrossRef]
45. Som, S.; Aggarwal, S. Effects of primary breakup modeling on spray and combustion characteristics of compression ignition engines. *Combust. Flame* **2010**, *157*, 1179–1193. [CrossRef]
46. Wu, H.; Ihme, M. Compliance of combustion models for turbulent reacting flow simulations. *Fuel* **2016**, *186*, 853–863. [CrossRef]
47. Wan, K.; Vervisch, L.; Xia, J.; Domingo, P.; Wang, Z.; Liu, Y.; Cen, K. Alkali metal emissions in an early-stage pulverized-coal flame: DNS analysis of reacting layers and chemistry tabulation. *Proc. Combust. Inst.* **2018**, *37*, 2791–2799. [CrossRef]
48. Butz, D.; Hartl, S.; Popp, S.; Walther, S.; Barlow, R.S.; Hasse, C.; Dreizler, A.; Geyer, D. Local flame structure analysis in turbulent CH_4/air flames with multi-regime characteristics. *Combust. Flame* **2019**, *210*, 426–438. [CrossRef]
49. Zirwes, T.; Zhang, F.; Habisreuther, P.; Hansinger, M.; Bockhorn, H.; Pfitzner, M.; Trimis, D. Identification of Flame Regimes in Partially Premixed Combustion from a Quasi-DNS Dataser. *Flow Turbul. Combust.* **2021**, *1006*, 373–404. [CrossRef]
50. Zhang, F.; Bonart, H.; Zirwes, T.; Habisreuther, P.; Bockhorn, H.; Zarzalis, N. Direct Numerical Simulation of Chemically Reacting Flows with the Public Domain Code OpenFOAM. In *High Performance Computing in Science and Engineering '16*; Springer: Berlin/Heidelberg, Germany, 2015; Volume 14, pp. 221–236.

51. Zirwes, T.; Zhang, F.; Denev, J.A.; Habisreuther, P.; Bockhorn, H. Automated Code Generation for Maximizing Performance of Detailed Chemistry Calculations in OpenFOAM. In *High Performance Computing in Science and Engineering '17*; Springer: Berlin/Heidelberg, Germany, 2018; pp. 189–204.
52. Zirwes, T.; Zhang, F.; Denev, J.A.; Habisreuther, P.; Bockhorn, H.; Trimis, D. Improved Vectorization for Efficient Chemistry Computations in OpenFOAM for Large Scale Combustion Simulations. In *High Performance Computing in Science and Engineering '18*; Springer: Berlin/Heidelberg, Germany, 2019; pp. 209–224.
53. Zirwes, T.; Zhang, F.; Habisreuther, P.; Denev, J.A.; Bockhorn, H.; Trimis, D. Implementation and Validation of a Computationally Efficient DNS Solver for Reacting Flows in OpenFOAM. In Proceedings of the 14th World Congress on Computational Mechanics, Virtual Congress, 11–15 January 2021.
54. Weller, H.G.; Tabor, G.; Jasak, H.; Fureby, C. A tensorial approach to computational continuum mechanics using object-oriented techniques. *Comput. Phys.* **1998**, *12*, 620–631. [CrossRef]
55. Weller, H.; Tabor, G.; Jasak, H.; Fureby, C. *OpenFOAM*; OpenCFD Ltd.: Bracknell, UK, 2017.
56. McDonald, P.W. The Computation of Transonic Flow Through Two-Dimensional Gas Turbine Cascades. In Proceedings of the ASME 1971 International Gas Turbine Conference and Products Show, Houston, TX, USA, 28 March–1 April 1971.
57. Maccormack, R.; Paullay, A. Computational efficiency achieved by time splitting of finite difference operators. In Proceedings of the 10th Aerospace Sciences Meeting, San Diego, CA, USA, 17–19 January 1972.
58. Zhang, F.; Zirwes, T.; Nawroth, H.; Habisreuther, P.; Bockhorn, H.; Paschereit, C.O. Combustion-Generated Noise: An Environment-Related Issue for Future Combustion Systems. *Energy Technol.* **2017**, *5*, 1045–1054. [CrossRef]
59. Zhang, F.; Zirwes, T.; Habisreuther, P.; Bockhorn, H. Effect of unsteady stretching on the flame local dynamics. *Combust. Flame* **2017**, *175*, 170–179. [CrossRef]
60. Zirwes, T.; Häber, T.; Zhang, F.; Kosaka, H.; Dreizler, A.; Steinhausen, M.; Hasse, C.; Stagni, A.; Trimis, D.; Suntz, R.; et al. Numerical Study of Quenching Distances for Side-Wall Quenching Using Detailed Diffusion and Chemistry. *Flow Turbul. Combust.* **2020**, *106*, 649–679. [CrossRef]
61. Zhang, F.; Zirwes, T.; Häaber, T.; Bockhorn, H.; Trimis, D.; Suntz, R. Near wall dynamics of premixed flames. *Proc. Combust. Inst.* **2021**, *38*, 1955–1964. [CrossRef]
62. Zirwes, T.; Zhang, F.; Habisreuther, P.; Hansinger, M.; Bockhorn, H.; Pfitzner, M.; Trimis, D. Quasi-DNS Dataset of a Piloted Flame with Inhomogeneous Inlet Conditions. *Flow Turbul. Combust.* **2019**, *104*, 997–1027. [CrossRef]
63. Zirwes, T.; Sontheimer, M.; Zhang, F.; Abdelsamie, A.; Pérez, F.E.H.; Stein, O.T.; Im, H.G.; Kronenburg, A.; Bockhorn, H. Assessment of Numerical Accuracy and Parallel Performance of OpenFOAM and its Reacting Flow Extension EBIdnsFoam. *Flow Turbul. Combust.* **2023**, *111*, 567–602. [CrossRef]
64. Martinez-Sanchis, D.; Sternin, A.; Santese, T.; Haidn, O.J. The role of turbulence in the characteristic velocity and length of rocket combustors. *Aerosp. Sci. Technol.* **2023**, *134*, 108158. [CrossRef]
65. Kee, R.; Coltrin, M.; Glarborg, P. *Chemically Reacting Flow: Theory and Practice*; Wiley: London, UK, 2005.
66. Winter, F.F.; Perakis, N.; Haidn, O.J. Emission imaging and CFD simulation of a coaxial single-element GOX/GCH_4 rocket combustor. In Proceedings of the 2018 Joint Propulsion Conference, Cincinnati, OH, USA, 9–11 July 2018.
67. Perakis, N.; Rahn, D.; Haidn, O.J.; Eiringhaus, D. Heat Transfer and Combustion Simulation of Seven-Element O_2/CH_4 Rocket Combustor. *J. Propuls. Power* **2019**, *35*, 1080–1097. [CrossRef]
68. Martinez-Sanchis, D.; Sternin, A.; Sternin, D.; Haidn, O.; Tajmar, M. Analysis of periodic synthetic turbulence generation and development for direct numerical simulations applications. *Phys. Fluids* **2021**, *33*, 125130. [CrossRef]
69. Shur, M.L.; Spalart, P.R.; Strelets, M.K.; Travin, A.K. Synthetic Turbulence Generators for RANS-LES Interfaces in Zonal Simulations of Aerodynamic and Aeroacoustic Problems. *Flow Turbul. Combust.* **2014**, *93*, 63–92. [CrossRef]
70. Morsbach, C.; Franke, M. Analysis of a Synthetic Turbulence Generation Method for Periodic Configurations. In *Direct and Large-Eddy Simulation XI*; Springer International Publishing: Berlin/Heidelberg, Germany, 2019; pp. 169–174.
71. Sternin, A.; Perakis, N.; Celano, M.P.; Haidn, O. CFD-analysis of the effect of a cooling film on flow and heat transfer characteristics in a GCH4/GOX rocket combustion chamber. In *Space Propulsion 2018*; 3AF: Sevilla, Spain, 2018.
72. Yeung, P.K.; Pope, S.B. Lagrangian statistics from direct numerical simulations of isotropic turbulence. *J. Fluid Mech.* **1989**, *207*, 531–586. [CrossRef]

energies

MDPI

Article

Investigating the Ignition and Stability Limits of Premixed Methane/Air Combustion in Micro-Channels

Almoutazbellah Kutkut [1], Mohsen Ayoobi [2], Marc E. Baumgardner [3] and V'yacheslav Akkerman [1,*]

[1] Center for Innovation in Gas Research and Utilization (CIGRU), Department of Mechanical and Aerospace Engineering, West Virginia University, Morgantown, WV 26506, USA; aak00010@mix.wvu.edu
[2] Division of Engineering Technology, Wayne State University, Detroit, MI 48202, USA; mohsen.ayoobi@wayne.edu
[3] Mechanical Engineering, Gonzaga University, Spokane, WA 99258, USA; baumgardner@gonzaga.edu
* Correspondence: vyacheslav.akkerman@mail.wvu.edu

Abstract: Understanding and improving the performance of miniature devices powered by micro-combustion have been the focus of continued attention of researchers recently. The goal of the present work is to investigate the behavior of premixed methane–air combustion in a quartz microreactor with an externally controlled wall temperature. Specifically, the impacts of the flow inlet velocity, the equivalence ratio, and the microreactor channel size are examined. This study is conducted by means of computational simulations, and the results are validated against prior experimental data, as well as by other similar studies in the literature. Utilizing simulation results with detailed chemistry, the present work provides more in-depth insight into a variety of phenomena, such as ignition, flame propagation, flames with repetitive extinctions and ignitions (FREI), and flame stabilization. In particular, the ignition, the flame span, and the FREI-related characteristics are scrutinized to understand the underlying physics of the flame stability/instability modes. It is shown that the flames appear stable at higher inlet velocities, while the FREI mode is detected at a lower inlet velocity, depending on the equivalence ratio and the channel size. The findings also explain how different operating conditions impact the flame characteristics in both stability modes.

Keywords: numerical simulations; microscale combustion; premixed combustion; lean and rich flames; stable flame; flames with repetitive extinctions and ignitions (FREI); flame span

1. Introduction

Recent developments in micro-manufacturing technologies have dramatically accelerated the versatility of micro-electromechanical systems (MEMS), unmanned aerial vehicles (UAV), and communication and sensing gadgets. All these devices necessitate compact, long-lasting, and instantly rechargeable power supplies capable of delivering power in the range of 10^{-3}~10^{2} W [1,2]. Motivated by the significantly higher chemical energy densities of hydrocarbons as compared to most advanced batteries, microscale combustion research has been actively pursued in an effort to maximize the operational lifespan of these devices [3,4]. To be specific, current cutting-edge batteries have an energy density of 0.05~0.265 kWh/kg, which is ~1/60th (for the upper energy density limit) of that of typical hydrocarbon fuel, such as methane (CH_4) [1,5]. Many physical and chemical processes impact micro-combustion systems; topics such as gas-phase phenomena, surface reactions, species transport, and thermal properties of the combustor material, as well as thermal and mass diffusion, convection, and radiation can all have a dramatic impact on a specific application [1,5–8].

In a micro- or mesoscale combustor, the combustion characteristics differ from those of a conventional "macro-combustor", which is mostly attributed to the differences in the surface area per unit volume ratio, and the thrust–weight ratio [9]. A combustion process becomes slightly less efficient as compared to macro-combustors due to the heat losses from the flame region to the combustor wall. On the other hand, the resulting power–weight ratio is higher for the micro-combustion devices due to the inverse relation of the characteristic length scale and the thrust–weight ratio [10]. Because of the inherently small heat release rate of a micro-combustion system, the effects of heat and radical losses dominate the propagation characteristics of a premixed flame and, as such, directly impact the flame stability and quenching. The literature on premixed microchannel combustion distinguishes three main flame stability modes, depending on the inlet flow velocity, namely, (i) a stationary stable flame; (ii) a flame with repetitive extinctions and ignitions (FREI); and (iii) a weak flame [11,12]. Among them, stationary stable flames are commonly observed at higher inlet velocities. The restrained preheating of unburned reactants emphasizes the flame anchoring at a position where the local flame speed is balanced by the flow rate. Decreasing the flow rate results in further upstream propagation of the flame front and, eventually, flame extinction in relatively colder regions. Subsequent to this extinction, however, once the unburned mixture arrives at a high wall temperature region, it re-ignites and again propagates upstream, thus resulting in the FREI regime. Further reduction of the inlet velocity generates "slowly" propagating flames, different from the conventional stable flames defined previously. The developed flames are stabilized by the transverse heat transfer, in contrast to the stable flame regime. Although these flames do not behave in the same way as conventional flames, the term "weak flames" is widely used in the literature on microchannel reactors [4,13–15].

Previous analytical, experimental, and numerical studies focused primarily on the stable and weak flame modes for various fuel mixtures and under various operating conditions. These studies provided insight to understand the concepts of low-temperature ignition (LTI), cool flames, and the influence of the negative temperature coefficient (NTC) region [4,12,16–21]. It has been shown that cool flames are usually generated in the unburned fuel mixture region, while another conventional flame propagates through the reacting mixture. Such a phenomenon is referred to as autoignition-assisted flame propagation [22]. Pizza et al. [23] studied the dynamics of fuel-lean premixed hydrogen/air flames in microchannels using three-dimensional (3D) numerical simulations, and observed that various flame modes depend on the inlet flow velocity and the channel size. Similar behavior was also observed in the planar, two-dimensional (2D) simulations of the same group [24].

The FREI phenomenon has been investigated for a methane/air mixture in a series of works [18,25–27]. Miyata et al. [26] studied the unsteady behavior of the oscillating flames and the FREI for methane/air premixed combustion in microchannels. It was observed that the flame stability is directly affected by the total heat transfer rate, the wall temperature gradient, and the wall thickness. For example, the FREI amplitude and period are directly related to the thickness of the channel wall and the wall temperature gradient. Norton et al. [28] identified the optimal flow velocity, optimal wall thermal conductivity, and the optimal combustor dimensions to achieve stable methane/air flames. Furthermore, the behavior and the stability modes of methane/air flames in microchannels were studied experimentally by Di Stazio et al. [19] over a wide range of equivalence ratios and flow inlet velocities. Different flame regimes were detected, including stable, weak, and FREI flames. They performed a frequency analysis of the CH* signals to demonstrate the periodic nature of the FREI phenomenon and adopted the concept of the Strouhal number to characterize the effect of the equivalence ratio variations on the FREI frequency.

Aligned with the aforementioned studies, the literature would still benefit from more comprehensive examinations targeted at understanding and characterizing the underlying combustion physics related to flame stability in microchannels. In the present work, premixed CH_4–air combustion in a microreactor is investigated numerically and validated experimentally. The main objective of this work is to scrutinize the combustion characteristics and flame behavior for various inlet velocities (0.1–0.6 m/s), equivalence ratios representing the fuel-lean ($\phi = 0.6$, 0.8), stoichiometric ($\phi = 1$), and fuel-rich ($\phi = 1.2$, 1.4) conditions, and channel diameters of 1.8 mm, 2.3 mm, and 3 mm, resulting in simulating a total of 34 unique cases with 2 to 5 days needed to complete each simulation using high-performance computers, depending on the operating conditions and the flame behavior. Varying these parameters, the flame can become stable or experience repeated extinctions and ignitions [11,21,29,30]. It is important to highlight that the models and the methods employed in this work have been accepted and acknowledged in the literature [24,27]. In the following sections, the simulation methodology is described. Then, the metrics, such as the ignition time, length, location, and the flame span, as well as the characteristics of the FREI are identified and studied for various operating conditions.

2. Methodology

2.1. Governing Equations

For a chemically reacting fluid, the governing equations describing, respectively, the balance of mass, momentum, species mass fractions, and energy, read

$$\frac{\partial \rho}{\partial t} + \nabla \cdot \left(\rho \vec{u}\right) = 0, \quad (1)$$

$$\frac{\partial}{\partial t}\left(\rho \vec{u}\right) + \nabla \cdot \left(\rho \vec{u}\,\vec{u}\right) = -\nabla P + \nabla \cdot (\overline{\tau}) + \rho \vec{g} + \vec{F}_B, \quad (2)$$

$$\frac{\partial}{\partial t}(\rho Y_i) + \nabla \cdot \left(\rho \vec{u} Y_i\right) = -\nabla \cdot \vec{J}_i + R_i, \quad (3)$$

$$\rho C_p \frac{\partial T}{\partial t} + (\rho C_p) \vec{u}\, \nabla T = -\nabla q - \rho \sum_{i=1}^{N} C_{p,i} Y_i V_i - \sum_{i=1}^{N} h_i \Omega_i. \quad (4)$$

Here, ρ is the fluid density; $\vec{u}$ is the flow velocity vector; P is the static pressure; $\overline{\tau}$ is the viscus stress tensor; $\vec{g}$ and $\vec{F}_B$ are the acceleration vectors associated with gravity and the external body force, respectively; C_p is the specific heat at constant pressure; q is the heat flux vector; and R_i is the rate of production in the chemical reactions for the *ith* species, with N being the total number of species. It should be noted that gravity and other external body forces of Equation (2) are neglected in this work. The mass fraction of each species Y_i is obtained by solving the diffusion equation for the *ith* species, while h_i, Ω_i, and V_i are the specific enthalpy, the net production rate, and the diffusion velocity, respectively, and $\vec{J}_i$ stands for the diffusion flux of the *ith* species caused by the gradients of the temperature and the species concentration; it is given by Fick's law of diffusion with respect to the concentration gradient:

$$\vec{J}_i = -\rho D_{i,m}\, \nabla(Y_i) - D_{T,i} \frac{\nabla T}{T}, \quad (5)$$

where $D_{i,m}$ is the mass diffusion coefficient of the *i*th species, while $D_{T,i}$ is the thermal diffusion coefficient. The numerical simulations were carried out by means of the commercial computational fluid dynamics (CFD) package Ansys FLUENT [31]. The simulation setup parameters are listed in Table 1.

2.2. Operating Conditions

A 2D axisymmetric computational domain represents a micro-channel of length 10.75 cm and a radius (half-width) of 1.15 mm, as illustrated in Figure 1. In addition, Table 2 provides a summary of all the simulated cases showing the chosen inlet velocities, the mixture regime (lean/rich/stoichiometric), and the equivalence ratios for a channel width of 2.3 mm.

Table 1. Main parameters of the computational model.

Parameters	Methods
Solver type	Pressure-based solver
Flow viscous model	Laminar
Species reaction (chemical kinetics)	Volumetric species transport, San Diego mechanism (46 species and 235 reactions) [32]
Discretization	Second-order upwind
Pressure–velocity coupling	COUPLE scheme
Transient formulation	First-order implicit
Boundary conditions	Inlet: velocity-inlet, Outlet: pressure-outlet External wall: temperature thermal conditions Axis: axisymmetric line
Initialization method	Standard

(a)

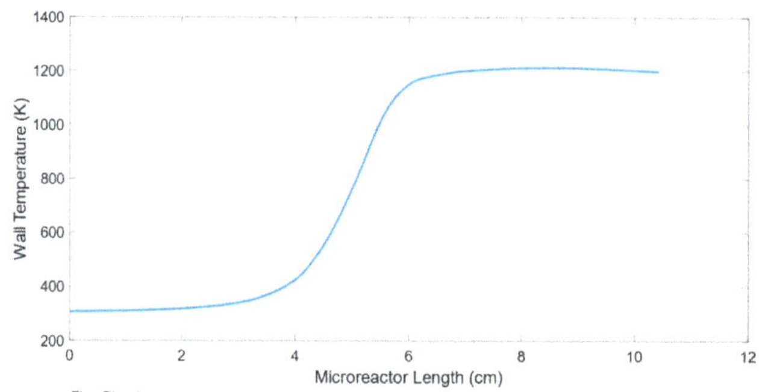

(b)

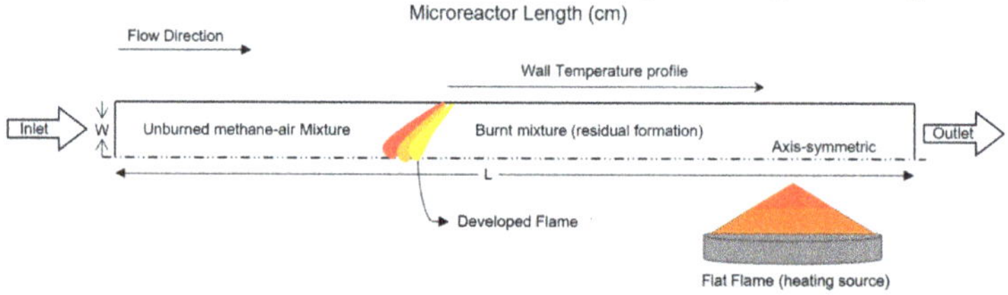

(c)

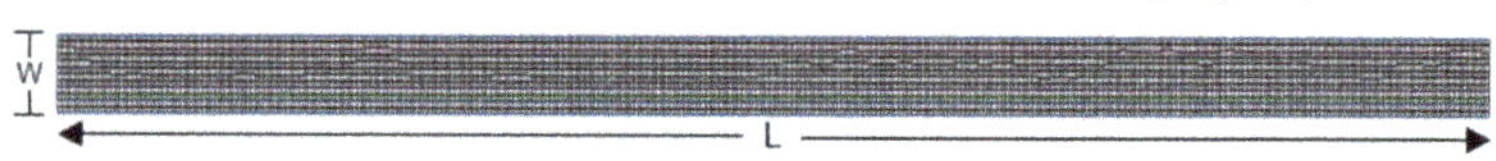

Figure 1. The imposed wall temperature profile along the tube (**a**), a schematic of the 2D computational domain (**b**), and the mesh layout showing the distribution of the square elements in the computational domain, where L is the channel length and W is the channel half width (**c**).

The 2D geometry was chosen for the following reasons: (i) considering the simulation time and the number of cases to be simulated, it would be extremely challenging to perform all the simulations using a 3D domain; (ii) although 3D simulations could provide more refined results and additional details, 2D results nevertheless provide solid and sufficient understanding of the primary combustion stages, especially for our case, where the flow in the channel has a 2D nature, i.e., exhibits a significantly axisymmetric behavior; (iii) due to the laminarity of the flow, the standard approach for such applications is 2D modelling, as used to be recognized and accepted in the literature [24,27]. Furthermore, in order to save time and computational resources, half of the tube was simulated by employing the axisymmetric boundary condition at the lower side of the computational domain.

Table 2. A summary of the simulated cases depicting the selected inlet velocities, the mixture regimes, and the equivalence ratios for a channel width of 2.3 mm.

Inlet Velocity (m/s)	Mixture Regime	Equivalence Ratio (ϕ)
0.10	Stoichiometric	1
0.15	Stoichiometric	1
0.20	Stoichiometric, lean, rich	0.6, 0.8, 1, 1.2, 1.4
0.24	Stoichiometric	1
0.30	Stoichiometric	1
0.40	Stoichiometric, lean, rich	0.6, 0.8, 1, 1.2, 1.4
0.50	Stoichiometric	1
0.60	Stoichiometric	1

The inlet and outlet extremes of the channel are open, and a temperature profile was imposed as a wall boundary condition similar to the experimental wall temperature measured by Baumgardner [33]. The wall temperature increased from 305 K at the inlet to around 1200 K at the outlet, and was externally controlled by a McKenna-style, flat-flame burner fueled by a H_2–air mixture. Such a configuration ensured the auto-ignition of the CH_4–air mixture in the microchannel. The fuel–air mixture entered the tube at a constant temperature (extracted from the temperature profile mentioned earlier), along with a no-slip boundary condition at the wall, with a uniform inlet velocity value. A square mesh of 0.1 mm in size (as shown in Figure 1c) and a time step of 10 µs were used for all the simulated cases. Preliminary trials in the setup of the model utilized herein showed that such a time step is sufficient in producing accurate simulation results that successfully predict the ignition characteristics. Additionally, this time step has been shown in the literature to be sufficiently small for capturing the microscale combustion phenomena, including the ones of interest in this study [27,34]. Furthermore, with the flame thickness being of the order of 1 mm, the chosen grid size ensured that the flame characteristics were resolved sufficiently within circa 15 cells. The grid resolution and its sufficiency to capture the flame details will be discussed in Section 2.3. The chemical kinetics were modelled by the San Diego mechanism with 46 species and 235 reactions [32], which is sufficient to describe the laminar flame dynamics at the microscale level and is able to successfully depict the ignition characteristics [35]. Table 3 summarizes all the species mass fractions at the inlet for the simulated cases.

Table 3. A summary of the reactants mass fractions for the simulated cases.

Equivalence Ratio (ϕ)		Reactants' Mass Fractions (y_x)		
		y_{CH4}	y_{O2}	y_{N2}
Fuel-lean	0.6	0.0337	0.2252	0.7411
	0.8	0.0445	0.2227	0.7328
Stoichiometric	1	0.0550	0.2202	0.7248
Fuel-rich	1.2	0.0655	0.2178	0.7167
	1.4	0.0754	0.2155	0.7091

2.3. Grid Independence Test

The computational domain was segregated into evenly sized square mesh elements, with three different grid sizes employed, namely, 0.075 mm (fine), 0.1 mm (moderate), and 0.133 mm (coarse). To compare the performance of these meshes, the maximum axial temperature and the total heat release rate (HRR_T) were monitored over a specific time, as shown in Figure 2. Here, HRR_T was computed numerically by integrating the heat release rate for all the chemical reactions over the computational domain at each time step. When this property peaks, it indicates the ignition and then the formation of a flame front. As depicted in Figure 2, the temperature for the three cases was identical from the starting point of the heating process until the mixture moved into the ignition phase; however, variations in the peak temperature were observable thereafter, especially for the coarse mesh. Nevertheless, the other two cases (the moderate and fine meshes) agree well, with almost the same ignition times and maximum temperatures. In a similar way, the HRR_T curves experienced a noticeable deviation upon varying the grid size. This is especially profound when approaching the ignition time, where the properties experience the highest gradients. The coarse mesh failed to predict the ignition time and the peak reaction rate, as well as the other two cases, whereas satisfactory agreement is seen between the moderate and fine meshes. It is noted that the HRR_T data were quite noisy; therefore, smoothing it was necessary to avoid any misleading data representation. Because of that, slight deviations produced by such a process are inevitable, especially where high noisiness points were detected; however, the time at which the peak of HRR_T occurred (indicating the ignition time, which is of significant importance in this comparison) was the same for both the 0.1 mm and 0.075 mm meshes. Recognizing the fact that the simulation time and required memory almost doubled as the mesh size was reduced from 0.1 to 0.075 mm, and accounting for the number of cases simulated in this work, the moderate mesh, 0.1 mm, was chosen for all the remaining simulation runs. It is noted that, in terms of the flame behavior, the simulation results are comparable with those found throughout the literature [19,29,33].

(a)

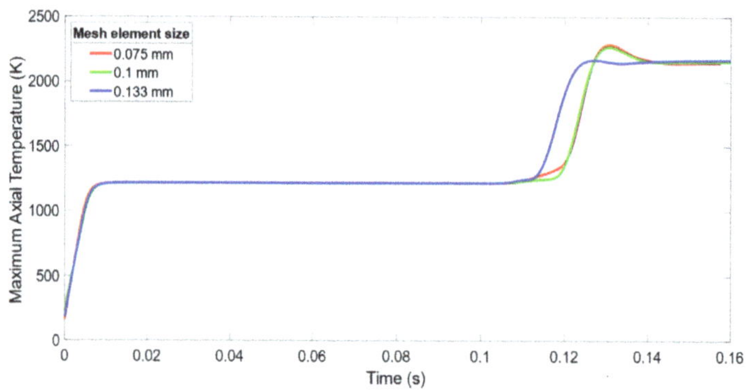

(b)

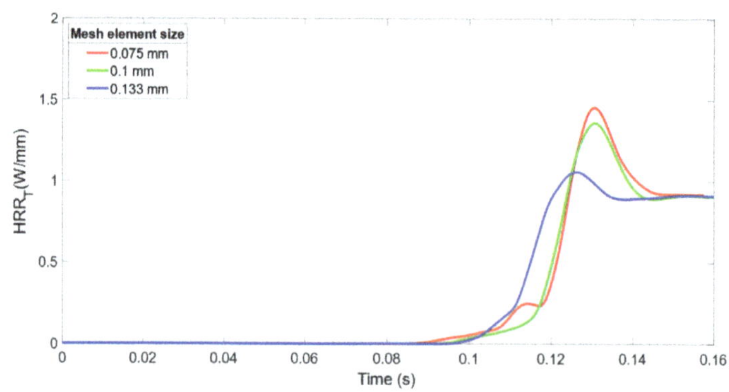

Figure 2. The grid independence tests with an inlet velocity of 0.4 m/s, a channel width of 2.3 mm, and a stoichiometric fuel/air mixture: the maximum axial temperature (**a**), and total heat release rate (HRRT) (**b**) vs. time.

2.4. Metrics

The purpose of this section is to describe the metrics that were defined to characterize and analyze the ignition and flame properties under various operating conditions.

2.4.1. Ignition Metrics: Time, Location, and Span

The ignition process was identified by tracking the time evolution of the transient maximum axial temperature, $T_{axial-max}$, obtained from the numerical simulation. As the mixture was heated through the walls, the chain-branching chemical reactions were initiated and eventually triggered ignition, thereby resulting in a V-shape flame front. Figure 3 further demonstrates the pre-ignition, ignition, and post-ignition features of a stable flame. Here, point (b) indicates the ignition with the peak temperature gradient. Figure 4 complements Figure 3 with the temperature contours to visualize the ignition process.

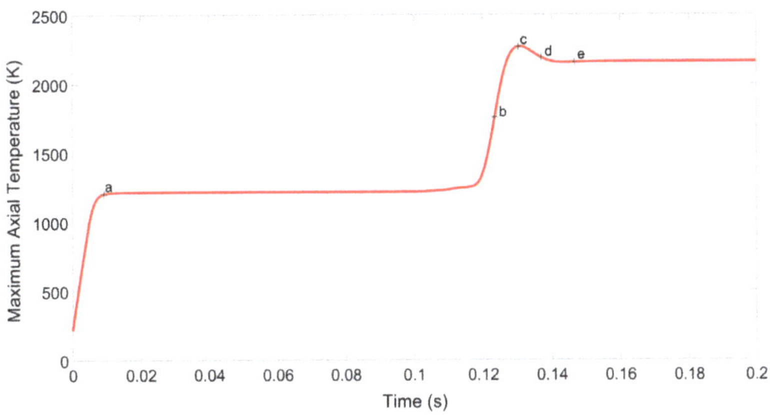

Figure 3. The maximum axial temperature vs. time for an inlet velocity of 0.4 m/s, a channel width of 2.3 mm, and a stoichiometric fuel/air mixture. The markers represent the specific time steps for further clarification in Figure 4.

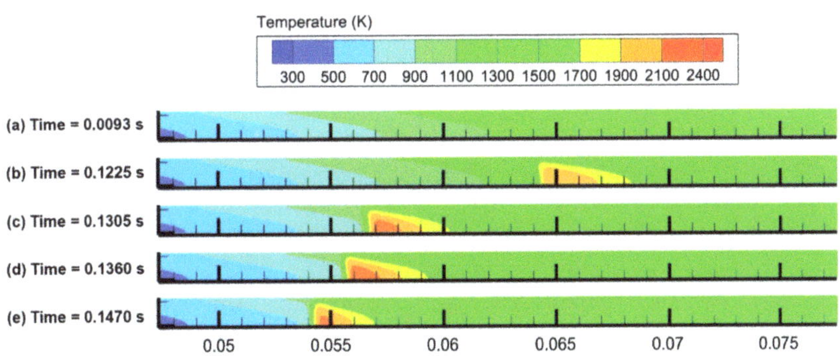

Figure 4. The temperature contours at the selected successive time steps for the ignition and propagating flame process with an inlet velocity of 0.4 m/s, a channel width of 2.3 mm, and a stoichiometric fuel/air mixture. The length is measured in meters.

In particular, the chain-branching reactions started at instant (a), as the fuel/air mixture was preheated, and went through time (b), thereby indicating the beginning of the ignition process. At instant (b), the maximum axial temperature experienced the highest gradient, where the flame front was formed, thus separating the unburnt and burnt gases.

State (b) determined the ignition time t_{ig} and the ignition span; see Figure 5. Moving forward from instant (b) to instants (c), (d), and (e) in Figures 3 and 4, the flame started propagating upstream until it either became stable or was extinguished, depending on the operating conditions, which are discussed in the next sections. The ignition time, t_{ig}, is defined as the time associated with the maximum gradient of the quantity $T_{axial-max}$:

$$t_{ig} = t_{\left(\frac{dT_{axial-max}}{dt}\right)_{max}}. \tag{6}$$

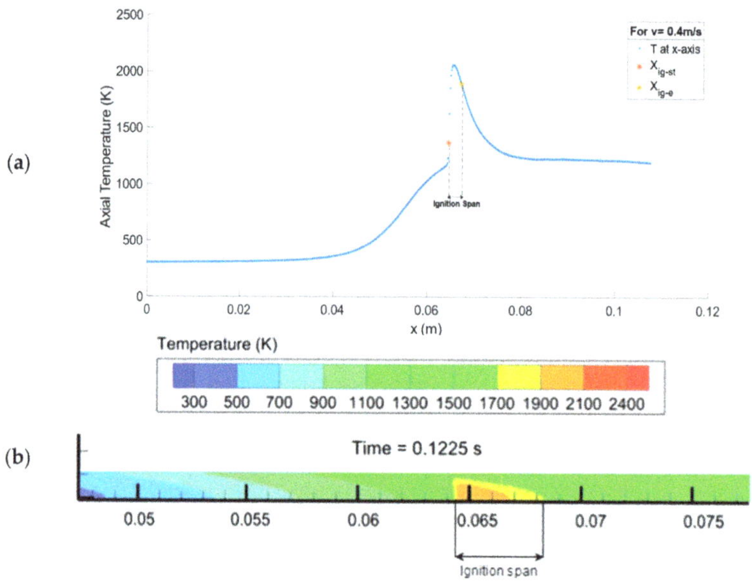

Figure 5. The axial temperature data at the time of ignition for an inlet velocity of 0.4 m/s, a channel width of 2.3 mm, and a stoichiometric fuel/air mixture. The length is measured in meters. (**a**) The temperature profile along the x-axis and (**b**) the temperature contour at the selected time step.

Similarly, the extinction time (if any) is defined as the time associated with the minimum gradient (maximum negative) of the quantity $T_{axial-max}$:

$$t_{ex} = t_{\left(\frac{dT_{axial-max}}{dt}\right)_{min}}. \tag{7}$$

It is worth mentioning that the extinction time parameter is defined and utilized only when the flame is not stable and there is an extinction; therefore, an additional parameter is introduced to describe the time instant at which the flame is stabilized and has a fixed location and span for the remainder of the simulation time:

$$t_{st} = t_{\left(\frac{dT_{axial-max}}{dt}=0\right)}. \tag{8}$$

Moreover, the ignition span was also examined to fully describe the ignition process. Using the axial temperature variation at the ignition time, the ignition span is defined as the location bounded by the peak positive temperature gradient, x_{ig-st}, and the peak negative temperature gradient, x_{ig-e}. The ignition span, Δx_{ig}, is then defined as the distance between these two locations (see Figure 5):

$$x_{ig-st} = x_{\left(\frac{dT}{dx}\right)_{max}}; t = t_{ig}, \tag{9}$$

$$x_{ig-e} = x_{\left(\frac{dT}{dx}\right)_{min}}; t = t_{ig}, \quad (10)$$

$$\Delta x_{ig} = x_{ig-st} - x_{ig-e}. \quad (11)$$

It should be noted that using HRR_T for the ignition phenomenon analyses is a more common approach in the literature than using the axial temperature [18,22,28,34,36]; however, in the present work, the HRR_T data were not directly implemented, as they typically produce noisy curves, which cannot be utilized to define such time-sensitive properties. Such noise is primarily attributed to the numerical artifacts resulting from the computational analysis; therefore, a smoothened HRR_T data trace was plotted against the axial temperature to validate our approach and to make sure that the axial temperature data are technically supported and can be used as a metric to predict the ignition properties and the flame characteristics. Figure 6 illustrates the comparison between these metrics. It can be seen that both curves had almost the same rate of change around the ignition time, and they concurred at the maximum point detected. In addition, both metrics demonstrate a comparable stabilized flame time, where both curves settle down to steady behavior, ultimately indicating the stable flame regime, as will be discussed further in Section 2.4.2.

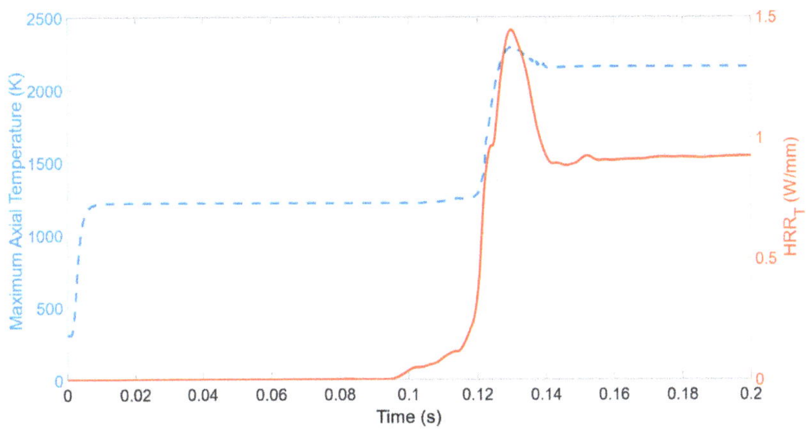

Figure 6. The maximum axial temperature and the smoothed heat release rate (HRR_T) data vs. time for an inlet velocity of 0.4 m/s, a channel width of 2.3 mm, and a stoichiometric fuel/air mixture. The secondary y-axis shows the smoothed data instead of the raw HRR_T data.

Moreover, to verify the reliability and functionality of the axial temperature metric, as well as its sufficiency to predict the ignition properties, the mass fraction of the CH radicals was monitored along with the axial temperature at the same points that have been defined earlier in Figure 3. Figure 7a shows both the axial temperature and the CH mass fraction at the time of ignition for an inlet velocity of 0.4 m/s; here, it is important to note that the location of the flame on the contour plots is clearly correlated to the formation of CH radicals. The highest CH concentrations were located at the upstream flame side, where the reaction occurred at the unburned fuel/air mixture region. In addition, as shown in Figure 7b, the maximum temperature gradient coincided with the location of highest concentration of the CH radical, which indicates the ignition location. These results demonstrate that the axial temperature data can provide the required accuracy to describe the ignition phenomena herein.

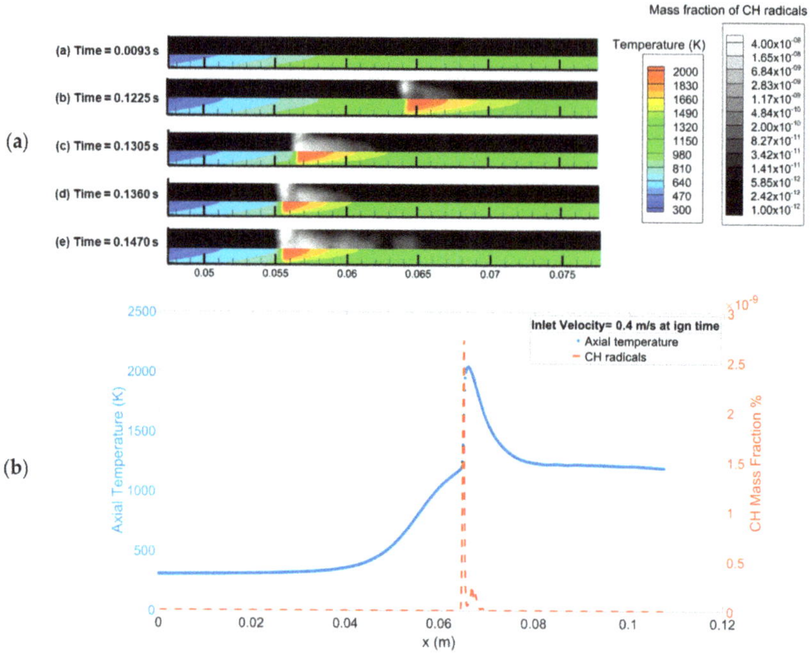

Figure 7. (**a**) The temperature (below) and the CH radical (above) contours at the selected successive time steps for the ignition and the propagating flame process with an inlet velocity of 0.4 m/s. (**b**) The axial temperature and the CH radicals vs. the axial distance at the ignition time, all at a channel width of 2.3 mm and with a stoichiometric fuel/air mixture.

2.4.2. Stable Flame Metrics: Location and Span

In the event of a stable mode, the flame propagated further upstream and sustained its shape, behavior, and location, starting from point (e) in Figure 3. Similar to the ignition process, the stable flame was also characterized by its location and span. Using the axial temperature profile at time instant t_{st}, the maximum and minimum temperature gradients were determined to mark the two extremes of the stable flame:

$$x_{sz-st} = x_{\left(\frac{dT}{dx}\right)_{max}}; t = t_{st}, \tag{12}$$

$$x_{sz-e} = x_{\left(\frac{dT}{dx}\right)_{min}}; t = t_{st}. \tag{13}$$

Here, x_{sz-st} indicates the start of the flame span, while x_{sz-e} stands for the ending point of the flame span; see Figure 8. In addition, the span of the stabilized flame is defined as the difference between these points:

$$\Delta x_{sz} = x_{sz-st} - x_{sz-e}. \tag{14}$$

2.4.3. FREI Metrics: Length, Period, Frequency, Ignition-to-Extinction Time, and Extinction-to-Ignition Time

In the event of FREI, the flame could not sustain itself due to unrestrained heat loss through the microreactor wall; however, after the flame extinguished, the supply of the fresh fuel–air mixture eventually reached the high-temperature zone downstream of the channel and ignited again. Such phenomenon is detected numerically and experimentally in the literature [27–29,34]. To better comprehend the FREI phenomena, a specific case with an inlet velocity of 0.2 m/s is presented in this section. Specifically, Figure 9 shows variations of the maximum axial temperature during the combustion process, while Figure 10 presents the spatial temperature contours corresponding to the selective time steps in Figure 9. The ignition process was exactly the same as that for the stable flame; however, in a FREI mode, as the flame front formed and propagated upstream, it started to diminish after point (d) until flame extinction occurred just before point (e). Due to the uninterrupted fuel–air feeding source, the mixture ignited again, and the process occurred repeatedly, as clarified in the rest of the points; see positions (f)–(h) in Figures 9 and 10.

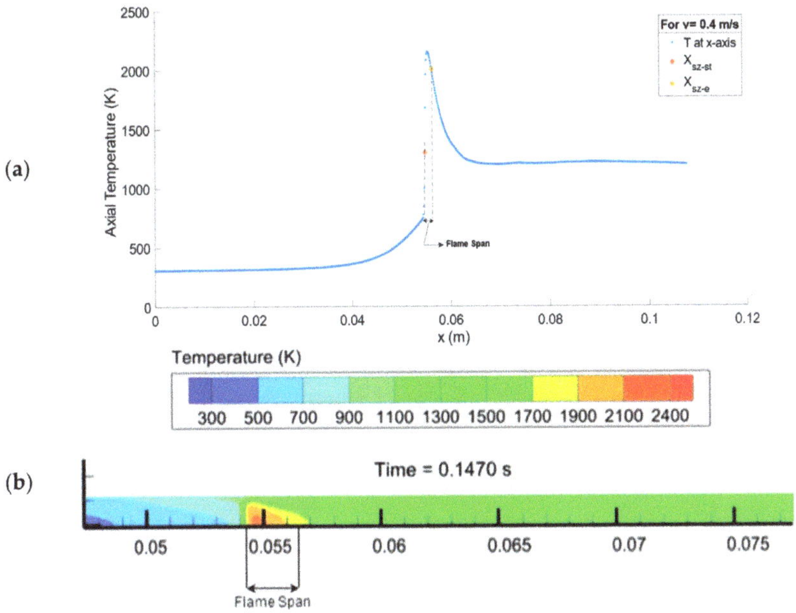

Figure 8. The axial temperature data for the stabilized flame for an inlet velocity of 0.4 m/s, a channel size of 2.3 mm, and a stoichiometric fuel/air mixture. The length is measured in meters. (**a**) The temperature profile along the x-axis and (**b**) the temperature contour at the selected time step.

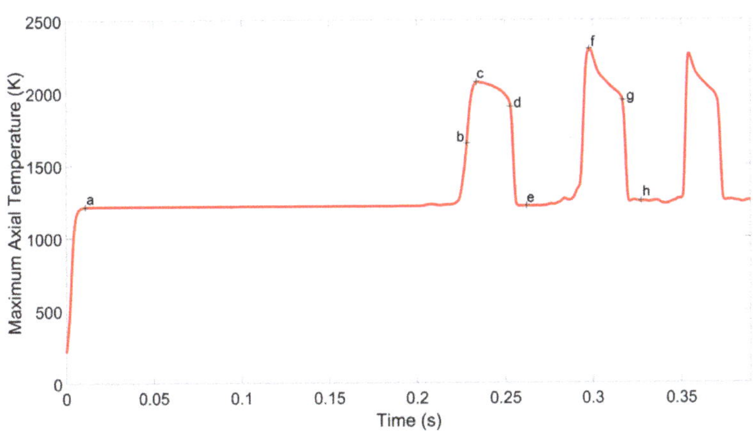

Figure 9. The maximum axial temperature vs. time for an inlet velocity of 0.2 m/s, a channel width of 2.3 mm, and a stoichiometric fuel/air mixture, in the case of the FREI mode. The markers represent the specific time steps for further clarification in Figure 10.

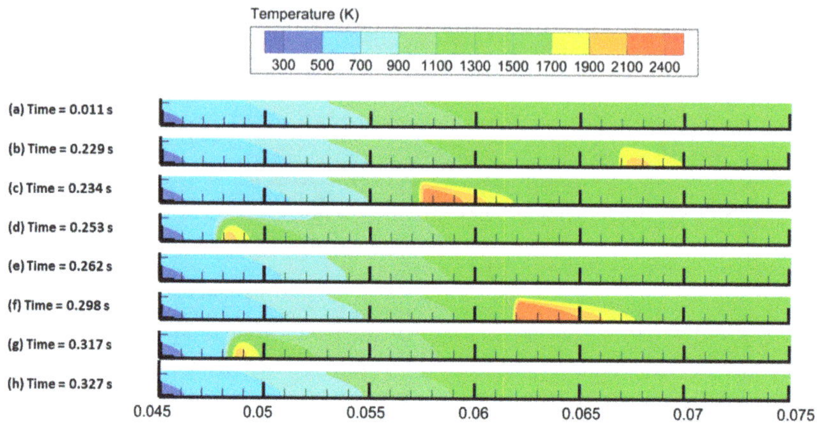

Figure 10. The temperature contour for a flame with a FREI mode, with an inlet velocity of 0.2 m/s, a channel width of 2.3 mm, and a stoichiometric fuel/air mixture. Points (a)–(h) correlate with the relevant markers in Figure 9.

It has historically been a challenge to characterize the FREI by means of simple parameters due to the intriguing behavior of flames under such circumstances. For example, the cycle-to-cycle variations will notably influence the simulation results. In the present work, several parameters were considered simultaneously to allow for a more comprehensive characterization, as follows:

- The FREI length: defined as the distance the flame travels along the channel before extinction;
- The FREI period (τ_{FREI}) defined as the time duration required for a flame to attain two successive ignition events [34];
- The FREI frequency: the reciprocal of the FREI period (τ_{FREI})
- The ignition-to-extinction time: defined as the time duration from the ignition time to the extinction time in the same FREI cycle [22];
- The extinction-to-ignition time: defined as the time duration from the extinction time to the ignition time of the subsequent cycle [22].

3. Results

In this section, first, the numerical results are validated by means of the experimental observations. Then, the impacts of the inlet velocity, the equivalence ratio ϕ, and the channel width on the flame behavior and the metrics, defined in the previous section, are scrutinized.

3.1. Comparison with Experimental Observations

3.1.1. Experimental Work Overview

The computational study of the present work is validated by an earlier experimental method [33], where an optically accessible microreactor with a controlled temperature profile was used in order to explore stoichiometric fuel/air mixtures of methane and other hydrocarbons, under velocities ranging from 0.2 m/s to 0.6 m/s. In this experimental work, two main flame modes were observed: (a) stable flames at higher inlet velocities and (b) unstable flames (FREI) at lower inlet velocities. The experimental arrangement is shown in Figure 11, in which the premixed fuel/air mixture flowed through a 2.3 mm diameter quartz tube, heated by a McKenna-style flat burner. The CH* chemiluminescence data were utilized to detect the flame positions, where the CH* signals were captured via long exposure images (with an associated optical filter). A CH* chemiluminescence profile for the FREI flame indicating the extinction and ignition limits was used to compare the numerical results with those of the experiment via the so-called FREI length metric. The maximum axial temperature, the local heat release rate (HRR_{axial}) at the ignition and extinction points, and the temperature contours for one FREI cycle represent such a comparative approach, as detailed in Sections 3.1.2 and 3.1.3.

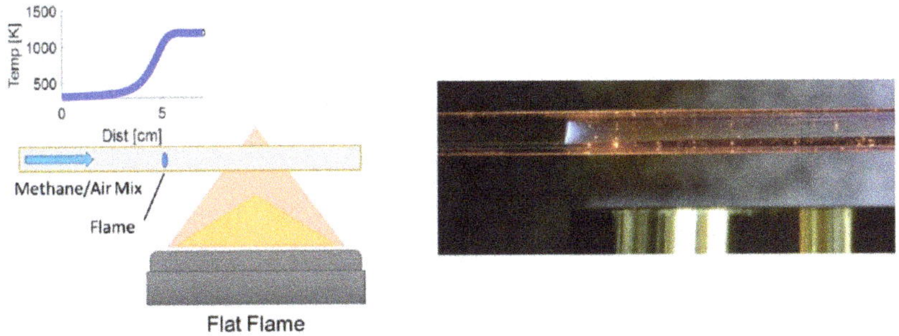

Figure 11. A schematic of the microreactor arrangement.

3.1.2. Comparison with Experimental Work: Axial Temperature

Figure 12 constitutes the processed image of CH* chemiluminescence extracted from the experiments [33], along with the temperature contours obtained from the simulation data for one FREI cycle at the same operating conditions. The maximum axial temperature plots for the corresponding time instants in the contour plots were also added to this figure for a more comprehensive comparison. As defined earlier, the maximum and the minimum gradient points for each plot explicated the span of the respective flame at that specific time instant. The experimental FREI length of 26 mm appears to be in excellent agreement with its numerical counterpart of 25.6 mm. Furthermore, the phenomenon of a double flame front was also detected in the first three instants (a–c). This double flame front occurs when a strong flame propagates upstream and a second, weaker flame also appears and travels further downstream. The double flame front was not readily visible in the methane experimental studies, but was clearly visible from the simulation, thus highlighting the essential role of the simulations in complementing the experimental observations. Such phenomenon was also explored numerically for a methane–air mixture by Nakamura et al. [18] and experimentally for a propane–air mixture by Baumgardner et al. [37]; however, this phenomenon is not the focus of this study, as it requires more simulations and post-processing.

On the other hand, there was an evident deviation for the FREI location, i.e., the experimental FREI location is slightly further downstream as compared to the numerical one. This difference can be attributed to two primary observations. First, the method of defining the ignition and extinction points when post-processing the numerical simulation results is slightly different than what was used in the experimental work; the FREI location and length were detected experimentally by means of the chemiluminescence profile of the CH^* radicals. This is unlike the present numerical approach utilizing the maximum axial temperature data. Second, there are inherent uncertainties in detecting the FREI location in the experimental setup, as for such a small-scale work environment, slight variations in the ambient conditions could potentially influence the readings and the outputs of the experiment.

3.1.3. Comparison with Experimental Work: Local Heat Release Rate (HRR_{axial})

The heat release rate at the microreactor axis is also a good indication of the presence of active reaction zones throughout the computational domain, hence, the HRR_{axial} peaks at a developed flame region starting from the time of ignition up to the time just before flame extinction. Such information facilitates a viable comparison with the experimental data. Figure 13 presents the variation of the axial heat release rate at the ignition and extinction times for the FREI mode (with an inlet velocity of 0.2 m/s). This therefore establishes the limits of the FREI cycle and the so-called FREI length. In addition, it can be seen that the heat release rate values for both the ignition and extinction points were not identical. It is shown that the upstream end of the FREI segments (i.e., the extinction location) is where the heat release peaked. This figure also provides the configurations of the CH* radicals at the same locations detected in the experiment. With a slight deviation between the two results, acceptable agreement is evident, thereby fulfilling the validity of the computational results. The uncertainty of the presented data is attributed to the same reasons that were mentioned previously in Section 3.1.2.

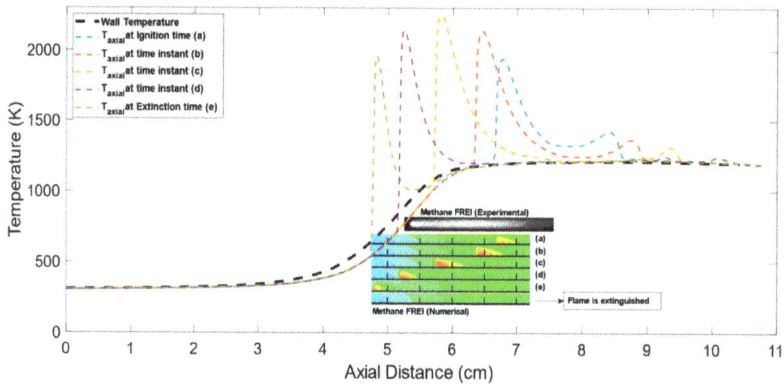

Figure 12. The axial temperature data along with the wall temperature profile and the temperature contour for a flame with a FREI mode with an inlet velocity of 0.2 m/s, a channel width of 2.3 mm, and a stoichiometric fuel/air mixture. Points (a)–(e) show the consecutive time instants of one cycle of the FREI mode. Also shown is the processed image of the chemiluminescence profile of CH* radicals for the actual FREI flames at 0.2 m/s; the start and end locations of each image/contour correspond to the respective experimental locations.

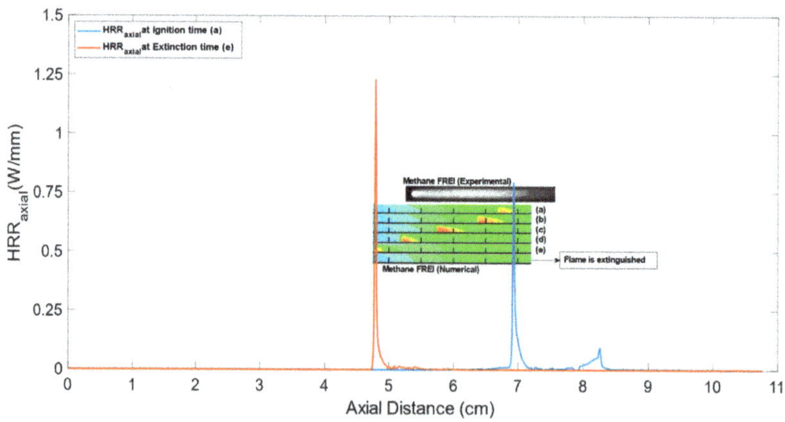

Figure 13. The axial heat release rate at the ignition and the extinction time of one cycle of the FREI mode and the temperature contour for a flame with a FREI mode with an inlet velocity of 0.2 m/s, a channel width of 2.3 mm, and a stoichiometric fuel/air mixture. Points (a)–(e) show the consecutive time instants of one cycle of the FREI mode. The processed image of the chemiluminescence profile of the CH* radicals for the actual FREI flames at 0.2 m/s is also shown; the start and end locations of each image/contour correspond to the respective experimental locations.

3.2. Impact of the Inlet Velocity

This section will illustrate how the variation of the inlet velocity can affect the combustion characteristics and metrics that were discussed in prior sections. Various inlet velocities ranging from 0.1 m/s to 0.6 m/s were used to form a total of eight (8) unique cases, all at a stoichiometric condition ($\phi = 1$) and a channel width of 2.3 mm. The results from these simulations show that the cases with the inlet velocities exceeding 0.24 m/s became stable, while those with lower inlet velocities experienced the FREI mode.

3.2.1. Impact of Inlet Velocity on Ignition Process

The ignition properties (as defined in Section 2.4.1) were determined for all the cases and are listed in Table 4. These results indicate that ignition occurred faster but also further downstream, which is attributed to a smaller residence time in the reactor at higher inlet velocities. This is consistent with the experimental and numerical findings in the literature [27,28,34,38]. Similarly, the ignition locations presented in Table 4 suggest an intuitive trend, i.e., ignition occurred further downstream for higher inlet velocities. Furthermore, the ignition span was larger for the cases with higher inlet velocities. This is explained by the fact that higher inlet velocities push the unburned fuel/air mixture further downstream and increase its concentration at a given time, as well as expand the high-temperature region (where the ignition is more likely to occur) at a faster rate. Under these circumstances, when a given differential volume (dx) ignites, it affects the surrounding area instantly, thus increasing the ignition span and the rate of heat release, resulting in an extended ignition surface (flame kernels). Accordingly, the HRR_T (at the ignition time) for the inlet velocities of 0.2 m/s, 0.4 m/s, and 0.6 m/s are 1.12 W/mm, 1.45 W/mm, and 1.64 W/mm, respectively.

Table 4. Ignition properties for various inlet velocities, a channel width of 2.3 mm, and a stoichiometric fuel/air mixture.

Inlet Velocity (m/s)	Ignition Time (s)	Ignition Location (mm)		Ignition Span (mm)
	t_{ig}	x_{ig-st}	x_{ig-e}	Δx_{ig}
0.10	0.4405	54.90	56.80	1.90
0.15	0.2430	64.00	66.20	2.20
0.20	0.2285	67.40	69.85	2.45
0.24	0.1625	67.50	69.80	2.30
0.30	0.1580	66.80	69.20	2.40
0.40	0.1195	64.50	67.90	3.40
0.50	0.0975	67.00	70.50	3.50
0.60	0.0835	74.10	78.00	3.90

To further examine the influence of the flow inlet velocity and the imposed wall temperature on the ignition span and location, Figure 14 describes these trends and illustrates the changes in the wall temperature difference at the ignition span limits. First, it can be seen that the largest wall temperature difference was detected for the lowest inlet velocity case (a FREI case), which also coincides with the closest ignition event to the channel inlet. It was noted that closer to the inlet, the wall temperature was still low and increasing axially, which exacerbates the flame instability, resulting in a FREI mode. In addition, it was found that the ignition location for both stability modes was affected by the inlet velocity, i.e., higher inlet velocities drive the unburnt fuel/air mixture further down the reactor before arriving to the required ignition temperature, which also directly affects the residence time. At the left-hand-side of the cutoff velocity, the flame front propagated upstream towards the inlet, where the wall temperature was low enough to maximize the wall heat losses and led to flame extinction. As for the stable flames (right-hand side of the cutoff velocity), the ignition span increased as the wall temperature profile became uniform and the wall temperature difference at the ignition span limits was minimal. This also reveals that at the uniform wall temperature region, the high temperature zone in the channel grew, providing an optimum environment for the flame to reside and stabilize. These observations assert that temperature is the key parameter to ensure the ignition process, while the ignition location is dominated by the inlet velocity.

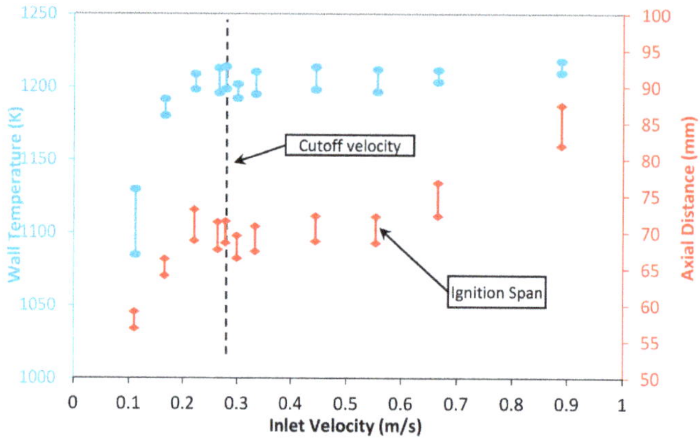

Figure 14. The ignition span along with the wall temperature at the span limits vs. the inlet velocity, for a channel width of 2.3 mm and a stoichiometric fuel/air mixture.

3.2.2. Impact of Inlet Velocity on Stable Flames

The stable flame metrics are presented in Table 5 (various inlet velocities exceeding 0.24 m/s at $\phi = 1$ and with a channel diameter of 2.3 mm). It is seen that an increase in the inlet velocity resulted in the formation of a wider flame span stabilized further downstream, consistent with the literature [28,34]. Overall, it can be inferred that with relatively higher inlet velocities, the combustion process was more stable—the wall temperature at the stable flame location was relatively high and the heat loss was therefore relatively low. This resulted in achieving a stable flame, where the flame speed balanced the flow velocity. Figure 15 shows how the axial temperature across the flame span varied. Similar to the ignition span, it can be concluded that the flame span was directly proportional to the temperature across the flame. This is explained by the fact that the propagating flame stabilized where the wall temperature was still high and the axial flame temperature difference across the flame span was minimal. Such conditions diminished the heat loss effect through the walls and sustained ideal conditions for the flame to stabilize. Additionally, the flame span did not change significantly with the inlet velocity; however, the temperature range across the flame span increased with higher inlet velocities (at higher inlet velocities, the stabilized flames reside further downstream with higher wall temperatures).

Table 5. The stabilization location and the flame span for the stable cases, all at a channel width of 2.3 mm and with a stoichiometric fuel/air mixture.

Inlet Velocity (m/s)	Flame Location (mm)		Flame Span (mm)
	x_{sz-st}	x_{sz-e}	Δx_{sz}
0.30	51.10	52.60	1.50
0.40	54.40	56.15	1.75
0.50	56.90	58.80	1.90
0.60	58.90	61.00	2.10

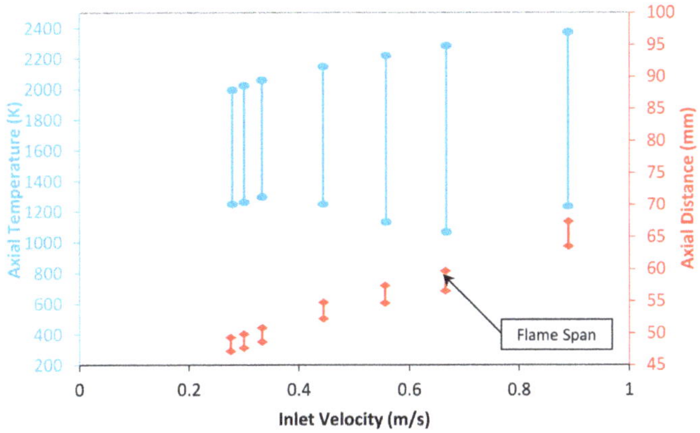

Figure 15. The flame span limits along with the maximum flame temperature range within the flame span region vs. the inlet velocity, for a channel size of 2.3 mm and a stoichiometric fuel/air mixture.

3.2.3. Impact of Inlet Velocity on the FREI Mode

The flames experiencing the FREI mode are mainly characterized by the FREI period (τ_{FREI}), adopted to further explain the impact of the inlet velocity on the FREI mode [34]. As illustrated in Table 6, τ_{FREI} increased for lower inlet velocities, which is explained by the larger flame residence time during each FREI cycle and the fact that the flame was created closer to the cold regions, where the wall temperature was below 750 K. This hindered the reignition event of the unburned mixture, causing the cyclic FREI behavior to slow down, thereby increasing the FREI period. This behavior was evident until reaching a point where the inlet flow velocity was low enough to develop the so-called weak flames, as previously mentioned in the Introduction.

Table 6. The FREI time for different inlet velocities, a channel width of 2.3 mm, and a stoichiometric fuel/air mixture.

Inlet Velocity (m/s)	0.10	0.15	0.20	0.24
τ_{FREI} (ms)	102.53	89.60	63.65	64.67

3.3. *Impact of Equivalence Ratio*

3.3.1. Impact of Equivalence Ratio on Ignition Process

In this section, the impact of the equivalence ratio, ϕ, is analyzed. For this purpose, five various equivalence ratios were chosen: two at the lean regime ($\phi = 0.6,\ 0.8$), stoichiometric ($\phi = 1$), and two others at the fuel-rich regime ($\phi = 1.2,\ 1.4$), all with a channel diameter of 2.3 mm. Each equivalence ratio ϕ was paired with two different inlet velocities of 0.4 m/s (representing the stable mode at $\phi = 1$) and 0.2 m/s (representing the FREI mode at $\phi = 1$) to examine the impact of ϕ on both stable and unstable flame modes. The simulation results show that the flame behavior was highly sensitive to such alterations, with the ignition time, ignition span, and flame temperature being affected significantly. Figure 16 compares the ignition progress for various equivalence ratios. It is evident that the fuel/air mixture is ignited slightly faster when the mixture was richer, due to an increased fuel concentration in the mixture; however, such an increase in the fuel concentration was limited—exceeding the flammability limit resulted in no ignition. It was also observed that the ignition was more delayed for the leaner cases. On the other hand, increasing the equivalence ratio further did not show substantial variation in the ignition time. For example, the ignition times of the rich cases of $\phi = 1.2$ and 1.4 were almost the same. It was also noted that the peak maximum temperature points for the stoichiometric and the rich cases were quite close, as well, while they were lower for the fuel-lean cases. Table 7 summarizes the ignition properties for various equivalence ratios under each inlet velocity.

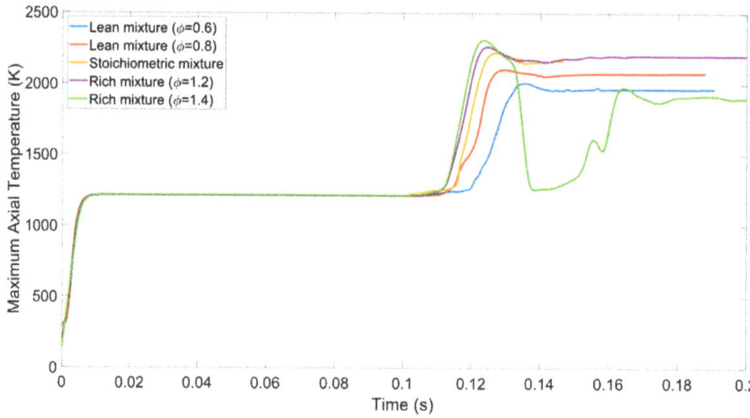

Figure 16. The maximum axial temperature profile vs. time for five cases of various equivalence ratio with an inlet velocity of 0.4 m/s and a channel width of 2.3 mm.

Table 7. The ignition properties for two inlet velocities (0.2 m/s and 0.4 m/s) and a range of equivalence ratios, all at a channel width of 2.3 mm.

Inlet Velocity	Equivalence Ratio (ϕ)		Ignition Time (s)	Ignition Location (mm)		Ignition Span (mm)
			t_{ig}	x_{ig-st}	x_{ig-e}	Δx_{ig}
0.4 m/s	Fuel-lean	0.6	0.1275	67.20	69.40	2.20
		0.8	0.1225	69.70	72.50	2.80
	Stoichiometric	1	0.1195	64.50	67.90	3.40
	Fuel-rich	1.2	0.1170	69.00	72.10	3.10
		1.4	0.1160	69.95	72.90	2.95
0.2 m/s	Fuel-lean	0.6	0.2390	62.20	64.45	2.25
		0.8	0.2305	60.05	62.45	2.40
	Stoichiometric	1	0.2285	67.40	69.85	2.45
	Fuel-rich	1.2	0.2215	61.00	62.95	1.95
		1.4	0.2195	60.05	62.05	2.00

For the ignition location, the stoichiometric mixture ignited farther upstream, unlike the fuel-lean and the fuel-rich cases that tended to creep downstream. In contrast, for the unstable mode, the ignition location for the stoichiometric case was the farthest from the inlet, whereas those of the richer and leaner cases were located upstream, closer to the inlet. Interestingly, the ignition span was the largest for the $\phi = 1$ condition for both stability modes. The following two subsections will further elaborate on the impacts of the equivalence ratio on the two flame modes comprehensively.

3.3.2. Impact of Equivalence Ratio on Stable Flames

Even though the variation of the equivalence ratio changed the main flame characteristics, the flame mode remained stable in most cases, except for the ultra-rich mixtures ($\phi = 1.4$), in which the flame became distinctly unstable (for the channel diameter of 2.3 mm). Specifically, after the flame was ignited, it moved upstream and extinguished, and then a weak flame appeared further downstream, where it resided and sustained its shape. This is especially evident in Figure 16, where the tendency of the transition between stable and FREI modes is higher for richer fuel mixtures. In addition, the inlet velocity at which such a transition was observed (cutoff velocity) decreased from the rich to the lean cases. For instance, at the stoichiometric condition, the cutoff velocity was 0.25 m/s, as compared to 0.4 m/s at the fuel-rich conditions. This observation is aligned with the experimental findings of Di Stazio et al. [19] at similar conditions.

Table 8 shows that the stable flame location was also affected by the equivalence ratio. For example, a lean ($\phi = 0.6$) flame stabilized further downstream, requiring more time to achieve stability, as is also illustrated in Figure 16; however, the stable flame location moved upstream when the fuel–air mixture was closer to the stoichiometric conditions. On the other hand, when the mixture became richer, the stabilized flame location shifted downstream again, even though the flame stabilized a bit faster than for the leaner cases. Such results are attributed to the higher fuel concentration. As for the flame span, the stoichiometric and lean cases were quite similar, 1.70~1.75 mm, whereas the flame span was as low as 1.45 mm for the rich fuel–air mixture.

Table 8. The stabilization location and the flame span for the stable cases at an inlet velocity of 0.4 m/s and a channel size of 2.3 mm.

Equivalence Ratio (ϕ)		**Flame Location (mm)**		**Flame Span (mm)**
		x_{sz-st}	x_{sz-e}	Δx_{sz}
Fuel-lean	0.6	56.51	58.23	1.72
	0.8	54.90	56.60	1.70
Stoichiometric	1	54.40	56.15	1.75
Fuel-rich	1.2	56.60	58.05	1.45

3.3.3. Impact of Equivalence Ratio on FREI Mode

The impact of the equivalence ratio on the FREI characteristics was analyzed using a channel diameter of 2.3 mm. In particular, Figure 17 depicts the maximum axial temperature data for five various equivalence ratios with the same inlet velocity of 0.2 m/s. It is shown that for $\phi = 0.8 \sim 1.2$, the FREI mode was maintained; however, the FREI characteristics, such as the FREI period and maximum axial temperature, were variable. For instance, it is evidently shown that the higher the equivalence ratio, the earlier the ignition occurs since the hot region (where the ignition event is initiated) developed at a faster rate for the richer cases as compared with the leaner ones. This also can be inferred from the peaks of the maximum axial temperature profiles, as the highest temperatures were detected at higher ϕ; see Figure 17. Additionally, the maximum axial temperature profiles for the first ignition cycle notably differed for each regime (the subsequent FREI cycles were affected by the burnt products left after the first extinction). Additionally, the lean case represents the largest period for a single FREI cycle among the three cases, where it took longer for the flame to extinguish (see Table 9). Increasing ϕ above a certain limit will eventually lead to flame quenching, due to a relative lack of the oxidizer in the mixture, and this is evident for the richest mixture here ($\phi = 1.4$). In this case, the flame did not sustain after the initial ignition. In contrast, for the leanest mixture ($\phi = 0.6$), the FREI mode was no longer seen, and the flame experienced a stable mode, indicating the very important role of the equivalence ratio in altering the flame stability limits. In some cases, therefore, the equivalence ratio can be used as a controlling factor to stabilize the flame.

Table 9. The FREI time for various equivalence ratios at an inlet velocity of 0.2 m/s and a channel width of 2.3 mm.

Equivalence Ratio (ϕ)	**0.8**	**1**	**1.2**	**1.4**
τ_{FREI} (ms)	66.85	63.65	59.10	N/A

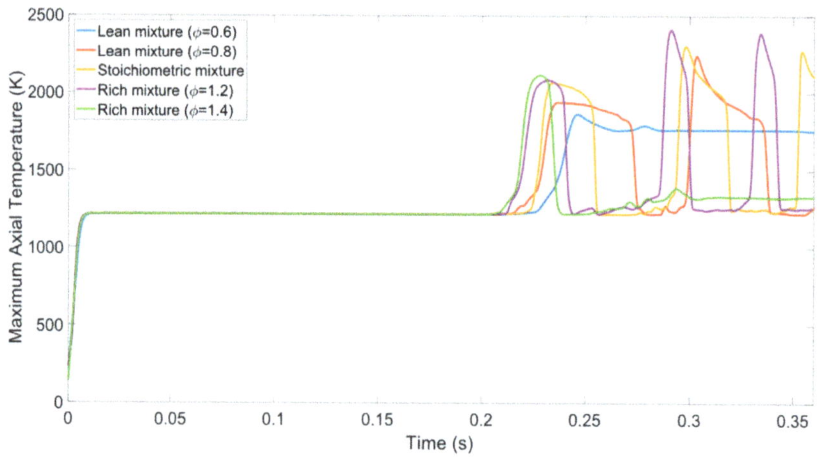

Figure 17. The maximum axial temperature profile vs. time for five cases of various equivalence ratios with an inlet velocity of 0.2 m/s and a channel width of 2.3 mm.

3.4. *Impact of Channel Width*

3.4.1. Impact of Channel Width on Stability

Three channel diameters (1.8 mm, 2.3 mm, and 3 mm) were used to assess the impact of the channel size on the flame behavior and stability modes. Varying the inlet velocity for each channel size at the stoichiometric conditions, the cutoff velocity was determined. It is shown that the cutoff velocity was lower for wider channels. To be specific, the cutoff velocities for the channel diameters of 1.8 mm, 2.3 mm, and 3.0 mm were found to be 0.30 m/s, 0.25 m/s, and 0.18 m/s, respectively. This is a good indication that it is more likely to observe instabilities, such as FREIs in narrower channels, while wider channels tend to be more stable at lower velocities. Similar analytical and experimental results have been reported for a propane/air mixture [39].

3.4.2. Impact of Channel Width on FREI Characteristics

This section examines the impact of the channel width on the FREI mode properties (e.g., the FREI frequency, the time from ignition to extinction, and the time from extinction to ignition). Figure 18a illustrates how the FREI frequency was affected by the inlet velocity at two different channel diameters of 1.8 mm and 2.3 mm. Specifically, the FREI frequency increased with the inlet velocity for both channels, arriving at a maximum frequency of 22.5 Hz for the narrower channel. Simultaneously, this behavior is explained by Figure 18b,c, where the FREI frequency is associated with the difference between the ignition-to-extinction and the extinction-to-ignition time intervals. In particular, it was concluded that the effect of the ignition-to-extinction time was minimal as compared to the extinction-to-ignition time. The reason for this is that the driving force for extinction is primarily the temperature difference between the reaction zone and the wall, which was relatively constant for all the FREI cases, whereas the wall temperature directly impacted the ignition timing. Furthermore, the wider the channel, the longer it takes for the flame to extinguish, thus diminishing the FREI frequency. These results confirm the findings of similar studies for methane/air combustion [22,40].

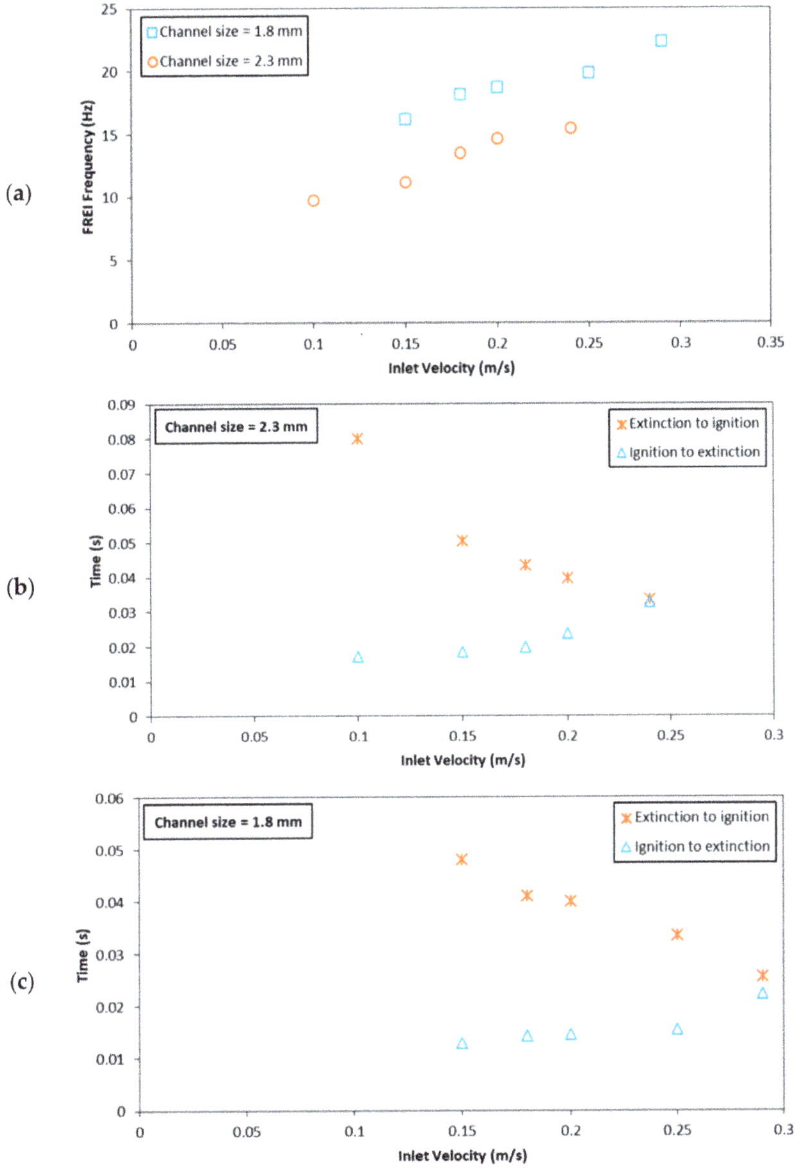

Figure 18. (**a**) The frequency of the FREI vs. the inlet velocity. (**b**,**c**) The time from ignition to extinction and extinction to ignition for two channel diameters, 2.3 mm and 1.8 mm, all with a stoichiometric fuel/air mixture.

4. Conclusions

In this work, methane-air premixed combustion in a microchannel with a controlled wall temperature profile was investigated numerically and validated experimentally. The combustion characteristics were examined for various inlet velocities (0.1 to 0.6 m/s), channel diameters (1.8 to 3.0 mm), and equivalence ratios (0.6 to 1.4), scrutinizing two distinct flame modes: stable flames and FREI. The outcomes from this work offer the following insights:

1. The flow rates can directly affect the behavior of the propagating flame and its stability conditions. Specifically, it was shown that at lower inlet velocities, the flames propagate further upstream towards the cold wall temperature region, increasing the likelihood for the flame to experience FREI.
2. At a given inlet velocity and channel size, the equivalence ratio can play a role in controlling the stability (e.g., a flame may lose its stability if going to an ultra-rich regime or an unstable flame may become stable if going to an ultra-lean regime, producing a weak, but stable flame).
3. At a given channel size and in the case of the FREI mode, the rich fuel/air mixture ignites faster than the other cases and it produces the longest FREI frequency.
4. At a given equivalence ratio, the cutoff velocity is higher for narrower channels, i.e., it is more likely for a flame to experience a FREI mode; the flame extinction occurs at a higher temperature as the channel narrows and the difference between the characteristic ignition and extinction points diminishes, increasing the chance of developing a FREI mode. This is directly associated with the competition between the heat release rate and the wall heat losses, impacting the temperature, the location, and the stability mode of the flame.
5. The FREI frequency is affected by the channel size and the inlet velocity, such that the frequency is higher for narrower channels at a given inlet velocity.

The findings of this work will be beneficial in (a) realizing the nature of the combustion process and flame propagation in a microscale environment, (b) constituting the stability limits of CH_4–air combustion in microreactors at various conditions, (c) selecting the optimum operating conditions for sustained microscale combustion, and (d) offering insight into stability mode transitions and the impacts of operating conditions.

Author Contributions: Conceptualization, M.A. and V.A.; data curation, A.K., M.A., M.E.B. and V.A.; formal analysis, A.K., M.A., M.E.B. and V.A.; funding acquisition, A.K., M.A., M.E.B. and V.A.; investigation, A.K., M.A., M.E.B. and V.A.; methodology, A.K., M.A., M.E.B. and V.A.; project administration, M.A. and V.A.; resources, V.A.; software, A.K.; supervision, M.A., M.E.B. and V.A.; validation, A.K. and M.A.; visualization, A.K.; writing—original draft, A.K.; writing—review and editing, M.A., M.E.B. and V.A. All authors have read and agreed to the published version of the manuscript.

Funding: This research was funded by the National Science Foundation (NSF) through CAREER Award #1554254 (V.A.), as well as by the West Virginia Higher Education Policy Commission, through Grant #HEPC.dsr.18.7 (V.A.).

Data Availability Statement: The data presented in this study are available on request from the corresponding author.

Conflicts of Interest: The authors declare no conflict of interest.

References

1. Ju, Y.; Maruta, K. Microscale combustion: Technology development and fundamental research. *Prog. Energy Combust. Sci.* **2011**, *37*, 669–715. [CrossRef]
2. Walther, D.C.; Ahn, J. Advances and challenges in the development of power-generation systems at small scales. *Prog. Energy Combust. Sci.* **2011**, *37*, 583–610. [CrossRef]
3. Dogwiler, U.; Mantzaras, J.; Benz, P.; Kaeppeli, B.; Bombach, R.; Arnold, A. Homogeneous ignition of methane-air mixtures over platinum: Comparison of measurements and detailed numerical predictions. In Proceedings of the International Symposium of Combustion, Boulder, CO, USA, 2–7 August 1998. [CrossRef]
4. Kamada, T.; Nakamura, H.; Tezuka, T.; Hasegawa, S.; Maruta, K. Study on combustion and ignition characteristics of natural gas components in a micro flow reactor with a controlled temperature profile. *Combust. Flame* **2014**, *161*, 37–48. [CrossRef]
5. Chigier, N.; Gemci, T. A review of micro propulsion technology. In Proceedings of the 41st Aerospace Sciences Meeting and Exhibit, Reno, Nevada, 6–9 January 2003. [CrossRef]
6. Buliński, Z.; Szczygieł, I.; Kabaj, A.; Krysiński, T.; Gładysz, P.; Czarnowska, L.; Stanek, W. Performance analysis of the small-scale α-type Stirling engine using computational fluid dynamics tools. *J. Energy Resour. Technol.* **2018**, *140*, 3. [CrossRef]
7. Yan, Y.; Liu, Y.; Li, H.; Huang, W.; Chen, Y.; Li, L.; Yang, Z. Effect of cavity coupling factors of opposed counter-flow microcombustor on the methane-fueled catalytic combustion characteristics. *J. Energy Resour. Technol.* **2019**, *141*, 2. [CrossRef]

8. Li, H.; Chen, Y.; Yan, Y.; Hu, C.; Fan, H.; Feng, S. Numerical study on heat transfer enhanced in a microcombustor with staggered cylindrical array for micro-thermophotovoltaic system. *J. Energy Resour. Technol.* **2018**, *140*, 11. [CrossRef]
9. Kaisare, N.; Vlachos, D. A review on microcombustion: Fundamentals, devices and applications. *Prog. Energy Combust. Sci.* **2012**, *38*, 321–359. [CrossRef]
10. Epstein, A.H. Millimeter-scale, micro-electro-mechanical systems gas turbine engines. *J. Eng. Gas Turbines Power* **2004**, *126*, 205–226. [CrossRef]
11. Maruta, K.; Kataoka, T.; Kim, N.; Minaev, S.; Fursenko, R. Characteristics of combustion in a narrow channel with a temperature gradient. *Proc. Combust. Inst.* **2005**, *30*, 2429–2436. [CrossRef]
12. Maruta, K.; Parc, J.K.; Oh, K.C.; Fujimori, T.; Minaev, S.S.; Fursenko, R.V. Characteristics of microscale combustion in a narrowheated channel. *Combust. Explos. Shock Waves* **2004**, *40*, 516–523. [CrossRef]
13. Kikui, S.; Kamada, T.; Nakamura, H.; Tezuka, T.; Hasegawa, S.; Maruta, K. Characteristics of n-butane weak flames at elevated pressures in a micro flow reactor with a controlled temperature profile. *Proc. Combust. Inst.* **2015**, *35*, 3405–3412. [CrossRef]
14. Nakamura, H.; Yamamoto, A.; Hori, M.; Tezuka, T.; Hasegawa, S.; Maruta, K. Study on pressure dependences of ethanol oxidation by separated weak flames in a micro flow reactor with a controlled temperature profile. *Proc. Combust. Inst.* **2013**, *34*, 3435–3443. [CrossRef]
15. Hori, M.; Nakamura, H.; Tezuka, T.; Hasegawa, S.; Maruta, K. Characteristics of n-heptane and toluene weak flames in a micro flow reactor with a controlled temperature profile. *Proc. Combust. Inst.* **2013**, *34*, 3419–3426. [CrossRef]
16. Watson, G.M.; Bergthorson, J.M. The effect of chemical energy release on heat transfer from flames in small channels. *Combust. Flame* **2012**, *159*, 1239–1252. [CrossRef]
17. Yamamoto, A.; Oshibe, H.; Nakamura, H.; Tezuka, T.; Hasegawa, S.; Maruta, K. Stabilized three-stage oxidation of gaseous n-heptane/air mixture in a micro flow reactor with a controlled temperature profile. *Proc. Combust. Inst.* **2011**, *33*, 3259–3266. [CrossRef]
18. Nakamura, H.; Fan, A.; Minaev, S.; Sereshchenko, E.; Fursenko, R.; Tsuboi, Y.; Maruta, K. Bifurcations and negative propagation speeds of methane/air premixed flames with repetitive extinction and ignition in a heated microchannel. *Combust. Flame* **2012**, *159*, 1631–1643. [CrossRef]
19. Di Stazio, A.; Chauveau, C.; Dayma, G.; Dagaut, P. Combustion in micro-channels with a controlled temperature gradient. *Exp. Therm. Fluid Sci.* **2016**, *73*, 79–86. [CrossRef]
20. Edalati-nejad, A.; Fanaee, S.A.; Ghodrat, M.; Khadem, J. Investigation of unsteady premixed micro/macro counterflow flames for lean to rich methane/air mixture. *J. Energy Resour. Technol.* **2021**, *143*, 052302. [CrossRef]
21. Kayed, H.; Mohamed, A.; Yehia, M.; Nemitallah, M.A.; Habib, M.A. Numerical investigation of auto-ignition characteristics in microstructured catalytic honeycomb reactor for CH4–air and CH4–H2–air mixtures. *J. Energy Resour. Technol.* **2019**, *141*, 082209. [CrossRef]
22. Akita, K.; Morii, Y.; Murakami, Y.; Nakamura, H.; Tezuka, T.; Maruta, K. Dynamics of FREI with/without cool flame interaction. *Proc. Combust. Inst.* **2023**, *39*, 1957–1965. [CrossRef]
23. Pizza, G.; Frouzakis, C.E.; Mantzaras, J.; Tomboulides, A.G.; Boulouchos, K. Three-dimensional simulations of premixed hydrogen/air flames in microtubes. *J. Fluid Mech.* **2010**, *658*, 463–491. [CrossRef]
24. Pizza, G.; Frouzakis, C.E.; Mantzaras, J.; Tomboulides, A.G.; Boulouchos, K. Dynamics of premixed hydrogen/air flames in mesoscale channels. *Combust. Flame* **2008**, *155*, 2–20. [CrossRef]
25. Nakamura, Y.; Yamashita, H.; Saito, K. A numerical study on extinction behavior of laminar micro-diffusion flames. *Combust. Theor. Model.* **2007**, *10*, 927–938. [CrossRef]
26. Miyata, E.; Fukushima, N.; Naka, Y.; Shimura, M.; Tanahashi, M.; Miyauchi, T. Direct numerical simulation of micro combustion in a narrow circular channel with a detailed kinetic mechanism. *Proc. Combust. Inst.* **2015**, *35*, 3421–3427. [CrossRef]
27. Ayoobi, M.; Schoegl, I. Numerical analysis of flame instabilities in narrow channels: Laminar premixed methane/air combustion. *Int. J. Spray Combust. Dyn.* **2017**, *9*, 155–171. [CrossRef]
28. Norton, D.G.; Vlachos, D.G. Combustion characteristics and flame stability at the microscale: A CFD study of premixed-methane/air mixtures. *Chem. Eng. Sci.* **2003**, *58*, 4871–4882. [CrossRef]
29. Di Stazio, A.; Chauveau, C.; Dayma, G.; Dagaut, P. Oscillating flames in micro-combustion. *Combust. Flame* **2016**, *167*, 392–394. [CrossRef]
30. Singh, S.; Veetil, J.E.; Kumbhakarna, N.; Velamati, R.K.; Kumar, S. Flame dynamics of premixed CH4/H2/air flames in a microchannel with a wall temperature gradient. *Combust. Theor. Model.* **2022**, *26*, 989–1013. [CrossRef]
31. Ansys | Engineering Simulation Software. Available online: https://www.ansys.com (accessed on 10 July 2023).
32. Chemical-Kinetic Mechanisms for Combustion Applications, San Diego Mechanism, Mechanical and Aerospace Engineering (Combustion Research), University of California at San Diego. Available online: http://combustion.ucsd.edu (accessed on 16 July 2023).
33. Baumgardner, M.E. Microreactor combustion of simple hydrocarbons. In Proceedings of the 11th U.S. National CombustionMeeting, Pasadena, CA, USA, 24–27 March 2019.
34. Pokharel, S.; Ayoobi, M.; Akkerman, V. Computational analysis of premixed syngas/air combustion in micro-channels: Impacts of flow rate and fuel composition. *Energies* **2021**, *14*, 4190. [CrossRef]
35. Zettervall, N.; Fureby, C.; Nilsson, E.J. Evaluation of chemical kinetic mechanisms for methane combustion: A review from a CFD perspective. *Fuels* **2021**, *2*, 210–240. [CrossRef]

36. Minaev, S.; Maruta, K.; Fursenko, R. Nonlinear dynamics of flame in a narrow channel with a temperature gradient. *Combust. Theor Model.* **2007**, *11*, 187–203. [CrossRef]
37. Baumgardner, M.E.; Harvey, J. Analyzing OH*, CH*, and C2* chemiluminescence of bifurcating FREI propane-air flames in a micro flow reactor. Combust. *Flame* **2020**, *221*, 349–351. [CrossRef]
38. Fan, Y.; Suzuki, Y.; Kasagi, N. Experimental study of micro-scale premixed flame in quartz channels. *Proc. Combust. Inst.* **2015**, *35*, 3429–3437. [CrossRef]
39. Schoegl, I.; Sauer, V.M.; Sharma, P. Predicting combustion characteristics in externally heated micro-tubes. *Combust. Flame* **2019**, *204*, 33–48. [CrossRef]
40. Bucci, M.; Di Stazio, A.; Chauveau, C.; Dayma, G.; Legros, G.; Dagaut, P.; Chibbaro, S. Numerical and experimental analysis of combustion in microchannels with controlled temperature. *Chem. Eng. Sci.* **2019**, *4*, 100034. [CrossRef]

energies

MDPI

Article

The Effects of Cracking Ratio on Ammonia/Air Non-Premixed Flames under High-Pressure Conditions Using Large Eddy Simulations

Chengming Wang [1], Haiou Wang [1,*], Kun Luo [1,2] and Jianren Fan [1,2]

[1] State Key Laboratory of Clean Energy Utilization, Zhejiang University, Hangzhou 310027, China; wangchengming@zju.edu.cn (C.W.); zjulk@zju.edu.cn (K.L.); fanjr@zju.edu.cn (J.F.)
[2] Shanghai Institute for Advanced Study, Zhejiang University, Shanghai 200120, China
* Correspondence: wanghaiou@zju.edu.cn

Abstract: Ammonia is a promising carbon-free fuel. However, one of the main challenges for ammonia combustion is the high level of NO emissions. In this study, simulations were conducted for ammonia/air laminar counterflow flames and turbulent non-premixed jet flames in the KAUST high-pressure combustion duct (HPCD) at a pressure of 5 bar, with two ammonia cracking ratios of 14% and 28%. The influence of ammonia cracking ratio on the flame structure and NO formation mechanism were examined. The laminar counterflow flame results showed that HNO is one of the most critical species related to NO formation and NO is mainly generated through the path of $NH_2 \rightarrow NH \rightarrow HNO \rightarrow NO$. For the turbulent flames, the flamelet/progress variable (FPV) approach was employed in the context of large eddy simulations (LES) for high-fidelity simulations. The simulation results were compared with the measured data with promising agreements, which proves the accuracy of the FPV method for the present flames. It was shown that with increasing cracking ratio, not only the flame reactivity is enhanced, but also the generation of NO is increased. The correlation between NO and HNO is weaker when compared to that between NO and radicals such as O, H and OH in the entire flame. Through the distribution of NO source terms, it was found that the NO source term has a higher absolute value in the upstream region and the absolute value rapidly decreases with increasing streamwise distance. The total NO source term is positive in the fuel-lean zone and shows negative values in the fuel-rich zone.

Keywords: cracked ammonia; elevated pressure; LES-FPV; NO formation; reaction path analysis

1. Introduction

Ammonia has been widely applied in the global industrial chain. Its production and transportation infrastructures are relatively mature [1,2]. Most importantly, as a carbon-free fuel, ammonia does not produce CO_2 in its chemical conversion process, greatly reducing greenhouse gas emissions [3]. However, ammonia combustion features a flame speed much lower than traditional hydrocarbon fuels or hydrogen, low calorific value [4] and large amounts of harmful nitrogen monoxide produced during combustion, which presents challenges for its industrial application [5].

Previous studies showed that using blend fuels such as NH_3/CH_4 or NH_3/H_2 can increase flame speed and improve heat release rate [6,7] compared with pure ammonia, thereby extending the flame stability limits [8–10]. Khateeb et al. [11] showed that in a laboratory-scale swirl burner NH_3/CH_4 mixtures could generate a larger stability range than pure methane and also reduce the tendency of flame flashback. Otomo et al. [12] found that adding H_2 to NH_3 can increase flame temperature and improve flame speed. Frigo et al. [13] demonstrated that ammonia could directly burn in internal combustion engines, but required a combustion promoter; the best carbon-free promoter is hydrogen, with a very high combustion speed and a wide combustion range.

Obtaining a mixture of ammonia and hydrogen via on-board (in situ) cracking of ammonia is an ideal choice to avoid the risks and costs of storing and distributing hydrogen [14]. Mei et al. [15] conducted experimental and numerical simulations on partially cracked NH_3/air mixtures under high-pressure conditions. They found that as the ammonia cracking ratio increases, chemical and transport effects dominate the flame speed, whereas the thermal effect plays a minor role. The formation of NO with an increase in the cracking ratio exhibits non-monotonic behavior. Shohdy et al. [16] investigated how the cracking ratio of ammonia affects the overall flame speed, flammable limits and NO_x concentration in a swirling premixed combustor. Mercier et al. [17] explored the impact of ammonia cracking levels on the performance and emissions of a single-cylinder spark ignition engine, finding that a very low cracking ratio of ammonia (10%) can significantly expand the engine's operating range. Kim et al. [18] employed a chemical reactor network to numerically study the NO_x emission characteristics of air-staged and fuel-staged combustion of partially cracked NH_3/air mixtures, revealing the impact of ammonia cracking ratio, residence time and staging strategy on NO_x emissions. It is clear that with partial ammonia cracking, there is a clear trade-off: high ammonia cracking ratios lead to high nitrogen oxide levels, while low ratios result in high unburned ammonia levels.

Recently, Tang et al. [19] at KAUST reported experimental results of ammonia/air non-premixed flames with different ammonia cracking ratios of 14% and 28%. Both laminar counterflow flames and turbulent jet flames were considered. The experiments provided detailed quantitative information about the temperature, mass fraction and mixture fraction of the flames. Later, Wang et al. [20] used the one-dimensional NO laser-induced fluorescence (NO-LIF) method combined with one-dimensional Raman spectroscopy to quantitatively obtain the NO mole fraction for these flames. The experimental configurations are featured by well-defined boundary conditions and the measurements of the species mass fractions, temperature and main pollutant NO mole fractions are comprehensive. Therefore, they are chosen as the target flames of the present work.

Abdelwahid et al. [21] proposed a PC-transport model based on principal component analysis (PCA) and deep neural networks (DNN) for simulating the above mentioned flames. The authors showed the potential of the proposed model for turbulent ammonia/air non-premixed combustion. On the other hand, the flamelet model represented by the flamelet/progress variable (FPV) method is based on the fast chemistry assumption. It is suggested that the thermochemical state in turbulent flames can be described by a few control variables [22]. The flamelet model has been widely used in large eddy simulations (LES) of turbulent premixed [23–26] and non-premixed [27–29] flames. However, it has been rarely used for the study of non-premixed NH_3/air flames with partial ammonia cracking.

In this context, the focus of the present work is on the flame structure and NO formation mechanism of turbulent ammonia/air non-premixed combustion with various ammonia cracking ratios, which was not covered by the study of Abdelwahid et al. [21]. The objectives are as follows. First, the relative importance of different NO formation pathways with various ammonia cracking ratios is examined in laminar counterflow flames. Second, the LES results of turbulent non-premixed jet flames using the FPV model are compared with experimental results, validating the accuracy of the numerical methods. The effects of the ammonia cracking ratio on the flame structure are studied. Finally, the mechanisms of NO formation under different ammonia cracking ratio conditions are explored.

2. Experimental and Numerical Setup

2.1. Experimental Details

In this work, the experimental configurations of laminar counterflow flames and turbulent non-premixed jet flames are considered. The experiments were conducted in the High-Pressure Combustion Duct (HPCD) at the KAUST Clean Combustion Research Center [30], with an operating pressure of 5 bar and inflow temperature of 294 K. The counterflow burner is horizontally installed inside the HPCD, composed of two identical nozzles with a diameter of 5 mm and a distance of 5 mm. Two cases with different ammonia cracking ratios were investigated. Specifically, the ammonia cracking ratio is 14% for CACF14, while it is 28% for CACF28. The fuel composed of an $NH_3/H_2/N_2$ mixture is injected from one nozzle and the air from the other, with a strain rate of 130 1/s [31].

The turbulent jet burner consists of a 0.57-m-long stainless steel tube, supplying the fuel mixture $NH_3/H_2/N_2$, surrounded by an air co-flow nozzle with a diameter of 0.25 m, as shown in Figure 1. The inner and outer diameters of the central jet tube are 4.58 mm and 6.35 mm, respectively. Again, two cases with different ammonia cracking ratios of 14% and 28% are considered in the experiments, which are denoted as CAJF14 and CAJF28, respectively. The Reynolds numbers of the turbulent flames are Re = 11,200 for both cases based on the jet velocity and nozzle diameter. Detailed experimental parameters for various cases are provided in Table 1.

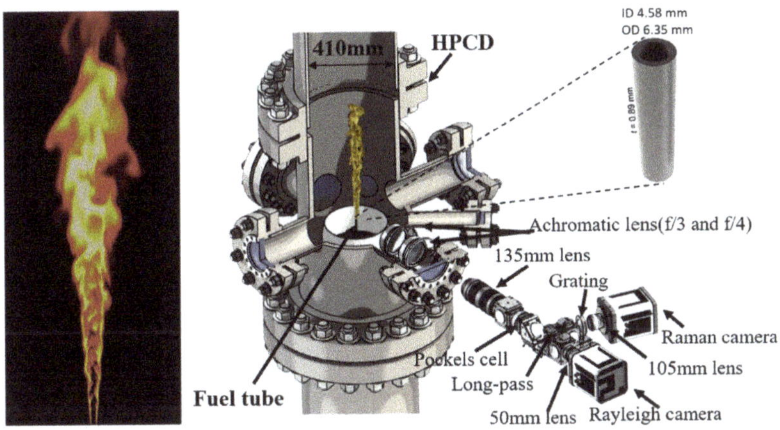

Figure 1. Instantaneous temperature cloud map of CAJF28 (**left**) and schematic diagram of KAUST jet burner in HPCD (**right**) [19].

Table 1. Operating condition parameters

Flame Type	Flame Case	NH_3	H_2	N_2	U_{fuel} (m/s)	U_{air} (m/s)
Laminar	CACF14	0.754	0.184	0.062	0.36	0.29
	CACF28	0.563	0.328	0.109	0.36	0.29
Turbulent	CAJF14	0.754	0.184	0.062	8.6	0.24
	CAJF28	0.563	0.328	0.109	10.1	0.24

2.2. Computational Setup

Figure 2 shows a schematic of the configurations for the laminar counterflow flames and turbulent jet flames. For the laminar counterflow flames, the length of the computational domain is the same as the distance between the two nozzles. The domain is discretized using 150 grids. The information of the composition and velocity in the two streams is provided in Table 1. The strain rate is 130 1/s, which is consistent with the experiment [19]. As for the turbulent jet flames, the length of the computational domain is 80D, which is sufficient to avoid the influence of the outlet boundary on the region of interest, and the diameter is 35D. The simulation uses a high-quality structured hexahedral grid with 258,000 cells. The grid is refined in the streamwise and radial directions to achieve a high resolution in the reaction zone. To make the turbulence at the inlet more realistic, a fully developed circular tube flow is used to generate the turbulence at the inlet. A fixed reference pressure with a zero-gradient boundary condition is set at the outlet of the burner.

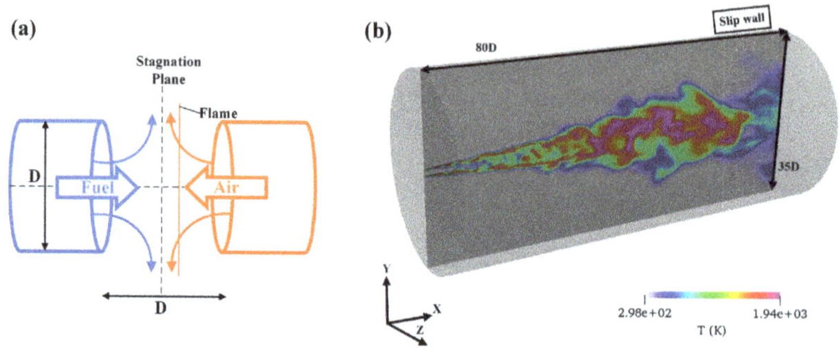

Figure 2. Computational domain and boundary conditions of (**a**) counterflow flame and (**b**) jet flame.

The one-dimensional simulations of laminar counterflow flames are performed using ANSYS Chemkin-PRO [32], in which the OPPDIF model developed by Kee et al. [33] was used. This model is based on a similarity solution for the governing equations, simplifying two-dimensional or three-dimensional flows mathematically to one dimension. Unity Lewis number transport model is used for the one-dimensional calculations. In the laminar flames, a focus is on the distributions of main species mass fractions, temperature and NO mole fractions. Four different mechanisms of Mathieu [7], Otomo [12], Okafor [34] and Jiang [35] are employed for the simulations. The Mathieu mechanism consists of 55 species and 275 elementary reactions, and accurately predicts the ammonia ignition delay time and laminar flame propagating velocity, as validated in refs. [36,37]. The Otomo mechanism includes 32 species and 213 elementary reactions, which was developed based on the mechanism of Song et al. [38], and can also predict the laminar flame propagating velocity and ignition delay times of NH_3 combustion very well [39]. The Okafor mechanism, featured by 59 species and 356 elementary reactions, is a reduced reaction mechanism developed for the combustion of CH_4/NH_3 flames optimized for laminar burning velocities [40]. It has been verified for both the pure NH_3/air flame and NH_3/CO/air flames [41]. The Jiang mechanism was derived from San Diego's short nitrogen mechanism [42] and consists of 19 species and 60 elementary reactions, which effectively reduces time consumption in reactive flow simulations while still adhering to acceptable accuracy thresholds.

The three-dimensional large-eddy simulations are carried out using a solver called FPVFoam, which is developed based on the open-source code OpenFOAM [43]. The PIMPLE algorithm is adopted to solve the transport equations. The second-order backward scheme is used for the unsteady term and the second-order Gauss linear is used for diffusion and convective terms. The time step is set to 2×10^{-6} s to keep the maximum Courant–Friedrich–Lewy (CFL) number less than 0.5 under all conditions. Only the Okafor mechanism is selected for the LES, which provides promising predictions for the turbulent flames. For both cases, the simulations ran for at least 20 flow-through times to reach a statistically steady state. The physical time for collecting statistical data exceeds 100 ms, equivalent to three flow-through times.

3. Numerical Methodology

3.1. Conservation Equations for LES

In the framework of large eddy simulation, the filtered equations of mass, momentum, mixture fraction, mixture fraction variance [44] and progress variable [45] are written as follows:

$$\frac{\partial \bar{\rho}}{\partial t} + \frac{\partial (\bar{\rho}\widetilde{u}_j)}{\partial x_j} = 0 \tag{1}$$

$$\frac{\partial (\bar{\rho}\widetilde{u}_i\widetilde{u}_i)}{\partial t} + \frac{\partial (\rho\widetilde{u}_i\widetilde{u}_j)}{\partial x_j} = -\frac{\partial \bar{p}}{\partial x_i} + \frac{\partial}{\partial x_j}\left[(\bar{\mu} + \mu_t)\left(\frac{\partial \widetilde{u}_i}{\partial x_j} + \frac{\partial \widetilde{u}_j}{\partial x_i} - \frac{2}{3}\delta_{ij}\frac{\partial \widetilde{u}_k}{\partial x_k}\right)\right] \tag{2}$$

$$\frac{\partial(\bar{\rho}\widetilde{Z})}{\partial t} + \frac{\partial\left(\bar{\rho}\widetilde{u}_i\widetilde{Z}\right)}{\partial x_i} = \frac{\partial}{\partial x_i}\left[\left(\bar{\rho}\widetilde{D} + \frac{\mu_t}{Sc_t}\right)\frac{\partial\widetilde{Z}}{\partial x_i}\right] \tag{3}$$

$$\frac{\partial\left(\bar{\rho}\widetilde{Z}''^2\right)}{\partial t} + \frac{\partial\left(\bar{\rho}\widetilde{u}_i\widetilde{Z}''^2\right)}{\partial x_i} = \frac{\partial}{\partial x_i}\left[\left(\bar{\rho}\widetilde{D} + \frac{\mu_t}{Sc_t}\right)\frac{\partial\widetilde{Z}''^2}{\partial x_i}\right] + 2\left(\bar{\rho}\widetilde{D} + \frac{\mu_t}{Sc_t}\right)\left(\frac{\partial\widetilde{Z}}{\partial x_i}\right)^2 - 2\bar{\rho}\widetilde{\chi} \tag{4}$$

$$\frac{\partial\bar{\rho}\widetilde{Y}_c}{\partial t} + \frac{\partial\bar{\rho}\widetilde{u}_j\widetilde{Y}_c}{\partial x_j} = \frac{\partial}{\partial x_j}\left[\left(\bar{\rho}\widetilde{D} + \frac{\mu_t}{Sc_t}\right)\frac{\partial\widetilde{Y}_c}{\partial x_j}\right] + \bar{\rho}\widetilde{\dot{\omega}_{Y_c}}, \tag{5}$$

where $\widetilde{\phi}$ and $\bar{\phi}$ represent the density-weighted Favre filtering and spatial filtering for a variable ϕ, respectively. ρ denotes the density, u_i the velocity component in the ith direction, p the pressure, and μ dynamic viscosity. μ_t is the subgrid eddy viscosity, which is closed by the dynamic Smagorinsky model [46]. δ_{ij} denotes the Kronecker delta function. The mixture fraction Z is calculated based on the Bilger's method [47] as follows:

$$Z = \frac{2(Y_C - Y_{C,o})/MW_C + 0.5(Y_H - Y_{H,o})/MW_H - (Y_O - Y_{O,o})/MW_O}{2\left(Y_{C,f} - Y_{C,o}\right)/MW_C + 0.5\left(Y_{H,f} - Y_{H,o}\right)/MW_H - \left(Y_{O,f} - Y_{O,o}\right)/MW_O}, \tag{6}$$

where Y and MW are the elemental mass fraction and atomic mass, respectively. The subscripts f and o represent fuel and oxidizer, respectively. Sc_t is the turbulent Schmidt number. $\dot{\omega}_{Y_c}$ is the source term of the progress variable, which is extracted from the flamelet library. In the equations, the molecular diffusivity is calculated via the unity Lewis number assumption $D = \lambda/(\rho c_p)$. χ represents the scalar dissipation rate. The filtered scalar dissipation rate is modeled according to the method in Ref. [48], which is composed of two parts, i.e., the resolved large-scale component and the unresolved sub-grid scale (SGS) component:

$$\widetilde{\chi} = \widetilde{D}|\nabla\widetilde{Z}|^2 + \frac{\mu_t}{Sc_t}\frac{C_\chi}{2\Delta^2}\widetilde{Z}''^2, \tag{7}$$

where Δ represents the filter width and C_χ is a model constant.

3.2. Flamelet/Progress Variable (FPV) Model

In the FPV approach, a flamelet library is required [45]. In this work, the library is constructed using FlameMaster [49] by solving one-dimensional steady-state counterflow flames. In this approach, all the thermochemical variables (including temperature, species mass fractions, thermodynamic and transport properties) can be parameterized using two control variables, i.e., the mixture fraction Z and the progress variable Y_c. The thermochemical scalars can be extracted from the flamelet library as follows:

$$\phi = \phi(Z, Y_c). \tag{8}$$

The progress variable Y_c is typically defined as the sum of the main species mass fractions of the products [50]. In this work, the progress variable is defined based on the mass fraction of H_2O. The normalized progress variable is calculated as follows:

$$C = \frac{Y_c - Y_c^{min}(Z)}{Y_c^{max}(Z) - Y_c^{min}(Z)}, \tag{9}$$

where $Y_c^{max}(Z)$ and $Y_c^{min}(Z)$ are the local maximum and minimum of the progress variable, respectively, which can be extracted from the flamelet library. With the normalized progress variable C, the flamelet library solutions can be parameterized as follows:

$$\phi = \phi(Z, C). \tag{10}$$

In the FPV model, the interactions of turbulence and combustion are considered by combining the thermochemical quantities with the joint probability density function (PDF) [44]:

$$\widetilde{\phi} = \int_0^1 \int_0^1 \phi(Z,C)\widetilde{P}(Z,C)dZdC. \tag{11}$$

Here, the δ-PDF is employed for the progress variable C, while the β-PDF is used for the mixture fraction Z. Therefore, all thermochemical scalars in the flamelet table, including fluid density, flame temperature and chemical species mass fractions, are represented in terms of $\widetilde{Z}$, $\widetilde{Z}''^2$ and $\widetilde{C}$:

$$\widetilde{\phi} = \widetilde{\phi}\left(\widetilde{Z}, \widetilde{Z}''^2, \widetilde{C}\right). \tag{12}$$

In the flamelet table, the coordinates for mixture fraction, mixture fraction variance and progress variable are discretized using 251, 7 and 251 uniformly spaced grid points, respectively. In total, over 350 flamelets were computed to construct the table.

3.3. The NO Prediction Model

The flamelet model is based on the fast chemistry assumption, where chemical reactions occur at a sufficiently fast rate compared to the turbulent Kolmogorov time scale [51]. However, the reaction rate of NO is relatively slow, and in actual turbulent flames, the concentration of NO often deviates far from its equilibrium state within a short residence time. Previous studies have shown that directly calculating NO using lookup tables in the flamelet model can lead to very serious errors [52,53].

In order to overcome the inherent defects of inaccurate prediction of NO in traditional flamelet models and predict the NO in the ammonia/air combustion accurately, the approach of Ihme et al. [53] is used, in which an extra transport equation is solved for NO. Specifically, the forward and backward NO reaction rates are modeled with the reaction rates taken from the flamelet library, respectively. The backward NO reaction rates directly related to NO concentration are corrected by NO obtained in the flow field.

$$\frac{\partial \bar{\rho} \widetilde{Y}_{NO}}{\partial t} + \frac{\partial \bar{\rho} \widetilde{u}_j \widetilde{Y}_{NO}}{\partial x_j} = \frac{\partial}{\partial x_j}\left(\left(\bar{\rho}\widetilde{D} + \frac{\mu_t}{Sc_t}\right)\frac{\partial \widetilde{Y}_{NO}}{\partial x_j}\right) + \bar{\rho}\widetilde{\omega_{\dot{NO}}} \tag{13}$$

$$\widetilde{\omega_{\dot{NO}}} = \widetilde{\omega_{\dot{NO}}^{+}} + \widetilde{Y}_{NO}\frac{\widetilde{\omega_{\dot{NO}}^{-}}}{\widetilde{Y}_{NO}^{FPV}} \tag{14}$$

In the above equations, $\widetilde{Y}_{NO}^{FPV}$ is the mass fraction of species NO obtained directly from the flamelet table. $\widetilde{\omega_{\dot{NO}}^{+}}$ and $\widetilde{\omega_{\dot{NO}}^{-}}$ are the filtered source terms of NO generation and consumption, respectively. To calculate the NO source term, all the reactions that involve NO are considered. The reactions involving NO in the Okafor mechanism are listed in Table 2.

Table 2. The reactions involving NO in the Okafor mechanism.

Sequence	Reaction	Sequence	Reaction
R220:	$N + NO = N_2 + O$	R255:	$NH_2 + HNO = NH_3 + NO$
R221:	$N + O_2 = NO + O$	R256:	$NH_2 + NO = N_2 + H_2O$
R222:	$N + OH = NO + H$	R257:	$NH_2 + NO = NNH + OH$
R224:	$N_2O + O = 2\ NO$	R263:	$NNH + O = NH + NO$
R228:	$HO_2 + NO = NO_2 + OH$	R269:	$NNH + NO = N_2 + HNO$
R229:	$NO + O + M = NO_2 + M$	R271:	$H + NO + M = HNO + M$
R230:	$NO_2 + O = NO + O_2$	R272:	$HNO + O = NO + OH$
R231:	$NO_2 + H = NO + OH$	R273:	$HNO + H = H_2 + NO$
R233:	$NH + O = NO + H$	R274:	$HNO + OH = NO + H_2O$
R237:	$NH + O_2 = NO + OH$	R275:	$HNO + O_2 = HO_2 + NO$
R240:	$NH + NO = N_2O + H$	R276:	$HNO + NO_2 = HONO + NO$
R241:	$NH + NO = N_2O + H$	R290:	$N_2H_2 + O = NH_2 + NO$
R242:	$NH + NO = N_2 + OH$		

4. Results and Discussions

4.1. 1D Laminar Counterflow Flame Simulation

4.1.1. Mechanism Comparison

Figure 3 shows the profiles of major species mass fractions and temperature in the mixture fraction space. It can be seen that the predictions of the four mechanisms almost completely overlap. The simulation results for the species mass fractions and temperature match the experimental results quite well. Some minor discrepancies are observed in the fuel-rich region for the H_2 mass fraction. This can be attributed to the high mass fraction of H_2 at the positions with large mixture fractions, leading to a pronounced preferential diffusion effect.

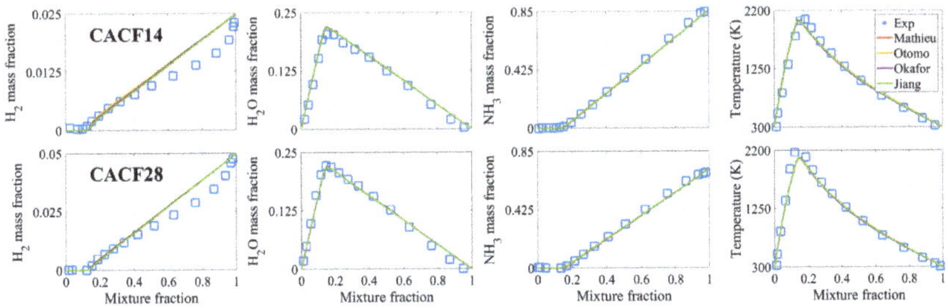

Figure 3. Measured mass fractions of major species and temperature [19] (symbols) as a function of mixture fraction in CACF14 and CACF28. The solid line is the simulation results using different chemical mechanisms.

Figure 4 shows the profiles of NO mole fractions in the mixture fraction space. The predictions generally agree with the experimental results, especially in the fuel-lean side. In the fuel-rich side, the NO mole fraction is under-predicted for all mechanisms. The experimental results show that the peak NO mole fraction of CACF14 is larger than that of CACF28, so that the Mathieu and Jiang mechanisms, where the peak mole fraction of CACF14 is less than that of CACF28, are excluded for further study. The Okafor mechanism has a peak NO mole fraction closer to the experimental measurement value under the current unity Lewis number assumption. Therefore, the Okafor mechanism is chosen for subsequent study.

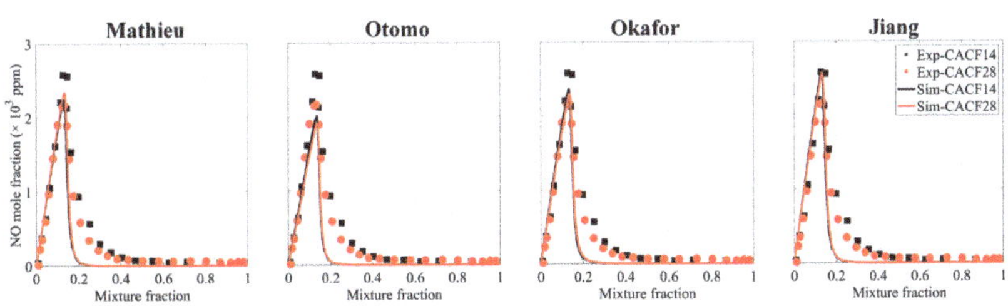

Figure 4. Measured mole fractions of NO [20] (symbols) in mixture fraction space of CACF14/28. The solid line is the simulation results using different chemical mechanisms.

4.1.2. Reaction Path Analysis

In order to explore the impact of ammonia cracking ratio on the formation of nitrogen-containing species, especially NO formation, this section performs nitrogen flow analysis, which is implemented by integrating the reactions across the entire field to calculate the nitrogen flow fraction between relevant species [54]. The nitrogen flow between species A and B is calculated by summing the average integrated reaction rates of all elementary reactions that contribute to A $\rightarrow$ B. Under the current conditions, almost 100% of NH_3

is converted to NH_2; thus, the nitrogen flow is normalized using the sum of the average integrated reaction rates of all elementary reactions that contribute to $NH_3 \rightarrow NH_2$.

Figure 5 presents the nitrogen flow diagrams of major paths for CACF14 and CACF28. It is worth noting that, for simplicity, the reaction paths in Figure 5 are not complete. As can be seen, the species directly related to NO formation are HNO, N and NH. Among them, the species that contributes the most to NO generation is HNO. The second most important species is N, while NH contributes the least to NO generation. As the cracking ratio increases, the amount of NO generated through N increases, but the amount of NO generated through species HNO and NH remains basically unchanged. It can also be seen that NO is mainly generated through the following four paths: $NH_2 \rightarrow HNO \rightarrow NO$ (path a), $NH_2 \rightarrow NH \rightarrow HNO \rightarrow NO$ (path b), $NH_2 \rightarrow NH \rightarrow NO$ (path c) and $NH_2 \rightarrow NH \rightarrow N \rightarrow NO$ (path d). On the other hand, the products of NO consumption are N_2O, NNH and N_2. As the cracking ratio increases, the amount of NO directly converted to N_2 increases, but the conversion to NNH and N_2O remains basically unchanged. NO is mainly consumed through $NO \rightarrow N_2O \rightarrow N_2$ (path e), $NO \rightarrow N_2O \rightarrow NNH \rightarrow N_2$ (path f), $NO \rightarrow N_2$ (path g) and $NO \rightarrow NNH \rightarrow N_2$ (path h).

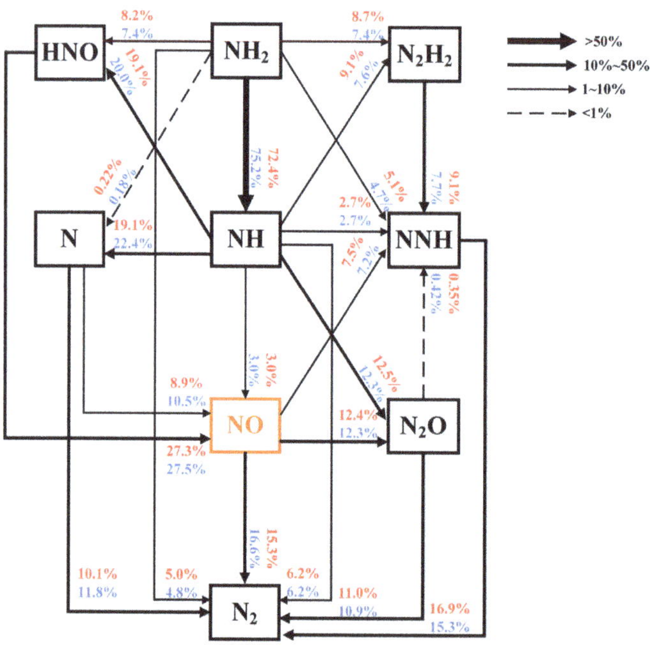

Figure 5. Nitrogen flow diagrams for CACF14/CACF28. The thickness of the arrows represents the relative size of the flow. The numbers on the arrows indicate the relative conversion rate, with CACF14 in red and CACF28 in blue.

In order to study the influence of the ammonia cracking ratio on the conversion of nitrogen-containing species and the relative magnitudes of the NO pathways, a quantitative analysis is conducted in terms of the conversion rate, which is defined as the ratio of the net outflow of nitrogen from species A to species B to the total net outflow of nitrogen from species A. Figure 6a–c present the conversion rates of NH_2, NH, HNO and N for various flames.

Figure 6a shows that most NH_2 is directly converted to NH, and the ratio of this conversion increases with the increasing cracking ratio. According to Figure 6b, NH is mostly converted to HNO and N. As the cracking ratio increases, the conversion ratio of NH to these two species increases, with the conversion to N being more significant.

Figure 6c shows that HNO is entirely converted to NO, while N is mostly converted to NO and N_2.

The relative importance of NO pathways is quantitatively assessed, as shown in Figure 6d. It is obvious that the most important pathway for NO generation is path b, followed by path d. As the ammonia cracking ratio increases, the generation of NO through paths b and d increases, while that through path a decreases. For path c, the change in the cracking ratio does not lead to significant change in NO formation.

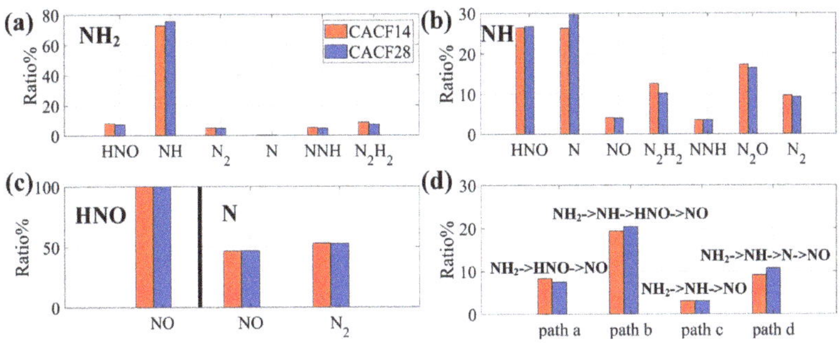

Figure 6. (**a**) The ratios of NH_2 converted to nitrogen-containing species in CACF14 (red) and CACF28 (blue); (**b**) ratios of NH converted to nitrogen-containing species; (**c**) ratios of HNO and N converted to nitrogen-containing species; (**d**) conversion rates of NO generated from NH_2 via four pathways: a–d.

Figure 7 shows the profiles of mole fractions for various radicals in the mixture fraction space. It is seen that the concentrations of HNO and NH in CACF28 are lower than those in CACF14, which is due to the higher hydrogen concentration and lower NH_i $(i = 1, 2, 3)$ concentration related to the higher cracking ratio in CACF28, which further leads to a lower concentration of HNO [20]. According to Figure 5, HNO is an important radical for NO generation [55]. Therefore, the mole fraction of NO in CACF28 is lower than that in CACF14, as shown in Figure 4. In addition, it can be seen that the concentrations of radicals such as H, OH and O in CACF28 are higher than those in CACF14, which is attributed to the higher H_2 concentration associated with the larger cracking ratio, so the reactivity of CACF28 is stronger than that of CACF14.

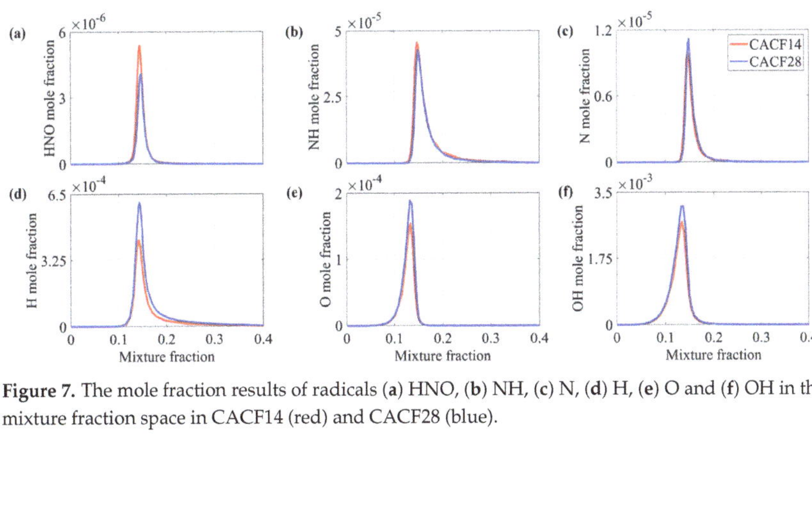

Figure 7. The mole fraction results of radicals (**a**) HNO, (**b**) NH, (**c**) N, (**d**) H, (**e**) O and (**f**) OH in the mixture fraction space in CACF14 (red) and CACF28 (blue).

4.2. 3D Turbulent Jet Flame Simulation

4.2.1. Simulation Verification

In this section, the LES results of the two turbulent flames, i.e., CAJF14 and CAJF28, are compared with the experimental results. Figure 8 shows mean and root-mean-square (RMS) values of the mixture fraction Z and progress variable Y_c at various streamwise locations. For the mean values, it can be observed that the simulation results demonstrate a favorable agreement with the measured data across the entire domain. However, the mean of Y_c is under-predicted near the peaks in the upstream region $X/D = 5$, which is more pronounced in the CAJF28 flame with a higher cracking ratio. This is attributed to the more prominent differential diffusion effects near the nozzle region [19,21]. It is evident that at the streamwise locations $X/D > 5$, turbulent diffusion overtakes molecular diffusion, leading to a good performance of simulations. For the RMS values of the mixture fraction, the peaks initially experience a slight increase along the streamwise direction and subsequently decrease as the flame progresses. It can be observed that for the progress variable, two RMS peaks are observed, corresponding to the lean and rich sides, as shown in the experiments. The RMS peak in the rich side becomes less pronounced with increasing streamwise distance, attributed to the more uniform distribution of jet temperature and composition due to combustion product entrainment. The RMS peak in the lean side gradually increases with the streamwise location.

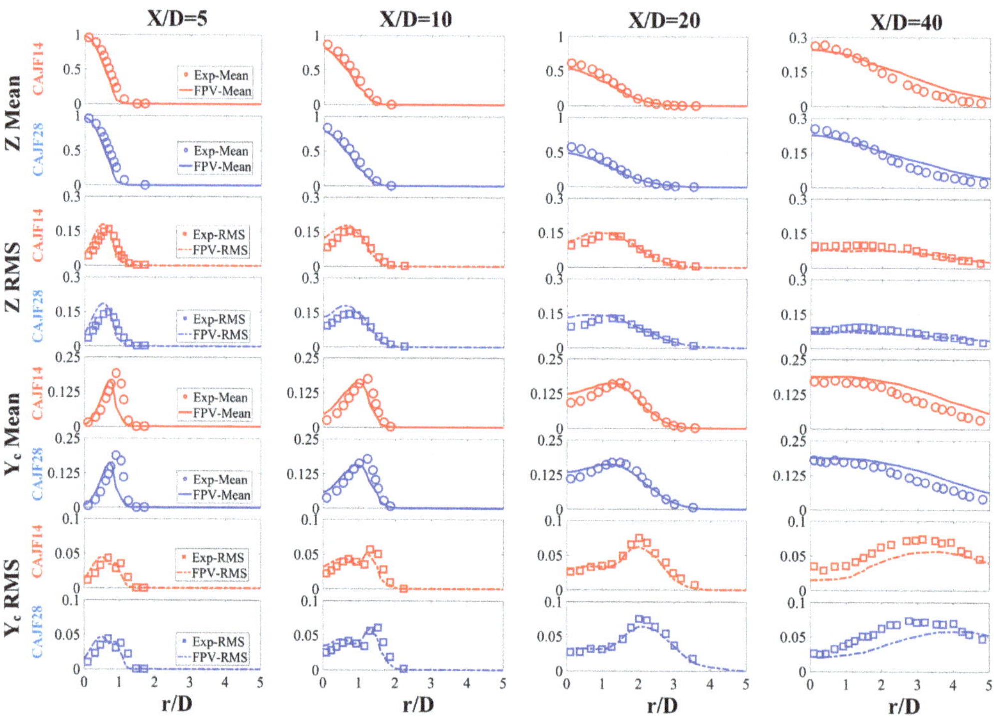

Figure 8. Radial distribution of time-averaged (circle points, solid lines) and RMS values (square points, dashed lines) of mixture fraction, progress variable at different streamwise locations (X/D = 5, 10, 20, 40) for CAJF14 (red) and CAJF28 (blue), as measured in experiments [19] (points) and computed with FPV (lines).

Figure 9 presents the mean and RMS profiles of Y_{NH_3}, Y_{H_2} and the temperature T. In terms of the mean values, similar to the mixture fraction, the measured and predicted values of Y_{NH_3} and Y_{H_2} exhibit a good match throughout the entire field. This indicates the satisfactory performance of the unit Lewis number transport model under the current operating conditions. The profile of temperature T resembles that of the progress variable. Both flames exhibit varying degrees of under-prediction for temperature in the upstream region X/D = 5, with this effect being more pronounced for CAJF28. Regarding the RMS values, the profiles of Y_{NH_3} and Y_{H_2} show a single peak, similar to the mixture fraction Z. The peak value increases first with the streamwise position and then decreases. The temperature T profile exhibits a double-peak distribution in the upstream, similar to that of the progress variable.

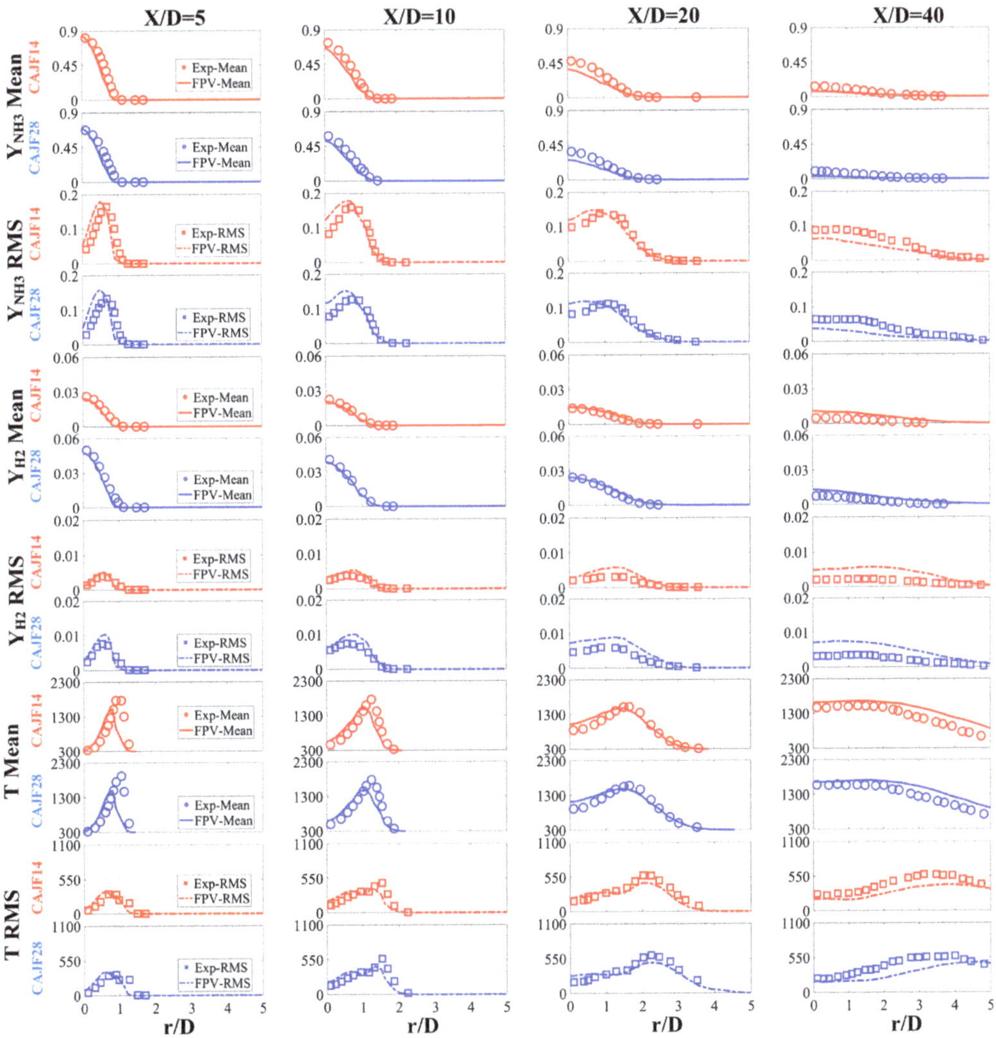

Figure 9. Radial distribution of time-averaged (circle points, solid lines) and RMS values (square points, dashed lines) of Y_{NH_3}, Y_{H_2} and temperature T at different streamwise locations (X/D=5, 10, 20, 40) for CAJF14 (red) and CAJF28 (blue), as measured in experiments [19] (points) and computed with FPV (lines).

Figure 10 shows the mean and RMS profile of X_{NO}. Notably, for the mole fraction of NO, the numerical method of the present wrok also produces satisfactory results, affirming the accuracy of the employed NO model. Nevertheless, in the downstream region at X/D = 40, both cases display varying degrees of over-prediction in the NO mole fraction. The peak mole fraction of NO appears at X/D = 5. With increasing streamwise distance due to the expansion of the jet, the radial position of the peak NO mole fraction gradually increases. Meanwhile, the peak mole fraction of NO decreases along the streamwise direction. The RMS values of X_{NO} also agree well with the experiment results.

In summary, both the mean and RMS values of the simulations successfully capture the characteristics of the experimental flame [19]. This confirms the rationality of the numerical methods and boundary conditions, providing a solid foundation for subsequent analyses.

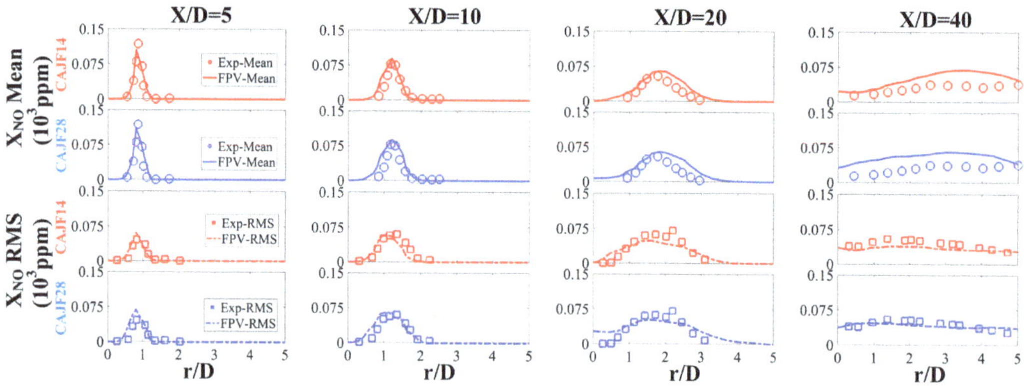

Figure 10. Radial distribution of time-averaged (circle points, solid lines) and RMS values (square points, dashed lines) of X_{NO} at different streamwise locations (X/D = 5, 10, 20, 40) for CAJF14 (red) and CAJF28 (blue), as measured in experiments [20] (points) and computed with FPV (lines).

4.2.2. Flame structure analysis

By examining the distributions of various quantities in mixture fraction space, a deeper understanding of the flame structure can be achieved. Figures 11 and 12 show the scatter plots of NO mole fraction and temperature, respectively, in mixture fraction space at two streamwise locations of X/D = 5 and X/D = 20. The conditional means and flamelet solutions with different scalar dissipation rates are also presented. It is shown that as the scalar dissipation rate increases, the mole fraction of NO in the steady flamelet branch increases while the temperature gradually decreases. This indicates that in both flames, the pathway of thermal NO is not dominant. It is also evident that with an increase in streamwise distance, the conditional mean values of the NO mole fraction approach the flamelet solutions with lower scalar dissipation rates.

In Figure 12, dashed lines illustrate the unsteady flamelet solutions for different scalar dissipation rates. It can be seen that at the streamwise location X/D = 5, a considerable number of data points are distributed over the unsteady flamelet solutions. However, as the streamwise distance increases, the level of local extinction decreases. By comparing CAJF14 and CAJF28, it is apparent that CAJF28 shows less local extinction compared to CAJF14. This is attributed to the elevated H_2 concentration resulting from a higher cracking ratio, which enhances the reaction activity and thus promotes the completeness of combustion.

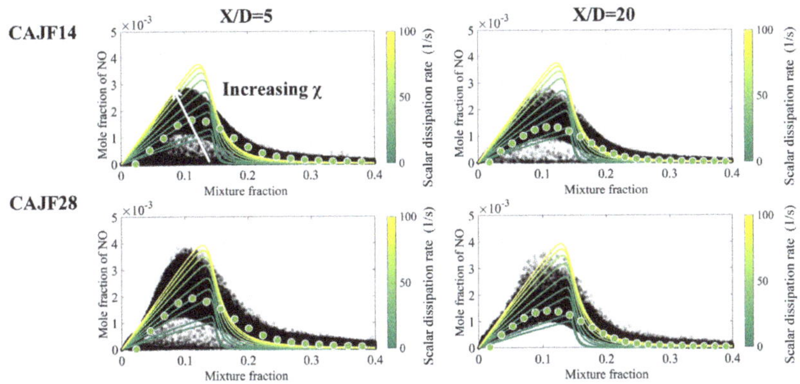

Figure 11. The figure of instantaneous NO mole fraction (black dots), NO conditional mean value (green dots) and steady flamelet solutions at different scalar dissipation rates (colored solid lines according to the value of scalar dissipation rate) versus mixture fraction at streamwise locations X/D = 5, X/D = 20 for CAJF14 and CAJF28. The arrows point towards the flamelets where the scalar dissipation rate increases.

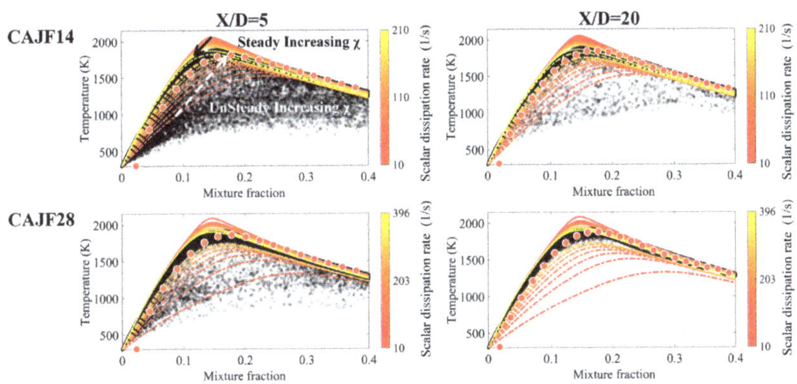

Figure 12. The figure of instantaneous temperature (black dots), temperature conditional mean value (red dots), steady flamelet solutions of temperature at different scalar dissipation rates (colored solid lines according to the size of scalar dissipation rate) and unsteady flamelet solutions (colored dashed lines according to the value of scalar dissipation rate) versus mixture fraction at streamwise locations X/D = 5 and X/D = 20 for CAJF14 and CAJF28. Arrows pointing towards the flamelets where the scalar dissipation rate increases; solid black arrow indicates steady flamelets and dashed white arrow indicates unsteady flamelets.

4.2.3. NO Formation Mechanism

Figure 13 displays the time-averaged distributions of NO mole fraction for CAJF14 and CAJF28. It can be seen that NO is primarily observed in a narrow region upstream near the stoichiometric mixture fraction Z_{st} where fuel and oxidizer meet, and a broader region downstream as the jet flame develops. In the downstream region, the peak NO concentration for CAJF28 occurs at a more upstream location and the peak concentration is higher. This means that when the cracking ratio increases, not only is the flame reactivity enhanced but also the generation of NO is increased for the turbulent flames. In addition, compared with Figure 4, it can be found that the impact of changes in the ammonia cracking ratio on NO production seems to be more significant in turbulent flames. This is because in the turbulent jet flame, there are complex interactions between turbulence and the formation of NO, which might result in a more pronounced difference of NO formation for the two turbulent flames with different ammonia cracking ratios.

In the analysis of laminar flames, it has been shown that intermediates such as HNO, NH and N are directly associated with NO generation. Figure 14 shows time-averaged distributions of key species in the turbulent jet flames. As can be seen, in the upstream region of the flames, similar to the laminar flames, a higher cracking ratio results in a lower NH concentration, which in turn leads to a reduced HNO level. As the streamwise distance increases, the magnitude of nitrogen-containing species concentrations decreases rapidly, indicating that the downstream regions approach the equilibrium state more closely than the upstream regions. Radicals such as OH are highly indicative of reaction zones and are commonly used to characterize turbulent flames [24], which also play a role in NO generation in ammonia flames [56,57]. Time-averaged distributions of O, H and OH concentrations are presented in Figure 14d–f. It can be observed that in the upstream regions of the flames, similar to the laminar flame results, the higher cracking ratio in CAJF28 results in higher radical concentrations compared to CAJF14. The differences in O, H and OH concentrations between the two flames remain distinct in the downstream regions. Furthermore, based on the distributions of radicals, it is concluded that the flame length of CAJF28 is shorter than that of CAJF14. This observation is consistent with the long exposure duration pictures obtained from the experimental results [20], indicating that with increasing cracking ratio, more intense combustion is achieved and the fuel is consumed more rapidly.

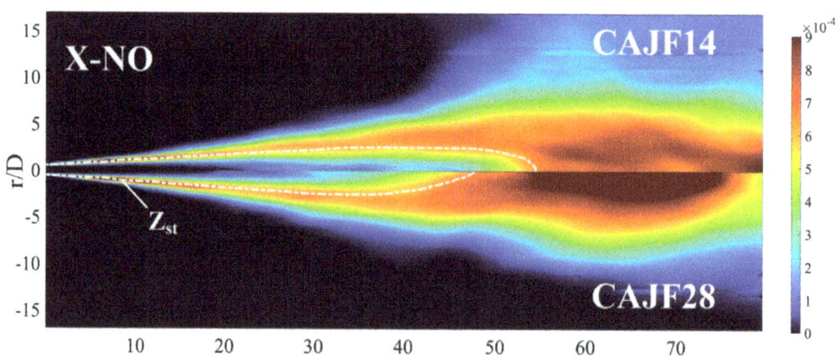

Figure 13. Time-averaged distributions of NO mole fraction for CAJF14 (**top**) and CAJF28 (**bottom**). The Zst isoline is represented by a white dotted line.

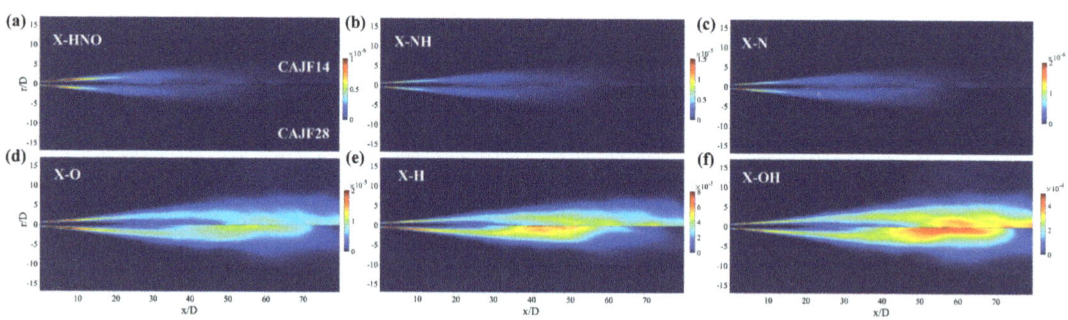

Figure 14. Time-averaged distributions of mole fractions for (**a**) HNO, (**b**) NH, (**c**) N, (**d**) H, (**e**) O and (**f**) OH in CAJF14 (**top**) and CAJF28 (**bottom**).

To gain a deeper understanding of NO formation in turbulent jet flames, two representative locations are chosen for the following analysis, i.e., $X/D = 5$ and 60. From Figure 13, it is evident that these two locations correspond to the regions where the NO mole fraction peaks. Figure 15 illustrates the scatter plots between important radical concentrations with the NO concentration at various locations. It is apparent that the correlation between NO and HNO is weaker when compared to that between NO and radicals such as O, H and OH, both in the upstream and downstream regions. Figure 15b–d indicate that the radicals O, H and OH exhibit a positive correlation with NO, confirming the a role that these radicals play in NO generation in ammonia flames.

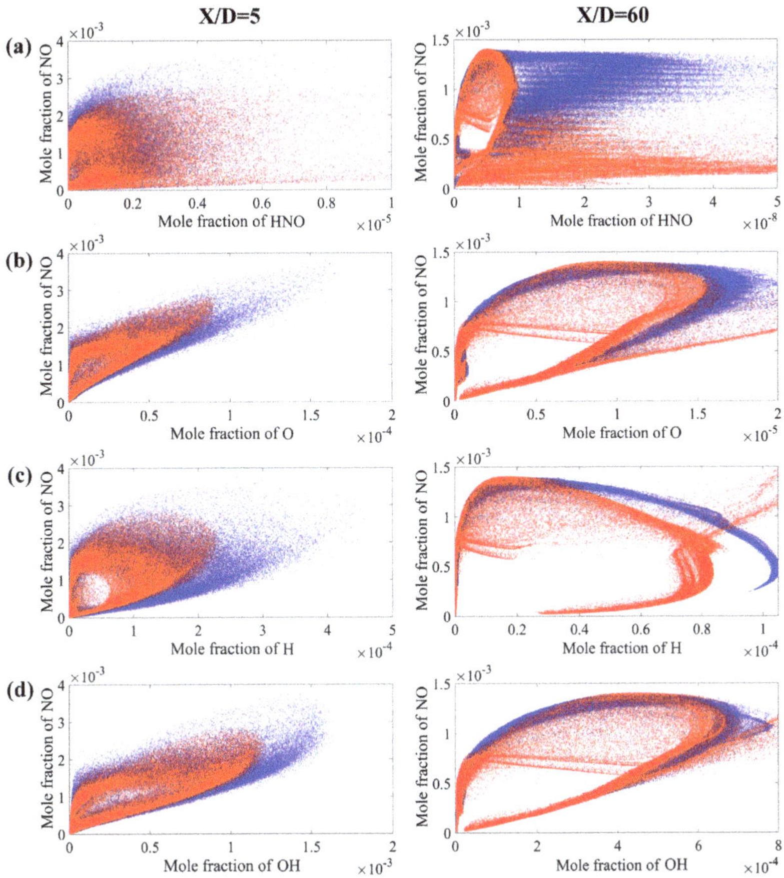

Figure 15. Scatter plots showing the distribution of mole fractions of NO (**a**), O (**b**), H (**c**) and OH (**d**) in the mixture fraction space, where CAJF14 is represented in red and CAJF28 in blue.

The following analysis focuses on the influence of varying ammonia cracking ratios on the distribution of NO source terms, the information of which is not available from the experiments. The combustion of NH_3/H_2 mixture mainly involves two mechanisms for NO formation. The first is the thermal NO mechanism, also known as the Zeldovich mechanism, which includes three reactions, i.e., $N + NO \rightarrow N_2 + O$, $N + O_2 \rightarrow NO + O$ and $N + OH \rightarrow NO + H$. The second mechanism is the fuel mechanism, i.e., $NH_i + OX \rightarrow NO + H_iX$, where OX is an oxygen-containing species [58]. Figure 16 shows the mean distributions of the fuel, thermal and total NO source terms of both flames. It can be seen that the NO source term has a higher absolute value in the upstream region, and the absolute value of the source term rapidly decreases with the increase in the streamwise distance. The total NO source term is positive in the fuel-lean zone and shows negative values in the fuel-rich zone. In addition, it can be observed that the fuel NO in most regions is negative, while the thermal NO is positive.

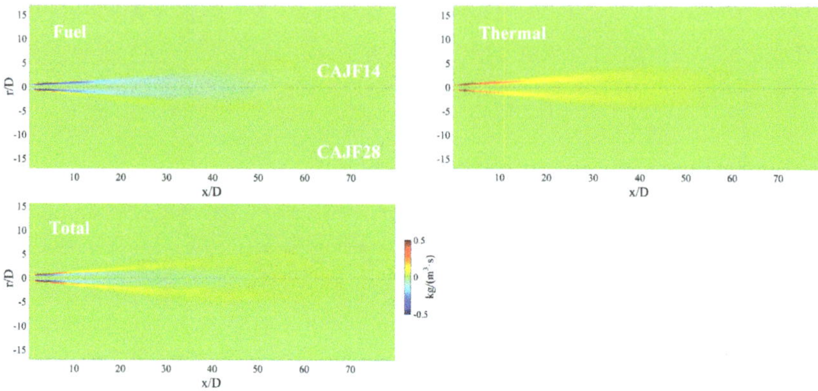

Figure 16. Time-averaged distributions of fuel, thermal and total NO source terms for CAJF14 (**top**) and CAJF28 (**bottom**).

5. Conclusions

In this study, simulations of laminar counterflow and turbulent non-premixed ammonia/air flames in HPCD were conducted. Two cases with different ammonia cracking ratios of 14% and 28% were investigated. The influence of changes in the ammonia cracking ratio on flame structure and NO formation mechanism under high-pressure conditions were examined. The main findings are summarized as follows:

- Through nitrogen flow analyses, the conversion of nitrogen-containing species and the NO pathways were examined. It was found that the species that contributes the most to NO generation is HNO, and the most important pathway for NO generation is $NH_2 \rightarrow NH \rightarrow HNO \rightarrow NO$.
- The simulation and experimental results agree quite well, and the simulations provided fairly accurate predictions for NO, proving the applicability of the FPV method in hydrogen–ammonia flames. The flame structure and NO formation of turbulent jet flames were examined. It was found that with increasing cracking ratio, not only is the flame reactivity enhanced but the generation of NO is also increased for the turbulent flames.
- The correlation between important radical concentrations and NO concentration was examined. We showed that the correlation between NO and HNO is weaker when compared to that of NO and radicals such as O, H and OH throughout the entire field. Moreover, the increased cracking ratio also results in a higher level of radical concentrations.
- The distribution of NO source terms were also explored. It was found that the NO source term has a higher absolute value upstream, and the absolute value rapidly decreases with the increase in the streamwise distance. The total NO source term is positive in the fuel-lean zone and shows negative values in the fuel-rich zone. In addition, the fuel NO in most regions is negative, while the thermal NO is positive.

Author Contributions: Conceptualization, C.W. and H.W.; methodology, C.W.; software, C.W.; validation, C.W. and H.W.; formal analysis, C.W.; investigation, C.W.; resources, C.W.; data curation, C.W. and H.W.; writing—original draft preparation, C.W. and H.W.; writing—review and editing, C.W., H.W., K.L. and J.F.; visualization, C.W.; supervision, H.W.; project administration, H.W., K.L. and J.F.; funding acquisition, H.W., K.L. and J.F. All authors have read and agreed to the published version of the manuscript.

Funding: This research was funded by the Natural Science Foundation of China (Grant Nos.: 52022091, 51976185, 51925603). The work was also supported by the Fundamental Research Funds for the Central Universities, China (Grant No: 2021FZZX001-10).

Data Availability Statement: The data that support the findings of this study are available from the corresponding author upon reasonable request.

Conflicts of Interest: The authors declare no conflict of interest.

References

1. Smith, C.; Hill, A.K.; Torrente-Murciano, L. Current and Future Role of Haber–Bosch Ammonia in a Carbon-Free Energy Landscape. *Energy Environ. Sci.* **2020**, *13*, 331–344. [CrossRef]
2. Yapicioglu, A.; Dincer, I. A Review on Clean Ammonia as a Potential Fuel for Power Generators. *Renew. Sustain. Energy Rev.* **2019**, *103*, 96–108. [CrossRef]
3. Awad, O.I.; Zhou, B.; Harrath, K.; Kadirgama, K. Characteristics of NH_3/H_2 Blend as Carbon-Free Fuels: A Review. *Int. J. Hydrogen Energy* 2022, *in press*. [CrossRef]
4. Hayakawa, A.; Goto, T.; Mimoto, R.; Arakawa, Y.; Kudo, T.; Kobayashi, H. Laminar Burning Velocity and Markstein Length of Ammonia/Air Premixed Flames at Various Pressures. *Fuel* **2015**, *159*, 98–106. [CrossRef]
5. Kobayashi, H.; Hayakawa, A.; Somarathne, K.D.K.A.; Okafor, E.C. Science and Technology of Ammonia Combustion. *Proc. Combust. Inst.* **2019**, *37*, 109–133. [CrossRef]
6. Li, J.; Huang, H.; Kobayashi, N.; He, Z.; Nagai, Y. Study on Using Hydrogen and Ammonia as Fuels: Combustion Characteristics and NO_x Formation: Hydrogen and Ammonia as Fuels. *Int. J. Energy Res.* **2014**, *38*, 1214–1223. [CrossRef]
7. Mathieu, O.; Petersen, E.L. Experimental and Modeling Study on the High-Temperature Oxidation of Ammonia and Related NOx Chemistry. *Combust. Flame* **2015**, *162*, 554–570. [CrossRef]
8. Li, S.; Zhang, S.; Zhou, H.; Ren, Z. Analysis of Air-Staged Combustion of NH3/CH4 Mixture with Low NOx Emission at Gas Turbine Conditions in Model Combustors. *Fuel* **2019**, *237*, 50–59. [CrossRef]
9. Xiao, H.; Lai, S.; Valera-Medina, A.; Li, J.; Liu, J.; Fu, H. Study on Counterflow Premixed Flames Using High Concentration Ammonia Mixed with Methane. *Fuel* **2020**, *275*, 117902. [CrossRef]
10. Kurata, O.; Iki, N.; Matsunuma, T.; Inoue, T.; Tsujimura, T.; Furutani, H.; Kobayashi, H.; Hayakawa, A. Performances and Emission Characteristics of NH3–Air and NH3CH4–Air Combustion Gas-Turbine Power Generations. *Proc. Combust. Inst.* **2017**, *36*, 3351–3359. [CrossRef]
11. Khateeb, A.A.; Guiberti, T.F.; Zhu, X.; Younes, M.; Jamal, A.; Roberts, W.L. Stability Limits and Exhaust NO Performances of Ammonia-Methane-Air Swirl Flames. *Exp. Therm. Fluid Sci.* **2020**, *114*, 110058. [CrossRef]
12. Otomo, J.; Koshi, M.; Mitsumori, T.; Iwasaki, H.; Yamada, K. Chemical Kinetic Modeling of Ammonia Oxidation with Improved Reaction Mechanism for Ammonia/Air and Ammonia/Hydrogen/Air Combustion. *Int. J. Hydrogen Energy* **2018**, *43*, 3004–3014. [CrossRef]
13. Frigo, S.; Gentili, R. Analysis of the Behaviour of a 4-Stroke Si Engine Fuelled with Ammonia and Hydrogen. *Int. J. Hydrogen Energy* **2013**, *38*, 1607–1615. [CrossRef]
14. Comotti, M.; Frigo, S. Hydrogen Generation System for Ammonia–Hydrogen Fuelled Internal Combustion Engines. *Int. J. Hydrogen Energy* **2015**, *40*, 10673–10686. [CrossRef]
15. Mei, B.; Zhang, J.; Shi, X.; Xi, Z.; Li, Y. Enhancement of Ammonia Combustion with Partial Fuel Cracking Strategy: Laminar Flame Propagation and Kinetic Modeling Investigation of $NH_3/H_2/N_2$/Air Mixtures up to 10 Atm. *Combust. Flame* **2021**, *231*, 111472. [CrossRef]
16. Shohdy, N.N.; Alicherif, M.; Lacoste, D.A. Transfer Functions of Ammonia and Partly Cracked Ammonia Swirl Flames. *Energies* **2023**, *16*, 1323. [CrossRef]
17. Mercier, A.; Mounaïm-Rousselle, C.; Brequigny, P.; Bouriot, J.; Dumand, C. Improvement of SI Engine Combustion with Ammonia as Fuel: Effect of Ammonia Dissociation Prior to Combustion. *Fuel Commun.* **2022**, *11*, 100058. [CrossRef]
18. Kim, N.; Lee, M.; Park, J.; Park, J.; Lee, T. A Comparative Study of NOx Emission Characteristics in a Fuel Staging and Air Staging Combustor Fueled with Partially Cracked Ammonia. *Energies* **2022**, *15*, 9617. [CrossRef]
19. Tang, H.; Yang, C.; Wang, G.; Krishna, Y.; Guiberti, T.F.; Roberts, W.L.; Magnotti, G. Scalar Structure in Turbulent Non-Premixed NH3/H2/N2 Jet Flames at Elevated Pressure Using Raman Spectroscopy. *Combust. Flame* **2022**, *244*, 112292. [CrossRef]
20. Wang, G.; Tang, H.; Yang, C.; Magnotti, G.; Roberts, W.L.; Guiberti, T.F. Quantitative Laser-Induced Fluorescence of NO in Ammonia-Hydrogen-Nitrogen Turbulent Jet Flames at Elevated Pressure. *Proc. Combust. Inst.* **2023**, *39*, 1465–1474. [CrossRef]
21. Abdelwahid, S.; Malik, M.R.; Al Kader Hammoud, H.A.; Hernández-Pérez, F.E.; Ghanem, B.; Im, H.G. Large Eddy Simulations of Ammonia-Hydrogen Jet Flames at Elevated Pressure Using Principal Component Analysis and Deep Neural Networks. *Combust. Flame* **2023**, *253*, 112781. [CrossRef]
22. Tang, J. Study and Application of NOx Prediction Model and Flamelet Tabulation Method for Aeroengine Combustion Chamber. Ph.D. Thesis, Northwestern Polytechnical University, Xi'an, China, 2018.
23. Tang, Y.; Raman, V. Large Eddy Simulation of Premixed Turbulent Combustion Using a Non-Adiabatic, Strain-Sensitive Flamelet Approach. *Combust. Flame* **2021**, *234*, 111655. [CrossRef]

24. An, Z.; Zhang, M.; Zhang, W.; Mao, R.; Wei, X.; Wang, J.; Huang, Z.; Tan, H. Emission Prediction and Analysis on CH4/NH3/Air Swirl Flames with LES-FGM Method. *Fuel* **2021**, *304*, 121370. [CrossRef]
25. Honzawa, T.; Kai, R.; Okada, A.; Valera-Medina, A.; Bowen, P.J.; Kurose, R. Predictions of NO and CO Emissions in Ammonia/Methane/Air Combustion by LES Using a Non-Adiabatic Flamelet Generated Manifold. *Energy* **2019**, *186*, 115771. [CrossRef]
26. Kai, R.; Tokuoka, T.; Nagao, J.; Pillai, A.L.; Kurose, R. LES Flamelet Modeling of Hydrogen Combustion Considering Preferential Diffusion Effect. *Int. J. Hydrogen Energy* **2022**, *48*, 11086–11101. [CrossRef]
27. Jiang, X.; Guo, J.; Wei, Z.; Quadarella, E.; Im, H.G.; Liu, Z. A Species-Weighted Flamelet/Progress Variable Model with Differential Diffusion Effects for Oxy-Fuel Jet Flames. *Combust. Flame* **2023**, *251*, 112674.
28. Gao, Z.; Wang, H.; Song, C.; Luo, K.; Fan, J. Large-Eddy Simulation of Hydrothermal Flames Using Extended Flamelet/Progress Variable Approach. *J. Supercrit. Fluids* **2020**, *163*, 104843. [CrossRef]
29. Wang, S.; Wang, H.; Luo, K.; Fan, J. The Effects of Differential Diffusion on Turbulent Non-Premixed Flames LO2/CH4 under Transcritical Conditions Using Large-Eddy Simulation. *Energies* **2023**, *16*, 1065. [CrossRef]
30. Boyette, W.R.; Guiberti, T.F.; Magnotti, G.; Roberts, W.L. Structure of Turbulent Nonpremixed Syngas Flames at High Pressure. *Proc. Combust. Inst.* **2019**, *37*, 2207–2214. [CrossRef]
31. Tang, H.; Yang, C.; Wang, G.; Guiberti, T.F.; Magnotti, G. Raman Spectroscopy for Quantitative Measurements of Temperature and Major Species in High-Pressure Non-Premixed $NH_3/H_2/N_2$ Counterflow Flames. *Combust. Flame* **2022**, *237*, 111840. [CrossRef]
32. Design, R. *CHEMKIN-PRO 18.2;* ANSYS, Inc.: San Diego, CA, USA, 2017.
33. Kee, R.J.; Miller, J.A.; Evans, G.H.; Dixon-Lewis, G. A Computational Model of the Structure and Extinction of Strained, Opposed Flow, Premixed Methane-Air Flames. *Symp. (Int.) Combust.* **1989**, *22*, 1479–1494. [CrossRef]
34. Okafor, E.C.; Naito, Y.; Colson, S.; Ichikawa, A.; Kudo, T.; Hayakawa, A.; Kobayashi, H. Experimental and Numerical Study of the Laminar Burning Velocity of CH4–NH3–Air Premixed Flames. *Combust. Flame* **2018**, *187*, 185–198. [CrossRef]
35. Jiang, Y.; Gruber, A.; Seshadri, K.; Williams, F. An Updated Short Chemical-kinetic Nitrogen Mechanism for Carbon-free Combustion Applications. *Int. J. Energy Res.* **2020**, *44*, 795–810. [CrossRef]
36. Cheng, M.; Wang, H.; Xiao, H.; Luo, K.; Fan, J. Emission Characteristics and Heat Release Rate Surrogates for Ammonia Premixed Laminar Flames. *Int. J. Hydrogen Energy* **2021**, *46*, 13461–13470. [CrossRef]
37. Chai, W.S.; Bao, Y.; Jin, P.; Tang, G.; Zhou, L. A Review on Ammonia, Ammonia-Hydrogen and Ammonia-Methane Fuels. *Renew. Sustain. Energy Rev.* **2021**, *147*, 111254. [CrossRef]
38. Song, Y.; Hashemi, H.; Christensen, J.M.; Zou, C.; Marshall, P.; Glarborg, P. Ammonia Oxidation at High Pressure and Intermediate Temperatures. *Fuel* **2016**, *181*, 358–365. [CrossRef]
39. Xu, L.; Chang, Y.; Treacy, M.; Zhou, Y.; Jia, M.; Bai, X.S. A Skeletal Chemical Kinetic Mechanism for Ammonia/n-Heptane Combustion. *Fuel* **2023**, *331*, 125830. [CrossRef]
40. Alnasif, A.; Mashruk, S.; Shi, H.; Alnajideen, M.; Wang, P.; Pugh, D.; Valera-Medina, A. Evolution of Ammonia Reaction Mechanisms and Modeling Parameters: A Review. *Appl. Energy Combust. Sci.* **2023**, *15*, 100175.
41. Han, X.; Wang, Z.; Costa, M.; Sun, Z.; He, Y.; Cen, K. Experimental and Kinetic Modeling Study of Laminar Burning Velocities of NH_3/Air, NH_3/H_2/Air, NH_3/CO/Air and NH_3/CH_4/Air Premixed Flames. *Combust. Flame* **2019**, *206*, 214–226. [CrossRef]
42. Petrova, M.; Williams, F. A Small Detailed Chemical-Kinetic Mechanism for Hydrocarbon Combustion. *Combust. Flame* **2006**, *144*, 526–544. [CrossRef]
43. Weller, H.G.; Tabor, G.; Jasak, H.; Fureby, C. A Tensorial Approach to Computational Continuum Mechanics Using Object-Oriented Techniques. *Comput. Phys.* **1998**, *12*, 620. [CrossRef]
44. Peters, N. Turbulent Combustion. *Meas. Sci. Technol.* **2001**, *12*, 2022–2022. [CrossRef]
45. Pierce, C.D.; Moin, P. Progress-Variable Approach for Large-Eddy Simulation of Non-Premixed Turbulent Combustion. *J. Fluid Mech.* **2004**, *504*, 73–97. [CrossRef]
46. Lilly, D.K. A Proposed Modification of the Germano Subgrid-scale Closure Method. *Phys. Fluids A Fluid Dyn.* **1992**, *4*, 633–635. [CrossRef]
47. Bilger, R.; Stårner, S.; Kee, R. On reduced mechanisms for methane-air combustion in nonpremixed flames. *Combust. Flame* **1990**, *80*, 135–149. [CrossRef]
48. Domingo, P.; Vervisch, L.; Veynante, D. Large-Eddy Simulation of a Lifted Methane Jet Flame in a Vitiated Coflow. *Combust. Flame* **2008**, *152*, 415–432. [CrossRef]
49. Pitsch, H. *A C++ Computer Program for 0-D and 1-D Laminar Flame Calculations*; RWTH: Aachen, Germany, 1998.
50. Ihme, M.; Shunn, L.; Zhang, J. Regularization of Reaction Progress Variable for Application to Flamelet-Based Combustion Models. *J. Comput. Phys.* **2012**, *231*, 7715–7721. [CrossRef]
51. Peters, N. Laminar Diffusion Flamelet Models in Non-Premixed Turbulent Combustion. *Prog. Energy Combust. Sci.* **1984**, *10*, 319–339. [CrossRef]
52. Coelho, P.J.; Peters, N. Unsteady Modelling of a Piloted Methane/Air Jet Flame Based on the Eulerian Particle Flamelet Model. *Combust. Flame* **2001**, *124*, 444–465. [CrossRef]
53. Ihme, M.; Pitsch, H. Modeling of Radiation and Nitric Oxide Formation in Turbulent Nonpremixed Flames Using a Flamelet/Progress Variable Formulation. *Phys. Fluids* **2008**, *20*, 055110. [CrossRef]

54. Wang, H.; Hawkes, E.R.; Savard, B.; Chen, J.H. Direct Numerical Simulation of a High Ka CH4/Air Stratified Premixed Jet Flame. *Combust. Flame* **2018**, *193*, 229–245. [CrossRef]
55. Okafor, E.C.; Somarathne, K.D.K.A.; Ratthanan, R.; Hayakawa, A.; Kudo, T.; Kurata, O.; Iki, N.; Tsujimura, T.; Furutani, H.; Kobayashi, H. Control of NOx and Other Emissions in Micro Gas Turbine Combustors Fuelled with Mixtures of Methane and Ammonia. *Combust. Flame* **2020**, *211*, 406–416. [CrossRef]
56. Somarathne, K.D.K.A.; Okafor, E.C.; Sugawara, D.; Hayakawa, A.; Kobayashi, H. Effects of OH Concentration and Temperature on NO Emission Characteristics of Turbulent Non-Premixed CH4/NH3/Air Flames in a Two-Stage Gas Turbine like Combustor at High Pressure. *Proc. Combust. Inst.* **2021**, *38*, 5163–5170. [CrossRef]
57. Somarathne, K.D.K.A.; Okafor, E.C.; Hayakawa, A.; Kudo, T.; Kurata, O.; Iki, N.; Kobayashi, H. Emission Characteristics of Turbulent Non-Premixed Ammonia/Air and Methane/Air Swirl Flames through a Rich-Lean Combustor under Various Wall Thermal Boundary Conditions at High Pressure. *Combust. Flame* **2019**, *210*, 247–261. [CrossRef]
58. Duynslaegher, C.; Jeanmart, H.; Vandooren, J. Flame Structure Studies of Premixed Ammonia/Hydrogen/Oxygen/Argon Flames: Experimental and Numerical Investigation. *Proc. Combust. Inst.* **2009**, *32*, 1277–1284. [CrossRef]

applied sciences

MDPI

Article

The Influence of Co-Firing Coal with Biomass Syngas on the Thermodynamic Parameters of a Boiler

Jin Wang, Qiaopeng Yao, Xiaoling Jin and Lei Deng *

State Key Laboratory of Multiphase Flow in Power Engineering, School of Energy and Power Engineering, Xi'an Jiaotong University, Xi'an 710049, China; 3122103009@stu.xjtu.edu.cn (J.W.); qiaopengyao@stu.xjtu.edu.cn (Q.Y.); j13778637284@stu.xjtu.edu.cn (X.J.)

* Correspondence: leideng@mail.xjtu.edu.cn; Tel./Fax: +86-29-8266-8703

Abstract: Biomass syngas can be considered as a supplementary fuel to partially substitute coal, which is beneficial to CO_2 emission reduction. For the case study, the influences of co-firing typical biomass syngas (gasification from palm, straw, and wood) with coal on the thermodynamic parameters of a 300 MW tangentially fired boiler are evaluated through a thermal calculation based on the principles of mass conservation, heat conservation, and heat transfer. The effects of boiler loads, biomass syngas species, and consumption rates are discussed. The results show that the introduction of biomass syngas weakens the radiative characteristics of the flame and reduces the furnace exit flue-gas temperature. As 3×10^4 m^3 h^{-1} of wood syngas is introduced, the decrement of thermal efficiency reaches 0.4%, while that of the coal consumption rate is 5.1%. The retrofitting of the boiler was not necessary and the corrosion of the low-temperature heating surface did not appear. The CO_2 annual emission reduction could achieve 0.001 to 0.095 million tons for palm syngas, 0.005 to 0.069 million tons for straw syngas, and 0.013 to 0.107 million tons for wood syngas with increasing biomass syngas consumption rates under the full load. Moreover, the main thermodynamic parameters changed more significantly under the low loads.

Keywords: coal-fired boiler; biomass syngas; co-firing; thermal calculation; thermodynamic parameter; peak regulation

1. Introduction

CO_2 emissions, which contribute to 76% of entire greenhouse gas emissions, can lead to a number of problems driven by global warming [1]. Moreover, power generation industries in China are facing tremendous pressure to mitigate CO_2 emissions, which should be lower than 550 g kW^{-1} h^{-1} [2]. Therefore, reducing CO_2 emissions has become a primary issue, especially emissions from the combustion of fossil fuels. Biomass can be adopted as an alternative energy resource to partially substitute coal in a boiler, which is a promising solution to reduce the emissions of CO_2 and NO_x, as well as improve the combustion characteristics of fuel owing to its renewability, CO_2 neutrality, and relatively high volatile-matter content [3–5].

Co-firing technologies include direct, indirect, and separate co-firing [6]. For direct co-firing, fuels are milled before entering the furnace. Although this method is a straightforward and economical option, it can create some challenges, such as the pre-treatment and feeding of biomass, fouling, and corrosion, which prevents it from being widely applied [7]. Separate co-firing can be adopted to increase the steam capacity of an existing boiler as a separate biomass-fired boiler is added to the entire system, which is costly and redundant [6]. For indirect co-firing, the biomass must be converted into syngas by a gasifier at first, and then the biomass syngas enters the furnace and is co-fired with the coal [8]. The interaction between alkali species from the biomass and the sulfur from the coal during co-firing might alleviate the ash deposition created by biomass combustion [9].

A high number of problems related to corrosion and deactivation can also be mitigated by indirect co-firing [7,10]. The distinct benefit of the indirect co-firing method is the preservation of some substances (containing K, Na, S, and Cl) that ordinarily contribute to the high-temperature corrosion of the water-cooled wall in the furnace. In addition, serious fouling of the heat exchangers produced by the inorganic components that contain K, Na, S, and Cl can be prevented [7]. Therefore, the indirect co-firing of coal with biomass syngas is an attractive scheme.

A great number of experimental [9,11–17] and numerical studies [8,18–22] have been performed on co-firing in a boiler. The researchers mainly focus on the impacts on NO_x emissions [12,19], slagging/ash deposition [9,13,14], combustion characteristics [15,16], and temperature distribution [8,18,19]. The variations in the main thermodynamic parameters, such as the radiant attenuation factor, furnace exit flue-gas temperature, thermal efficiency, and coal combustion rate, are not fully evaluated. Additionally, the thermal calculations of the principles of mass conservation, heat conservation, and heat transfer are precise and helpful in understanding the design, operation, and retrofit of boilers through the variations in the thermodynamic parameters [2]. However, only a few studies [23,24] (our previous studies) have been conducted to investigate the influence of co-firing on the thermodynamic parameters through thermal calculations, to our knowledge, but not for co-firing with the biomass syngas, which is beneficial to the reduction in CO_2 emissions. Moreover, it is significant to study the indirect co-firing method under different loads, especially a low load, in conjunction with the current deep-peaking issue. Biomass syngas, especially syngas from the gasification of woody biomass, contain low levels of nitrogen, virtually no sulfur, and have a relatively high heating value, which can lead to an apparent reduction in NO_x emissions [12]. Thus, it is beneficial to study the effect of co-firing as the boiler load alters.

In this study, three typical types of biomass syngases (from the gasifications of palm, straw, and wood) are chosen to investigate the impact of co-firing on the thermodynamic parameter variations in a 300 MW tangentially coal-fired boiler through thermal calculations. The influences of boiler loads, biomass syngas species, and consumption rates are discussed in detail. The solely coal-fired condition is employed as a reference to be compared with the other conditions. It is expected to provide a reference for the design, operation, and retrofit of boilers under co-firing conditions.

2. Thermal Calculation

The primary thermodynamic parameters, including the radiant attenuation factor, thermal efficiency, exhaust gas temperature (θ_{ex}, °C), and acidic dew point, were obtained through thermal calculations. Then, the heat distribution and effectiveness (including efficiency and reliability) of the boiler were evaluated. Since the 0-dimensional model has been employed in many previous studies [25–27], a semiempirical method based on it, proposed by the former Soviet Union and Russia, was adopted in this study [28,29]. It was widely applied for heat transfer calculations in the furnace, the basic equations of which were created according to the principles of heat conservation and heat transfer, as the heat transfer and combustion of the fuels occurred simultaneously in the furnace. By the principle of heat conservation, the heat absorption from the flue gas was considered to be equal to the enthalpy decreasing from the theoretical combustion temperature to the furnace exit flue-gas temperature. By the principle of heat transfer, the heat transfer calculation is equal to radiant heat transfer since the level of radiant heat transfer accounts for more than 95% of the total heat transfer in the furnace, which can be conducted through the Stephan–Boltzmann law. Since the heat obtained from heat conservation is equal to that from radiant heat transfer, some main thermodynamic parameters can be evaluated [28]. The similitude and differential theorems are adopted to build the model. The flowcharts of the thermal calculation procedures applied in this study are presented in Figure 1 [23]. First, the values of exhaust gas temperature (for the boiler), furnace exit flue-gas temperature (for the furnace), and outlet gas temperature (for the convective heating surfaces) should be assumed. Then, an estimated value of these thermodynamic parameters can be obtained according to the principles of mass conservation, heat conservation, and heat transfer. Finally, it is of vital importance to note that iterative calculations are required until the discrepancy between the estimated and fictitious exhaust gas temperatures of the boiler falls within a suitable range (±1 °C in this study). Based on the above procedures, a calculation program was proposed by our group [24]. The accuracy of the program can be verified since it has been applied in lots of manufacturers and research institutes [23,24]. The calculation methods of thermodynamic parameters described below are based on 1 kg of coal and the corresponding volume of biomass syngas, represented by x (m^3 kg^{-1}), which is correlated to the volume fraction of biomass syngas and the fuel compositions under the standard condition.

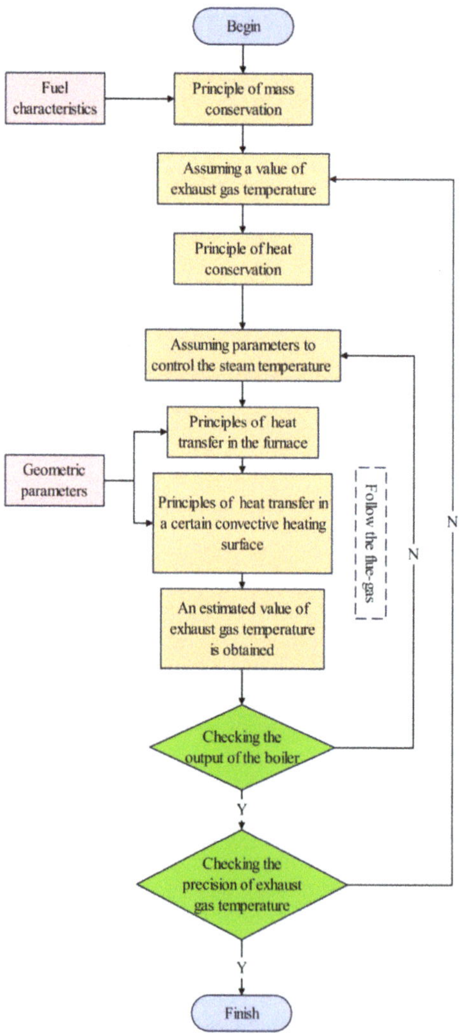

(a)

Figure 1. *Cont.*

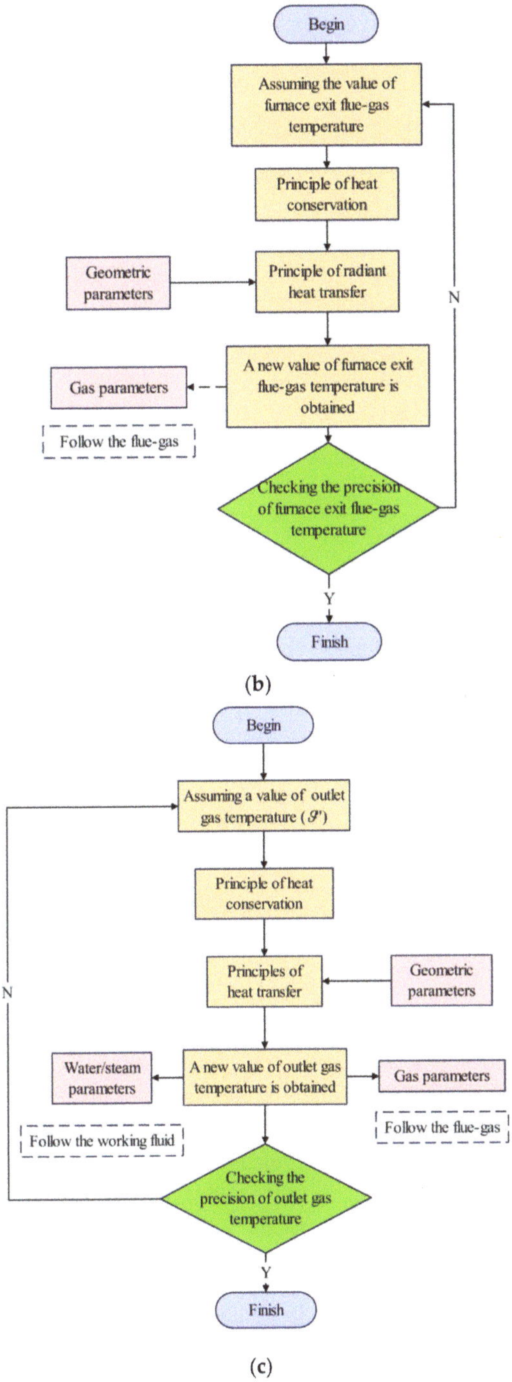

Figure 1. Flowcharts of the thermal calculation procedures [23]. (**a**) The boiler. (**b**) The furnace. (**c**) The convective heating surfaces.

2.1. Mass Conservation

To evaluate whether the retrofitting of boilers is necessary after the introduction of biomass syngas, the total air V_a and flue-gas flow rates V_f ($m^3 s^{-1}$) under the standard condition should be evaluated. The calculation methods are listed in Equations (1) and (2) [23, 28].

$$V_a = B_c\left(\beta_k V^0 + x\right) \tag{1}$$

$$V_f = B_c\left(V_y^0 + (\alpha_o - 1)V^0 + \frac{\rho d_{H_2O}}{18/22.4}(\alpha_o - 1)V^0\right) \tag{2}$$

where B_c ($kg\ s^{-1}$) represents the calculation coal consumption rate, the definition of which would be displayed in Section 2.2. The excess air coefficients at the inlet and outlet of the air heater are represented by β_k and α_o, respectively, β_k is equal to 1.49 in this study, the air density represented by ρ in Equation (2) is equal to 1.29 kg m^{-3} under the standard condition, and the ratio of water vapor mass to air mass (d_{H_2O}) is around 0.01.

The theoretical air volume represented by V^0 ($m^3\ kg^{-1}$) can be obtained by Equation (3), while that of flue gas (V_y^0, $m^3\ kg^{-1}$) is determined by Equation (4) [28].

$$V^0 = \frac{4}{45}C_{ar} + \frac{1}{30}(S_{ar} - O_{ar}) + \frac{50}{189}H_{ar} \tag{3}$$

$$V_y^0 = \frac{7}{375}C_{ar} + \frac{7}{1000}S_{ar} + \frac{1}{125}N_{ar} + \frac{79}{100}V^0 + V_{H_2O}^0 \tag{4}$$

where C_{ar}, S_{ar}, H_{ar}, O_{ar}, and N_{ar} (%) are the contents of carbon, sulfur, hydrogen, oxygen, and nitrogen on the as-received basis, respectively. $V_{H_2O}^0$ ($m^3\ kg^{-1}$) represents the theoretical water vapor volume, which is mainly determined by the components of fuels. It is confirmed from Equations (3) and (4) that V^0 and V_y^0 can be calculated as the fuel components are known.

The water vapor volume fraction (r_{H_2O}) is the ratio of the total water vapor volume (V_{H_2O}, $m^3\ kg^{-1}$) to the total flue-gas volume (V_y, $m^3\ kg^{-1}$), which can be defined as [28].

$$r_{H_2O} = \frac{V_{H_2O}}{V_y} = \frac{V_{H_2O}^0 + \frac{\rho d_{H_2O}}{18/22.4}(\alpha - 1)V^0}{V_y} \tag{5}$$

where α represents the excess air coefficient of the furnace exit (about 1.10–1.35 in this study). V_y can be obtained by [28].

$$V_y = V_y^0 + (\alpha - 1)V^0 + \frac{\rho d_{H_2O}}{18/22.4}(\alpha - 1)V^0 \tag{6}$$

2.2. Heat Conservation

Heat conservation, which is the balance between the heat entering and leaving the boiler envelope, is calculated to obtain the coal consumption rate and thermal efficiency of boilers [23,24,28]. The coal consumption rate B ($kg\ s^{-1}$) is defined as

$$B = 100Q_{av}/(\eta Q_r) \tag{7}$$

The total available heat of working fluid represented by Q_{av} (kW) is decided by the mass flow rate of heated steam and enthalpies of heated steam and feed water, which is closely related to boiler loads. η (%) is the thermal efficiency, which can be calculated by Equation (8) as the heat losses are given.

$$\eta = 100 - q_2 - q_3 - q_4 - q_5 - q_6 \tag{8}$$

where q_3 and q_4 (%) are the heat losses owing to incomplete combustion of biomass syngas and coal, respectively, q_5 and q_6 are the heat losses that are related to the radiation and convection, and ash residue, respectively. q_3, q_4, q_5, and q_6 are primarily determined by the fuel characteristics, boiler structure, and boiler capacity. q_2 is the heat loss owing to the exhaust gas, the value of which can be obtained by

$$q_2 = \left(I_{ex} - \alpha_o I_c^0\right)(100 - q_4)/Q_r \tag{9}$$

where I_c^0 and I_{ex} (kJ kg^{-1}) represent the enthalpies of cold air and exhaust gas, respectively. The heat entering the boiler envelope represented by Q_r (kJ kg^{-1}) is a combination of the sensible heat i_s and lower heating value $Q_{net,ar}$ (LHV) of the mixed fuels (kJ kg^{-1}), which can be calculated by

$$Q_r = Q_{net,ar} + i_s = Q_{net,ar}^{coal} + xQ_{net,ar}^{syngas} + i_s^{coal} + i_s^{syngas} \tag{10}$$

where i_s^{coal} and i_s^{syngas} are the sensible heats of coal and biomass syngas, while the LHV of coal and biomass syngas are represented by $Q_{net,ar}^{coal}$ (kJ kg^{-1}) and $Q_{net,ar}^{syngas}$ (kJ m^{-3}), respectively. It should be noted that i_s^{syngas} is correlated to the environment temperature.

Thus, the calculation coal consumption rate B_c mentioned in Equations (1) and (2) can be obtained by

$$B_c = B(1 - q_4/100) \tag{11}$$

2.3. Heat Transfer Calculations in the Furnace

The furnace exit flue-gas temperature (T_1'', K) is defined as [26].

$$T_1'' = T_a Bo^{0.6}/\left(M\widetilde{Bu}^{0.3} + Bo^{0.6}\right) \tag{12}$$

where Bo, M, and $\widetilde{Bu}$ are dimensionless coefficients. M is related to the fuel characteristics, combustion mode, and temperature distribution. The theoretical combustion temperature (T_a, K) depends on the available heat of fuels (Q_l, kJ kg^{-1}) that can be obtained by Equation (13) [23].

$$Q_l = Q_r\left(\frac{100 - q_3 - q_4 - q_6}{100 - q_4}\right) + Q_k \tag{13}$$

where Q_k (kJ kg^{-1}) represents the sensible heat of combustion air.

The radiant heat transfer, which is affected by the suspensions of triatomic gases and solid substances in the furnace, makes the primary contribution to the heat transfer of boilers. The radiative characteristics can be described with radiant attenuation factor (k, m^{-1} MPa^{-1}), including that of triatomic gases (k_q), soot ($k_{C,q}$), fly ash ($k_h\mu_h$), and char ($k_K\mu_K$), which is calculated by [28].

$$k = k_q + m_q k_{C,q}\frac{xQ_{net,ar}^{syngas}}{Q_{net,ar}} + (k_h\mu_h + k_K\mu_K)\frac{Q_{net,ar}^{coal}}{Q_{net,ar}} \tag{14}$$

where m_q is a dimensionless coefficient which is correlated to the proportion of the luminous area.

The heat absorption of flue gas (Q_{fl}, kW) is calculated by [28].

$$Q_{fl} = \varphi B_c\left(Q_l - I_{fl}''\right) \tag{15}$$

where φ is the heat retention factor (about 0.9–1.0), and the enthalpy of flue gas at the furnace exit is represented by I_{fl}'' (kJ kg^{-1}).

2.4. Heat Transfer Calculations of the Convective Heating Surfaces

The heat transfer of the convective heating surfaces, consisting of convective heat transfer and radiant heat transfer from the fluegas and flame, can be evaluated through Equations (16)–(18) [28].

K (W m^{-2} K^{-1}) represents the heat transfer coefficient, which can be determined by

$$K = 1/\left(\frac{1}{\xi(\alpha_d + \alpha_f)} + \varepsilon + \frac{1}{\alpha_2}\right) \tag{16}$$

where ξ is the utilization factor considering imperfect sweeping by flue gas, ε (m^2 K W^{-1}) is the fouling factor. The heat transfer coefficients in the flue-gas side include that of radiation and convection, which are represented by α_f and α_d (W m^{-2} K^{-1}), respectively. In the working fluid side, only convective heat transfer happens, while α_2 is the corresponding coefficient.

The temperature difference of heat transfer (Δt, K) is obtained by

$$\Delta t = \psi_t \frac{\Delta t_l - \Delta t_s}{\ln \frac{\Delta t_l}{\Delta t_s}} \tag{17}$$

where ψ_t is the temperature correction factor which is correlated to the terms of specific arrangement, Δt_l and Δt_s (K) are the larger one, and the smaller one of initial and final temperature differences between hot fluid and cold fluid, respectively.

The heat transfer quantity (Q_{tr}, W) can be determined by

$$Q_{tr} = KH\Delta t \tag{18}$$

where H is the area of convective heating surfaces (m^2).

2.5. Acidic Dew Point

To evaluate whether the corrosion would happen in the furnace, the acidic dew point (t_1, °C) should be calculated after coupling coal with biomass syngas, which can be obtained by [23,29].

$$t_1 = t_w + \beta \sqrt[3]{S_{ar,c}} / 1.05^{\alpha_{ash} A_{ar,c}} \tag{19}$$

the water dew point is represented by t_w (°C), β is correlated to the excess air coefficient at the furnace exit, and α_{ash} is related to fly ash (around 0.75–0.95). When the components and LHV of fuels are given, the converted contents of sulfur and ash represented by $S_{ar,c}$ and $A_{ar,c}$ (%), respectively, can be calculated.

2.6. CO_2 Annual Emission Reduction

The CO_2 annual emission reduction (m_r, t) can be determined by [24].

$$m_r = m_{t0} - m_t \tag{20}$$

where m_{t0} (t) is the CO_2 annual emission under the boiler maximum continuous rating (BMCR), and m_t (t) is the CO_2 annual emission, which is calculated by [24].

$$m_t = \frac{M_i \times W}{10^6} \tag{21}$$

The CO_2 annual emission is the product of CO_2 emission intensity (M_i, g kW^{-1} h^{-1}) and generation capacity (W, kW h). M_i can be calculated by [24].

$$M_i = \eta_{ce} C_{ar} B_c \frac{33 \times 10^2}{25P} \tag{22}$$

where P (MW) represents the power generation, and η_{ce} (%) is the combustion efficiency determined by the structure, capacity, and operation of boilers, which is equal to 0.99 in this study.

3. Case Study

A subcritical 300 MW tangentially fired boiler, the type of which is SG-1025/17.50-M749, is applied for the case study. The intricate boiler structure can be seen from Figure 2. The main design parameters of the boiler under BMCR, 75% THA (turbine heat acceptance), 50% THA, and 30% THA are shown in Table 1. The design parameters are adopted under (BMCR), with the main steam output of 1025 t h^{-1}, and the outlet temperatures of both superheated steam and reheated steam of 540 °C. The height, width, and depth of the furnace are 50,868, 14,020, and 12,350 mm, respectively. The wall reheater, platen superheaters, and high-temperature superheaters are sequentially arranged in the upper part of the boiler, while the horizontal convective superheater and economizer are installed in the tail flue vibration from top to bottom. The chosen biomass syngases in this study are palm syngas, straw syngas, and wood syngas from previous studies [30–32]. The biomass syngas can be obtained through the gasification of the biomass by a gasifier. The processes consist of desiccation, volatile release, oxidation, and reduction. Then, the tar and dust removal are required at high temperatures. Finally, the biomass syngas should be cooled at 450 °C before entering the furnace since the temperature of biomass syngas should be controlled below 500 °C [2]. The biomass syngas are injected into the furnace at the bottom level of the reburn zone in this study since NO emissions were at their lowest value in this condition according to the results in our previous study [8]. The details about the coal and biomass syngases are listed in Table 2.

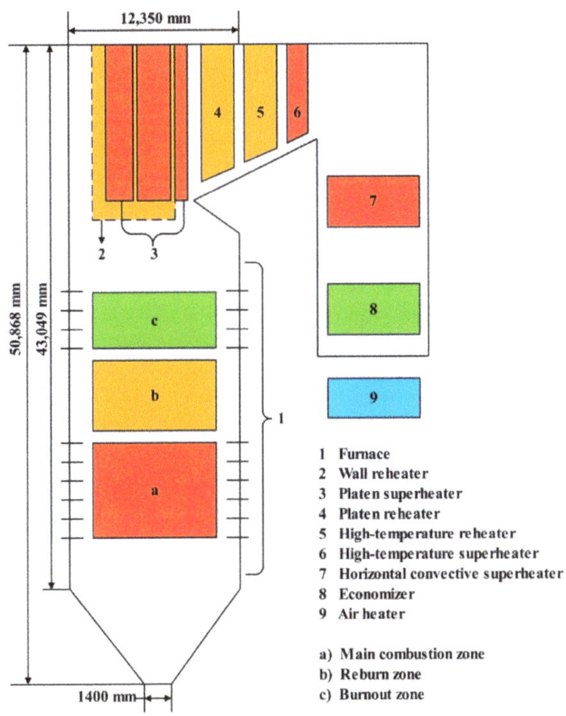

Figure 2. Boiler structure.

Table 1. Main design parameters of the boiler.

	Value			
	BMCR	**75% THA**	**50% THA**	**30% THA**
Power generation (MW)	330	225	150	113
Main steam output (t h^{-1})	1025	662	441	335
Main steam pressure (MPa)	17.47	14.45	9.79	6.23
Outlet temperature of superheated steam (°C)	540	540	540	540
Reheated steam output (t h^{-1})	880.0	582.4	395.6	313.7
Outlet temperature of reheated steam (°C)	540	540	540	535
Outlet pressure of reheated steam (MPa)	3.68	2.44	1.63	1.25
Desuperheating water output (t h^{-1})	0	0	0	0
Desuperheating water temperature (°C)	193	176	160	140
Excess air coefficient of furnace exit	1.25	1.25	1.25	1.35
Exhaust gas temperature (°C)	114	108	108	106
Thermal efficiency (%)	93.88	94.13	92.61	92.14

Table 2. Fuel characteristics of coal (as-received basis) and biomass syngases.

Fuel	**Ultimate Analysis, wt.%**					**Proximate Analysis, wt.%**			**LHV**
	C_{ar}	H_{ar}	O_{ar}	N_{ar}	S_{ar}	A_{ar}	M_{ar}	FC_{ar}	V_{ar}	
Coal	47.49	3.25	4.97	0.80	0.98	36.21	6.30	36.95	20.54	18,460 kJ kg^{-1}
	Reference				Volume fraction of components (%)					
					H_2	CO	CO_2	N_2	CH_4	
Wood Syngas	[30] (pp. 4196–4205)				17.80	20.30	8.30	51.90	1.70	5300 kJ m^{-3}
Straw Syngas	[31] (pp. 0025–0028)				5.60	17.89	10.86	55.60	10.05	3906 kJ m^{-3}
Palm Syngas	[32] (pp. 0491–0501)				9.60	25.30	8.20	55.70	1.20	4800 kJ m^{-3}

Since the fuel burnt in the boiler changes from coal to biomass syngas and coal, the thermal calculations are performed with the verification method. The comparisons of thermodynamic parameters for various biomass syngas consumption rates are conducted under a full load. The biomass syngas consumption rates coupled with coal vary from 0 to 5×10^4 m^3 h^{-1}. Moreover, the comparisons of thermodynamic parameters between burning coal alone and co-firing coal with 3×10^4 m^3 h^{-1} of wood syngas (taking the safety and economy into consideration) are performed with different boiler loads. The results of thermal calculations are similar to that the boiler manufacturer provided, which means the thermodynamic parameters calculated by our developed program are reliable and valid. The following assumptions are made after the introduction of biomass syngas:

(1) The excess air coefficient remains unchanged at the exits of the furnace and convective heating surfaces;
(2) The temperatures of the cold air and hot air, and the proportions of primary and secondary air remain unchanged;
(3) No variations happen in the fuel characteristics, boiler structure, and boiler capacity.

4. Results and Discussion

4.1. Effect of Co-Firing on Radiant Heat Transfer

The primary thermodynamic parameters of the boiler under BMCR are shown in Figure 3. The heating value of the fuel, the quantity, and the specific heat of the combustion products determine the theoretical combustion temperature, which represents the highest temperature that the combustion products can achieve [28]. The available heat of fuels is correlated to the volume, specific heat, and theoretical combustion temperature if the fuel is burnt completely and adiabatically. It can be seen from Figure 3a that the available heat would rise monotonically, as the consumption rates and heating value of biomass syngas increase. The result is the same as that in our previous study [23]. Compared with the other two types of biomass syngases, the increment of available heat can reach 10.8% after introducing wood syngas with 5×10^4 m^3 h^{-1}. It can be noted from Figure 3b that the theoretical combustion temperature increases with ascending heating values, while the trend reverses with increasing consumption rates. The result is similar to that in the study of Wang et al. [2]. The reason for the variations in volume and specific heat might be different compositions and quantities of combustion products. The variation in the straw syngas consumption rate has the most pronounced influence on the theoretical combustion temperature, the highest decrement of which is 3.8%.

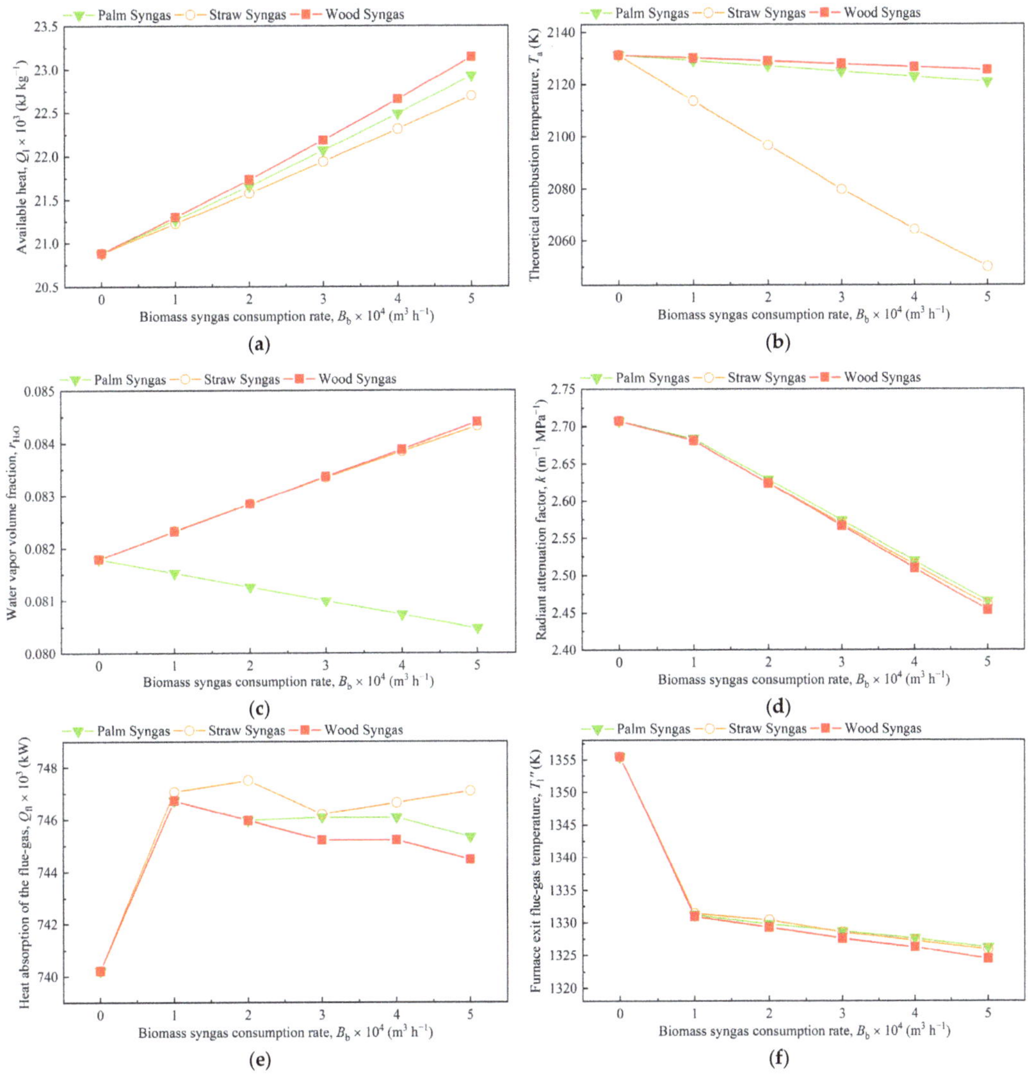

Figure 3. Furnace thermodynamic parameters for various biomass syngas consumption rates (BMCR). (**a**) Available heat of fuel; (**b**) theoretical combustion temperature; (**c**) water vapor volume fraction in flue gas; (**d**) radiant attenuation factor of flue gas; (**e**) heat absorption of flue gas; (**f**) furnace exit flue-gas temperature.

It is believed that the radiant attenuation factor of flue gas (k, m^{-1} MPa^{-1}) can be applied to represent the radiative characteristics of flame [28]. The water vapor volume fraction rises with increasing biomass syngas consumption rates after the introductions of straw and wood syngases, while that decreases after the introduction of palm syngas (see Figure 3c). The radiant attenuation factor is related to that of triatomic gases (k_q), soot ($k_{C,q}$), fly ash ($k_h\mu_h$), and char ($k_K\mu_K$) according to Equation (14). The $k_{C,q}$ for a coal-fired boiler can be disregarded, while $k_K\mu_K$ would not change owing to the unchanged coal characteristics [28]. The k_q rises as a result of the rising volume fraction of CO_2, while $k_h\mu_h$ decreases due to the decreasing fly ash concentration after the introduction of biomass syngas. Consequently, the k descends as the biomass syngas consumption rate rises according to Figure 3d.

The heat absorption of flue gas (Q_{fl}, kW) is determined by the average temperature and radiative characteristics of the flame, and the size and shape of the furnace [24]. The average temperature of the flame decreases due to the falling theoretical combustion temperature. Compared with the solely coal-fired condition, the heat absorption ascends after introducing the biomass syngas. For the introduction of wood syngas, the heat absorption descends with rising biomass syngas consumption rates. However, the heat absorption for introducing palm and straw syngases does not present a monotonous trend (see Figure 3e), since the complex combustion products produced by burning mixed fuels can contribute to radiant heat transfer in numerous ways [28]. Since the difference between the available heat and heat absorption of flue gas ascends, the enthalpy of flue gas, which reflects the variations of the volume, specific heat, and temperature in flue gas, rises as the biomass syngas consumption rate increases. It can be seen that T_1'' declines with ascending biomass syngas consumption rates in Figure 3f. T_1'' decreases by 2.1% for palm syngas and straw syngas, while that for wood syngas can reach 2.3%, as 5×10^4 m^3 h^{-1} of biomass syngas is entrained.

To evaluate the influence of boiler load variation on the thermodynamic parameters after the introduction of biomass syngas, the consumption rate of the biomass syngas would not change (3×10^4 m^3 h^{-1} in this study) because of the difficulty of adjusting the load of the gasifier. The comparisons of the primary thermodynamic parameters between burning coal alone and co-firing coal with biomass syngas (3×10^4 m^3 h^{-1} of wood syngas) under different boiler loads are revealed in Figure 4. The results demonstrate that the available heat and water vapor volume fraction ascend after coupling the wood syngas, while the radiant attenuation factor and furnace exit flue-gas temperature descend after the introduction of syngas. Meanwhile, the influence of wood syngas introduction on the theoretical combustion temperature and heat absorption is weakened. For the solely coal-fired condition, the available heat declines with falling boiler loads, while that for the wood syngas introduction shows a reverse trend. It could be confirmed that the introduction of wood syngas affects the available heat more significantly under a low load (see Figure 4a). The syngas from the gasification of the woody biomass contains low levels of nitrogen, a high volume fraction of combustible materials, and a relatively high heating value, which would result in a higher water vapor volume fraction after the introduction of wood syngas (see Figure 4c). As the consumption rate of biomass syngas remains unchanged, operating the boiler under lower loads means a higher consumption rate percentage of biomass syngas than that of coal. Thus, the water vapor volume fraction increases more dramatically under low boiler loads. The variation in the water vapor volume fraction would affect that of the radiant attenuation factor, which can be seen in Figure 4d. The difference ratios are 4.6% for the water vapor volume fraction and 12.1% for the radiant attenuation factor under 30% THA. The other three thermodynamic parameters decrease when the boiler load descends (see Figure 4b,e,f).

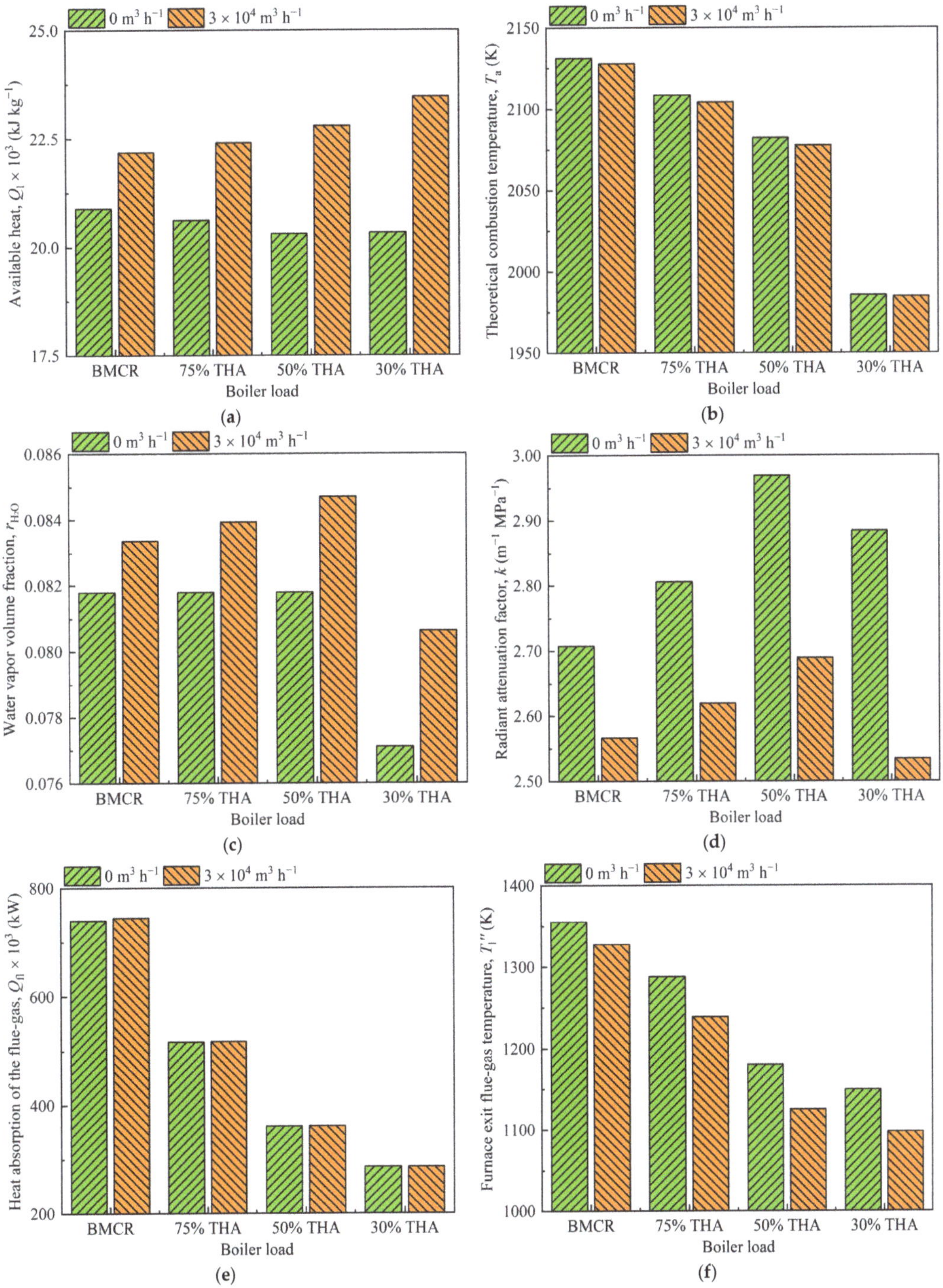

Figure 4. Furnace thermodynamic parameters comparison under various boiler loads (wood syngas, 450 °C). (**a**) Available heat of fuel; (**b**) theoretical combustion temperature; (**c**) water vapor volume fraction in flue gas; (**d**) radiant attenuation factor of flue gas; (**e**) heat absorption of flue gas; (**f**) furnace exit flue-gas temperature.

4.2. Effect of Co-Firing on Heat Transfer of the Convective Heating Surfaces

The heat transfer coefficient (K), temperature difference (Δt), and heat transfer quantity (Q_{tr}) of a certain convective heating surface with various biomass syngas consumption rates under BMCR are displayed in Figure 5. K is affected by many factors, such as the radiative characteristics of the flue gas, and the physical characteristics and velocity of the flue gas and working fluid. Since the convective heat transfer on the working fluid side is scarcely affected, K is primarily impacted by the heat transfer (including convective and radiant heat transfer) on the flue-gas side. The inflection point occurs when only the coal is burned, the result is similar to that in our previous study [24]. The heat transfer coefficient decreases (for the palm and wood syngases) or increases (for the straw syngas) with ascending biomass syngas consumption rates (see Figure 5a). A possible explanation might be the alterations of the temperature, component, and quantity. The temperature difference is influenced by the inlet and outlet temperatures of the working fluid and flue gas of a certain heating surface. The temperature of the working fluid could be barely impacted. Although T_1'' declines with rising biomass syngas consumption rates, the flue-gas temperature of the low-temperature superheater rises slightly. Thus, Δt climbs moderately in Figure 5b. Q_{tr} is related to K, Δt, and geometric parameters according to Equation (18), while the geometric parameters remain unchanged. The variations in the heat transfer quantity are displayed in Figure 5c. It can be seen from Figure 6 that all of these three thermodynamic parameters decrease when the boiler load declines, while the introduction of the wood syngas could not lead to a dramatic variation.

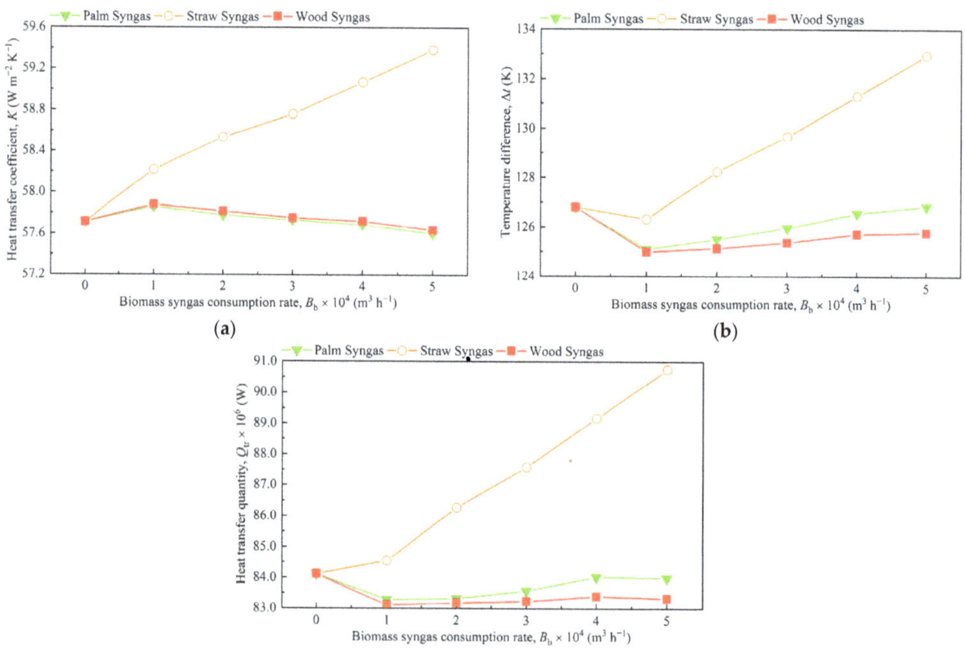

(a) (b) (c)

Figure 5. Some parameters related to heat transfer of a low-temperature superheater under various biomass syngas consumption rates (BMCR). (**a**) Heat transfer coefficient; (**b**) temperature difference; (**c**) heat transfer quantity.

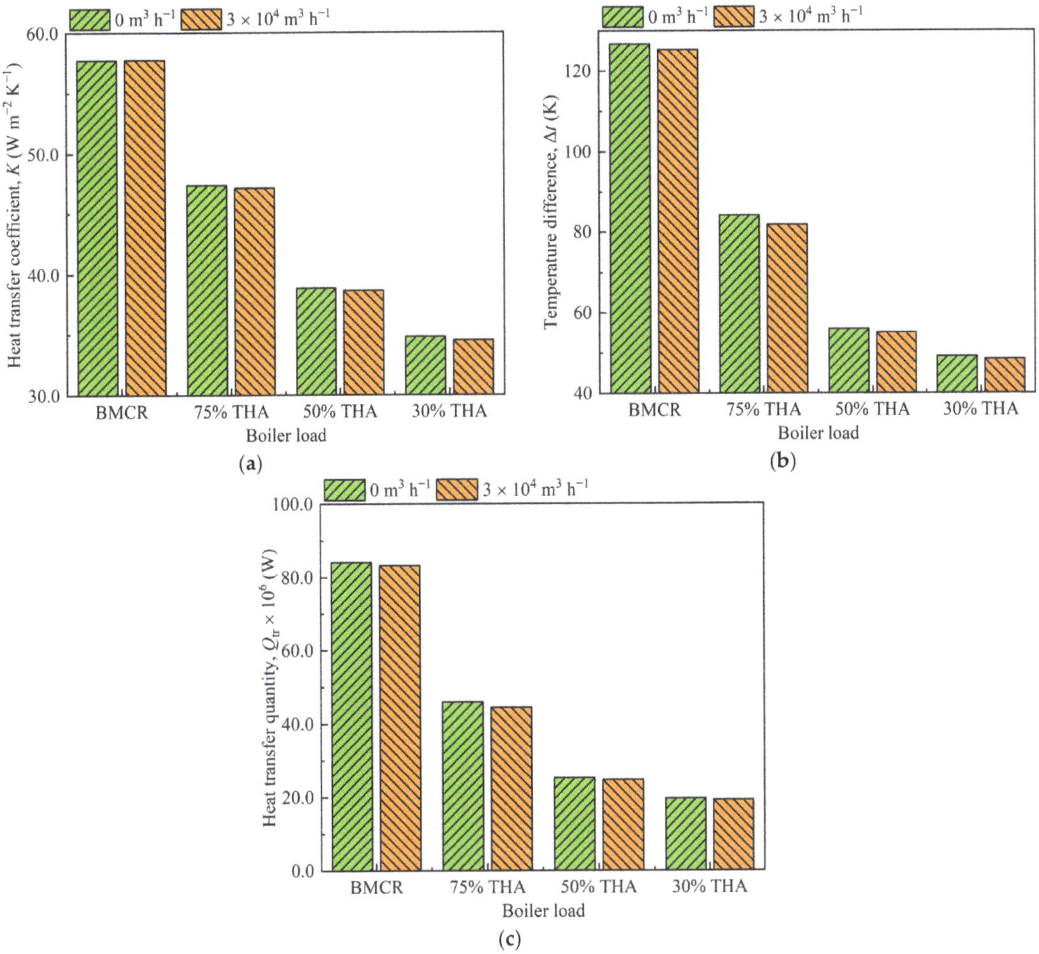

Figure 6. Some parameters comparison related to heat transfer of a low-temperature superheater under various boiler loads (wood syngas, 450 °C). (**a**) Heat transfer coefficient; (**b**) temperature difference; (**c**) heat transfer quantity.

4.3. Effect of Co-Firing on Thermal Efficiency

Figure 7 reveals the results of the exhaust gas temperature, thermal efficiency, and coal consumption rate through heat balance calculations. As the biomass syngas is entrained, the alterations of the component, velocity, volume, and physical characteristics of the flue gas, which happen in the processes of convective and radiant heat transfer in the boiler, would result in the exhaust gas temperature variation according to Figure 7a. The exhaust gas temperature increases with increasing biomass syngas consumption rates. The highest variations of the exhaust gas temperature for various consumption rates under a full load are 9 °C for the wood syngas, 10 °C for the palm syngas, and 14 °C for the straw syngas. It can be inferred from Equations (8) and (9) and the assumption made in Section 3 that the thermal efficiency (η, %) is mainly determined by the heat losses owing to the exhaust gas (q_2, %), while the other heat losses remain unchanged. The growth of the exhaust gas temperature would lead to the growth of q_2, and then the thermal efficiency would descend with increasing biomass syngas consumption rates (see Figure 7b). A combination of the available heat of the working fluid and the thermal efficiency would reduce the coal consumption rate, as the biomass syngas consumption rate increases (see Figure 7c). As

3×10^4 m^3 h^{-1} of wood syngas is entrained, the thermal efficiency is reduced by 0.4%, while the coal consumption rate is declined by 5.1%. The variations of the exhaust gas temperature and thermal efficiency under various coupling percentages of the fuels between this current study and the previous studies [23,24] are quite different due to the different fuels co-firing with the coal, while it could be confirmed from these three studies [23,24] that the coal consumption rate would significantly descend with increasing coupling percentages of the fuels.

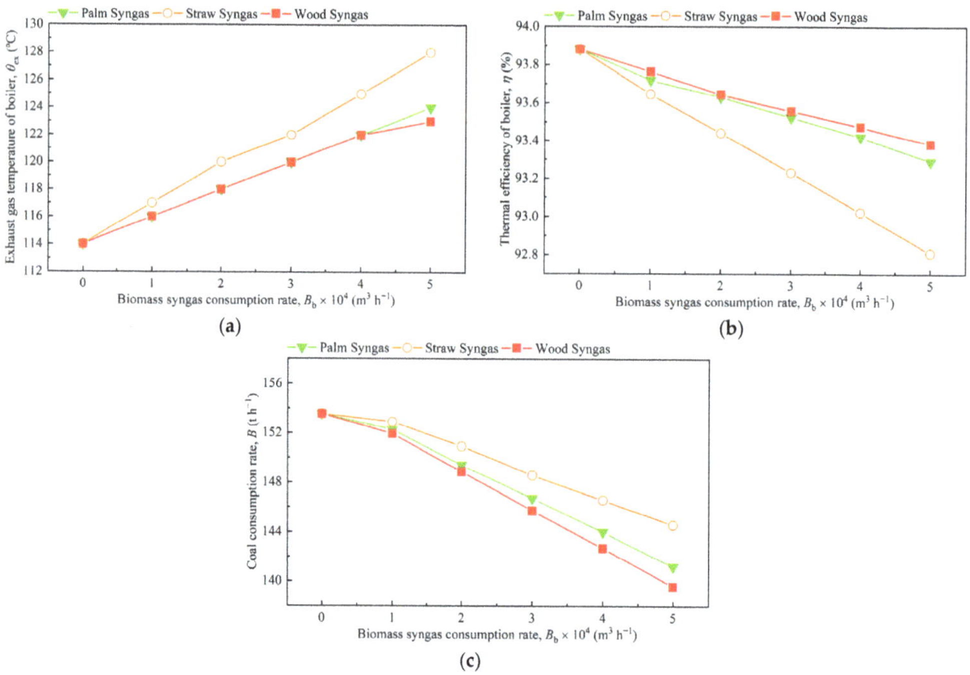

Figure 7. Exhaust gas temperature, thermal efficiency and coal consumption rate under various biomass syngas consumption rates (BMCR). (**a**) Exhaust gas temperature of boiler; (**b**) thermal efficiency of boiler; (**c**) coal consumption rate.

The alterations of the exhaust gas temperature, thermal efficiency of the boiler, and coal consumption rate under various boiler loads after introducing 3×10^4 m^3 h^{-1} of the wood syngas are shown in Figure 8. Coupling wood syngas with the coal would lead to the growth of the exhaust gas temperature (see Figure 8a), but reduce the thermal efficiency of the boiler and the coal consumption rate (see Figure 8b,c). The alterations of the exhaust gas temperature and the thermal efficiency are not monotonous as the boiler load is changed. However, the increment (for the exhaust gas temperature) or the decrement (for the thermal efficiency and the coal consumption rate) rises monotonously with declining boiler loads. When the boiler operates under 30% THA, the exhaust gas temperature is increased by 7.5%, while η and B are decreased by 0.5% and 13.2%, respectively.

4.4. *Effect of Co-Firing on Operational Safety*

It can be determined from Equation (19) that the acidic dew point depends on the water dew point, excess air coefficient, and converted contents of sulfur and fly ash [23]. The introduction of biomass syngas brings a large increment of the volume fraction of hydrogen, and thus the content of the water vapor rises. Therefore, the water dew point ascends with the growth of the biomass syngas consumption rate. The contents of both sulfur and fly ash decrease as the biomass syngas is entrained. A combination of these factors mentioned above causes the acidic dew point to climb (for the straw and wood syngases) or decline (for the palm syngas), the variation in which is within ±1 °C (see Figure 9a). It can be clearly seen from Figures 7a and 9a that the exhaust gas temperature has a greater value than the acidic dew point. Therefore, the low-temperature heating surface would not be corroded after the introduction of biomass syngas.

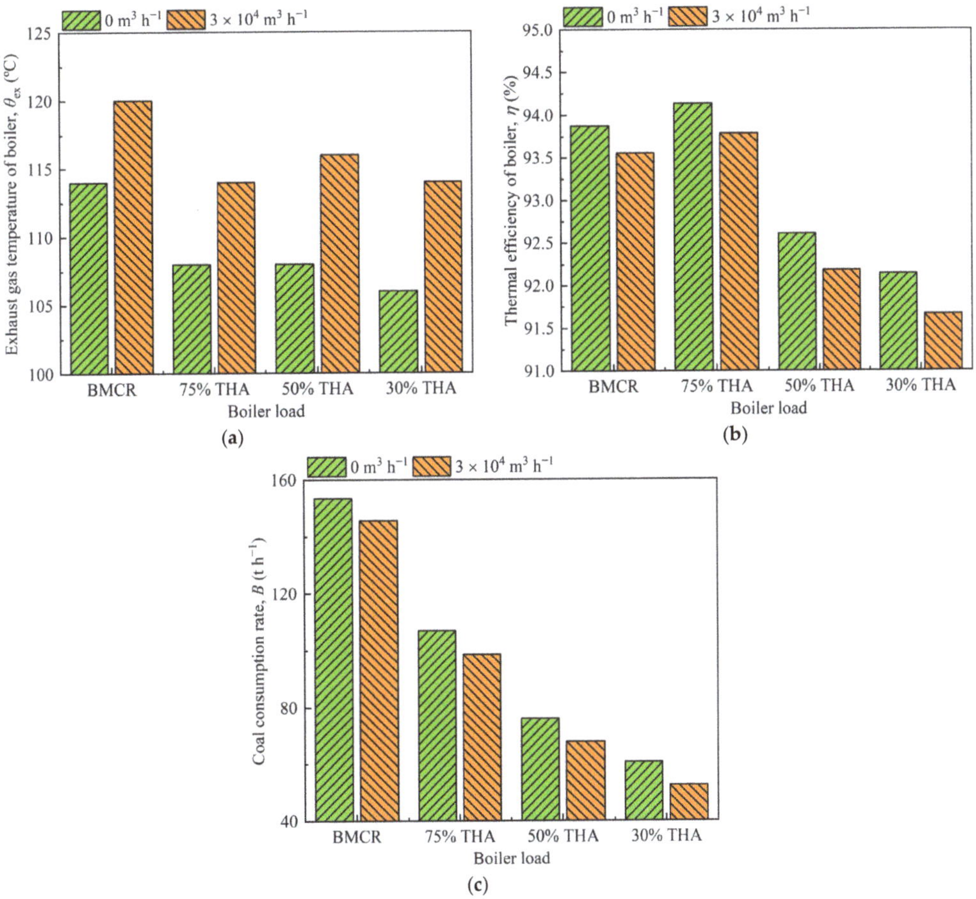

Figure 8. Exhaust gas temperature, thermal efficiency, and coal consumption rate comparison under various boiler loads (wood syngas, 450 °C). (**a**) Exhaust gas temperature of boiler; (**b**) thermal efficiency of boiler; (**c**) coal consumption rate.

Figure 9b,c show the total air and flue-gas flow rates evaluated under the standard condition for various biomass syngas consumption rates. These two parameters are influenced by the coal consumption rate and theoretical volumes of the air and flue gas [23,28]. After coupling biomass syngas with coal, the theoretical volumes of the air and flue gas ascend with increasing biomass syngas consumption rates, while the coal consumption rate descends. Therefore, the variations in the total air and flue-gas flow rates are within ±4.0% and ±7.6%, respectively. Consequently, the retrofitting of the boiler is not required.

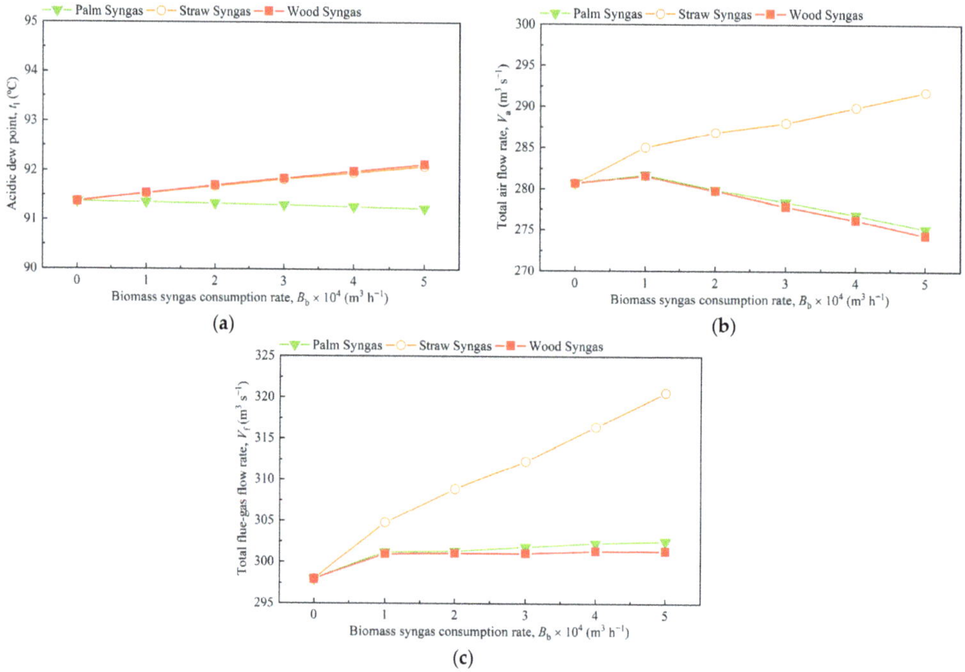

Figure 9. Acidic dew point, total air flow rate, and total flue-gas flow rate with various biomass syngas consumption rates (BMCR, standard condition). (**a**) Acidic dew point; (**b**) total air flow rate of air heater inlet; (**c**) total flue-gas flow rate of air heater outlet.

It could be implied from Figures 8a and 10a that although the acidic dew point would rise slightly after the introduction of wood syngas under various boiler loads, the acidic dew point is still lower than the exhaust gas temperature. Consequently, no corrosion occurs on the low-temperature heating surface even under a low load. Compared with the standard condition, the total air and flue-gas flow rates do not alter dramatically as 3×10^4 m^3 h^{-1} of wood syngas is entrained. However, the total air and flue-gas flow rates decline significantly with decreasing boiler loads (see Figure 10b,c).

4.5. Effect of Co-Firing on CO_2 Emissions

The CO_2 emissions are related to the content of carbon and the calculation coal consumption rate [24]. Therefore, the CO_2 emissions descend gradually due to the drop of the calculation coal consumption rate with ascending biomass syngas consumption rates. Then, the CO_2 emission intensity of power generation and the CO_2 annual emission (m_t, t) decline according to Equations (21) and (22), the result is displayed in Figure 11a. Consequently, an obvious growth of the CO_2 annual emission reduction (m_r, t) for different biomass syngas consumption rates under BMCR can be seen in Figure 11b. The results in Figure 11a,b are the same as those in our previous study [24], which could confirm that co-firing the fuels, such as biomass syngas and hydrogen-derived fuel, with the coal is significantly beneficial to CO_2 emission reduction. The CO_2 annual emission reduction equals 0.001 to 0.095 million tons for the palm syngas, 0.005 to 0.069 million tons for the straw syngas, and 0.013 to 0.107 million tons for the wood syngas.

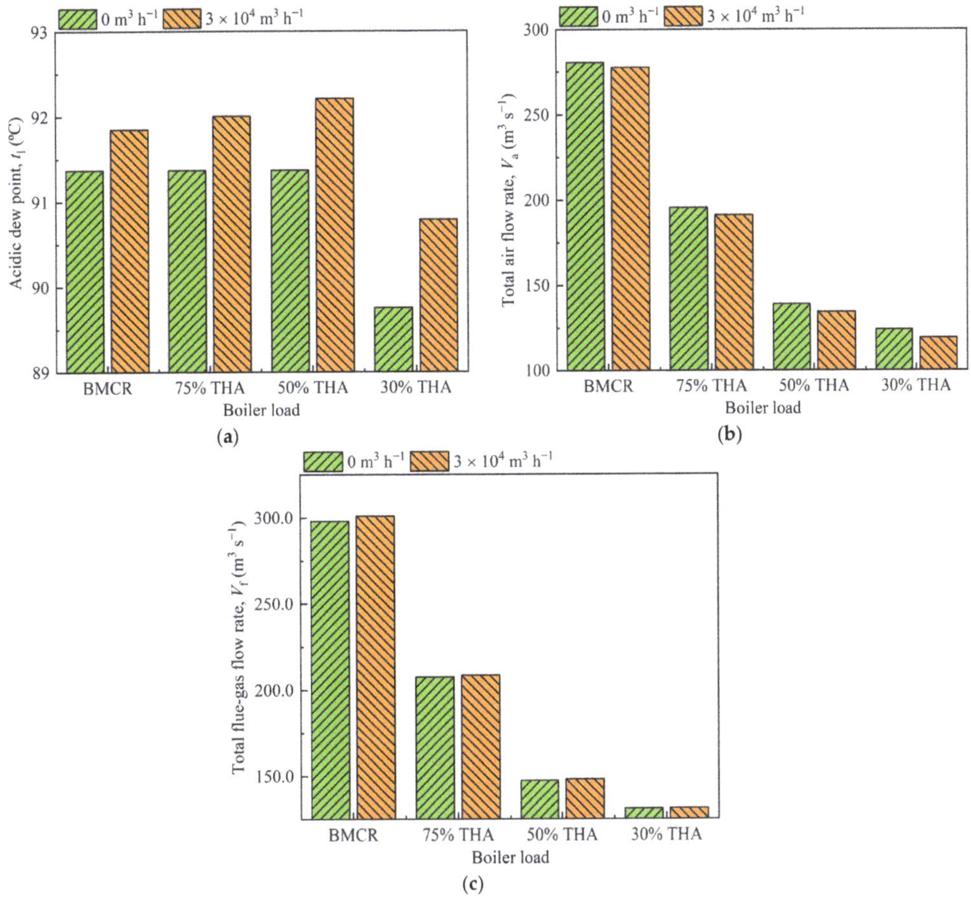

Figure 10. Acidic dew point, total air flow rate, and total flue-gas flow rate comparison with various boiler loads (wood syngas, 450 °C, standard condition). (**a**) Acidic dew point; (**b**) total air flow rate of air heater inlet; (**c**) total flue-gas flow rate of air heater outlet.

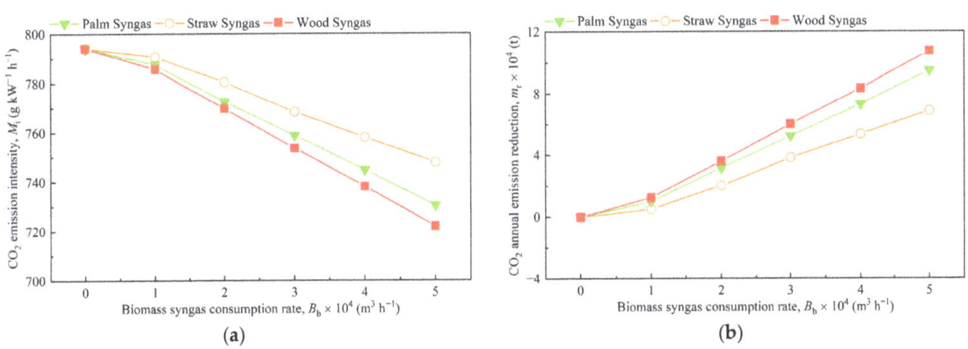

Figure 11. CO_2 emission intensity and CO_2 annual emission reduction for various biomass syngas consumption rates (BMCR). (**a**) CO_2 emission intensity; (**b**) CO_2 annual emission reduction.

As 3×10^4 m^3 h^{-1} of wood syngas is entrained, the CO_2 emission intensity decreases and the CO_2 annual emission reduction rises compared with the standard condition. It can be seen from Figure 12a that the decrement of the CO_2 emission intensity increases with descending boiler loads, which could achieve 13.2% under 30% THA. It is apparent from Figure 12b that the CO_2 annual emission reduces more under a low load, while the introduction of wood syngas would exacerbate the reduction.

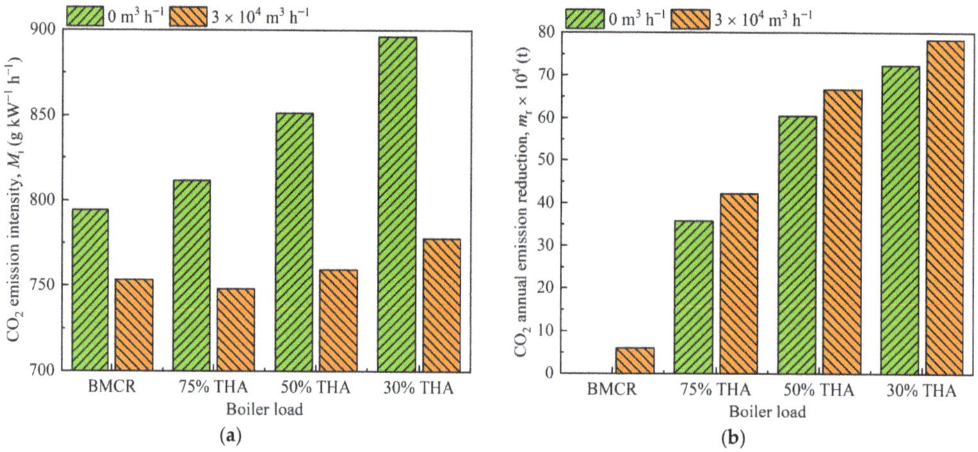

Figure 12. CO_2 emission intensity and CO_2 annual emission reduction comparison under various boiler loads (wood syngas, 450 °C). (**a**) CO_2 emission intensity; (**b**) CO_2 annual emission reduction.

4.6. Effect of Co-Firing on the Boiler during Peak Regulation

Figure 13 displays the variations of some thermodynamic parameters under different boiler loads. The result in Figure 13a suggests that the radiant attenuation factor decreases more significantly after coupling wood syngas with coal under a low load. The radiant attenuation factor increases gradually and then decreases when the boiler load descends, while the inflection point occurs under 50% THA. As the boiler load is reduced, the heat transfer coefficient declines dramatically in Figure 13b. It could be easily concluded from the result in Figure 13c that the thermal efficiency of the boiler does not alter monotonously with decreasing boiler loads. The thermal efficiency rises (from 100% BMCR to 75% THA) and then decreases (from 75% THA to 30% THA) when the boiler load declines. Co-firing coal with wood syngas would reduce the thermal efficiency slightly. Although the introduction of wood syngas would increase the acidic dew point, the difference between the acidic dew point and the exhaust gas temperature can be enlarged, which means the boiler would operate more safely. The differences between these two parameters are higher than 20 K under various boiler loads after coupling wood syngas with the coal. It could be revealed from Figures 13d and 8a. What could be significantly concluded from Figure 13e is that co-firing coal with the wood syngas under a low load would increase the CO_2 annual emission reduction apparently, which is beneficial for the environment.

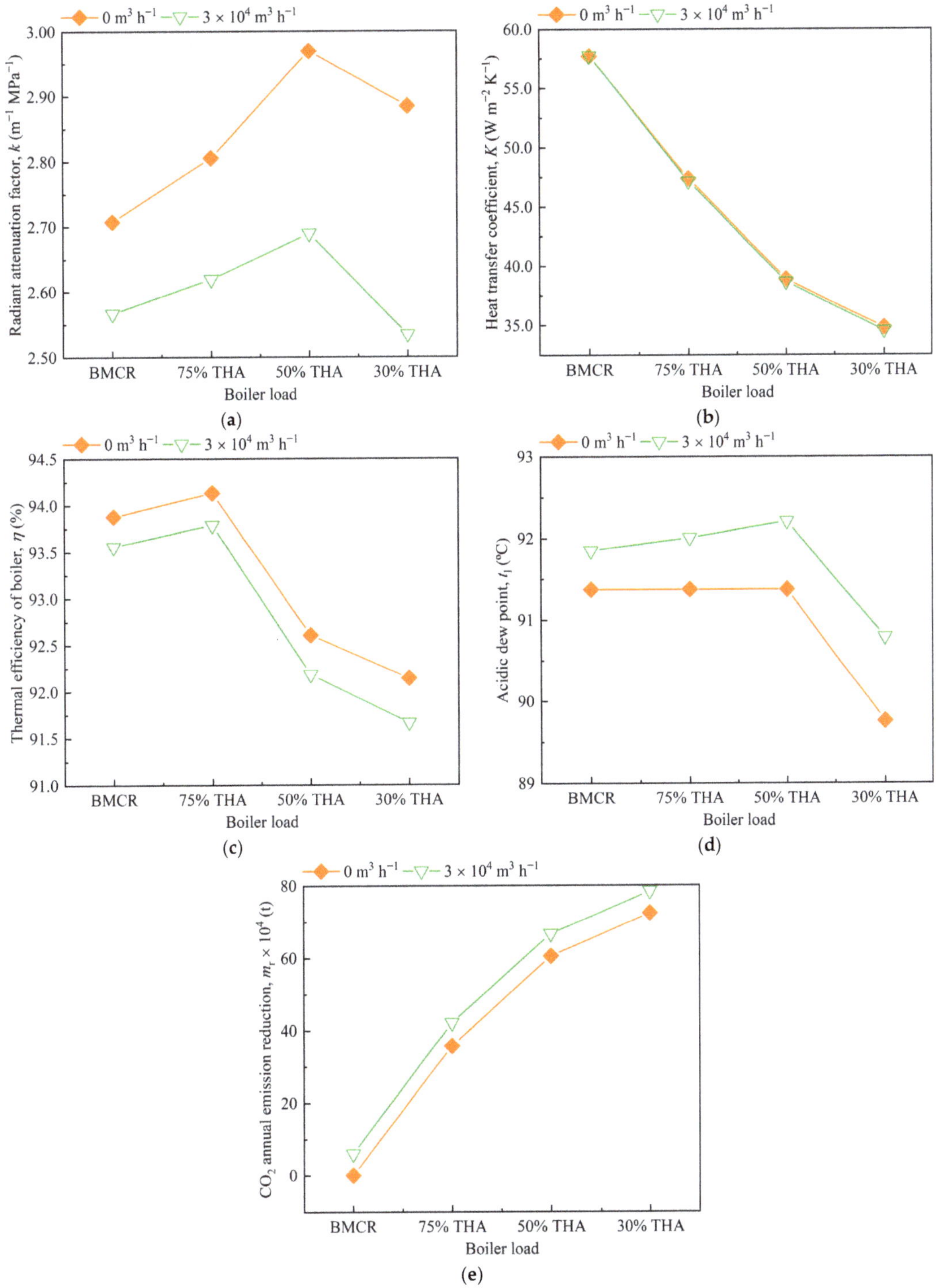

Figure 13. Some thermodynamic parameters variations under different boiler loads (wood syngas, 450 °C). (**a**) Radiant attenuation factor; (**b**) heat transfer coefficient; (**c**) thermal efficiency of boiler; (**d**) acidic dew point; (**e**) CO_2 annual emission reduction.

5. Conclusions

The thermal calculations are conducted to investigate the effect of co-firing coal with biomass syngas on the thermodynamic parameters in a 300 MW tangentially fired boiler in this study. The biomass syngas is applied as a supplementary fuel entering the furnace with various consumption rates under BMCR. In addition, the comparisons of thermodynamic parameters between two typical conditions are performed under different boiler loads. The conclusions are listed as follows:

(1) The introduction of biomass syngas weakens the average temperature in the furnace and the radiative characteristics of flame, which leads to a reduction of furnace exit flue-gas temperature. The alterations of these thermodynamic parameters are monotonically related to the biomass syngas consumption rate;
(2) The heat absorption of flue gas would be enhanced after the introduction of biomass syngas. Coupling palm syngas and wood syngas with coal would increase the heat transfer on the convective heating surfaces, while that for the straw syngas would be weakened. Thus, a combination of these two parameters leads to an increment in the exhaust gas temperature of the boiler and a decrement in the thermal efficiency of the boiler;
(3) The variations of acidic dew point, total air, and flue-gas flow rates are not monotonous with increasing biomass syngas consumption rates. The variations are within ± 1 °C, $\pm 4.0\%$, and $\pm 7.6\%$, respectively. The difference between the acidic dew point and exhaust gas temperature ascends after the introduction of biomass syngas. Therefore, the retrofitting of the boiler is not necessary and the corrosion of a low-temperature heating surface would not appear. The coal consumption rate descends dramatically after coupling;
(4) The CO_2 annual emission reduction increases dramatically with increasing biomass syngas consumption rates under BMCR, which equals 0.001 to 0.095 million tons for palm syngas, 0.005 to 0.069 million tons for straw syngas, and 0.013 to 0.107 million tons for wood syngas. Although the variations are not monotonous, the introduction of wood syngas enlarges the difference of these thermodynamic parameters between two different consumption rates of wood syngas as the boiler load decreases.

Author Contributions: Conceptualization, J.W., Q.Y., X.J. and L.D.; Data Curation, J.W., Q.Y. and L.D.; Investigation, J.W., Q.Y., X.J. and L.D.; Resources, L.D.; Supervision, L.D.; Writing—Original Draft, J.W., Q.Y. and X.J.; Writing—Review and Editing, L.D. All authors have read and agreed to the published version of the manuscript.

Funding: This research was funded by the National Natural Science Foundation of China, grant number [51406147].

Institutional Review Board Statement: Not applicable.

Informed Consent Statement: Not applicable.

Data Availability Statement: The data that support the findings of this study are available from the corresponding author upon reasonable request.

Conflicts of Interest: The authors declare no conflict of interest.

Nomenclature

A_{ar}	content of ash on the as-received basis (%)
$A_{ar,c}$	converted content of ash (%)
B	coal consumption rate (kg s^{-1})
B_c	calculation coal consumption rate (kg s^{-1})
Bo	Boltzmann number
$\widetilde{Bu}$	effective value of Bouguer number
C_{ar}	content of carbon on the as-received basis (%)
d_{H_2O}	the ratio of water vapor mass to air mass

FC_{ar}	content of fixed carbon on the as-received basis (%)
H	area of convective heating surfaces (m^2)
H_{ar}	content of hydrogen on the as-received basis (%)
i_s	sensible heat of mixed fuels (kJ kg^{-1})
i_s^{coal}	sensible heat of coal (kJ kg^{-1})
i_s^{syngas}	sensible heat of biomass syngas (kJ kg^{-1})
I_c^0	enthalpy of cold air (kJ kg^{-1})
I_{ex}	enthalpy of exhaust gas (kJ kg^{-1})
I''_{fl}	enthalpy of flue gas at the furnace exit (kJ kg^{-1})
k	radiant attenuation factor of flue gas (m^{-1} MPa^{-1})
$k_{C,q}$	radiant attenuation factor of soot (m^{-1} MPa^{-1})
$k_h\mu_h$	radiant attenuation factor of fly ash (m^{-1} MPa^{-1})
$k_K\mu_K$	radiant attenuation factor of char (m^{-1} MPa^{-1})
k_q	radiant attenuation factor of triatomic gases (m^{-1} MPa^{-1})
K	heat transfer coefficient of convective heating surfaces (W m^{-2} K^{-1})
m_q	a dimensionless coefficient correlated to the proportion of luminous area in the flame
m_r	CO_2 annual emission reduction (t)
m_t	CO_2 annual emission (t)
m_{t0}	CO_2 annual emission under BMCR (t)
M	a dimensionless coefficient related to fuel characteristics, combustion mode, and temperature distribution
M_{ar}	content of moisture on the as-received basis (%)
M_i	CO_2 emission intensity of power generation (g kW^{-1} h^{-1})
N_{ar}	content of nitrogen on the as-received basis (%)
O_{ar}	content of oxygen on the as-received basis (%)
P	power generation (MW)
q_2	heat loss owing to the exhaust gas (%)
q_3	heat loss owing to incomplete combustion of gaseous fuels (%)
q_4	heat loss owing to incomplete combustion of solid fuels (%)
q_5	heat loss owing to radiation and convection (%)
q_6	heat loss owing to ash residue (%)
Q_1	available heat of fuel (kJ kg^{-1})
Q_{tr}	heat transfer quantity of convective heating surfaces (W)
Q_{fl}	heat absorption of flue gas (kJ kg^{-1})
Q_k	sensible heat of combustion air (kJ kg^{-1})
$Q_{net,ar}$	lower heating value of mixed fuels (kJ kg^{-1})
$Q_{net,ar}^{coal}$	lower heating value of coal (kJ kg^{-1})
$Q_{net,ar}^{syngas}$	lower heating value of biomass syngas (kJ m^{-3})
Q_r	heat entrained into the boiler envelope (kJ kg^{-1})
Q_{av}	total available heat of working fluid (kW)
r_{H_2O}	water vapor volume fraction
S_{ar}	content of sulfur on the as-received basis (%)
$S_{ar,c}$	converted content of sulfur (%)
t_1	acidic dew point (°C)
t_w	water dew point (°C)
T''_1	furnace exit flue-gas temperature (K)
T_a	theoretical combustion temperature (K)
V^0	theoretical air volume (m^3 kg^{-1})
V_a	total air flow rate of air heater inlet (m^3 s^{-1})
V_f	total flue-gas flow rate of air heater outlet (m^3 s^{-1})
$V_{H_2O}^0$	theoretical water vapor volume (m^3 kg^{-1})
V_{H_2O}	total water vapor volume (m^3 kg^{-1})
V_y	total flue-gas volume (m^3 kg^{-1})
V_y^0	theoretical flue-gas volume (m^3 kg^{-1})
V_{daf}	content of volatile matter on a dry ash-free basis (%)
W	generation capacity of the boiler (kW h)
x	volume of biomass syngas to 1 kg of coal under standard conditions (m^3 kg^{-1})

Δt	temperature difference of heat transfer (K)
Δt_l	the larger one of initial and final temperature differences between hot and cold fluid (K)
Δt_s	the smaller one of initial and final temperature differences between hot and cold fluid (K)
α	excess air coefficient of furnace exit
α_{ash}	a dimensionless coefficient related to fly ash
α_o	excess air coefficient of the air heater outlet
α_d	convective heat transfer coefficient of flue gas (W m^{-2} K^{-1})
α_2	convective heat transfer coefficient of working fluid (W m^{-2} K^{-1})
α_f	radiant heat transfer coefficient of flue gas (W m^{-2} K^{-1})
ξ	utilization factor considering imperfect sweeping by flue gas
ε	fouling factor (m^2 K W^{-1})
β	a dimensionless coefficient that depends on the excess air coefficient at the furnace exit
β_k	excess air coefficient of the air heater inlet
η	thermal efficiency of the boiler (%)
η_{ce}	combustion efficiency
φ	heat retention factor
θ_{ex}	exhaust gas temperature (°C)
ρ	air density (kg m^{-3})
ϑ''	outlet gas temperature (°C)
ψ_t	temperature correction factor
BMCR	boiler maximum continuous rating
THA	turbine heat acceptance
LHV	lower heating value

References

1. Ziegler, C.; Morelli, V.; Fawibe, O. Climate Change and Underserved Communities. *Physician Assist. Clin.* **2019**, *4*, 203–216. [CrossRef]
2. Wang, Y.; Deng, L.; Chang, G.; Liu, H.; Lyu, K.; Nie, X.; Zhang, G. Influence of biomass gas parameters on coupled coal-fired biomass generation. *Therm. Power Gener.* **2021**, *50*, 34–40. (In Chinese) [CrossRef]
3. Long, J.; Deng, L.; Che, D. Analysis on organic compounds in water leachate from biomass. *Renew. Energy* **2020**, *155*, 1070–1078. [CrossRef]
4. Olivieri, A.; Ravelli, S. Cogasification of Coal and Biomass in an Integrated Gasification Combined Cycle Power Plant: Effects on Thermodynamic Performance and Gas Composition. *J. Energy Eng.* **2020**, *146*, 04020071. [CrossRef]
5. Roni, M.S.; Chowdhury, S.; Marufuzzaman, M.; Mamun, S.; Lein, W.; Johnson, S. Biomass co-firing technology with policies, challenges, and opportunities: A global review. *Renew. Sustain. Energy Rev.* **2017**, *78*, 1089–1101. [CrossRef]
6. Neville, A. Biomass Cofiring: A Promising New Generation Option. *Power* **2011**, *155*, 52–56.
7. Kalisz, S.; Pronobis, M.; Baxter, D. Co-firing of biomass waste-derived syngas in coal power boiler. *Energy* **2008**, *33*, 1770–1778. [CrossRef]
8. Deng, L.; Ma, S.; Jiang, J.; Tie, Y.; Zhang, Y.; Zhu, Z.; Belošević, S.; Tomanović, I.; Che, D. Numerical Investigation on Cofiring Characteristics of Biomass Syngas and Coal in a 660-MW Tower Boiler. *J. Energy Eng.* **2022**, *148*, 04022014. [CrossRef]
9. Robinson, A.L.; Junker, H.; Baxter, L.L. Pilot-Scale Investigation of the Influence of Coal−Biomass Cofiring on Ash Deposition. *Energy Fuels* **2002**, *16*, 343–355. [CrossRef]
10. Badour, C.; Gilbert, A.; Xu, C. Combustion and air emissions from co-firing a wood biomass, a Canadian peat and a Canadian lignite coal in a bubbling fluidised bed combustor. *Can. J. Chem. Eng.* **2012**, *90*, 1170–1177. [CrossRef]
11. Yao, X.; Zhou, H.; Xu, K.; Chen, S.; Ge, J.; Xu, Q. Systematic study on ash transformation behaviour and thermal kinetic characteristics during co-firing of biomass with high ratios of bituminous coal. *Renew. Energy* **2020**, *147*, 1453–1468. [CrossRef]
12. Harding, N.S.; Adams, B.R. Biomass as a reburning fuel: A specialized cofiring application. *Biomass Bioenergy* **2000**, *19*, 429–445. [CrossRef]
13. Sefidari, H.; Ma, C.; Fredriksson, C.; Lindblom, B.; Wiinikka, H.; Nordin, L.O.; Wu, G.; Yazhenskikh, E.; Müller, M.; Öhman, M. The effect of co-firing coal and woody biomass upon the slagging/deposition tendency in iron-ore pelletizing grate-kiln plants. *Fuel Process. Technol.* **2020**, *199*, 106254. [CrossRef]
14. Ma, W.; Zhou, H.; Zhang, J.; Zhang, K.; Liu, D.; Zhou, C.; Cen, K. Behavior of Slagging Deposits during Coal and Biomass Co-combustion in a 300 kW Down-Fired Furnace. *Energy Fuels* **2018**, *32*, 4399–4409. [CrossRef]
15. Luo, Z.; Xu, D.; Ma, Y.; Cheng, Q. Experimental Study on Co-Firing of Coal and Brewery Wastewater Sludge. *Appl. Sci.* **2020**, *10*, 7589. [CrossRef]

16. Unchaisri, T.; Fukuda, S.; Phongphiphat, A.; Saetia, S.; Sajjakulnukit, B. Experimental Study on Combustion Characteristics in a CFB during Co-firing of Coal with Biomass Pellets in Thailand. *Int. Energy J.* **2019**, *19*, 101–114.
17. Hou, S.; Huang, W.; Lin, T. Co-combustion of Fast Pyrolysis Bio-oil Derived from Coffee Bean Residue and Diesel in an Oil-fired Furnace. *Appl. Sci.* **2017**, *7*, 1085. [CrossRef]
18. Chen, C.; Wu, X.; Zhao, L. Simulation of Coal and Biomass Cofiring with Different Particle Density and Diameter in Bubbling Fluidized Bed under O_2/CO_2 Atmospheres. *J. Combust.* **2018**, *2018*, 6931483. [CrossRef]
19. Szufa, S.; Piersa, P.; Junga, R.; Błaszczuk, A.; Modliński, N.; Sobek, S.; Marczak-Grzesik, M.; Adrian, Ł.; Dzikuć, M. Numerical modeling of the co-firing process of an in situ steam-torrefied biomass with coal in a 230 MW industrial-scale boiler. *Energy* **2023**, *263*, 125918. [CrossRef]
20. Liu, Q.; Zhong, W.; Yu, A.; Wang, C.-H. Modelling the co-firing of coal and biomass in a 10 kW_{th} oxy-fuel fluidized bed. *Powder Technol.* **2022**, *395*, 43–59. [CrossRef]
21. Zhou, M.; Wang, S.; Luo, K.; Fan, J. Three-dimensional modeling study of the oxy-fuel co-firing of coal and biomass in a bubbling fluidized bed. *Energy* **2022**, *247*, 123496. [CrossRef]
22. Dong, L.; Alexiadis, A. Simulation of char burnout characteristics of biomass/coal blend with a simplified single particle reaction model. *Energy* **2023**, *264*, 126075. [CrossRef]
23. Deng, L.; Wang, Y.; Wu, S.; Che, D. Utilization of combustible waste gas as a supplementary fuel in coal-fired boilers. *Int. J. Energy Res.* **2018**, *42*, 1677–1692. [CrossRef]
24. Yao, Q.; Li, R.; Wang, Y.; Li, Y.; Zhang, L.; Deng, L.; Che, D. Feasibility Analysis of Coupling Hydrogen-Derived Fuel on a Coal-Fired Boiler for Power Generation. *Energy Fuels* **2023**, *37*, 477–491. [CrossRef]
25. Blokh, A.G.; Zhuravlev, Y.A.; Gorb, E.I.; Talibdzhanov, Z.S. Heat transfer in furnaces of high-power boilers. *Heat Transf. Res.* **1993**, *25*, 418–424.
26. Antonovskii, V.I. Heat Transfer in Steam Boiler Furnaces. A Retrospective View on the Development of a Standard Calculation Method. *Therm. Eng.* **2004**, *51*, 738–749.
27. Alekhnovich, A.N. Thermal efficiency coefficient of furnace shields as applied to standard method for thermal calculation of boilers. *Teploenergetika* **2009**, *9*, 23–29.
28. Che, D. *Boilers: Theory, Design and Operation*; Xi'an Jiaotong University Press: Xi'an, China, 2008.
29. Kuznetsov, N.; Mitor, V.; Karasina, E. *Thermal Calculation of Boiler Units (Standard Method)*; Energiya Press: Moscow, Russia, 1973.
30. Hsi, C.-L.; Kuo, J.-T. Estimation of fuel burning rate and heating value with highly variable properties for optimum combustion control. *Biomass Bioenergy* **2008**, *32*, 1255–1262. [CrossRef]
31. Song, Q. Numerical Simulation and Combustion Analysis of Biomass Gas and Coal Co-firing Process. *J. Henan Sci. Technol.* **2016**, *15*, 25–28. (In Chinese)
32. Atnaw, S.M.; Sulaiman, S.A.; Yusup, S. Syngas production from downdraft gasification of oil palm fronds. *Energy* **2013**, *61*, 491–501. [CrossRef]

applied sciences

MDPI

Article

The Effects of Parameter Settings on Triggering Time and Climb Rate during Lean-Premixed Combustion Thermoacoustic Oscillations

Chengfei Tao [1,2,*], Rongyue Sun [1,*], Yichen Wang [1], Liang Zhang [1], Jiangming Ye [1] and Shaohua Liang [1]

1 College of Energy and Power Engineering, Nanjing Institute of Technology, Nanjing 211167, China
2 MayAir Technology (China) Co., Ltd., Nanjing 211167, China
* Correspondence: taochengfei@njit.edu.cn (C.T.); sunrongyue@njit.edu.cn (R.S.)

Abstract: This study theoretically explored the effects of parameter settings on thermoacoustic oscillations with a low-order model. Three factors were explored—combustor length, inlet gas temperature and thermal power. The research findings indicate that optimizing the parameter settings can yield better thermoacoustic oscillation suppression results. The sound pressure amplitude decreased from 3.2×10^5 Pa to 2.1×10^5 Pa as the combustor length increased from 1.2 m to 6.0 m. The triggering time increased from 0.32 s to 0.91 s when the combustion chamber length increased. The climb rate declined from 23.38×10^5 Pa/s to 3.75×10^5 Pa/s when the combustor length was elongated. The sound pressure amplitude decreased from 3.44×10^5 Pa to 2.4×10^5 Pa as the gas temperature rose from 0 to 100 °C. The triggering time and climb rate variation tendency were similar when the gas temperature increased—both declined as the gas temperature rose. The sound pressure amplitude experienced a slight fluctuation when the thermal power rose. However, the triggering time decreased from 0.26 s to 0.043 s when the thermal power improved. The climb rate increased from 18.72×10^5 Pa/s to 27.65×10^5 Pa/s when the thermal power rose. The oscillation frequency presented was completely different in three cases that had different wavelengths and oscillation intensities. The triggering time and climb rate fluctuated extensively in varying conditions, and the above two factors were interrelated and contradictory to each other when thermoacoustic oscillation was excited. This study explored parameters' effects on triggering time and climb rate, thereby providing references for constructing a model-based control system for thermoacoustic oscillation feedback control.

Keywords: thermoacoustic instability; triggering time; premixed flame; climb rate; sound pressure amplitude

1. Introduction

Unsteady thermoacoustic oscillations, also referred to as combustion instability, have emerged as a common problem in modern gas turbine burners and have hindered the development of lean-premixed combustors [1]. Their oscillation amplitude can reach sufficiently high levels to interfere with gas turbine operations and, in some extreme cases, can lead to the failure of a combustion system due to excessive structural vibrations. With the extensive use of lean-premixed combustors in the modern gas turbine industry, the suppression of thermoacoustic oscillations is always crucial for the progression of gas turbine developments. Thus, understanding how to predict and control thermoacoustic oscillations has become important [2]. Mastering and controlling thermoacoustic oscillations has become crucial for future renewable energy power systems [3]. Generally, there are two ways to control thermoacoustic oscillations: active control and passive control [4–6]. Acoustic dampers are widely applied as a passive control tool to stabilize combustion systems. However, this method cannot respond to the dynamic changes of real operating conditions and tends to be effective only within certain narrow-range frequencies [7]. Compared with passive control approaches, active control, especially in the closed-loop-feedback form, is more flexible and has promotional prospects [8]. Therefore, active control technology is expected to more efficiently suppress thermoacoustic oscillations. This has become a current research hotspot.

Active control is generally based on feedback control technologies, either by modulating the acoustic field in the combustion chamber or modulating the unsteady heat-release rate of flames [9]. Currently, reduced-order models of thermoacoustic oscillations are mainly used to investigate the feedback control of combustion instability [10]. Li et al. [11] studied the feedback control of combustion instabilities with a low-order combustor model and revealed that the open-loop transfer function needed for controller design varies with the oscillation level. Gelbert et al. [12] explored the feedback control of unstable thermoacoustic modes in an annular Rijke tube; a thermoacoustic network model was constructed and used to derive low-order models for modal control of the system. Morgans et al. [13,14] proved that model-based control is a good method for studying thermoacoustic oscillations in combustion chambers. Different controllers were implemented in simulations using a low-order thermoacoustic model. Although active control, which is based on these models, has significant potential, under different working conditions, the triggering time and climb rate of thermoacoustic oscillations are different [15,16], which also leads to control system failure or poor control effectiveness. Therefore, it is necessary to further explore the triggering time and climb rate of thermoacoustic oscillations under different conditions. Accordingly, to develop a robust feedback control system based on low-order thermoacoustic oscillation models, exploring the influence of different combustor parameters on the triggering time and climb rate is crucial.

The triggering time and climb rate of combustion thermoacoustic oscillations are related to many parameters, and relevant scholars have conducted extensive research on this topic. Zhang et al. [17] studied the thermodynamic properties of a thermoacoustic swirl combustor and revealed that oscillation mode switching depends on the equivalence ratio and methane flow rate. Du et al. [18] investigated the effects of different heat source distributions on a Rijke tube; their experimental results show that optimum spacing exists to minimize the starting power of thermoacoustic oscillations and growth rates and maximize the sound pressure amplitude. Furthermore, it has also been proven that parameters such as the flame equivalence ratio and flow field play a major role in the excitation of thermoacoustic oscillations [19]. Zhao et al. [20] explored the effects of the flame equivalence ratio and fuel flow rate on nonlinear thermoacoustic instability in a swirl combustor, and mode switching between higher harmonic and fundamental oscillation was observed. It has also been found that the fuel type and air/fuel supply method affect the triggering and growth of thermoacoustic oscillations [21–24]. The influence of geometric parameters on thermoacoustic oscillations cannot be ignored. Jo et al. [25] evaluated the damping capacity according to geometrics and the number of resonators using a Rijke tube, and the decay time of thermoacoustic oscillations was measured to quantify the damping capability in the time domain. Therefore, by illuminating the influence of parameter settings on the triggering time and climb rate of thermoacoustic oscillations, combined with using low-order models to construct a model-based control system [26–28], a better design for active control systems for thermoacoustic oscillations can be achieved for real burners.

The analytical models that have been used by researchers to study the thermoacoustic oscillation phenomenon contain the G-equation method [29,30] or the n-τ model [31]. In this study, only linear acoustic vibrations were expected in the combustor chamber; the nonlinear effects were supposed to be related to the flame heat-release rate, which exhibited saturation characteristics. To predict limit cycle thermoacoustic oscillations, scholars establish the flame's G-equation by linking the flame surface with flame heat release. The empirical approach is to use a parameter-dependent n-τ model. Therefore, this study intended to construct a low-order model of thermoacoustic oscillations with the n-τ model, aiming to explore the temporal signal characteristics of thermoacoustic oscillation under different key parameter settings. Through analyzing the differences in the triggering time and climb rate of thermoacoustic oscillations, this study will provide references to better construct a model-based control system for thermoacoustic oscillation feedback control.

In this work, the triggering times and climb rates of thermoacoustic oscillations are theoretically studied. In order to present the impact of changes in the combustion parameters, three main factors of the combustion system were researched—burner length, inlet air temperature and thermal power. Five burner lengths were measured in this study. The inlet air temperature for combustion ranged from 0 to 100 °C. The thermal power of the burner was also set to five different values. Under these parameters, the triggering times and climb rates of thermoacoustic oscillations were recorded and studied. A low-order model of thermoacoustic oscillation was constructed in MATLAB software 2021R1. This research aimed to reveal the parameter settings of a lean-premixed combustor to trigger and grow thermoacoustic oscillations to better achieve the feedback control effect of thermoacoustic oscillations in modern gas turbine engines. This study found that the triggering time and climb rate affect the safety of a combustion system; therefore, the present investigation examines the effect of different parameter settings on thermoacoustic oscillations to better understand the evolution properties of triggering time and climb rate. This study provides a theoretical basis for the prevention of combustion thermoacoustic oscillation. At the same time, it can also be seen that an incorrect burner parameter design will lead to difficulties in suppressing thermoacoustic oscillations.

2. Materials and Methods

2.1. Design of the Lean-Premixed Combustor

The theoretical low-order model in this research was conducted with a model lean-premixed combustor from our previous research [32,33]. Detailed geometrical characteristics and critical design parameters of the combustor are displayed in Figure 1. In Figure 1, τ denotes the time delay, the reflection coefficient for the chocked boundary condition is defined as R_u and the reflection coefficient of the open-ended downstream is defined as R_d. Acoustic waves in the combustion chamber are denoted as f and g, with $p(t)$ representing the sound pressure signal. In this research, the mixture's gas temperature was set as 0, 25, 50, 75 and 100 °C. The length of the whole burner was set as 1.2, 2.4, 3.6, 4.8 and 6.0 m. The thermal power of the combustor was set as 0.013, 0.026, 0.039, 0.052 and 0.065 kW. The geometrical parameters of the burner are presented in Table 1.

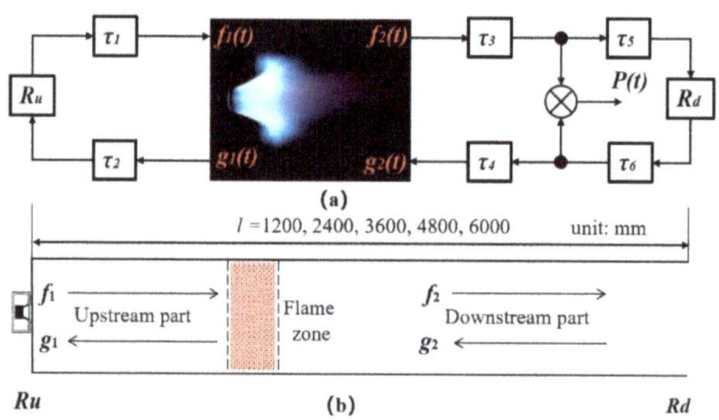

Figure 1. Low-order thermoacoustic oscillation model of the lean-premixed gas turbine combustor. (**a**) Time delay and reflection coefficient of the low-order model. (**b**) Acoustic waves in the open-ended combustion chamber.

Table 1. Geometric parameters of the lean-premixed combustor.

Geometries of the Combustor	
Burner inlet length (mm)	550
Diameter (mm)	114
Burner total length (mm)	1200, 2400, 3600, 4800, 6000
Swirl section length (mm)	150

2.2. Low-Order Model of Thermoacoustic Oscillation

The self-excited frequency (264 Hz) was below the cut-off frequency (1501 Hz, 1.2 m); thus, the combustor can be treated as one-dimensional. Figure 1a presents the low-order thermoacoustic oscillation model of the lean-premixed gas turbine combustor. The low-order thermoacoustic oscillation model of the combustor was modeled with MATLAB Simulink software using the method proposed by Dowling [34] and Fabian [35], and the governing conservation equations of the wave equation are as follows:

$$p(x,t) = \overline{p} + \left[f\left(t - \frac{x}{\overline{c}+\overline{u}}\right) - g\left(t + \frac{x}{\overline{c}-\overline{u}}\right) \right] \tag{1}$$

$$u(x,t) = \overline{u} + \frac{1}{\overline{\rho c}}\left[f\left(t - \frac{x}{\overline{c}+\overline{u}}\right) - g\left(t - \frac{x}{\overline{c}-\overline{u}}\right) \right] \tag{2}$$

In Equations (1) and (2), $p(x,t)$ and $u(x,t)$ denote the total pressure and velocity fluctuation, $f(t)$ and $g(t)$ denote traveling waves in the combustion chamber, $\overline{p}$ and $\overline{u}$ denote the average pressure and velocity value, $\overline{c}$ denotes the acoustic velocity in the combustion chamber and x denotes the relative position of traveling waves in the combustion chamber. The links between mean flow variables u and acoustic variables p can be described as in our previous research [33]:

$$\frac{p'}{p} \propto O(\gamma\overline{M})\frac{u'}{u} \tag{3}$$

where gamma (γ) is the adiabatic exponent of gas and $\overline{M}$ stands for the mean Mach number. In Figure 1b, R_u is the reflection coefficient of the upstream part of the combustor, and R_d is the reflection coefficient of the downstream part of the combustor. $P(t)$ is the sound pressure signal in the combustion chamber. The lean-premixed flame heat release transfer function was modeled as the n-τ equation, where n is the interaction index or combustion efficiency and τ is a time delay; the travel times of the acoustic waves f and g are denoted as time delays $\tau_i = x_i/c$ (i=1, 2, . . .,6); and x_i denotes the travelling distance of the acoustic waves. These considerations lead to the following transfer function:

$$q'(s) = \frac{k_f}{1+T_f s}e^{-s\tau_f}u_1'(s) \tag{4}$$

In Equation (4), k_f is the static gain, T_f is the time constant of the low pass element, τ_f is the travel dead time and u_1 is the acoustic velocity upstream of the flame. In this paper, parameters for modeling are summarized in Table 2. The equivalence ratio of the propane flame was maintained at 1.0, and air for combustion was 0.105 MPa with a relative humidity of 50%. The lower heat value of propane is 46.56 kJ/kg, and the heat capacity of air is 1.009 kJ/(kg K). As displayed in Figure 2, the low-order model was implemented in MATLAB. A linear flame transfer function was used, and the n-τ model was filtered by a first-order filter to capture the flame response [36,37].

$$H(s) = \frac{\omega_c}{\omega_c + s}a_f e^{-\tau_f s} \tag{5}$$

where a_f is the gain and τ_f indicates the time delay. $\omega_c = 2\pi f_c$ denotes the cut-off frequency of the filter. The reflection coefficient R_u for the chocked boundary condition is defined as 0.95. The reflection coefficients of the open-ended downstream are sensitive to geometric dimensions and sound speed. We defined the reflection coefficient of the open-ended downstream, R_d, as Equation (6) [38]:

$$R_d = \left[-\frac{l_d^2}{16c^2} - 1 + \frac{0.3 l_d}{c} \right] / \left[-\frac{l_d^2}{16c^2} + 1 + \frac{0.3 l_d}{c} \right] \quad (6)$$

Table 2. Parameter settings in this study.

Parameters	Values
Fuel type	Propane, gas purity: 99.995%
Thermal power (kW)	0.013, 0.026, 0.039, 0.052, 0.065
Lower heat value for propane (kJ/kg)	46.56
Equivalence ratio (Φ)	0.90
Atmospheric pressure (MPa)	0.105
Gas mixture temperature (°C)	0, 25, 50, 75, 100
Burner lengths (mm)	1.2, 2.4, 3.6, 4.8, 6.0

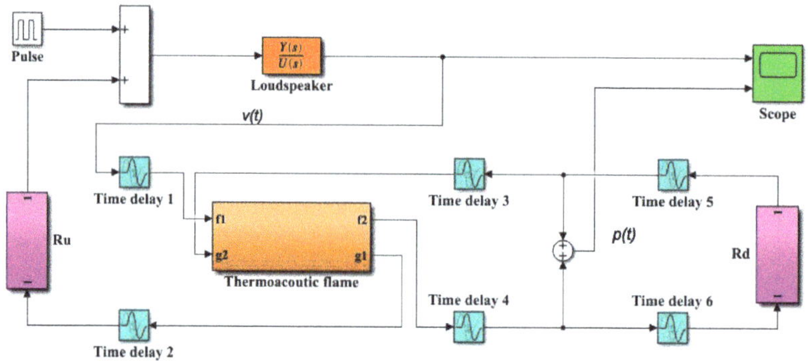

Figure 2. Low-order thermoacoustic oscillation model in MATLAB Simulink.

3. Results

3.1. *Effects of the Combustor Length*

Figure 3a–e show the variation tendency of the sound pressure amplitude in a combustion chamber. When the combustor length ranges from 1.2 to 6.0 m, the upstream part of the combustor remains constant at 0.55 m. It can be inferred from Figure 3 that the sound pressure amplitude gradually declines as the length of the combustor reduces. The oscillation curve becomes sparse, which denotes that the oscillation frequency also experienced a gradually decreasing trend. The triggering time of thermoacoustic oscillations varies significantly while the combustor length changes. In Figure 3a, at the length of 1.2 m, thermoacoustic oscillation emerges within 0.5 s; however, at the length of 6.0 m, Figure 3e shows that oscillation emerges after 1 s. Both sound pressure amplitude and oscillation frequency decline as the length grows, accompanied by the gradually increasing trigger time. Figure 3 also indicates that there is a significant difference in the climb rate during the excitation processes.

The main reason for the sound pressure amplitude decreasing can be attributed to the variation in sound wavelengths. At the same time, there is a significant change in the length of the sound waves. In Figure 3, as the chamber length rises, the oscillation frequency gradually declines. This is because, as the length of the combustion chamber increases, the oscillation period of thermoacoustic oscillations is also extended. According to the equation for the difference between the oscillation period and frequency, $f = 1/T$, it can be seen that the increase in the oscillation period will lead to a decrease in oscillation frequency. In the sound wavelength calculation equation, $\lambda = v/f$, v denotes the acoustic velocity and f denotes the oscillation frequency. The equation indicated that the sound wavelength λ increased as the oscillation frequency f declined. As the sound wavelength λ increased, the relative location where high amplitude oscillations are easily triggered also changed. Then, the intensity of thermoacoustic oscillation decreased.

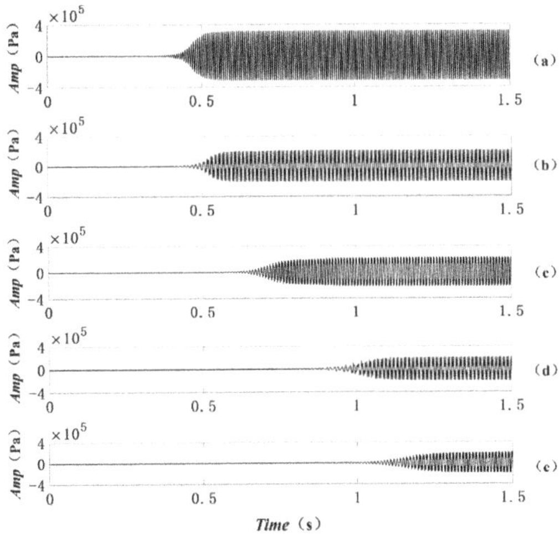

Figure 3. Effects of different combustor lengths on the sound pressure amplitude in the combustion chamber; the combustor length ranged from 1.2 to 6.0 m. (**a**) 1.2 m length. (**b**) 2.4 m length. (**c**) 3.6 m length. (**d**) 4.8 m length. (**e**) 6.0 m length.

The detailed values of sound pressure amplitude and triggering time under different combustor lengths are displayed in Figure 4. As the combustor length increased from 1.2 m to 6.0 m, the thermoacoustic oscillation triggering time increased from 0.32 s to 0.91 s, while the sound pressure amplitude decreased from 3.2×10^5 Pa to 2.1×10^5 Pa. The sound pressure amplitude experienced a rapid descent, while the chamber length rose from 1.2 m to 2.4 m. However, the downward trend slowed down when the chamber length rose from 2.4 m to 6.0 m. The change trend of the triggering time curve was relatively consistent as the chamber length rose. The reason why the triggering time changed is closely related to the increase in the acoustic wave oscillation frequency; as $f = 1/T$, the oscillation period increased. Due to the extension of the oscillation period and chamber length, the coupling process between the sound waves and flame was extended. As a result, the triggering time of oscillation was prolonged.

To accurately evaluate the impact of parameter settings on the triggering of thermoacoustic oscillations, we introduced the climb rate as an indicator of thermoacoustic oscillations. This method is similar to that in our previous research on calculating the decay time of thermoacoustic oscillations [32]. Figure 5 shows the calculation process of the climb rate, $C = \frac{(P_{max} - P_{max/e})}{\tau}$, where C denotes the thermoacoustic oscillation climb rate, P_{max} denotes the maximum value of the sound pressure amplitude, $P_{max/e}$ denotes the maximum value of the sound pressure amplitude, e is the natural constant and equals 2.71828 in this research and τ is the climbing time in which $P_{max/e}$ climbs to P_{max}. The unit for the thermoacoustic oscillation climb rate is 10^5 Pa/s.

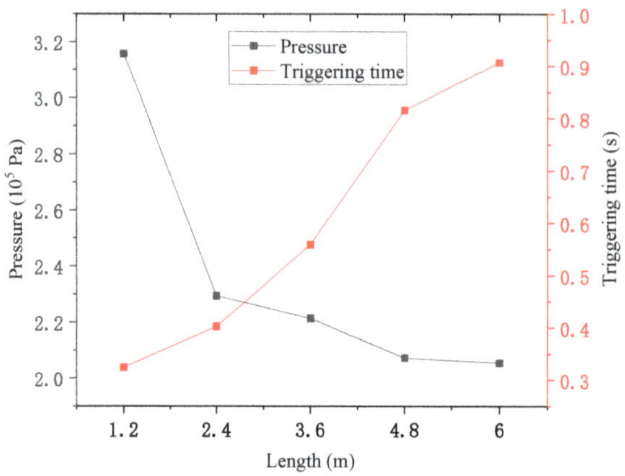

Figure 4. Effects of different combustor lengths on the sound pressure amplitude and triggering time; the combustor length ranged from 1.2 to 6.0 m.

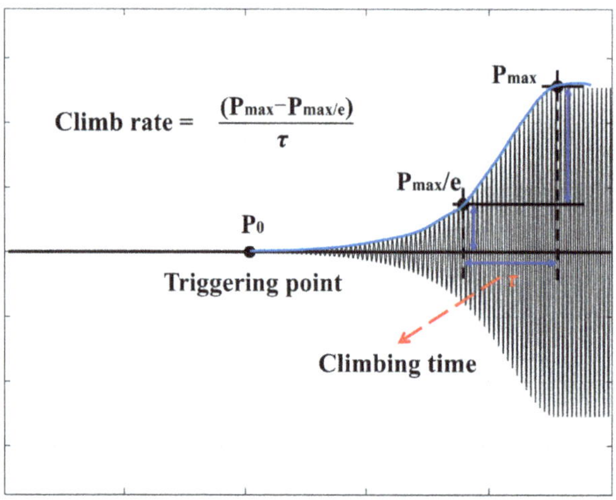

Figure 5. Schematic diagram for calculating the climb rate of thermoacoustic oscillation in this research.

Figure 6 shows the climb rate of thermoacoustic oscillation as the combustor length increased from 1.2 m to 6.0 m. Consistent with the amplitude variation trend of thermoacoustic oscillation, the climb rate dropped from 23.38 $\times$ 10^5 Pa/s to 3.75 $\times$ 10^5 Pa/s as the combustor length rose. This means that the geometric structure of the combustor was closely related to the excitation speed of the thermoacoustic oscillations. It is important to understand how to effectively control the acceleration rate during the thermoacoustic oscillation excitation process. This is related to whether the active controller can quickly suppress the thermoacoustic oscillation that occurs in gas turbine combustion chambers. As shown by the results in Figure 6, a reasonable combustor length design can not only meet the combustion heat volume but can also slow down the excitation speed of thermoacoustic oscillation.

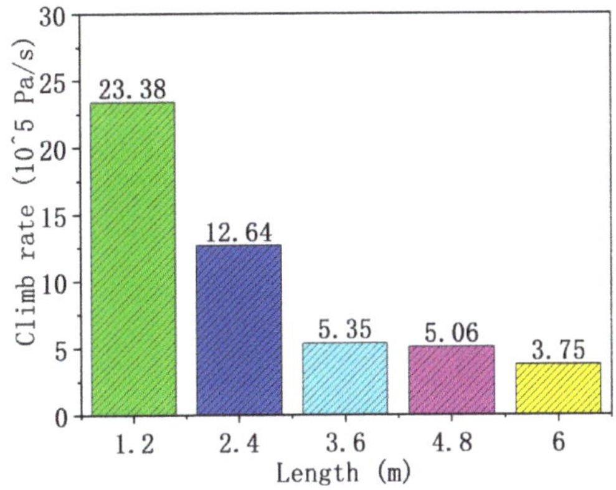

Figure 6. Effects of different combustor lengths on the thermoacoustic oscillation climb rate; the combustor length ranged from 1.2 to 6.0 m.

3.2. Effects of the Gas Temperature

Figure 7a–e show the variation tendency of sound pressure amplitude in the combustion chamber while the gas temperature ranged from 0 to 100 °C. This study selected five gas temperatures, each increasing by 25 °C. In this part, the combustor length and thermal power remain constant. It can be inferred from Figure 7 that the sound pressure amplitude gradually declined as the gas temperature increased from o to 100 °C. Different from the changes in combustor length shown in Figure 4, the oscillation curve showed a linear change. The trend of oscillation frequency change is also different from that in Figure 4; the oscillation frequency experienced a slowly increasing trend as the gas temperature rose. The triggering time of thermoacoustic oscillation varied obviously when the inlet gas temperature changed. In Figure 7a, at the temperature point of 0 °C, thermoacoustic oscillation emerged within 0.3 s; however, at the temperature point of 100 °C, Figure 7e illustrates that thermoacoustic oscillation emerged after 1.1 s.

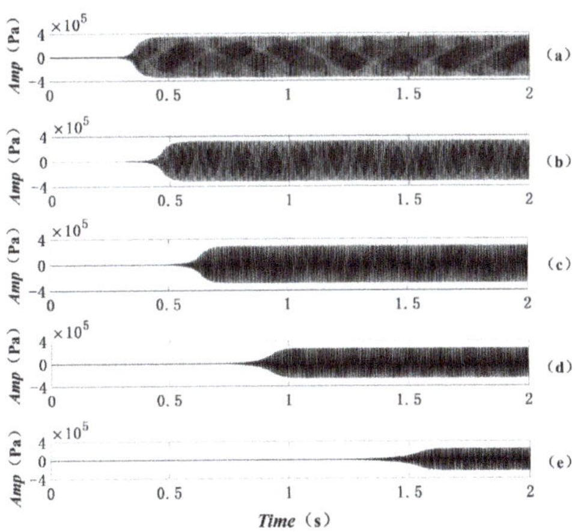

Figure 7. Effects of different gas temperatures on the sound pressure amplitude in the combustion chamber; the gas temperature ranged from 0 to 100 °C. (**a**) 0 °C. (**b**) 25 °C. (**c**) 50 °C. (**d**) 75 °C. (**e**) 100 °C.

The detailed values of the sound pressure amplitude and triggering time under different gas temperatures are displayed in Figure 8. It can be seen that the amplitude of thermoacoustic oscillation decreased but the triggering time was prolonged. As the inlet premixed gas temperature increased from 0 to 100 °C, the thermoacoustic oscillation triggering time increased from 0.26 s to 1.13 s, while the sound pressure amplitude decreased from 3.44×10^5 Pa to 2.4×10^5 Pa. The sound pressure amplitude experienced a linear descent while the gas temperature increased from 0 to 100 °C. This is because the triggering of thermoacoustic oscillations is related to the Rayleigh criterion. The oscillation frequency f slowly improved as the inlet gas temperature increased, and the phase difference between the flame heat-release rate and sound wave changed to $\Delta p = \omega t + \varphi$, where ω denotes the circular frequency, and the relationship is $\omega = 2\pi f$. The symbol φ stands for the initial phase value. As the thermoacoustic oscillation frequency f gradually increases, the circular frequency ω also starts to rise; this leads to an increase in the phase difference, causing a decrease in the oscillation amplitude.

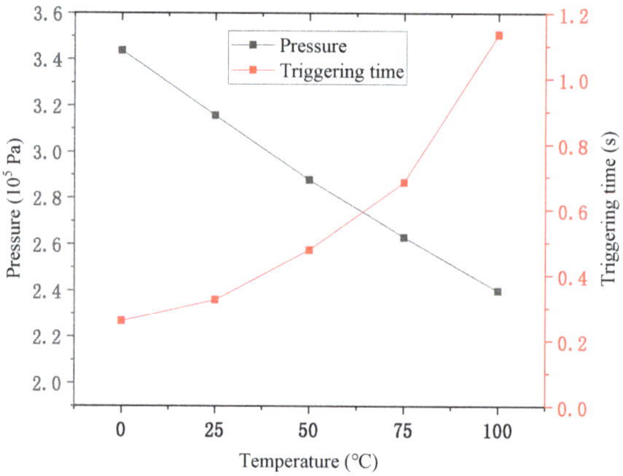

Figure 8. Effects of different gas temperatures on the sound pressure amplitude and triggering time; the gas temperature ranged from o to 100 °C.

Figure 9 shows the climb rate of thermoacoustic oscillations as the gas temperature increased from 0 to 100 °C. Consistent with the amplitude variation trend of thermoacoustic oscillation, the climb rate dropped from 24.01×10^5 Pa/s to 11.06×10^5 Pa/s as the temperature rose. This means that the thermodynamic characteristics of the working fluid at the burner inlet were closely related to the excitation speed of thermoacoustic oscillations. Compared with the results in Figure 8, where, as the temperature increased, the slope of the curve for the triggering time of the thermoacoustic oscillation increased, the climb rate in Figure 9 declined. For lean-premixed combustors, the duration of the thermoacoustic oscillation triggering time and the climb rate are contradictory indexes. Therefore, it is necessary to optimize the parameter settings to obtain a reasonable thermoacoustic oscillation triggering time and climb rate, thereby improving the stability and safety of the system [39]. The decrease in the climb rate of thermoacoustic oscillations is related to the decrease in local temperature differences in the combustion chamber.

3.3. Effects of Thermal Power

Figure 10a–e show the sound pressure amplitude in the combustion chamber. When the thermal power of the combustor ranged from 0.013 to 0.065 kW, the chamber length and inlet gas temperature of the combustor remained constant. In the beginning, it can be inferred from Figure 10 that the sound pressure amplitude gradually increased when the thermal bellows were 0.052 kW. However, the sound pressure amplitude gradually decreased when the thermal power reached 0.065 kW. Additionally, the sparsity of the oscillation curve remained almost unchanged, which denotes that the oscillation frequency maintained a relatively stable value. The triggering time of thermoacoustic oscillation varies significantly when thermal power changes. In Figure 10a, at the thermal power of 0.013 kW, thermoacoustic oscillation emerged within 0.3 s; however, at the thermal power of 0.065 kW, Figure 10e shows that thermoacoustic oscillation emerged within 0.05 s. Figure 10 also indicates that there was a remarkable difference in the climb rate during the thermoacoustic oscillation excitation processes. The reason why the thermoacoustic oscillation amplitude did not significantly increase may be related to the intake pressure and burner heat dissipation. After the increase in thermal power, the degree of heat dissipation of the combustor increases. The oscillation frequency is almost unchanged; this may be attributed to the slight change in the acoustic wavelength in the combustion chamber.

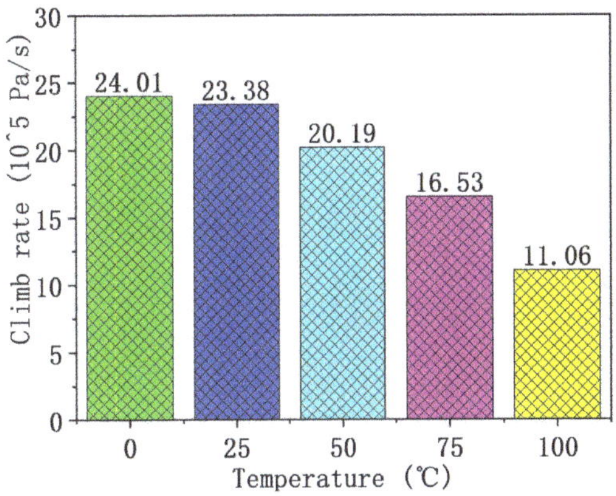

Figure 9. Effects of different gas temperatures on the thermoacoustic oscillation climb rate; the gas temperature ranged from 0 to 100 °C.

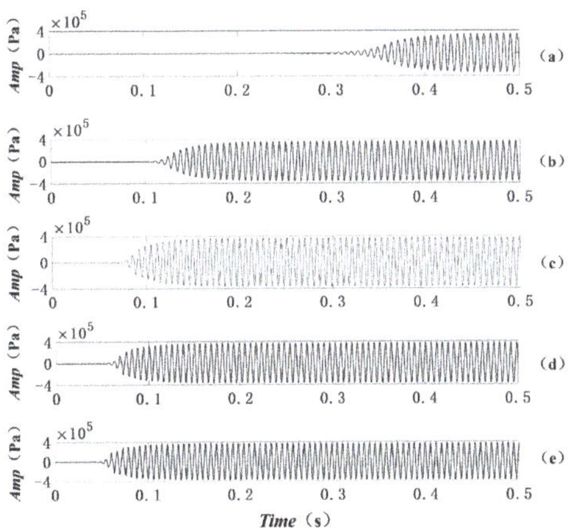

Figure 10. Effects of different thermal powers on the sound pressure amplitude in the combustion chamber; the thermal power ranged from 0.013 to 0.065 kW. (**a**) 0.013 kW. (**b**) 0.026 kW. (**c**) 0.039 kW. (**d**) 0.052 kW. (**e**) 0.065 kW.

The detailed values of the sound pressure amplitude and triggering time under different thermal powers are displayed in Figure 11, and the triggering time is shortened. As the combustor thermal power increased from 0.013 to 0.065 kW, the thermoacoustic oscillation triggering time decreased from 0.26 s to 0.043 s. The triggering time of thermoacoustic oscillations is related to the Rayleigh criterion. The triggering time declined as thermal power increased, which caused the increase in the flame heat-release rate, leading to easier excitation of thermoacoustic oscillations. However, when the thermal power reached

0.052 kW, the decreasing trend of the thermoacoustic oscillation triggering time became slower. This may be because the heat dissipation effect of the combustor was greater than the influence of the flame heat-release rate. In this study, adiabatic effects were not considered; this may have resulted in a slight decrease in the thermoacoustic oscillation amplitude in the combustion chamber.

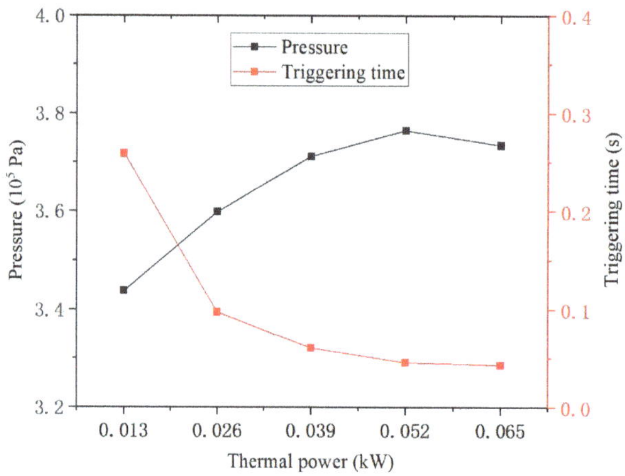

Figure 11. Effects of thermal power on the sound pressure amplitude and triggering time; the thermal power ranged from 0.013 to 0.065 kW.

Figure 12 shows the climb rate of thermoacoustic oscillations as the thermal power of the combustor increased from 0.013 to 0.065 kW. Contrary to the trend of the changes in the excitation time of thermoacoustic oscillations, the climb rate rose from 18.72 × 10^5 Pa/s to 27.65 × 10^5 Pa/s as the thermal power rose. However, the trend of the curve changes in both indexes was relatively consistent, ranging from significant to slow changes. Comparing this discovery with the research results in Sections 3.1 and 3.2, it can be found that it is crucial to reasonably set parameters for the effective active control of lean-premixed thermoacoustic oscillations. There are several areas that can be optimized to achieve the best thermoacoustic oscillation active control effect. If the amplitude, triggering time and climb rate of thermoacoustic oscillations are not well studied, the selected active controller may not meet specific oscillation conditions, leading to the failure of the control system and serious consequences.

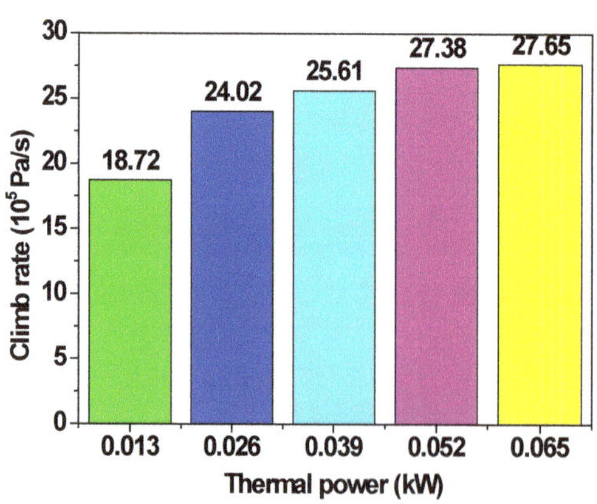

Figure 12. Effects of different thermal powers on the thermoacoustic oscillation climb rate; the thermal power ranged from 0.013 to 0.065 kW.

4. Conclusions

We researched the suppression effects of thermoacoustic oscillation variation under different parameter settings, which brings significant challenges to the design of active control systems. In this study, the influence of three different types of parameters on lean-premixed combustion thermoacoustic oscillation was theoretically investigated. The effects among three parameters were investigated: the burner length, the inlet air temperature and the thermal power.

The research results indicate that an elongated combustor length or improved gas temperature could alleviate the oscillation intensity. However, improved thermal combustion power cannot effectively suppress thermoacoustic instability. The oscillation frequency not only fluctuated with the parameter settings but also altered the process of thermoacoustic coupling. Meanwhile, some unreasonable designs, such as excessively short burner designs, can lead to stronger thermoacoustic oscillations.

The triggering time and climb rate were key factors that must be considered when designing an active control system for thermoacoustic oscillations. The triggering time and climb rate fluctuated extensively among different conditions, and the above two factors are interrelated and contradictory to each other when a thermoacoustic oscillation is excited. It is necessary to optimize the triggering time and climb rate of thermoacoustic oscillations in future research to achieve better active control effects in gas turbine combustors. Based on the research results in this article, the design of active control systems for thermal acoustic oscillations can be improved. For thermal acoustic oscillations with short triggering times and fast climb rates, it is necessary to adopt active control algorithms with fast response actions and small delay times and to use high-frequency actuators as much as possible.

Author Contributions: Conceptualization, C.T.; data curation, L.Z. and J.Y.; formal analysis, S.L.; methodology, Y.W.; supervision, C.T. and R.S.; visualization, C.T.; writing—original draft, C.T. All authors have read and agreed to the published version of the manuscript.

Funding: This research was funded by the Natural Science Foundation of Jiangsu Province (No. BK20230699), the China Postdoctoral Science Foundation (No. 2023M731492), Scientific Research Start-up Foundation funding of High-level Introduction Talents of Nanjing Institute of Technology (No. YKJ202107) and the Natural Science Research Project in Colleges and Universities in Jiangsu Province of China (No. 23KJB470010).

Institutional Review Board Statement: Not applicable.

Informed Consent Statement: Not applicable.

Data Availability Statement: Data are contained within the article.

Acknowledgments: We thank Ling Chen in MayAir Technology (China) Co., Ltd., for her support of this study.

Conflicts of Interest: Chengfei Tao was employed by the company MayAir Technology (China) Co., Ltd. The remaining authors declare that the research was conducted in the absence of any commercial or financial relationships that could be construed as a potential conflict of interest.

References

1. Huang, Y.; Yang, V. Dynamics and stability of lean-premixed swirl-stabilized combustion. *Prog. Energy Combust. Sci.* **2009**, *35*, 293–364. [CrossRef]
2. Poinsot, T. Prediction and control of combustion instabilities in real engines. *Proc. Combust. Inst.* **2017**, *36*, 1–28. [CrossRef]
3. Candel, S. Combustion dynamics and control: Progress and challenges. *Proc. Combust. Inst.* **2002**, *29*, 1–28. [CrossRef]
4. Zhang, M.X.; Li, J.X.; Cheng, W.W. Active control of thermoacoustic instability using microsecond plasma discharge. *J. Appl. Phys.* **2020**, *127*, 033301. [CrossRef]
5. Zhang, Y.M.; Huang, L.X. Electroacoustic control of Rijke tube instability. *J. Sound Vib.* **2017**, *409*, 131–144. [CrossRef]
6. Guan, Y.; He, W.; Murugesan, M.; Li, Q.; Liu, P.J.; Li, L.K.B. Control of self-excited thermoacoustic oscillations using transient forcing, hysteresis and mode switching. *Combust. Flame* **2019**, *202*, 262–275. [CrossRef]
7. Zhao, D.; Li, X.Y. A review of acoustic dampers applied to combustion chambers in aerospace industry. *Prog. Aerosp. Sci.* **2015**, *74*, 114–130. [CrossRef]

8. Zhao, D. A review of active control approaches in stabilizing combustion systems in aerospace industry. *Prog. Aerosp. Sci.* **2018**, *97*, 35–60. [CrossRef]
9. Rubio-Hervas, J.; Zhao, D.; Reyhanoglu, M. Nonlinear feedback control of self-sustained thermoacoustic oscillations. *Aerosp. Sci. Technol.* **2015**, *41*, 209–215. [CrossRef]
10. Illingworth, S.J.; Morgans, A.S.; Rowley, C.W. Feedback control of flow resonances using balanced reduced-order models. *J. Sound Vib.* **2011**, *330*, 1567–1581. [CrossRef]
11. Li, J.X.; Morgans, A.S. Feedback control of combustion instabilities from within limit cycle oscillations using H_∞ loop-shaping and the v-gap metric. *Proc. R. Soc. A* **2015**, *472*, 20150821. [CrossRef]
12. Gelbert, G.; Moeck, J.P.; Paschereit, C.O.; King, R. Feedback control of unstable thermoacoustic modes in an annular Rijke tube. *Control Eng. Pract.* **2012**, *20*, 770–782. [CrossRef]
13. Morgans, A.S.; Dowling, A.P. Model-based control of combustion instabilities. *J. Sound Vib.* **2007**, *299*, 261–282. [CrossRef]
14. Morgans, A.S.; Stow, S.R. Model-based control of combustion instabilities in annular combustors. *Combust. Flame* **2007**, *150*, 380–399. [CrossRef]
15. Zalluhoglu, U.; Kammer, A.S.; Olgac, N. Delayed feedback control laws for Rijke tube thermoacoustic instability, synthesis, and experimental validation. *IEEE Trans. Control Syst. Technol.* **2016**, *24*, 1861–1868. [CrossRef]
16. Li, X.Y.; Zhao, D.; Yang, X.L.; Wang, S.H. Unity maximum transient energy growth of heat-driven acoustic oscillations. *Energy. Convers. Manag.* **2016**, *116*, 1–10. [CrossRef]
17. Zhang, Z.G.; Zhao, D.; Ni, S.L.; Sun, Y.Z.; Wang, B.; Chen, Y.; Li, G.N.; Li, S. Experimental characterizing combustion emissions and thermodynamic properties of a thermoacoustic swirl combustor. *Appl. Energy* **2019**, *235*, 463–472. [CrossRef]
18. Du, M.L.; Li, F.; Yang, L. Effects of different heat source distribution on a Rijke tube. *Appl. Acoust.* **2019**, *146*, 66–75. [CrossRef]
19. Pires, J.M.; Fernandes, E.C. Combined effect of equivalence ratio and velocity gradients on flame stability and emission formation. *Fuel* **2018**, *222*, 800–809. [CrossRef]
20. Zhao, H.; Li, G.N.; Zhao, D.; Zhang, Z.G.; Sun, D.K.; Yang, W.M.; Li, S.; Lu, Z.L.; Zheng, Y.Q. Experimental study of equivalence ratio and fuel flow rate effects on nonlinear thermoacoustic instability in a swirl combustor. *Appl. Energy* **2017**, *208*, 123–131. [CrossRef]
21. Zhang, B.; Liu, H.; Liu, X.; Liu, H. Prediction Method of Swirling Flame Lean Blowout Based on Flame Image Morphological Features. *Appl. Sci.* **2023**, *13*, 3173. [CrossRef]
22. Shchepakina, E.A.; Zubrilin, I.A.; Kuznetsov, A.Y.; Tsapenkov, K.D.; Antonov, D.V.; Strizhak, P.A.; Yakushkin, D.V.; Ulitichev, A.G.; Dolinskiy, V.A.; Hernandez Morales, M. Physical and Chemical Features of Hydrogen Combustion and Their Influence on the Characteristics of Gas Turbine Combustion Chambers. *Appl. Sci.* **2023**, *13*, 3754. [CrossRef]
23. Pan, D.; Ji, C.; Zhu, T. Characterization of Nonlinear Responses of Non-Premixed Flames to Low-Frequency Acoustic Excitations. *Appl. Sci.* **2023**, *13*, 6237. [CrossRef]
24. Du, Y.; Zhang, Y.; Li, X.; Zhang, J.; Da, Y.; Jia, Y.; Che, D. Self-Excited Thermoacoustic Instability Behavior of a Hedge Premixed Combustion System with an Asymmetric Air/Fuel Supply or Combustion Condition. *Appl. Sci.* **2023**, *13*, 11463. [CrossRef]
25. Jo, S.; Choi, Y.; Kim, H.J. Evaluation of the damping capacity according to the geometric and the number of resonator with thermal environment using a Rijke tube. *Aerosp. Sci. Technol.* **2019**, *88*, 1–8. [CrossRef]
26. Li, S.; Li, Q.T.; Tang, L.; Yang, B.; Fu, J.Q.; Clarke, C.A.; Jin, X.; Ji, C.Z.; Zhao, H. Theoretical and experimental demonstration of minimizing self-excited thermoacoustic oscillations by applying anti-sound technique. *Appl. Energy* **2016**, *181*, 399–407. [CrossRef]
27. Li, X.Y.; Zhao, D.; Li, J.W.; Xu, Y.S. Experimental evaluation of anti-sound approach in damping self-sustained thermoacoustic oscillations. *J. Appl. Phys.* **2013**, *114*, 204903. [CrossRef]
28. Yuan, X.C.; Glover, K. Model-based control of thermoacoustic instabilities in partially premixed lean combustion—A design case study. *Int. J. Control* **2013**, *86*, 2052–2066. [CrossRef]
29. Dowling, A.P. A kinematic model of a ducted flame. *J. Fluid Mech.* **1999**, *394*, 51–72. [CrossRef]
30. Semlitsch, B.; Orchini, A.; Dowling, A.P.; Juniper, M.P. G-equation modelling of thermoacoustic oscillations of partially premixed flames. *Int. J. Spray Combust.* **2017**, *9*, 260–276. [CrossRef]
31. Hield, P.A.; Brear, M.J.; Jin, S.H. Thermoacoustic limit cycles in a premixed laboratory combustor with open and choked exits. *Combust. Flame* **2009**, *256*, 1683–1697. [CrossRef]
32. Tao, C.F.; Zhou, H. Effects of different preheated CO_2/O_2 jet in cross-flow on combustion instability and emissions in a lean premixed combustor. *J. Energy Inst.* **2020**, *93*, 2334–2343. [CrossRef]
33. Tao, C.F.; Zhou, H. Effects of different active control strategies on combustion instability decay time, actuator voltage, and damping ratio. *J. Appl. Phys.* **2020**, *128*, 054902. [CrossRef]
34. Dowling, A.P. Nonlinear self-excited oscillations of a ducted flame. *J. Fluid Mech.* **1997**, *346*, 271–290. [CrossRef]
35. Fabian, J.; Christopher, G.W.; Thomas, L.; Abel, D. Robust Output Model Predictive Control of an Unstable Rijke Tube. *J. Combust.* **2012**, *2012*, 927345.
36. Dowling, A.P.; Morgans, A.S. Feedback control of combustion oscillations. *Annu. Rev. Fluid Mech.* **2005**, *37*, 151–182. [CrossRef]
37. Li, J.X.; Yang, D.; Luzzato, C.; Morgans, A.S. Open Source Combustion Instability Low Order Simulator for Longitudinal Modes. 2014. Available online: http://www.oscilos.com/ (accessed on 5 September 2023).

38. Jarmolowitz, F.; Abel, D. Modeling and model predictive control of a thermoacoustically unstable burner. *IFAC Proc.* **2010**, *43*, 68–73. [CrossRef]
39. Wang, J.; Rubini, P.; Qin, Q. A Porous Media Model for the Numerical Simulation of Acoustic Attenuation by Perforated Liners in the Presence of Grazing Flows. *Appl. Sci.* **2021**, *11*, 4677. [CrossRef]

energies

MDPI

Article

A Data-Based Hybrid Chemistry Acceleration Framework for the Low-Temperature Oxidation of Complex Fuels

Sultan Alqahtani [1,2], Kevin M. Gitushi [1] and Tarek Echekki [1,*]

1 Department of Mechanical and Aerospace Engineering, North Carolina State University, Raleigh, NC 27695, USA; ssqahtani@kku.edu.sa (S.A.); kmgitush@ncsu.edu (K.M.G.)
2 Department of Mechanical Engineering, King Khalid University, Abha 61421, Saudi Arabia
* Correspondence: techekk@ncsu.edu

Abstract: The oxidation of complex hydrocarbons is a computationally expensive process involving detailed mechanisms with hundreds of chemical species and thousands of reactions. For low-temperature oxidation, an accurate account of the fuel-specific species is required to correctly describe the pyrolysis stage of oxidation. In this study, we develop a hybrid chemistry framework to model and accelerate the low-temperature oxidation of complex hydrocarbon fuels. The framework is based on a selection of representative species that capture the different stages of ignition, heat release, and final products. These species are selected using a two-step principal component analysis of the reaction rates of simulation data. Artificial neural networks (ANNs) are used to model the source terms of the representative species during the pyrolysis stage up to the transition time. This ANN-based model is coupled with C_0–C_4 foundational chemistry, which is used to model the remaining species up to the transition time and all species beyond the transition time. Coupled with the USC II mechanism as foundational chemistry, this framework is demonstrated using simple reactor homogeneous chemistry and perfectly stirred reactor (PSR) calculations for n-heptane oxidation over a range of composition and thermodynamic conditions. The hybrid chemistry framework accurately captures correct physical behavior and reproduces the results obtained using detailed chemistry at a fraction of the computational cost.

Keywords: chemistry reduction; data-based hybrid chemistry model; artificial neural network; principal component analysis

1. Introduction

Establishing a mechanistic separation between faster pyrolysis and subsequent slower oxidation in complex hydrocarbon fuels forms the basis for the HyChem approach to high-temperature oxidation (HTO) [1–3]. The pyrolysis stage features a common set of fuel fragments, whose global chemistry model is coupled with simpler C_0–C_4 foundational chemistry [1]. In contrast to HTO, the low-temperature oxidation (LTO) of these fuels is characterized by the initial formation of fuel-specific intermediates [4] (e.g., for alkanes, these include alkyl, alkylperoxy, and alkyl hydroperoxide radicals). The modeling of the chemistry of these intermediates represents an important challenge in the development of new detailed mechanisms for complex fuels.

In a recent study, we proposed a data-based hybrid chemistry approach for the chemistry acceleration of complex fuels in HTO [5]. In this approach, a set of representative species, instead of pre-selected fuel fragments, are used to model the pyrolysis stage. These species "shadow" the progress of the fuel fragments and can be modeled beyond the pyrolysis stage using simpler foundational C_0–C_2 or C_0–C_4 chemistry [5–7]. The representative species are selected using principal component analysis (PCA), a traditional tool for chemistry reduction [8], and their hybrid chemistry is modeled using artificial neural networks (ANNs) during pyrolysis and simpler foundational chemistry during oxidation.

Recently, Alqahtani [9] proposed an extension of the same approach for the LTO chemistry of complex hydrocarbon fuels. He showed that a set of representative species can be selected for different fuels. One of the key strategies in the development of HyChem [1–3,10–14] and the data-based hybrid chemistry approach [5,7] is based on modeling fuel and fragments during the pyrolysis stage, during which fuel consumption occurs. The fast pyrolysis followed by the slower oxidation of fragments during HTO in complex hydrocarbon fuels was observed by You et al. [15]. Unlike HTO, in LTO, fuel is consumed primarily by oxidation and this consumption occurs in two stages. The first stage, which corresponds to an exponential rise in OH and HO_2 radicals, is characterized by a build-up of fuel-specific intermediates, as mentioned above, a moderate increase in temperature, and oxygenated intermediates like CO, CH_2O, CH_3CHO (acetaldehyde), CH_2CO (ketene), and other heavier carbonyls. The slow oxidation of these intermediates is then followed by a second-stage ignition [16]. Similarly to HTO [5], where PCA is used to identify fragments in the context of representative species, key intermediates that track both stages of fuel consumption can be selected as representative species using PCA. These representative species can then be modeled using ANNs. ANNs have been successfully used in combustion for the tabulation of chemical reactions and chemistry reduction [17–21]. Additionally, as a regression tool for species and temperature chemical source terms, ANNs have been used to accelerate the evaluation of chemistry [22–27].

This study extends the earlier work of Alqahtani in several ways. First, we validate the hybrid chemistry model from data generation to a posteriori simulations by constructing ANN models for the reactions of the representative species. Second, the hybrid model is demonstrated using two canonical reactor models that are commonly used for chemistry reduction. They include ignition with homogeneous chemistry and the use of the perfectly stirred reactor (PSR). Data clustering is also implemented to couple these two reactor models during the training of the ANN. The present study's ultimate goal is to extend the hybrid chemistry strategy for complex fuel HTO [5] to complex fuel LTO. The resulting approach addresses the specific requirements for implementing hybrid chemistry for LTO and implements strategies for combining training data from different reactor models. We explore this feasibility using n-heptane, which was one of the three fuels considered in Ref. [5].

2. Methodology

To demonstrate the proposed hybrid chemistry approach, we contrast the temporal evolutions of fuel-specific intermediates and other species to temperature during the homogeneous ignition of n-heptane during LTO. They are obtained from the simulations of a simple reactor model based on homogeneous chemistry at constant volume, using Senkin [28] and the detailed mechanism proposed by Mehl et al. [29]. Figure 1 shows the temperature and species mole fraction profiles during n-heptane LTO based on the initial conditions of 700 K, 20 atm, and a mixture equivalence ratio of 0.5. The species shown in Figure 1 account for some of the higher species concentrations during all phases of LTO. Figure 1d shows the fuel-derived intermediates, while Figure 1b,c show other intermediate species with lower chemical complexity, including the usual species found in the fragment list for high-temperature chemistry combined with CH_2O, HO_2, and H_2O_2. These species exhibit the mechanistic phases of the ignition process, including the two stages of ignition. More importantly, the simpler species account for the bulk of the composition within the mixture, while the fuel-derived intermediates (e.g., $C_7H_{15}O_2$, C_7H_{14}, $C_7H_{14}O$, and C_7H_1OOH for n-heptane) account for much smaller fractions of the composition of the original fuel at their peaks between the two ignition stages. Toward the start of the second stage, the fuel-derived intermediates are depleted at a faster rate than other species, such as CO, H_2, CH_2O, and C_2H_4. These results support our proposed hybrid chemistry approach. Tracking the initial stages of fuel oxidation using a common set of simpler hydrocarbons and other species, which also track the bulk of the mixture, can potentially delineate the various stages of fuel oxidation and account for the bulk of the atoms in the mixture.

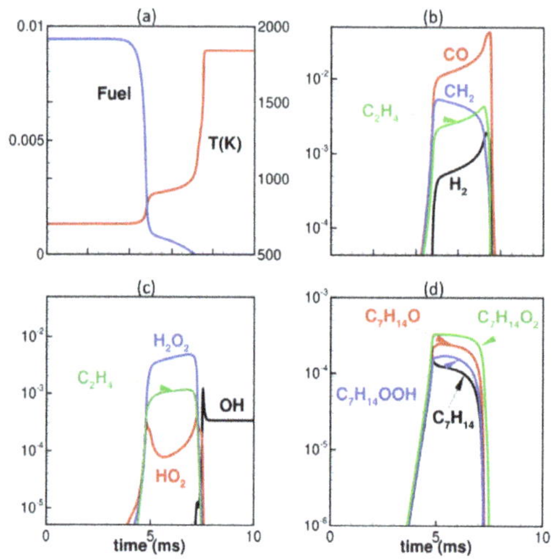

Figure 1. Temperature and species mole fraction profiles during the two-stage LTO of n-heptane, where $\phi = 0.5$, $T = 700$ K, and $p = 20$ atm. (**a**) temperature and fuel mole fraction, (**b**) CO, H_2, C_2H_4 and CH_2 mole fractions, (**c**) OH, HO_2, H_2O_2 and C_2H_4 mole fractions, (**d**) C_7H_{14}, $C_7H_{14}O$, $C_7H_{14}O_2$ and $C_7H_{14}OOH$ mole fractions.

As an extension of our recent work on HTO [5], we seek to couple a foundational chemistry model with a chemical model for complex fuel LTO. Using ANNs, we model the reaction rates of representative species, which are used to track the stages of fuel oxidation. These species tend to consist of simpler hydrocarbons and other intermediates or oxidation species. The remaining species are modeled entirely by foundational chemistry, using the species equations described in Section 2.1. As a data-based framework, the proposed hybrid model is designed to be adaptive to a set of conditions that is determined by the data on which the model is trained.

2.1. Database Generation

We generate a database of chemical data using the simple reactor model of homogeneous chemistry at constant pressure and steady PSR. The mixture conditions used for these reactors consist of equivalence ratios of 0.8, 1, and 1.2 and initial temperatures of 700 to 860 K in increments of 20 K. Additionally, 50 residence times that are evenly spaced on a log scale from −9 to 0 are used for the steady PSR. The lumped mechanism by Brunialti et al. [30] is used as the detailed chemistry mechanism, which consists of 538 species and 2824 reactions. The governing equations for the simple reactor are listed below.

- The energy equation

$$\frac{dT}{dt} = -\frac{\sum_{k=1}^{N} h_l \,\hat{\omega}_k \,\hat{W}_k}{\rho \bar{c}_p} \tag{1}$$

- The species equation for $k = 1, \ldots, N$

$$\frac{dY_k}{dt} = \frac{\hat{\omega}_k \hat{W}_k}{\rho}, \tag{2}$$

where T is the temperature, h_k, $\hat{\omega}_k$, and $\hat{W}_k$ are the enthalpy, reaction rate, and molecular weight of the kth species, respectively, ρ is the mixture density, $\bar{c}_p$ is the mixture average specific heat at constant pressure, and Y_k is the concentration of the kth species.

The governing equations for the steady PSR are listed below.

- The enthalpy equation

$$0 = \frac{h_{in} - h}{\tau} \tag{3}$$

- The species equation for $k = 1, \ldots, N$

$$0 = \frac{Y_{k,in} - Y_k}{\tau} + \frac{\hat{\omega}_k \hat{W}_k}{\rho}, \tag{4}$$

where h is the total enthalpy (chemical and sensible) of the mixture, τ is the residence time, and the subscript in denotes the inlet conditions to the PSR.

2.2. Representative Species Selection

Representative species are selected using a principal component analysis (PCA) of the simulation data [5], using the chemical reaction rates of all species in detailed or skeletal mechanisms. PCA is a powerful technique for identifying the importance of the chemical reactions involved in a mechanism [8].

PCA is implemented as an eigendecomposition of the covariance matrices of the species reaction rate data. It results in a vector of principal components (PCs), $\mathbf{\Psi}$, which represents a linear transformation of the centered species reaction rate vector $\mathbf{\Omega}$.

$$\mathbf{\Psi} = \mathbf{Q}^{\mathrm{T}} \ \mathbf{\Omega}. \tag{5}$$

where $\mathbf{Q}$ is the eigenvector of the species reaction rate covariance matrix. The jth PC can be expressed in terms of species reaction rate as follows:

$$\Psi_j = \sum_{k=1}^{N} q_{k,j} \, \hat{\omega}_k^* \tag{6}$$

where N is the total number of species/reaction rates, $\hat{\omega}_k^*$ is the centered (i.e., yielding a zero mean across training data) kth species reaction rate, and $q_{k,j}$ are the (k, j) elements of $\mathbf{Q}$.

The PCs are then ordered based on their eigenvalues, such that the bulk of the data variance is represented by the leading PCs. For a given PC, we also order the species reaction rates in Equation (6), based on the magnitudes of their coefficients $q_{k,j}$. From this re-ordering of the PCs and the species reaction rates for each PC, we identify the corresponding representative species. In Ref. [5], as well as here, we implement PCA in two steps. The two-step procedure is adopted to progressively select the representative species. A set of conditions is imposed on the retained principal components (PCs). The first condition is satisfied by accounting for more than 99% of the data variance. The second condition, i.e., the cutoff condition, is set at 90% of the data variance. The representative species are selected from the list of species that are part of the C_0–C_4 foundational chemistry. Once the representative species are selected, a model for their chemistry is constructed based on the available data.

2.3. Data Clustering

The database of chemical reactions used is a multidimensional and multiscale system with complex nonlinear behavior. Although it is theoretically possible to employ a single ANN to deal with such a complex problem, it is not a good choice because of poor computational efficiency and accuracy. And while a very deep and wide network can be used to approximate the entire dynamics, this may lead to higher computational costs. As a result, it is necessary to cluster data to improve accuracy and lower computational costs [17,18,31,32]. Self-organizing maps (SOMs) [33] and K-means [34] are well-established clustering techniques for clustering chemical data [17–19]. Local principal component analysis (LPA) [35] is another approach used to cluster data with similar chemical compositions.

While clustering techniques are not within the scope of this study, we use the clustering approach implemented in the PCAfold 2.0 software [36]. This software implements a zero-neighborhood clustering technique, which clusters data by separating close-to-zero observations into one or two clusters. This feature of clustering close-to-zero data is very important since the mass fractions and reaction rates of most representative species are zero, if not close to zero, during the early stages of oxidation. To account for the contributions of all representative species, zero-neighborhood clustering is carried out in the principal component (PC) spaces of the reaction rates of both the homogeneous reactor and PSR, where zero PCs correspond to regions where mass fractions and reaction rates are close to zero.

Before applying zero-neighborhood clustering, we separate the data into two categories based on the first and second stages of ignition. The first stage of ignition in LTO can be identified by the "plateau" of the slope of the CO time history [16]. In this study, this plateau occurs around the time when the CH_2O mass fraction reaches its maximum peak. After clustering the data into the two stages of ignition, we apply zero-neighborhood clustering on each stage of ignition.

The parameters of zero-neighborhood clustering are the offset percentage from zero, the number of clusters, and the option to group data close to zero into one or two different clusters. In this work, the number of clusters is set to four for both stages of ignition and data close to zero are clustered into two different bins. This results in a total of eight clusters. The offset percentage from zero at which splits are performed is set to 2.5 and 1 for the first and second stages of ignition, respectively. The clustering is performed as follows [36]:

(a) Using the offset percentage from zero ϵ_0, we obtain the offset ε

$$\varepsilon = \frac{\epsilon_0 \cdot |\max(PC_{obs}) - \min(PC_{obs})|}{100}, \tag{7}$$

where PC_{obs} is the reaction rate PCs;

(b) The cluster borders are obtained using the offset ε as

$$BC^- = \begin{cases} \mathcal{F}(\min(PC_{obs}), & -\varepsilon, & \frac{n_{clusters}+1}{2}), & \text{if } |\min(PC_{obs})| > |\max(PC_{obs})| \\ \mathcal{F}(\min(PC_{obs}), & -\varepsilon, & n_{clusters}/2), & \text{otherwise} \end{cases}, \tag{8}$$

$$BC^+ = \begin{cases} \mathcal{F}(\varepsilon, & \max(PC_{obs}), & n_{clusters}/2), & \text{if } |\min(PC_{obs})| > |\max(PC_{obs})| \\ \mathcal{F}(\varepsilon, & \max(PC_{obs}), & \frac{n_{clusters}+1}{2}), & \text{otherwise} \end{cases}, \tag{9}$$

where BC^- and BC^+ are the negative and positive borders of a cluster, $n_{clusters}$ is the set number of clusters, $||$ is the absolute value operator, and $\mathcal{F}$ is the function to evenly space numbers over a specified interval (*linspace*);

(c) Data are then clustered based on where they fall along the negative or positive borders.

The sensitivities of the offset percentage and the number of clusters are not in the scope of this study; however, these could be explored in future works.

2.4. Artificial Neural Network Architecture

In the present study, an ANN-based regression is developed to determine the reaction rates of the representative species in terms of the temperature and mass fractions of the fuel: O_2, H_2O_2, and CH_2O. These scalars, which are part of the representative species, are adequate markers of each of the stages of ignition and they have high correlations with the remaining representative species. The temperature and these five species are the inputs to the ANNs. ANNs are developed for each cluster.

For each cluster, two sets of ANNs are developed. The first set of ANNs is only used to predict each of the inputs. This is conducted to further improve the accuracy of the hybrid scheme during the a posteriori time integration of the governing equations. Since the ANN models are coupled with foundational chemistry, small errors in the predictions of temperature and species mass fractions propagate over time and become larger, resulting in the inaccurate prediction of all representative species and the chemical system. By training an ANN for each input, we eliminate the complexity associated with multidimensional outputs and independently optimize each ANN's prediction accuracy and computational efficiency. The second set consists of one ANN for predicting the reaction rates of all remaining species. The data are scaled for training using the equations below.

- Input scaling

$$TS^* = \frac{TS_n^\lambda - 1}{\lambda} \tag{10}$$

where

$$TS_n = \frac{TS - \min(TS)}{\max(TS) - \min(TS)} \quad and \quad \lambda = 0.1 \tag{11}$$

- Reaction rate scaling

$$\hat{\omega}_n = 2\frac{\hat{\omega} - \min(\hat{\omega})}{\max(\hat{\omega}) - \min(\hat{\omega})} - 1 \tag{12}$$

Each input is modeled by a single ANN comprising two hidden layers with 20 neurons, while the remaining representative species are modeled by a single ANN comprising three hidden layers with 20 neurons. Figure 2 shows an example of the network architectures for CH_2O and remaining species prediction. The transfer function connecting a given neuron to neurons in a previous layer is the hyperbolic tangent function. The ANN weights and biases are updated during the training process by minimizing the selected loss function, which is the mean squared error (MSE). For each input ANN, loss function minimization is optimized by a Levenberg–Marquardt backpropagation algorithm in Matlab [37] and the training is completed over 5000 epochs. The ANN modeling of the remaining species is optimized using the Adam optimizer in PyTorch [38] over 100,000 epochs. During the first 20,000 epochs, a learning rate of 0.005 is used with a weight decay of 10^{-6}, after which the learning rate and weight decay are decreased to 10^{-5} and 10^{-9}, respectively. The training data used to build the hybrid model are divided randomly into three sets: a training set, a test set, and a validation set with the proportions of 70/20/10%, respectively. The training is carried out on AMD Ryzen Threadripper ROI 5975WX CPU for a total time of about 1.5 h, of which about 1 h accounts for training the input ANNs and 30 min account for training the remaining representative species ANNs. A K-nearest neighbor (K-NN) algorithm [39] is used to classify each of the clusters and identify which ANN to use for the prediction of the representative species. This training is carried out in less than 2 min. The classifier uses the same inputs as the ANNs.

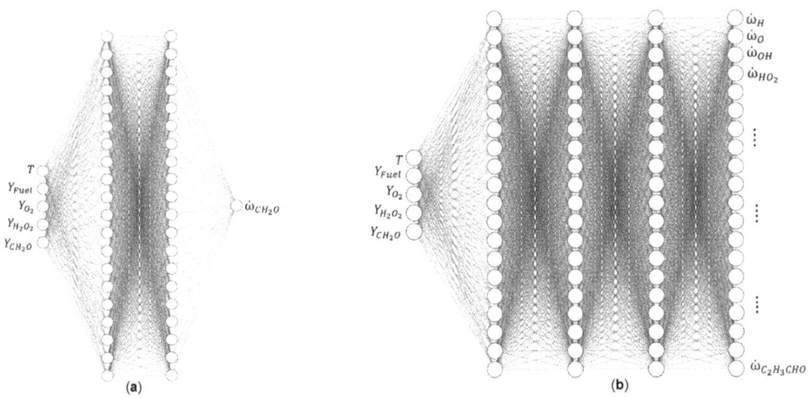

Figure 2. Illustration of the ANN architectures: (**a**) ANN architecture for CH_2O prediction; (**b**) ANN architecture for the remaining reaction rate prediction.

2.5. Summary of the Hybrid Chemistry Model

After generating the data, selecting the representative species, clustering the data, and modeling the reaction rates using ANNs, we couple the ANN-regression models with foundational chemistry, where the ANNs model the representative species and foundational chemistry models all other remaining species. To illustrate the implementation of this hybrid scheme, we use the simple reactor model of homogeneous chemistry at constant pressure, as described in Section 2.1. The governing equations for this system of N species in foundational chemistry and a subset of N_r representative species are listed below.

- The energy equation

$$\frac{dT}{dt} = -\frac{\sum_{k=1}^{N} h_k\, \hat{\omega}_k\, \hat{W}_k}{\rho \bar{c}_p} \tag{13}$$

- The representative species equation for $k = 1, \ldots, N_r$

$$\frac{dY_k}{dt} = \frac{\hat{\omega}_k \hat{W}_k}{\rho} \text{ where } \hat{\omega}_k = \begin{cases} \hat{\omega}_{k,\mathrm{h}}, & \text{if } t < \tau_{\mathrm{transition}} \\ \hat{\omega}_{k,\mathrm{f}}, & \text{otherwise} \end{cases}, \tag{14}$$

- Remaining species equation for $k = 1, \ldots, N - N_r$

$$\frac{dY_k}{dt} = \frac{\hat{\omega}_{k,\mathrm{f}} \hat{W}_k}{\rho}. \tag{15}$$

where T is the temperature, h_k, $\hat{\omega}_k$, and $\hat{W}_k$ are the enthalpy, reaction rate, and molecular weight of the kth species, respectively, ρ is the mixture density, $\bar{c}_p$ is the mixture average specific heat at constant pressure, and Y_k is the concentration of the kth species.

In the above equations, the subscript "h" corresponds to a global or ANN model for the representative species that is used up to a transition point and "f" corresponds to the foundational chemistry model. In the present study, that transition point also can be tagged as the transition time $\tau_{\mathrm{transition}}$ during fuel oxidation. An adequate model may be needed up to the onset of the second ignition stage. Beyond that transition time, the representative species reaction rates are modeled with foundational chemistry. Meanwhile, all remaining species are modeled with foundational chemistry. The hybrid model for representative species chemistry attempts to capture the roles of the initial stages of fuel oxidation, including the formation of fuel-specific intermediates, which are not captured by foundational chemistry.

For more complex reactor models, a classifier may be used to decide which chemistry model to use for a given mixture condition. These classifiers, such as the pattern recognition network (PRN) used in Ref. [5], are also trained on detailed chemistry simulation data. Hybrid models for the reaction rates of the representative species can be determined via regression using ANNs, as shown in [5–7]. Alternatively, they can also be determined using global reaction rates, similar to the forms proposed for fragments in HyChem [1–3].

3. Results and Discussion

3.1. Selected Representative Species

Table 1 lists the representative species obtained using the two-step PCA approach. From 538 species in the detailed mechanism, 24 species are retained using the two-step PCA. The two-step PCA achieves a 95.5% reduction. The major products of combustion, CO_2 and H_2O, along with CO, are among the selected species. Low-temperature radicals, such as H_2O_2, HO_2, and CH_2O, are also retained. Other species that are dominant and highly reactive in HTO or the second stage of ignition, such as H_2 and HCCO, are also among the retained species. Additionally, the two-step PCA also selects the species H, O, OH, HO_2, H_2, H_2O, H_2O_2, CH_3, HCO, CH_2O, CO, CO_2, and CH_3CHO, which are formed in the oxidation of any fuel and can be used as universal markers of oxidation reactions [40], as well as OH, HO_2, CH_2O, and CO, which can be used to predict the IDTs of many fuels [41].

This demonstrates that the two-step PCA retains key species that adequately mark both stages of ignition during LTO. A total of 26 thermochemical scalars are modeled, including 24 representative species, temperature, and fuel.

Table 1. List of the selected representative species.

Fuel	n-C_7H_{16}
Selected Species (24)	H, O, OH, HO_2, H_2, H_2O, H_2O_2, O_2, CH_3, HCO, CH_2O, CH_3O, CO, CO_2, C_2H_3, C_2H_4, C_2H_5, HCCO, CH_2CO, CH_3CO, CH_2CHO, CH_3CHO, C_3H_6, and C_2H_3 CHO

3.2. Data Clusters

The results of the clustering technique are shown in Figures 3 and 4 for the homogeneous reactor and Figures 5 and 6 for the steady PSR. In Figure 3, the clustered mass fractions (top row) and the reaction rates (bottom row) are shown for the fuel, O_2, H_2O_2, and CH_2O. Figure 4 shows the clusters during both stages of ignition. We can observe that Clusters 2, 3, and 4 contain all zero and close-to-zero reaction rates in the first stage of ignition, while Cluster 1 mostly contains reaction rates with larger magnitudes. This illustrates the importance of the clustering approach. Since most of the close-to-zero reaction rates present during the slow and early stages of oxidation are modeled separately from the large reaction rates, better prediction accuracy is achieved for the early stages of ignition compared to modeling the entire first stage of ignition. Similarly, for the second ignition stage, Cluster 7 contains all zero reaction rates, corresponding to equilibrium. Cluster 8 contains the high-reactivity data of the high-temperature ignition.

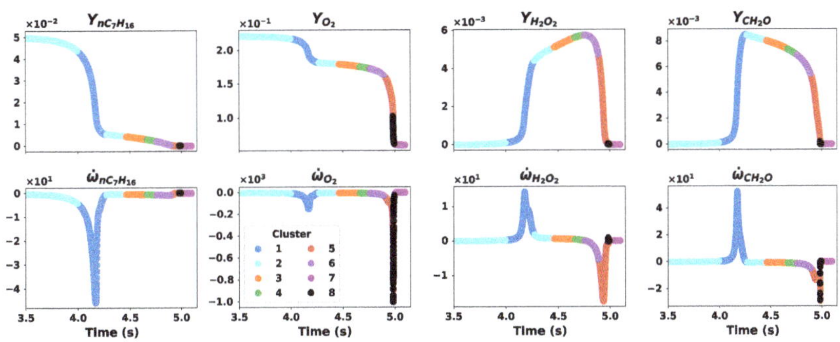

Figure 3. Homogeneous data clustering: species mass fraction and reaction rate temporal profiles during the oxidation of n-heptane, where $\phi = 0.8$, $T = 700$ K, and $p = 20$ atm.

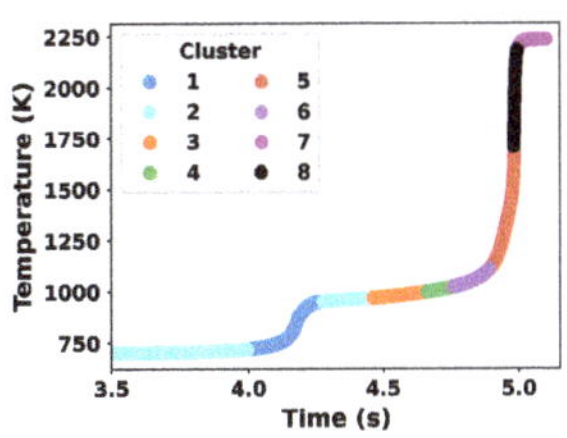

Figure 4. Homogeneous data clustering: temperature temporal profile during the oxidation of n-heptane, where $\phi = 0.8$, $T = 700$ K, and $p = 20$ atm.

Figure 6 shows the clustered mass fractions (top row) and reaction rates (bottom row) for the fuel, O_2, H_2O_2, and CH_2O and Figure 6 shows the clustered temperature. Unlike the homogeneous reactors where eight clusters are identified, here, data are only found in Clusters 2 and 7. This is due primarily to the initial separation of the data into two categories based on the stages of ignition, as described in Section 2.3. Additionally, all data after ignition are included in Cluster 7. The reaction rates of the PSRs are 1 to 2 orders of magnitude smaller than those of the homogeneous reactors; therefore, they are close to zero, which is why they fall into Cluster 7.

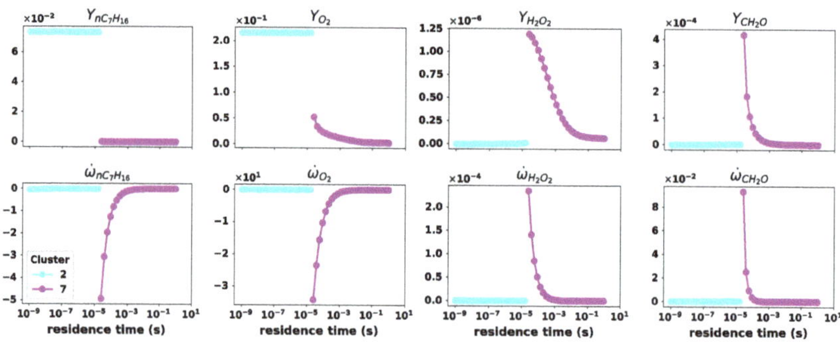

Figure 5. PSR data clustering: species mass fraction and reaction rate temporal profiles, where $\phi = 1.2$, $T = 780$ K, and $p = 20$ atm.

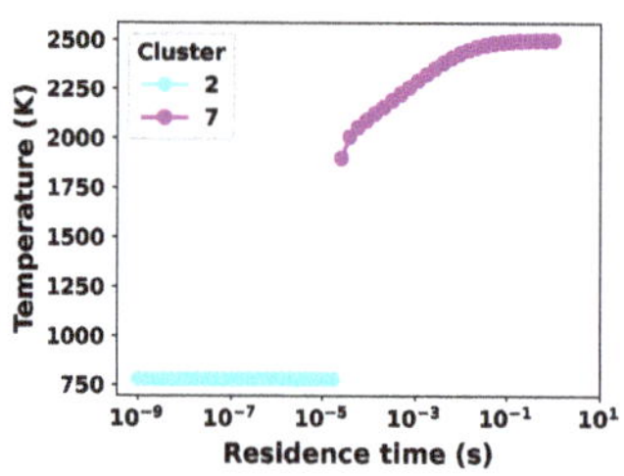

Figure 6. PSR data clustering: temperature temporal profile, where $\phi = 1.2$, $T = 780$ K, and $p = 20$ atm.

3.3. *ANN Training and Reaction Rate Prediction*

We investigate the generalization of the ANNs by predicting test and training datasets. The scatter plot in Figure 7 shows good correlation between the actual Cantera and ANN-predicted reaction rates. Here, the models are trained using data obtained by solving Equations (1)–(4) with the initial temperatures of 700, 740, 780, 820, and 860 K for equivalence ratios of 0.8, 1, and 1.2 and tested using the initial temperatures of 720, 760, 800, and 840 K for the same equivalence ratios. Figure 8 shows the evolution of the loss functions as a function of the epochs or iterations for the first four clusters. Training for Clusters 5 to 8 also exhibits similar behavior. At around 20,000 iterations, the loss function becomes somewhat "steady" and no longer decreases. Decreasing the learning rate and weight decay improves the network's training and decreases the loss function. Table 2 shows the training losses.

The homogeneous reactor ANN predictions of reaction rates are shown in Figure 9, with an initial temperature of 700 K and an equivalence ratio of 1. The PSR ANN predictions of reaction rates are shown in Figures 10 and 11, with initial temperatures of 740 K and 700 K and equivalence ratios of 1 and 0.8. The ANN predictions accurately match the Cantera solutions. Despite the reactors displaying two different behaviors, the ANNs in each cluster accurately learn and predict each reactor's behavior.

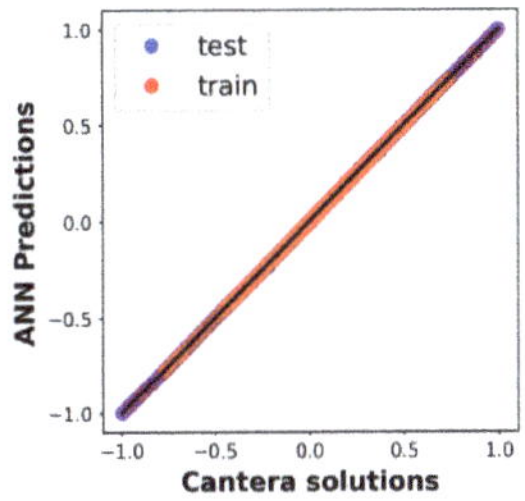

Figure 7. Training (red symbols) and test (blue symbols) data scatter plot of Cantera and ANN-predicted reaction rates. The black line represents the line of best fit.

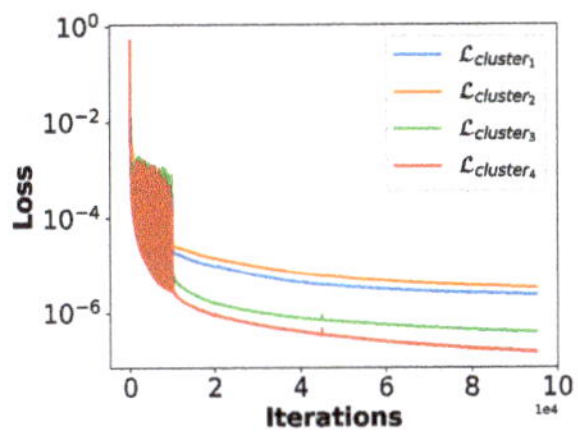

Figure 8. Training losses of clusters for the first stage of ignition.

Table 2. Values of the loss functions of the obtained models for the training dataset.

Cluster	Loss
1	2.4×10^{-6}
2	3.4×10^{-6}
3	3.9×10^{-7}
4	1.5×10^{-7}
5	2.6×10^{-6}
6	2.1×10^{-7}
7	4.6×10^{-6}
8	1.3×10^{-6}

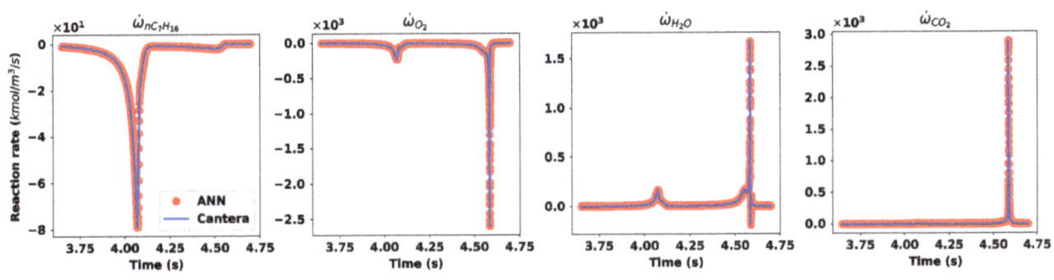

Figure 9. Comparison of the ANN-predicted and Cantera reaction rates for the homogeneous reactor, with an initial temperature of 700 K and an equivalence ratio of 1.

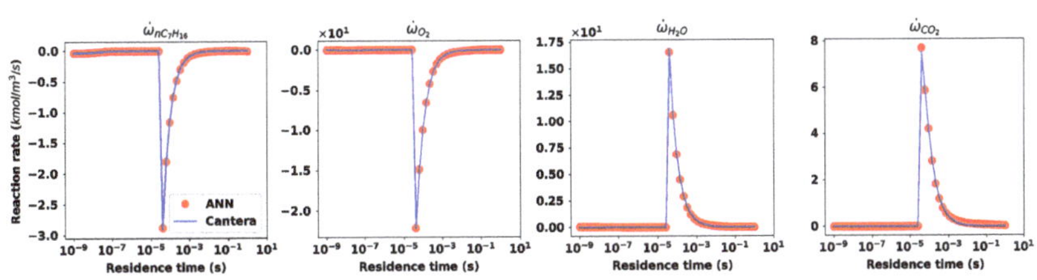

Figure 10. Comparison of the ANN-predicted and Cantera reaction rates for the PSR, with an initial temperature of 740 K and an equivalence ratio of 1.

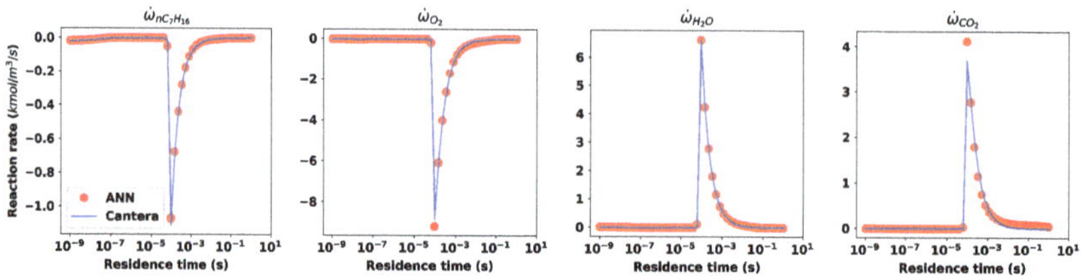

Figure 11. Comparison of the ANN-predicted and Cantera reaction rates for the PSR, with an initial temperature of 700 K and an equivalence ratio of 0.8.

3.4. Hybrid Chemistry Model

We demonstrate the hybrid chemistry approach by comparing a posteriori simulations to detailed chemistry data using the homogeneous reactor. The foundational chemistry is based on the C_0–C_4 USC Mech-II mechanism [42], with 111 species and 784 reactions. The governing Equations (13)–(15) are integrated using Cantera [43]. To demonstrate the hybrid chemistry scheme, we show the results of simulations with initial temperatures of 740 K and 820 K and an equivalence ratio of 1. These two sets of conditions are included in the training data. We also show results with an initial temperature of 720 K and an equivalence ratio of 0.8 to demonstrate the model's ability to interpolate. Figures 12–14 show comparisons of the temporal profiles of the detailed mechanism calculations (solid black) and those of the hybrid scheme (dotted red) for the temperature and select species mass fractions under these initial conditions.

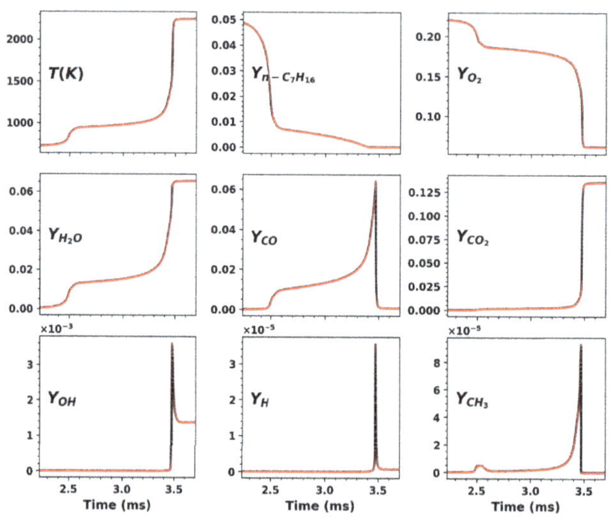

Figure 12. Comparison of temperature and species mass fraction profiles during the oxidation of n-heptane, where ϕ = 0.8, T = 720 K, and p = 20 atm. The hybrid model results are shown by the dotted red lines and those of the detailed mechanism are shown by the solid black lines.

The figures show that the timings of the two stages of ignition are accurately captured by the hybrid chemistry scheme. However, we do not consider fuel-specific intermediates nor the very initial stages of fuel oxidation, which include H abstraction and some of the chain propagation reactions that also involve other key radicals, such as OH and CH_3 [4]. Nonetheless, the profiles of these radicals are well predicted and match the detailed chemistry calculations. Regardless, if the goal is to capture the ignition delay, heat release, and products, the hybrid scheme's performance is very good. More importantly, the scheme results in a computation speed increase of up to 9 times that of detailed chemistry calculations. Additionally, there is a potential to save memory since the hybrid model tracks fewer species (111) and reactions (784) compared to the detailed mechanism (538 species and 2824 reactions). Further acceleration may be achieved with a further reduction in the number of representative species and foundational chemistry as well.

Overall, the proposed hybrid scheme can capture ignition delay times (IDTs) and the extent of temperature increases (i.e., the rates of heat release) at each stage and can be used to determine the final products of combustion.

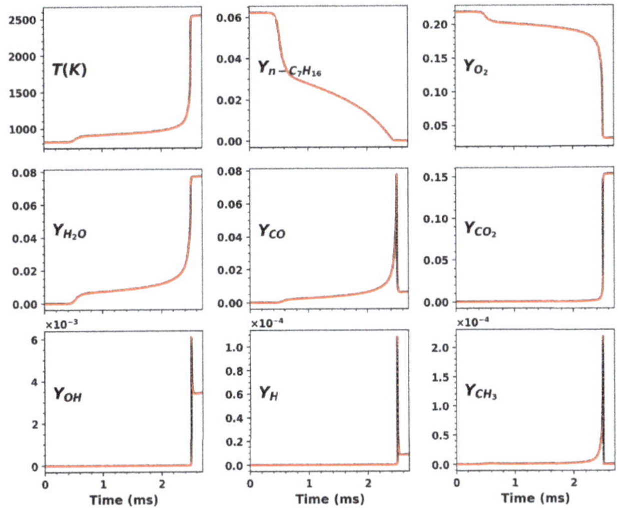

Figure 13. Comparison of temperature and species mass fraction profiles during the oxidation of n-heptane, where $\phi = 1$, $T = 820$ K, and $p = 20$ atm. The hybrid model results are shown by the dotted red lines and those of the detailed mechanism are shown by the solid black lines.

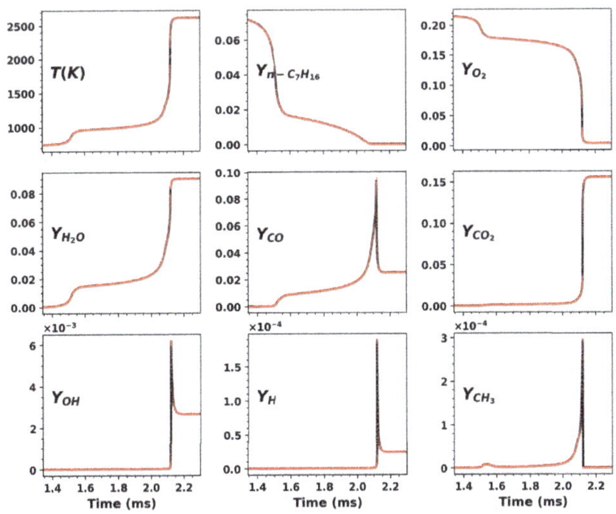

Figure 14. Comparison of temperature and species mass fraction profiles during the oxidation of n-heptane, where $\phi = 1.2$, $T = 740$ K, and $p = 20$ atm. The hybrid model results are shown by the dotted red lines and those of the detailed mechanism are shown by the solid black lines.

4. Conclusions

In the present study, we demonstrate the feasibility of a data-based hybrid chemistry scheme for LTO, based on modeling the chemistry of representative species and coupling this chemistry with simpler foundational chemistry. In total, 24 representative species are selected using two-step PCA carried on reaction rate data obtained from simulations using detailed chemistry with 538 species and 2824 reactions. Among the 24 representative species are 14 species that are formed during the oxidation of any fuel [40] and 4 species (OH, HO_2, CH_2O, and CO) that can be used for IDT prediction of many fuels [41]. ANNs are used to model the representative species reaction rates and are coupled with smaller C_0–C_4 foundational chemistry with 111 species and 784 reactions to model the remaining species. To improve the computational efficiency and prediction accuracy of the ANNs, the reaction rate data are clustered into eight clusters such that data close to zero are in the same cluster. A validation of this approach via comparisons to detailed chemistry calculations for n-heptane shows that the model results match well with detailed chemistry calculations. The model also correctly captures the timings of the stages of ignition, as well as IDT, heat release, and major species and radicals. With a calculation speed increase of up to 9 times that of detailed chemistry, the hybrid scheme can accelerate the chemistry integration process. Similarly, a speed increase of 1 order of magnitude is achieved for HTO [5].

Future work will aim to achieve aggressive reductions in the numbers of representative species by refining the PCA-based procedure or using other reduction methods, such as directed related graphs (DRGs) [44–46] or global pathway selection (GPS). Ref. [47] identified a smaller number of representative species in a hybrid model based on desired criteria for accuracy in prediction chemistry. The sensitivity of the hybrid scheme to the choice of foundational chemistry will also be investigated to establish prediction accuracy and computational acceleration across different foundational chemistry sizes. We will also extend the scheme using neural ordinary differential equations (NODEs) as these have been shown to have better prediction accuracy and generalization [48–51]. Additionally, improving the criteria for determining the transition from ANN regression to foundational chemistry could further enhance the model and these criteria could be used to implement PRN classifiers. The choice of transition can determine how accurate the final products are because, beyond this point, only foundational chemistry is used; thus, having accurate mixture compositions before the transition is important.

Author Contributions: Conceptualization, S.A., K.M.G. and T.E.; methodology, K.M.G., S.A. and T.E.; software, K.M.G. and S.A.; validation, K.M.G. and S.A.; formal analysis, K.M.G. and S.A.; writing—original draft preparation, K.M.G., S.A. and T.E.; writing—review and editing, K.M.G., S.A. and T.E.; supervision, T.E.; project administration, T.E.; funding acquisition, T.E. All authors have read and agreed to the published version of the manuscript.

Funding: This research was funded by King Abdullah University of Science and Technology, under grant number 4351-CRG9.

Data Availability Statement: The data presented in this study are available upon reasonable request from the corresponding author.

Acknowledgments: Sultan Alqahtani would like to acknowledge the support of King Khalid University in Abha, Saudi Arabia.

Conflicts of Interest: The authors declare no conflicts of interest and that they have no known competing financial interests or personal relationships that could have appeared to influence the work reported in this paper.

References

1. Wang, H.; Xu, R.; Wang, K.; Bowman, C.; Hanson, R.; Davidson, D.; Brezinsky, K. A physics-based approach to modeling real-fuel combustion chemistry-I. evidence from experiments, and thermodynamic, chemical kinetic and statistical considerations. *Combust. Flame* **2018**, *193*, 502–519. [CrossRef]

2. Xu, R.; Wang, K.; Banerjee, S.; Shao, J.; Parise, T.; Zhu, Y.; Wang, S.; Movaghar, A.; Lee, D.; Zhao, R.; et al. A physics-based approach to modeling real-fuel combustion chemistry-II. reaction kinetic models of jet and rocket fuels. *Combust. Flame* **2018**, *193*, 520–537. [CrossRef]
3. Tao, Y.; Xu, R.; Wang, K.; Shao, J.; Johnson, S.; Movaghar, A.; Han, X.; Park, J.-W.; Lu, T.; Brezinsky, K.; et al. A physics-based approach to modeling real-fuel combustion chemistry-III. reaction kinetic model for JP10. *Combust. Flame* **2018**, *198*, 466–476. [CrossRef]
4. Zádor, J.; Taatjes, C.; Fernandes, R. Kinetics of elementary reactions in low-temperature autoignition chemistry. *Prog. Energy Combust. Sci.* **2011**, *37* 371–421. [CrossRef]
5. Alqahtani, S.; Echekki, T. A data-based hybrid model for complex fuel chemistry acceleration at high temperatures. *Combust. Flame* **2021**, *223*, 142–152. [CrossRef]
6. Ranade, R.; Alqahtani, S.; Farooq, A.; Echekki, T. An ANN based hybrid chemistry framework for complex Fuels. *Fuel* **2019**, *241*, 625–636. [CrossRef]
7. Ranade, R.; Echekki, T. An extended hybrid chemistry framework for complex hydrocarbon fuels. *Fuel* **2019**, *251*, 4276–4284. [CrossRef]
8. Brown, N.; Li, G.; Koszykowski, M. Mechanism reduction via principal component analysis. *Int. J. Chem. Kin.* **1996**, *29*, 393–414. [CrossRef]
9. Alqahtani, S. Machine Learning Methods for Chemistry Reduction in Combustion. Ph.D. Thesis, North Carolina State University, Department of Mechanical and Aerospace Engineering, Raleigh, NC, USA, 2020.
10. Xu, R.; Wang, H.; Davidson, D.F.; Hanson, R.K.; Bowman, C.T.; Egolfopoulos, F.N. Evidence supporting a simplified approach to modeling high-temperature combustion chemistry. In Proceedings of the 10th US National Meeting on Combustion, College Park, MD, USA, 23–24 April 2017.
11. Xu, R.; Chen, D.; Wang, K.; Tao, Y.; Shao, J.K.; Parise, T.; Zhu, Y.; Wang, S.; Zhao, R.; Lee, D.J.; Egolfopoulos, F.N. HyChem model: Application to petroleum-derived jet fuels. In Proceedings of the 10th US National Meeting on Combustion, College Park, MD, USA, 23–24 April 2017; Volume 69, pp. 70–77.
12. Xu, R.; Chen, D.; Wang, K.; Wang, H. A comparative study of combustion chemistry of conventional and alternative jet fuels with hybrid chemistry approach. In Proceedings of the 55th AIAA Aerospace Sciences Meeting, Grapevine, TX, USA, 9–13 January 2017; p. 0607.
13. Wang, K.; Xu, R.; Parise, T.; Shao, J.K.; Lee, D.J.; Movaghar, A.; Davidson, D.F.; Hanson, R.K.; Wang, H.; Bowman, C.T.; et al. Combustion kinetics of conventional and alternative jet fuels using a hybrid chemistry (HyChem) approach. In Proceedings of the 10th US National Combustion Institute Meeting, the Combustion Institute, College Park, MD, USA, 23–24 April 2017.
14. Wang, K.; Xu, R.; Parise, T.; Shao, J.K.; Davidson, D.F.; Hanson, R.K.; Wang, H.; Bowman, C.T. Evaluation of a hybrid chemistry approach for combustion blended petroleum and bio-derived jet fuels. In Proceedings of the 10th US National Combustion Institute Meeting, the Combustion Institute, College Park, MD, USA, 23–24 April 2017.
15. You, X.; Egolfopoulos, F.N.; Wang, H. Detailed and simplified kinetic models for n-dodecane oxidation: The role of fuel cracking in aliphatic hydrocarbon combustion. *Proc. Combust. Inst.* **2009**, *32*, 403–410 [CrossRef]
16. Choudhary, R.; Boddapati, V.; Clees, S.; Biswas, P.; Shao, J.; Davidson, D.F.; Hanson, R.K. Towards HyChem modeling of kinetics of distillate fuels in the NTC regime. In Proceedings of the Spring Technical Meeting of the Western States Section of the Combustion Institute, Stanford, CA, USA, 21–22 March 2022.
17. An, J.; He, G.; Luo, K.; Qin, F.; Liu, B. Artificial neural network based chemical mechanisms for computationally efficient modeling of hydrogen/carbon monoxide/kerosene combustion. *Int. J. Hydrogen Energy* **2020**, *45*, 29594–29605. [CrossRef]
18. Haghshenas, M.; Mitra, P.; Santo, N.D.; Schmidt, D.P. Acceleration of chemical kinetics computation with the learned intelligent tabulation (LIT) method. *Energies* **2021**, *14*, 7851. [CrossRef]
19. Nguyen, H.-T.; Domingo, P.; Vervisch, L.; Nguyen, P.-D. Machine learning for integrating combustion chemistry in numerical simulations. *Energy AI* **2021**, *5*, 100082. [CrossRef]
20. Zhou, L.; Song, Y.; Ji, W.; Wei, H. Machine learning for combustion. *Energy AI* **2022**, *7*, 100128. [CrossRef]
21. Chen, J-Y.; Blasco, J. A.; Fueyo, N.; Dopazo, C. An economical strategy for storage of chemical kinetics: Fitting in situ adaptive tabulation with artificial neural networks. *Proc. Combust. Inst.* **2000**, *28*, 115–121. [CrossRef]
22. Blasco, J.A.; Fueyo, N.; Dopazo, C.; Ballester, J. Modelling the temporal evolution of a reduced combustion chemical system with an artificial neural network. *Combust. Flame* **1998**, *113*, 38–52. [CrossRef]
23. Blasco, J.A.; Fueyo, N.; Larroya, J.C.; Dopazo, C.; Chen, Y.J. A single-step time-integrator of a methane–air chemical system using artificial neural networks. *Comput. Chem. Eng.* **1999**, *23*, 1127–1133. [CrossRef]
24. Ihme, M.; Marsden, A.L; Pitsch, H. Generation of optimal artificial neural networks using a pattern search algorithm: Application to approximation of chemical systems. *Neural Comput.* **2008**, *20*, 573–601. [CrossRef]
25. Chatzopoulos, A.K.; Rigopoulos, S. A chemistry tabulation approach via rate-controlled constrained equilibrium (RCCE) and artificial neural networks (ANNs), with application to turbulent non-premixed $CH_4/H_2/N_2$ flames. *Proc. Combust. Inst.* **2013**, *34*, 1465–1473. [CrossRef]
26. Franke, L.L.; Chatzopoulos, A.K.; Rigopoulos, S. Tabulation of combustion chemistry via Artificial Neural Networks (ANNs): Methodology and application to LES-PDF simulation of Sydney flame L. *Combust. Flame* **2017**, *185*, 245–260. [CrossRef]

27. Sinaei, P.; Tabejamaat, S. Large eddy simulation of methane diffusion jet flame with representation of chemical kinetics using artificial neural network. *Proc. Inst. Mech. Eng. Part E J. Process Mech. Eng.* **2017**, *231*, 147–163. [CrossRef]
28. Lutz, A.E.; Kee, R.J.; Miller, J.A. *SENKIN: A FORTRAN Program for Predicting Homogeneous Gas Phase Chemical Kinetics with Sensitivity Analysis*; Sandia National Laboratories: Livermore, CA, USA, 1988.
29. Mehl, M.; Pitz, W.J.; Westbrook, C.K.; Curran, H.J. Kinetic modeling of gasoline surrogate components and mixtures under engine conditions. *Proc. Combust. Inst.* **2011**, *33*, 193–200. [CrossRef]
30. Brunialti, S.; Zhang, X.; Faravelli, T.; Frassoldati, A.; Sarathy, S.M. Automatically generated detailed and lumped reaction mechanisms for low- and high-temperature oxidation of alkanes. *Proc. Combust. Inst.* **2023**, *39*, 335-344. [CrossRef]
31. Barwey, S.; Prakash, S.; Hassanaly, M.; Raman, V. Data-driven classification and modeling of combustion regimes in detonation waves. *Flow Turbul. Combust.* **2021**, *106*, 1065–1089. [CrossRef]
32. Ranade, R.; Li, G.; Li, S.; Echekki, T. An efficient machine-learning approach for pdf tabulation in turbulent combustion closure. *Combust. Sci. Tech.* **2021**, *193*, 1258–1277. [CrossRef]
33. Kohonen, T.; Hynninen, J.; Kangas, J.; Laaksonen, J. SOM_PAK: The self-organizing map program package. *Tech. Rep. Citeseer* **1996**, *31*, 1996.
34. MacQueen, J. Some methods for classification and analysis of multivariate observations. In *Fifth Berkeley Symposium on Mathematical Statistics and Probability*; University of California Press: Berkeley, CA, USA, 1967; Volume 1, pp. 281–297.
35. D'Alessio, G.; Parente, A.; Stagni, A.; Cuoci, A. Adaptive chemistry via pre-partitioning of composition space and mechanism reduction. *Combust. Flame* **2020**, *211*, 68–82. [CrossRef]
36. Zdybal, K.; Armstrong, E.; Parente, A.; Sutherland, J.C. *PCAfold 2.0-Novel tools and algorithms for low-dimensional manifold assessment and optimization. SoftwareX* **2023**, *23*, 101447. [CrossRef]
37. MathWorks Inc. Matlab Version: 9.13.0 (r2022b). 2022. Available online: https://www.mathworks.com (accessed on 1 December 2022).
38. Paszke, A.; Gross, S.; Massa, F.; Lerer, A.; Bradbury, J.; Chanan, G.; Killeen, T.; Lin, Z.; Gimelshein, N.; Antiga, L.; et al. PyTorch: An Imperative Style, High-Performance Deep Learning Library. In *Advances in Neural Information Processing Systems*; Curran Associates, Inc.: Kottayam, India, 2019; Volume 32, pp. 8024–8035.
39. Pedregosa, F.; Varoquaux, G.; Gramfort, A.; Michel, V.; Thirion, B.; Grisel, O.; Blondel, M.; Prettenhofer, P.; Weiss, R.; Dubourg, V.; Vanderplas, J.; Passos, A.; Cournapeau, D.; Brucher, M.; Perrot, M.; Duchesnay, E. Scikit-learn: Machine learning in Python. *J. Mach. Learn. Res.* **2011**, *12*, 2825–2830.
40. Buras, Z.; Hansen, N.; Taatjes, C.A.; Sheps, L. Prospects and Limitations of Predicting Fuel Ignition Properties from Low-Temperature Speciation Data. *Energy Fuels* **2022**, *36*, 3229–3238. [CrossRef]
41. Buras, Z.J.; Safta, C.; Zádor, J.; Sheps, L. Simulated production of OH, HO_2, CH2O, and CO_2 during dilute fuel oxidation can predict 1st-stage ignition delays. *Combust. Flame* **2020**, *216*, 472–484. [CrossRef]
42. Wang, H.; You, X.; Joshi, A.; Davis, S.; Laskin, A.; Egolfopoulos, F.; Law, C.K. USC Mech Version II. High-Temperature Combustion Reaction Model of H_2/CO/C_1-C_4 Compounds, Tech. Rep. (May, 2007). Available online: https://ignis.usc.edu:80/Mechanisms/USC-Mech%20II/USC_Mech%20II.htm (accessed on 15 May 2021).
43. Goodwin, D. G.; Moffat, H. K.; Schoegl, I.; Speth, R. L.; Webber, B. W. Cantera: An Object-Oriented Software Toolkit for Chemical Kinetics, Thermodynamics, and Transport Processes. Version 2.6.0. 2022. Available online: https://www.cantera.org (accessed on 20 November 2023).
44. Lu, T.F.; Law, C.K. A directed relation graph method for mechanism reduction. *Proc. Combust. Inst.* **2005**, *30*, 1333–1341. [CrossRef]
45. Lu, T.F.; Law, C.K. Linear time reduction of large kinetic mechanisms with directed relation graph: N-heptane and iso-octane. *Combust. Flame* **2006**, *144*, 24–36. [CrossRef]
46. Lu, T.F.; Law, C.K. On the applicability of directed relation graphs to the reduction of reaction mechanisms. *Combust. Flame* **2006**, *146*, 472–483. [CrossRef]
47. Gao, X.; Yang, S.; Sun, W. A global pathway selection algorithm for the reduction of detailed chemical kinetic mechanisms. *Combust. Flame* **2016**, *167*, 238–247. [CrossRef]
48. Dikeman, H.E.; Zhang, H.; Yang, S. Stiffness-reduced neural ODE models for data-driven reduced-order modeling of combustion chemical kinetics. In Proceedings of the AIAA SCITECH 2022 Forum, San Diego, CA, USA, 3–7 January 2022; p. 0226.
49. Owoyele, O.; Pal, P. Chemnode: A neural ordinary differential equations framework for efficient chemical kinetic solvers. *Energy AI* **2022**, *7*, 100118. [CrossRef]
50. Lee, K.; Parish, E.J. Parameterized neural ordinary differential equations: Applications to computational physics problems. *Proc. R. Soc. A Math. Phys. Eng. Sci.* **2021**, *477*, 2253. [CrossRef]
51. Ji, W.; Richter, F.; Gollner, M.J.; Deng, S. Autonomous kinetic modeling of biomass pyrolysis using chemical reaction neural networks. *Combust. Flame* **2022**, *240*, 111992. [CrossRef]

energies

MDPI

Review

A Review on Micro-Combustion Flame Dynamics and Micro-Propulsion Systems

David M. Dias [1], Pedro R. Resende [2,3] and Alexandre M. Afonso [3,4,*]

[1] Department of Mechanical Engineering—FEUP, University of Porto, 4200-465 Porto, Portugal; engineer.david.dias@gmail.com
[2] proMetheus, Escola Superior de Tecnologia e Gestão, Instituto Politécnico de Viana do Castelo, 4900-347 Viana do Castelo, Portugal; pedroresende@estg.ipvc.pt
[3] CEFT—Transport Phenomena Research Center, Department of Mechanical Engineering—FEUP, University of Porto, 4200-465 Porto, Portugal
[4] ALiCE, Associate Laboratory in Chemical Engineering, Faculty of Engineering, University of Porto, 4200-465 Porto, Portugal
* Correspondence: aafonso@fe.up.pt

Abstract: This work presents a state-of-the-art review of micro-combustion flame dynamics and micro propulsion systems. In the initial section, we focus in on the different challenges of micro-combustion, investigating the typical length and time scales involved in micro-combustion and some critical phenomena such as flammability limits and the quenching diameter.We present an extensive collection of studies on the principal types of micro-flame dynamics, including flashback, blow-off, steady versus non-steady flames, mild combustion, stable flames, flames with repetitive extinction, and ignition and pulsatory flame burst. In the final part of this review, we focus on micropropulsion systems, their performance metrics, conventional manufacturing methods, and the advancements in Micro-Electro-Mechanical Systems manufacturing.

Keywords: micro-scale combustion; flame dynamics; micro-propulsion; flame types

1. Introduction

Current space agencies are making substantial investments in micro and meso satellites like CubeSat to advance communication networks and control systems, catering to the growing need for improved connectivity [1,2]. These control systems and the micro and meso satellites are integrated with Micro Electro-Mechanical Systems (MEMS), devices that depend on external power sources, typically in the form of batteries. There is some research which aims to explore alternative power generators at smaller scales, especially with the higher energy densities usually offered by hydrocarbon fuels through micro-combustion [1,3]. However, the use of power generated through combustion at the micro scale poses some challenges, such as sustaining a stable flame at smaller scales, due to the restricted residence time within micro combustors and the amplified heat losses due to the surface-to-volume ratios characteristic of micro-scale devices [1–5]. A review exploring the current challenges and advancements in flame stability and residence time in micro combustion techniques was presented by Nauman et al. [5], while the effect of combustion parameters on the length and flame structure was recently reviewed by Jabar and Al-Fahham [4].

In this review paper, we present a review of the progress in micro-combustion flame dynamics and micro-propulsion systems. In the initial section, we focus om the different challenges within micro-combustion, investigating the typical length and time scales involved in micro-combustion. In the next section, we present a review of the types of micro-flame dynamics, and in the final part of this review, we focus on micro-propulsion systems, their performance metrics, conventional manufacturing methods, and the advancements in Micro-Electro-Mechanical Systems manufacturing.

2. Definition of Micro-Scale Combustion

The definition of combustion in Webster's Dictionary [6] is "rapid oxidation generating heat, or both light and heat; also, slow oxidation accompanied by relatively little heat and no light". For the purpose of this review, we shall limit the definition to only the rapid oxidation portion, as most actual combustion devices fall within this category.

This definition emphasizes the inherent significance of chemical reactions in combustion. It also underlines the significance of combustion: burning converts energy held in chemical bonds to heat, which can be used in a variety of ways. In essence, combustion occurs when there is a chemical interaction between a combustible, a fuel, and an oxidizer under appropriate circumstances.

More relevant for the context of this review is combustion in microsystems, or "micro-scale" combustion. Generally speaking, there are three different length scales that have been used in previous research to define micro-scale combustion [1]:

i. Definition based on Physical length: A generally used characteristic to distinguish between micro-scale and meso-scale combustion is that, when the length scale of the combustor is smaller than 1 mm, the combustion is referred to as micro-combustion. On the other hand, if the physical length scale is greater than 1 mm but around 1 cm, the combustion is referred to as mesoscale combustion, which is usually utilized in the development of micro-engines [7];
ii. Definition based on the Flame quenching diameter: consists of using the quenching diameter as the largest diameter of a cylindrical tube that will just quench (extinguish) the flame front of a particular fuel–air mixture [8], which is a function of the mixture's composition and the wall properties (temperature/surface reactivity) [1], and the reference length scale for the flame [9]. Consequently, if the combustor dimensions are smaller (larger) than the quenching diameter, the combustion is referred to as micro-scale (mesoscale). This definition is more sensible in terms of the physical flame regimes and is preferred by researchers for studying the fundamental aspects of micro-combustion;
iii. Definition based on the Device Scale: Researchers sometimes use the relative size of a device for its intended purpose compared to conventional large-scale devices as a means of defining meso and micro-scale combustion. This approach offers a third way of defining these types of combustion (often utilized by researchers who aim to design micro-thrusters for particular applications).

It is important to keep in mind that most flows become laminar when reduced from the macro to the micro scale, and as a result, the numerical analysis of the flows is simplified when comparing them with turbulent flows.

3. Challenges within Microcombustion

The design and manufacture of micro burners, which have dimensions at the millimeter or submillimeter level, pose unique challenges when compared to conventional combustors, which make it not trivial to obtain a stable flame, as documented in Hossain and Nakamura [10], Ju and Maruta [1], Nakamura et al. [2], Resende et al. [11], and Lee et al. [12]. These challenges extend to combustion in microscale environments, where the effects of viscosity are amplified in small channels, the physical residence time of mixed gases is shortened, and the ratio of area to volume of the combustion chamber increases sharply [13,14]. These factors have direct or indirect effects on the flame and the heat release of chemical energy, thereby leading to significant differences in combustion characteristics between micro and conventional combustors [15]:

1. SHORT RESIDENCE TIME: Micro-combustion, with its small chamber volume, is limited by short residence times due to diffusion and reaction times [16]. For example, the reaction time for hydrogen or hydrocarbon species with air or oxygen is typically 0.1 to 1 ms [17], aligning with small-scale conditions [18]. The Damköhler number (Da), which relates residence time to combustion time, assesses fuel combustion com-

pleteness [19,20]. Da is defined as Da $= \tau_{res}/\tau_c$. When Da is less than 1, extinction will occur via incomplete combustion, which reduces combustion and power generation efficiency [21];

2. High heat loss ratio: Micro-combustion poses a challenge due to increased heat losses related to high surface-to-volume ratios within the combustion chamber. To address this, understanding the causes of these losses and developing effective mitigation strategies is crucial. Heat loss has been quantified by Kang et al. [15] using the following equation:

$$\rho_q = \frac{Q}{m} = \frac{\overline{\rho_q''}}{\bar{\rho}} = \left(\frac{S_{ch}}{V_{ch}}\right)$$

Here, ρ_q represents heat loss density, $\overline{\rho_q}''$ is the average heat loss density, S_{ch} is the inner surface area of the combustor, and V_{ch} is the chamber volume. As the combustion chamber size decreases in the micro-scale, the surface-to-volume ratio (S/V) increases substantially, leading to amplified heat loss in the combustor. For conventional burners, S/V is approximately 3 to 5 m^{-1}, while it reaches 500 m^{-1} in micro combustors [22], accelerating heat dissipation from the inner wall;

3. The stability of micro-combustion depends on reactions, heat transfer, and the interaction between burned and unburned gases [23,24]. Factors like flow fields, temperature, and species distribution in microchannels can cause combustion instabilities such as flame blowout, oscillations, and asymmetric distribution [25,26]. High heat loss from combustor walls can lead to non-self-sustaining combustion, including thermal or radical extinction in small chambers [27]. Radical extinction occurs when active radicals are absorbed by the inner wall, hindering their reactivity. External wall heating is proposed to enhance flame stability [28]. The increased surface-to-volume ratio in micro-combustion intensifies wall heat losses and radical collisions, affecting flame extinction and micro power systems [29,30]. A review of the advancements in flame stability and residence time in micro-combustion technologies was performed by Nauman et al. [5]. This review compares different methods for improving the stability of a wide range of features in various operational settings, from the use of porous media to increasing the radiation temperature of the burner, using backflow zones formed due to optimized chamber construction encouraging flame anchoring and increasing combustion efficiency, increasing the wall temperature, and using catalytic assisted combustion to speed up the reaction time and residence time.

4. Typical Length and Time Scales

Micro-combustion, akin to its conventional-scale counterpart, encompasses a plethora of physical and chemical processes that exert direct influence on the observed phenomena. These processes comprise gas-phase and surface reactions, molecular transport, thermal and mass diffusion, as well as convection and radiation effects [1]. In Table 1 we list some important length and time scales which strongly affect the phenomena of micro-scale and mesoscale combustion:

Small-scale combustion is influenced by length and time scales [1], as shown in Table 2. For instance, when the combustor size is close to the flame thickness or quenching diameter, flame extinction or instability can occur due to wall heat loss and/or radical quenching. Moreover, there is limited fuel–oxidizer mixing. The coupling of wall heat loss and auto-ignition at isolated hot spots causes steady/unsteady flame streets to appear. Furthermore, decreasing the combustor structure leads to multiple flame regimes and strong flame–wall coupling, expanding the flammability limit. Higher wall temperatures result in shorter ignition times and the possibility of weak flames or flameless combustion. Flame instability arises when the ignition time approaches the flow residence time and the combustor size nears the quenching limit. The Biot number governs the coupling between wall temperature and flame, while the Fourier number affects the wall temperature distribution.

Table 1. Typical length and time scales in micro-scale combustion, adapted from Ju et al. [1].

Time Scales	Length Scales
Flow residence time, τ_{res}	Combustor scales [1], d
Characteristic combustion time, τ_c	Combustor structure scales [2], d_s
Diffusion time scale of gas-phase, $\tau_\alpha = d^2/\alpha_f$	Flame thickness, δ_f
Diffusion time scale of solid-phase, $\tau_s = {d_s}^2/\alpha_s$	Quenching diameter [3], $d_0 = n\delta_r$
Time scale of heat loss [4], τ_h	Mass diffusion length, δ_D
Time scale of acoustic wave [5], τ_a	Thermal diffusion length, δ_{ff}
Ignition time scale, τ_{ig}	Mean free path, λ_{fp}
Characteristic flow time, τ_f	-

[1] Characteristic size of combustor inner diameter. [2] Characteristic length scale of the structure (solid phase). [3] The quenching diameter, which is contingent upon the wall temperature, is typically several times (n) the flame thickness. In the case of a wall temperature corresponding to room temperature, n is approximately 15. [4] Is a function of convective and radiative heat losses. [5] Depends both on the span-wise and streamwise dimensions of the combustor.

In small-scale combustion, non-equilibrium transport effects become important at reduced sizes (Knudsen number > 0.01), influencing ignition and extinction. Micro-scale combustion exhibits incomplete combustion and thermal–diffusional instability when the flow residence time approaches the characteristic combustion time. The Lewis number effect and solid phase thermal transport significantly impact flame stability boundaries. Spinning and pulsating combustion occur for low mixture Lewis numbers. Wall confinement and high temperatures introduce strong acoustic wave effects. Understanding these phenomena requires the analysis of relative length and time scales or the consideration of unsteady temperature distribution in the solid phase [1].

Table 2. Non-dimensional parameters vs. micro-scale combustion phenomena from Ju et al. [1].

Non-Dimensional Numbers	Phenomena
$d/\delta_f \sim O(10)$ or $d/d_0 \sim 1$	Flame extinction due to heat loss to the wall Wall catalytic/quenching effects, flame street & cells
$\tau_c/\tau_s = O(1)$, $d/d_0 \sim 1$	Flame-wall thermal coupling, multiple flame regime
$\tau_{ig}/\tau_{res} = O(1)$, $d/d_0 \sim 1$	Extinction and reignition instability, Weak flame Flameless combustion
$Bi = \tau_\alpha/\tau_h >> 1$	Small wall temperature change
$Fo = \tau_c/\tau_s >> 1$	Steady-state temperature distribution in solid phase
$Kn = \lambda/d > 0.01$	Non-equilibrium transport effect
$Da = \tau_{res}/\tau_c$ or $\tau_D/\tau_c \sim 1$	Flame extinction due to incomplete combustion
$Le = \delta_\alpha/\delta_D \sim 11$	Pulsating instability via flame structure coupling
$\tau_a/\tau_c = O(1)$	Acoustic instability

The typical lengths, flame structures, and time scales occurring in micro-combustion were recently reviewed by Jabar and Al-Fahham [4], reporting many experimental and theoretical ways to measure flame sizes inside combustion chambers, with their main focus being on fuel components and their mixing ratios.

4.1. Flammability Limits in Micro-Combustion and the Quenching Diameter

The standard flammability limit defines the boundaries for lean or rich flame propagation limits in a one-dimensional, infinite, planar, and unstretched flame. Extensive research [9,31–34] has determined that radiating species like H_2O and CO_2 contribute to these limits.

In micro-scale combustion, heat loss occurs not only due to radiation but also due to convection and conduction losses to the channel wall. By normalizing the convective heat loss by the chemical heat release, researchers found it to be proportional to the square of the ratio of flame thickness to the channel diameter. As such, assuming a constant flame thickness for a given mixture, the flame speed is influenced by the channel diameter [1].

Further analysis from Figure 1a indicates that the propagation limit corresponds to a limiting tube diameter, d_0, also known as the quenching diameter. Below this critical diameter, the flame fails to propagate in the channel (Ju et al. [1]).

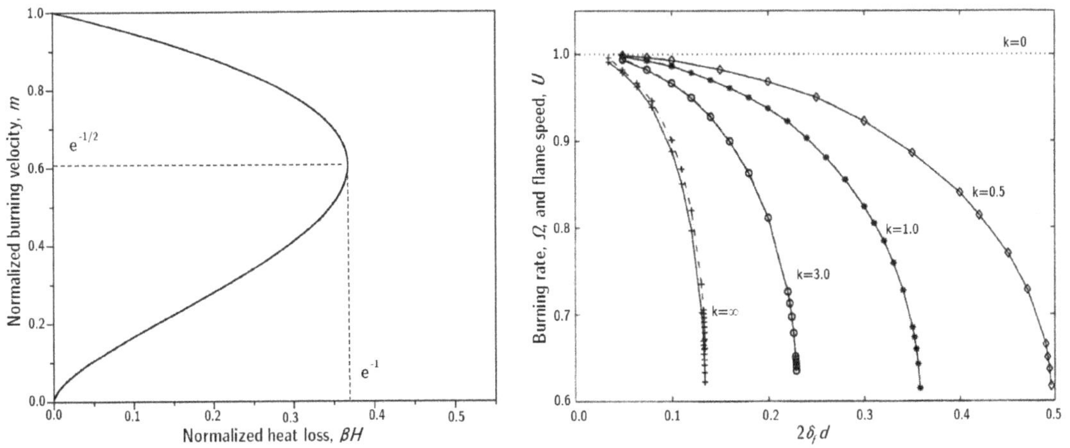

(a) (b)

Figure 1. (**a**) The dependence of the normalized burning velocity on the normalized radiative heat loss of a one-dimensional planar flame [1]; (**b**) burning rate, solid line, and normalized flame propagation speed U_{fl}, dash-dotted line, plotted against the ratio of flame thickness to channel width, d, for selected values of reduced heat transfer coefficient, k, within a quiescent, two-dimensional channel flow [35].

The size of a flame depends on how easily heat spreads (thermal diffusivity) and how fast the flame burns (flame speed), which, in turn, is influenced by the amount of fuel and its temperature. This means that the quenching diameter, in which the flame can no longer sustain itself, is determined by the fuel concentration and the properties of the fuel itself.

Figure 1b shows how the flame speed in a quiescent channel depends on its width (d) and the convective heat loss (k), which is proportional to the Nusselt number (Nu), as mentioned previously. When the wall is adiabatic ($k = 0$), there is no heat loss from the flame to the wall, so the flame speed does not depend on channel width (Nu $= 0$). With increasing heat loss, the flame speed decreases until there is a critical channel width in which the flame can not propagate [35].

At a high level of convective heat transfer ($k = \infty$), the flame extinguishes at $U = 0.6$, resulting in a quenching channel width of $2\delta_f/d_0 = 0.13$, or using the convention set in Table 2, $d_0/\delta_f = 15$, consistent with theory and experiments. Near-wall flame quenching occurs within 6% of the channel radius in a cold-wall scenario. Additionally [35], the flow velocity and direction also affect the quenching diameter; flames are more vulnerable to heat losses when the flow moves from the unburned to the burned side, and the quenching distance increases with higher flow rates. These findings match experimentally reported values [36].

4.2. Classification of Flame Types

Diffusion and premixed flames are two different kinds of micro combustion flames. In a premixed flame, combustion takes place after the fuel and oxidizer have been combined.

The fuel and oxidizer in a diffusion flame are kept apart and only mixed as they move closer to the reaction zone [37].

The two subtypes of premixed flames are turbulent premixed flames and laminar premixed flames, as shown in Figure 2a,b. A laminar premixed flame is one in which the fuel and oxidizer mixture flows uniformly and without turbulence. In a turbulent premixed flame, on the other hand, the mixture of fuel and oxidizer is turbulent, meaning that it is characterized by random fluctuations in velocity and pressure [38].

(a) (b) (c) (d)

Figure 2. Differences in premixed and diffusion flames: (**a**) Laminar premixed flame. (**b**) Turbulent premixed flame. (c) Laminar diffusion flame. (**d**) Turbulent diffusion flame [39].

Conversely, the separation of the fuel and oxidizer is what distinguishes a diffusion flame. When approaching the reaction zone, where combustion occurs, the fuel and oxidizer are only briefly mixed. Because of natural molecular diffusion, which is fueled by the concentration gradient of the two species, the fuel and oxidizer in a diffusion flame mix together [40]. Diffusion flames can also be classified into two types: laminar diffusion flames and turbulent diffusion flames, as shown in Figure 2c,d. The numerical studies of these type of diffusion flames at micro scales was studied by Resende et al. [11], where the authors extensively analyzed the importance of operating conditions, critical parameters, and the conjugate heat transfer/heat re-circulation.

Premixed and diffusion flames each have benefits and drawbacks. Premixed flames are commonly more efficient because they offer a more uniform mixture of fuel and oxidizer, which promotes faster combustion and less pollution. They may also have a smaller operating range and be more prone to explosions. While diffusion flames are typically more stable and capable of a wider range of operations, they may also be less effective and more polluting [37].

In the context of this review, we limited the analysis to premixed flames exclusively, and all mixtures were, as such, previously mixed. Therefore, it is imperative to note that, in the upcoming subsection, any discussion concerning the characteristics of micro-flames will pertain solely to premixed flames.

5. Classification of Micro-Flame Behaviours

The microflame is a complex phenomenon that can exhibit different behaviors, which are influenced by multiple factors, such as the input velocity of the fuel–oxidizer mixture, the heat flow rate supplied to the mixture, the thermal conduction coefficient of the burner walls (i.e., boundary conditions), and the burner geometry, among others. Thus, the present subsection aims to present a review of the possible behaviors of flames in microchannels, based on previous scientific research.

5.1. Flashback and Blow-Off

Premixed flames can experience situations where the reactant supply velocity exceeds or falls short of the flame speed, resulting in two possible scenarios for flame extinction [41,42], as observed in Figure 3:

i. Flashback: This occurs when the velocity of the reactants supplied to the burner chamber is lower than the flame speed. As a result, the flame propagates upstream towards the fuel injection system, burning the reactants in a zone increasingly upstream, possibly self-extinguishing (Combustion instability-induced flashback is connected to the phenomenon of the instantaneous axial flow velocity experiencing significant reductions, sometimes even reaching negative values, during large-amplitude oscillations [43]).
ii. Blow-off: In this scenario, the reactants' supply velocity is higher than the flame speed. The flame moves along the channel, burning the reactants in a zone increasingly downstream until it "blows off" (possibly self-extinguishing) when the fuel flow rate exceeds a certain limit.

(a) (b) (c) (d)

Figure 3. Flashback and blow-off for flames with fixed swirl: (**a**) flashback; (**b**) before flashback; (**c**) stable flame; (**d**) after blow-off [42].

The research paper by Fruzza et al. [44] studies the flashback of different H_2/CH_4 mixtures and uses 2D unsteady simulations with detailed kinetics and considers the fundamental role of the combustor wall by considering the Conjugate Heat Transfer. Different flashback regimes were studied for different H_2 contents which highlights the significant difference between the critical flashback velocity and laminar flame speed at high H_2 contents. This work was recently extended by Fruzza et al. [45], in which the authors coupled the numerical simulations with stochastic sensitivity analysis based on Generalized Polynomial Chaos, which allowed for the construction of a set of comprehensive maps for flashback velocities and burner temperatures for the parametric space of H_2 content, equivalence ratio, and slit width.

Recently, Cai et al. [46] performed an experimental study on the the the flame dynamics of CH_4-air within a micro-combustor with a flame holder and preheating channels. In addition to stable flame results, different types of blow-off dynamics were also presented, including anomalous blow-off dynamics at lower Reynolds numbers and fuel ultra-lean conditions, as observed in the stability maps presented in Figure 4B. It is found that, in these anomalous blow-off dynamics, the middle of the flame in the vertical direction presents a constricted shape, and that a flame with local extinction and reignition occurs periodically (see Figure 4C).

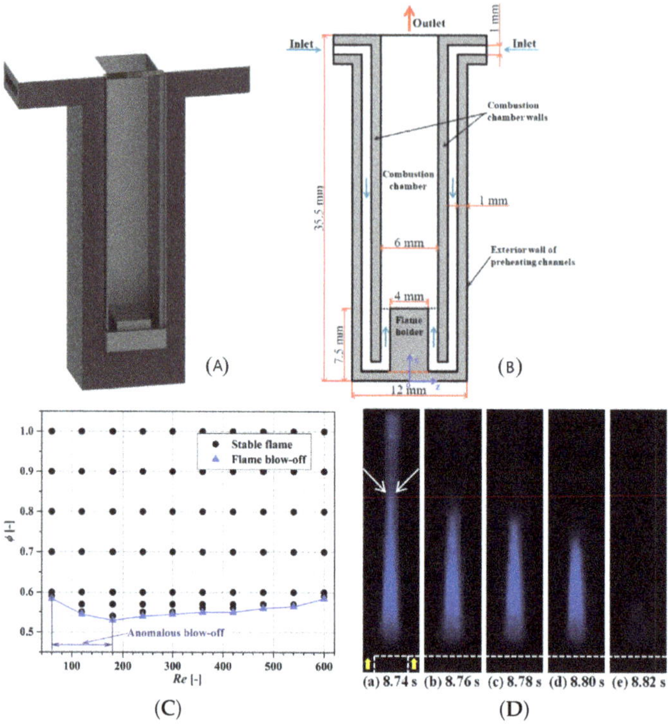

Figure 4. Flame stability and blow-off dynamics for micro-combustion with a flame holder and preheating channels: (**A**) three-dimensional view of the burner; (**B**) cross-section drawn; (**C**) stability map; (**D**) blow-off dynamics [46].

5.2. Steady vs. Non-Steady Flames

Combustion can be classified as steady or non-steady/unsteady, which can be latter divided into periodic, quasi-periodic, and chaotic. Non-steady flames show fluctuations in their behavior, such as changes in their burning rate or flame shape that happen over time, in contrast to steady flames, which keep a consistent burning pace [47].

As illustrated in Figure 5, non-steady flames can exhibit different types of dynamics, including periodic, quasi-periodic, and chaotic behavior [48]:

i. Periodic flames, which are characterized by fluctuations in their behavior that occur periodically over time, in a constant pattern, Figure 5a;
ii. Quasi-periodic flames, in which the oscillations present irregular periodicity and recurrence are observed, Figure 5b;
iii. Chaotic combustion, which can occur when steady combustion waves lose stability and transition to non-steady chaotic burning, exhibiting highly irregular and unpredictable behavior, Figure 5c.

The transition from steady to non-steady burning can be predicted using the nonlinear stability theory [47].

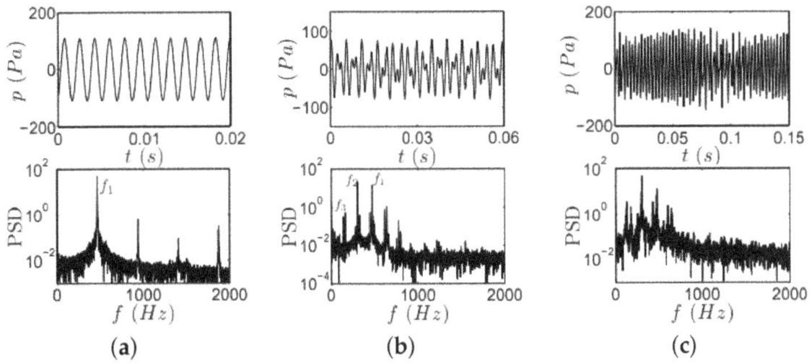

Figure 5. Time-spectrum (**up**) and power series (**down**) for different non-steady combustion modes in ducted premixed flames: (**a**) periodic; (**b**) quasi-periodic; (**c**) haotic [49].

5.3. Mild Combustion

Mild combustion, also called flameless combustion, is a combustion mode that is distinguished by a relatively small temperature rise and an incomplete but considerable fuel conversion. It is generally observed at low inflow velocities and exhibits an axisymmetric shape. This mode has been identified in numerous research studies, and is typically characterized by the lack of a distinct temperature peak, as documented in [50]. It is a potentially beneficial flame type for applications requiring a weak heat source in microsystems.

On the other hand, the combustion process is contingent upon two critical factors [51, 52]: the reactants must be preheated above the temperature at which they will self-ignite, and enough inert combustion products must be incorporated into the reaction area to saturate the flame. Furthermore, it is noteworthy to mention High Temperature Air Combustion, a technique that often leads to mild combustion scenarios, characterized by the synergistic utilization of highly preheated air and exhaust gas recirculation (EGR) in furnaces, resulting in high thermal efficiency and low NO_x emissions [53,54].

The phenomenon of interest was first observed in micro-scale systems in Maruta et al. [55]. The authors conducted a comprehensive investigation into the combustion characteristics of such systems employing experimental, numerical, and analytical techniques. Specifically, a narrow cylindrical quartz channel with a fixed temperature gradient was utilized, possessing an inner diameter of $2\,[\mathrm{mm}]$. For sufficiently low velocities in the range of $2\text{–}3\,[\mathrm{cm/s}]$, it was observed that a sharp temperature gradient did not form between the chamber's walls and the gas phase. The article by Pizza et al. [56] explored the combustion modes that occur in the three-dimensional combustion of premixed hydrogen and air in open cylindrical microtubes with the wall temperature controlled by gradually increasing it from the temperature of the incoming mixture T_{in} to a final value of $960\,[\mathrm{K}]$ over the initial $1/20$ length of the domain. This was achieved by applying a hyperbolic tangent function to smoothly ramp up the wall temperature. The researchers identified a particular type of mild combustion in both narrow and wider tubes with diameters of $1.0\,[\mathrm{mm}]$ and $1.5\,[\mathrm{mm}]$, respectively, when using the lowest air inlet velocities, which were $0.5\,[\mathrm{cm}\cdot\mathrm{s}^{-1}] \leq U_{\mathrm{in}} \leq 1.0\,[\mathrm{cm}\cdot\mathrm{s}^{-1}]$. The maximum temperature increase ranged from $1.5\,[\mathrm{K}]$ and $2.5\,[\mathrm{K}]$ (for U_{in} of $0.5\,[\mathrm{cm}\cdot\mathrm{s}^{-1}]$ to $1.0\,[\mathrm{cm}\cdot\mathrm{s}^{-1}]$, respectively).

The findings were consistent with the antecedent investigations conducted by the authors in their prior 2D study [57], wherein the phenomenon of mild combustion was also observed for low inlet velocities. The authors examined the dynamics of premixed hydrogen/air flames in a microchannel that is confined between two parallel, infinitely wide plates, and maintained a constant length-to-height aspect ratio of $L/h = 10\,[-]$.

5.4. Stable Flames

Upon initiation of combustion, the flame undergoes movement either upstream or downstream before stabilizing at a particular location on the burner, as reported numerically by several studies [58,59] and experimentally observed in Maruta et al. [55]. This constitutes the simplest form of micro-flame behavior and is illustrated schematically in Figure 6.

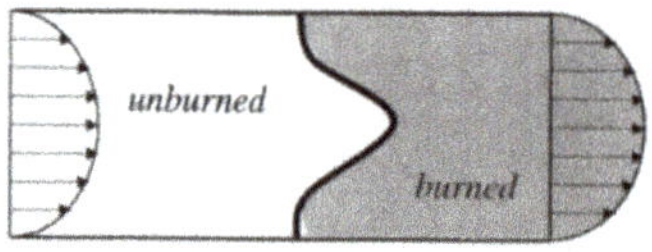

Figure 6. Schematic representation of a stable micro-flame [35].

Typically, flames of this type exhibit a reduction in flame temperature as inlet velocity increases, which can impact heat transfer to the wall [60]. Additionally, distinct velocity profiles can influence flame propagation and structure [61]. As seen in Norton et al. [62], an increase in inlet velocity necessitates a greater distance for incoming gases to preheat, leading to downstream stabilization of the flame. This stabilization results in higher wall temperatures at the flame location and lower heat losses, ultimately leading to increased reaction rates and elevated flame temperatures, exceeding the adiabatic flame temperature [60].

This phenomenon, which can exhibit symmetric and asymmetric characteristics (or both, depending on the axis, if in 3D settings), has been observed both numerically, as in the work of Lamione et al. [63], experimentally, and analytically in multiple studies such as Lewis and Von Elbe [34], Kurdyumov et al. [64,65], Dogwiller et al. [66], and Tsai [67]. The previously mentioned articles by Pizza et al. [56,57] and the experimental observations by Maruta et al. [50,55] also corroborate these findings since the phenomenon was also identified in these studies.

Cai et al. [46] obtained stable flame dynamics of CH4–air within a micro-combustor with a flame holder and preheating channels for non fuel ultra-lean conditions. The flame stability map is quite high, showing the importance of preheating the incoming unburned mixture. The results demonstrate that, in cases with a stable flame, the increasing Reynolds and equivalence ratios pushed the flame downstream, leading to a significant increase in the flame heigh, presenting a remarkable anchored effect of the flow recirculation, as observed Figure 7.

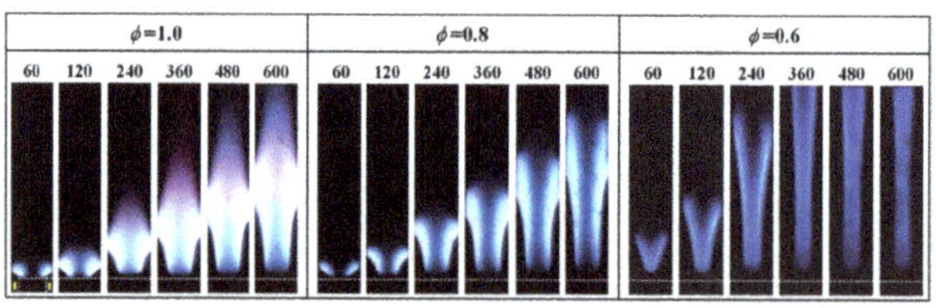

Figure 7. Stable flame dynamics of CH4–air within a micro-combustor with a flame holder for different Reynolds numbers and equivalence ratios [46].

5.5. FREI: Flames with Repetitive Extinction and Ignition

Flames with repetitive extinction and ignition (FREI) refer to a specific type of flame behavior characterized by the repeated extinguishing and reignition of the flame, and this phenomenon is observed in various systems, including micro flow reactors and lean premixed hydrogen–air combustion [1,68,69].

In the region characterized by the FREI phenomenon, a distinctive behavior of the flame is apparent: it ignites in a downstream region, propagates towards the unburned gases, and eventually extinguishes in an upstream region. Over the course of simulation, it becomes evident that the flame consistently occupies the same region, with the phenomenon recurring cyclically. Following ignition, the flame is transported through these vortices to an upstream position, ultimately leading to its cessation.

Maruta et al. [55] first captured experimentally FREI flames in a straight micro-channel. These flames exhibit cyclic behavior with luminescence starting at ignition, moving upstream to extinction, and then reigniting after a delay. This cyclic process, illustrated in Figure 8a contrasts with the stable flame behavior shown in Figure 8b. Furthermore, other experimental investigations established that the dynamics of FREI also manifest in curved mesoscale channels [70] and later in mesoscale cylindrical combustors featuring a backward-facing step [71].

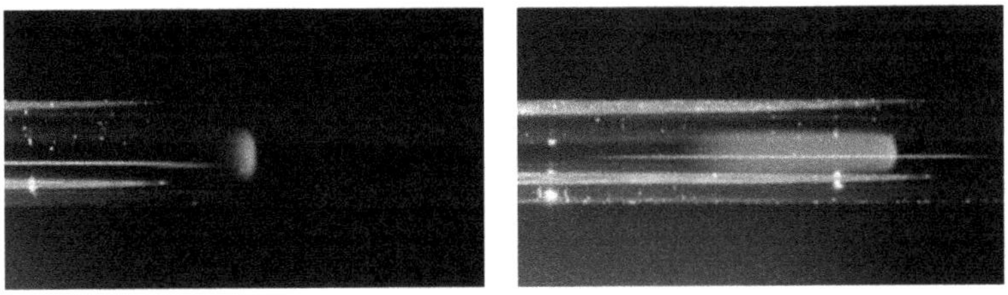

(a) (b)

Figure 8. Direct photographs of [55]: (**a**) stable flame; (**b**) FREI.

Numerous advancements in the computational field have been achieved by probing this phenomenon and utilizing lean hydrogen/air premixed flames, allowing for a better understanding of the FREI phenomena. Minaev et al. [72] provided a theoretical interpretation of FREI. They analyzed the non-stationary behavior of near-limit premixed flames in a microchannel with a temperature gradient. Their theoretical work resulted in a 1D nonlinear evolutionary equation for the flame front, explaining the flame stabilization, nonlinear oscillations, and repetitive extinction–ignition phenomena observed in the experiments. Later, Jackson et al. [73] numerically investigated FREI dynamics in a thermally active semi-infinite channel. They found stable solutions at specific flow rates, and instability leading to oscillations or extinction–reignition flames at higher flow rates, in alignment with experimental observations. The studies by Pizza et al., as mentioned earlier, observed FREI dynamics in planar micro combustors with prescribed wall temperatures at various channel heights [57] and also in 3D cylindrical microtubes [56]. Miyata et al. [74] performed direct numerical simulation to investigate FREI characteristics in a narrow circular channel, where they varied the wall temperature gradient and the inflow temperature in their study.

In a subsequent study, Alipoor and Mazaheri [68] scrutinized the FREI phenomenon through a singular simulation, affording a more intricate analysis of alternating extinction and ignition. In pursuit of this, contour maps of hydroxide radical (OH) concentrations were examined along the channel for various time instances, as shown in Figure 9. Furthermore, an investigation into heat release rate profiles, fluid and wall temperatures, the and molar concentrations of pertinent species central to the reaction was undertaken. Subsequently, flame bifurcation was evaluated from a hydrodynamic standpoint utilizing flow velocity vectors.

Subsequently, the same authors [75] delved into flames featuring equivalence ratios ranging from 0.5 up to the stoichiometric condition of $\phi = 1$, in microchannels with widths spanning 0.4 to 1 mm. The accrued findings illustrated that augmentations to the equivalence ratio and burner width increased the amplitude of the FREI phenomenon while reducing its frequency. Furthermore, this examination illuminated that the phenomenon engenders fluid recirculation zones and flame bifurcation due to the presence of heavier species. These species exhibit an elevated heat absorption capacity, thereby inducing a temperature reduction.

Later, Alipoor and Mazaheri [76] conducted a comprehensive study concerning these flames within a heated microchannel possessing a hyperbolic temperature profile. Within this inquiry, three distinct flame behaviors were observed, including the FREI phenomenon. This phenomenon manifested under low velocities near the lower flammability limit. In pursuit of a deeper comprehension of the phenomenon's causality, diverse simulations were performed, manipulating three key parameters: inlet velocity, equivalence ratio, and microchannel width. Elevating the velocity led to an escalated reaction intensity, consequently inducing rapid flame front extinction. Increasing the equivalence ratio from 0.5 to 0.6 resulted in an enhanced flame speed due to the augmented burning rate of the flame front, thus validating its accelerated extinction.

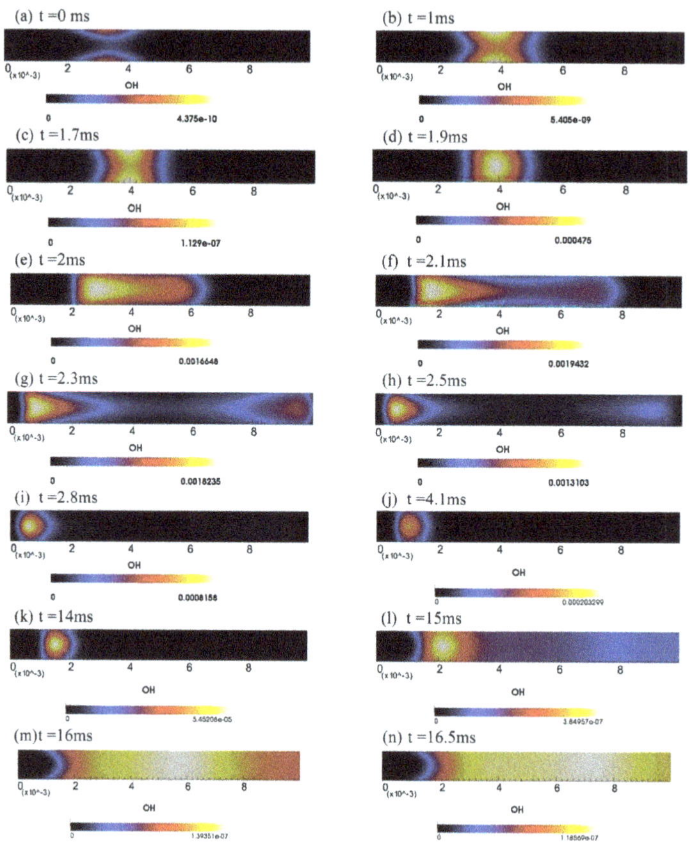

Figure 9. Repetitive extinction-ignition dynamics using hydroxide mass fraction (Y_{OH}) [68].

Beyond utilizing hydrogen as a fuel, the prospect of employing alternative fuels in microcombustion studies is viable. The FREI phenomenon was also observed in the research conducted by Biswas [77]. In this instance, a CH4/air mixture was employed, with ignition achieved through inserting a spark into the burner. The combustion of this mixture under stoichiometric conditions was examined within both a linear microchannel and a microchannel characterized by a contraction (convergent–divergent).

Additionally, other computational investigations have been conducted to enhance our comprehension of the FREI phenomenon across diverse fuel types. Nakamura et al. [78], Tsuboi et al. [79], and Miyata et al. [74] undertook analyses involving CH_4 and air mixtures. Initially, they employed a one-dimensional model, subsequently validated against their experimental findings. Subsequently, they exclusively utilized direct numerical simulations to scrutinize the impact of boundary conditions—such as inlet temperature and wall temperature gradient—on the flame dynamics. The combustion characteristics of n-heptane–air were subject to experimental and numerical inquiry by Yamamoto et al. [80], wherein they also noted the manifestation of FREI dynamics.

5.6. PFB: Pulsatory Flame Burst

The phenomenon known as Pulsatory Flame Burst (PFB) is a specific flame behavior characterized by burst-like flame pulsation due to flow and combustion interactions. Currently, PFB only occurs in undulating geometries with premixed hydrogen and air, and the flame bursts involve the ignition, propagation, bifurcation (splitting), and extinction of the flame in a cyclic manner. Unlike FREI, where the flame extinguishes upstream, in PFB, a new flame emerges before the current one extinguishes. This prevents complete extinction, marking a significant difference.

PFB, though initially observed without distinction from FREI, was first identified in the course of micropropulsion studies conducted with wavy geometries [81,82], involving various inlet velocities and equivalence ratios, using a mixture of H_2 and air. The geometry used is similar to our research, and it incorporated different wall conditions (hyperbolic and linear temperature profiles, as well as adiabatic conditions). Under the linear profile and adiabatic conditions, some simulations exhibited the phenomenon of PFB. This suggests that the microchannel geometry recirculates combustion energy, preventing flame extinction by flame lifting at higher speeds. Regarding propulsion capacity, it was found that increasing both the inlet velocity and equivalence ratio contribute to higher propulsion indicators. This underscores the significance of PFB, as pulsating flames exhibit substantial peaks in specific impulses.

Only in a subsequent study by Resende et al. [83], was PFB accurately identified and named. This investigation delved into stable and pulsatory flame burst behaviors within an undulating geometry, utilizing premixed hydrogen and air (with an equivalence ratio of $\phi = 1.0$) and applying a linear temperature profile along the wall. The findings revealed that, at a lower inlet velocity (4 [m/s]), the flame maintained stability, while at higher velocities, it exhibited pulsatory burst dynamics. The interplay between fluid dynamics and combustion response proved crucial, particularly due to vortices generated by the nonlinear burner geometry. With increased inlet velocity, the heat release rate transmitted through the vortices diminished, leading to ignition delay. This was evident through a reduced pulsatory burst frequency and a heightened peak heat release rate, albeit not substantial enough to amplify the maximum temperature amplitude. In Figures 10 and 11 the results of a typical PFB flame are presented, as reported in [83].

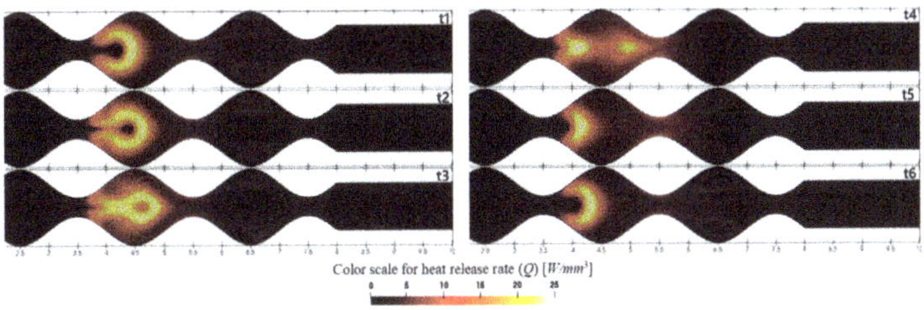

Figure 10. Streamlines of the flow field and heat release rate distribution for the pulsatory flame burst dynamics with $U_{in} = 6$ [m/s], with 10 ms intervals [83].

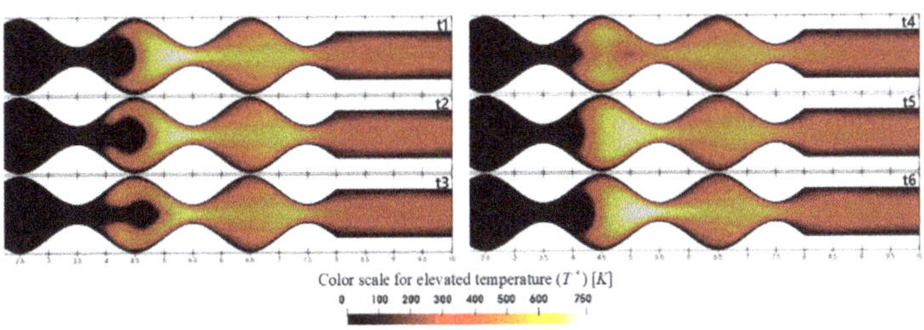

Figure 11. Streamlines of the flow field and temperature distribution for the pulsatory flame burst dynamics with $U_{in} = 6$ [m/s], with 10 ms intervals [83].

6. Numerical Investigations in Microcombustion

Li et al. [84] conducted numerical investigations into the combustion characteristics of hydrogen/air premixed flames in microtubes and parallel-plate microchannels with hydraulic diameters less than one millimeter. The results showed that microtubes lead to higher flame and wall temperatures, in comparison to parallel-plate microchannels. In subsequent studies, Li et al. [60] performed a similar investigation using methane as the fuel, finding higher temperatures in parallel-plate microchannels. Moreover, the upper flammability limit of the hydrogen/air mixture was higher than that of methane due to hydrogen's faster burning rate.

The wall thickness of the burner, flame thickness, and hydraulic diameter are of the same order of magnitude in microcombustion chambers. Therefore, heat transfer through the wall and heat recirculation have a significant impact on combustion sustainability and its characteristics. Ronney [85] demonstrated that, for burners with counterflow zones and heat recirculation, heat conduction through the wall in the direction of flow has a crucial effect on flammability limits, particularly in flows with low Reynolds numbers.

To investigate the effect of heat recirculation on flammability limits, Kim et al. [86] experimentally studied the use of Swiss-roll microburners with channel widths smaller than the extinction diameter for propane/air combustion. Here it was found that a stable propane/air flame could be observed in channels with dimensions smaller than the quenching diameter for a wide range of inlet velocities and equivalence ratios, primarily due to heat recirculation. Kuo and Ronney [87] also numerically investigated heat recirculation in a Swiss-roll microburner and suggested that heat conduction along the burner wall, dividing the reactant flow from the reaction product flow, reduces the reaction temperature and extends extinction limits.

Recently, the use of catalyst-containing surfaces [88,89] has been studied for potential application in Swiss-roll burners. In Pizza et al. [90,91], it was found that instabilities in hydrogen/air premixed flames in parallel-plate burners could be avoided through the use of catalysts. Zhou et al. [92] experimentally and numerically tested three microburners with copper, alumina, and quartz catalysts, confirming the advantages of catalyst utilization in combustion.

Flow recirculation zones, as well as low-velocity zones achievable through combustion chamber design, are highly effective techniques for flame stabilization. One common technique involves the insertion of a backward-facing step in the burner walls. In this regard, Yang et al. [93] experimentally investigated the enhancement of a thermophotovoltaic microcombustion system's efficiency using a backward-facing step in the combustion chamber. They concluded that the step insertion is useful for controlling the flame position and extending flammability limits. An experimental study on wall temperature and radiation heat flux from cylindrical microburners with a backward-facing step (as observed in Figure 12), conducted by Li et al. [94], revealed that optimal efficiency occurs at an equivalence ratio of 0.8, regardless of the burner's dimensions and flow velocity. Furthermore, it was found that the backward-facing step is helpful in stabilizing flame positions. Khandelwal et al. [95] experimentally investigated flame stabilization in a microtube with three steps and found that flammability limits increase with the number of steps inserted in the reaction zone. Faramarzpour et al. [96] conducted a numerical study on the effects of microburner geometry and inlet velocity on flame position and radiation efficiency. The results showed that the presence of the backward-facing step stabilizes the flame at high velocities, increases the average outer wall temperature, and anchors the flame closer to the burner inlet.

Fan et al. [97] explored the impact of obstacles (bluff body) on hydrogen/air premixed flames. They investigated triangular and circular obstacles, concluding that the insertion of obstacles increases the combustion flammability limits. Yang et al. [98] studied hydrogen/air premixed flames in a convergent–divergent microtube, represented in Figure 13. They simplified the geometry to two dimensions and used experimental temperature distribution data on the outer surface of the microtube as the thermal boundary conditions in numerical simulations. Their findings showed stable combustion for equivalence ratios ranging from 0.6 to 2.2 and inlet velocities from 3.4 m/s to 41.4 m/s. Comparing these results with a regular microtube, they observed broader flammability limits in the convergent–divergent microtube. Later, Biwas et al. [99] numerically investigated the extinction and flame propagation behaviors of methane/air premixed flames in straight and convergent–divergent microchannels with diameters ranging from 1 to 10 mm. They examined the effects of microchannel diameter, equivalence ratio, inlet velocity, and exit-to-throat area ratio. Adiabatic wall conditions with conjugate heat transfer were considered. Their results indicated that flames extinguish more readily in convergent–divergent microchannels than in straight microchannels under adiabatic conditions.

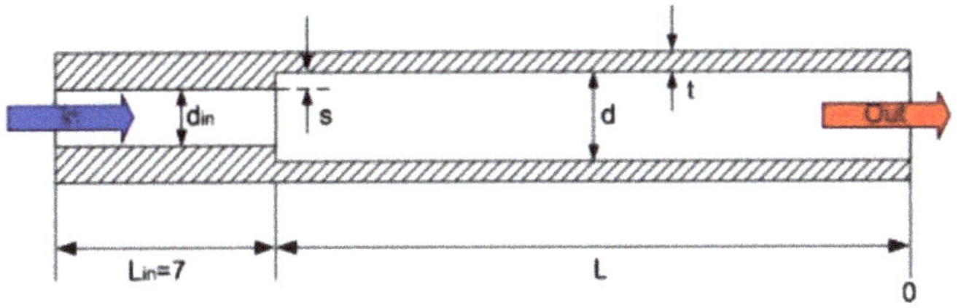

Figure 12. Example of a backward-facing step insertion in a microcombustor, used in Li et al. [94].

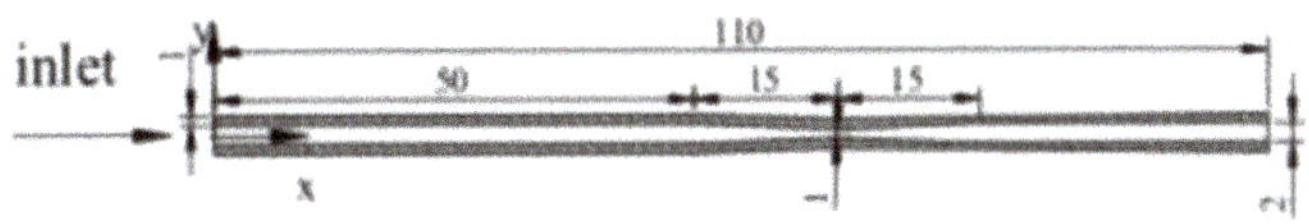

Figure 13. Schematics of the microburner used in Yang et al. [98].

Rao et al. [100] presented a numerical work on a micro-combustor with a slotted bluff body, with a flame-tip opening, observing that the presence of the central bluff adversely affects the combustion characteristics, leading to reduced average combustion efficiency and exhaust gas temperature. In order to minimize this adverse effect, Rao et al. [100] introduced a deflector downstream of the bluff body, that significantly increased the temperature in the central region due to a secondary flame. Zhang et al. [101] used a similar deflector strategy to try to improve combustion stability, using a slotted blunt body with a front baffle instead of the downstream position. Zhang et al. [101] confirmed the synergistic effectiveness of a blunt body slot and front baffle, with the front baffle increasing the regulation of the fuel supply while decreasing the pressure loss in the micro combustor.

Abbaspour and Alipoor [102] conducted numerical investigations into hydrogen/air premixed flame characteristics and behaviors in heated convergent–divergent microtubes. Results showed that the convergent–divergent section led to decreased flame temperature due to increased flow velocity and convective heat transfer, resulting in oscillating flames. However, this geometry also contributed to flame stability. When the flame stabilized near the outlet section, reduced flame wall instability and increased upper flammability limits were observed due to enhanced heat recirculation in the convergent–divergent section.

Zhao et al. [103] used a heat-recirculating approach to perform an in-depth investigation into the thermal performance, entropy generation, and pollutant emissions of a micro-combustor fuelled by blended H_2 and ammonia to increase flammability in atmospheric micro-combustion, as observed in Figure 14. They show that the thermal performance of the micro-combustor depends significantly on the inlet velocity and inlet equivalence ratio, while the impact of blending ratio between the hydrogen and ammonia is relatively minimal.

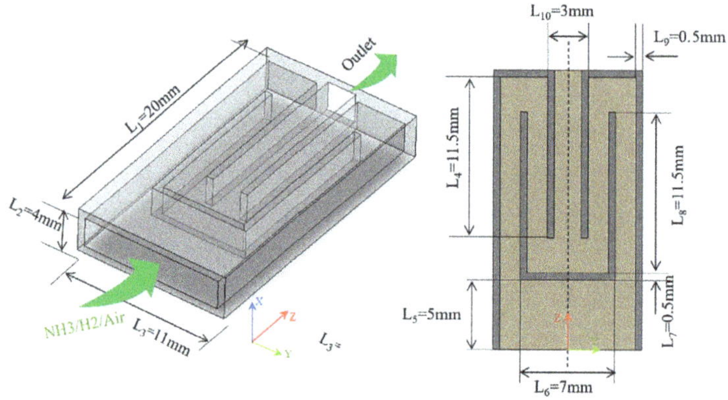

Figure 14. Microcombustor with heat recirculating approach [103].

The use of flow recirculation zones is also interesting in some applications of micro-combustion within micro-thermophotovoltaic systems, as in the recent work of Li et al. [104] regarding the comparison between the performance of the traditional micro combustor and the a modified microcombustor with twisted tape inserted, as observed in Figure 15a. The results show that the inclusion of the twisted tape in the new micro-thermophotovoltaic systems reduces the radial thermal resistance and increases the gas residence time, when compared with the classical system. He et al. [105] presented other modification to the micro-thermophotovoltaic system using internal spiral fines, and showed that internal spiral fins can effectively improve the energy efficiency and energy output, but with high pressure loss. The optimal value for the ratio of the length of the fin and the outer diameter was suggested to be 5/2, while the optimal ratio between the internal spiral fin pitch distance and the outer diameter should be greater than 9.42/4. Cai et al. [3] presented a comprehensive review on the recent advances on stabilization of micro-thermophotovoltaic systems, exploring the thermal effects, pollutant emission, and energy conversion.

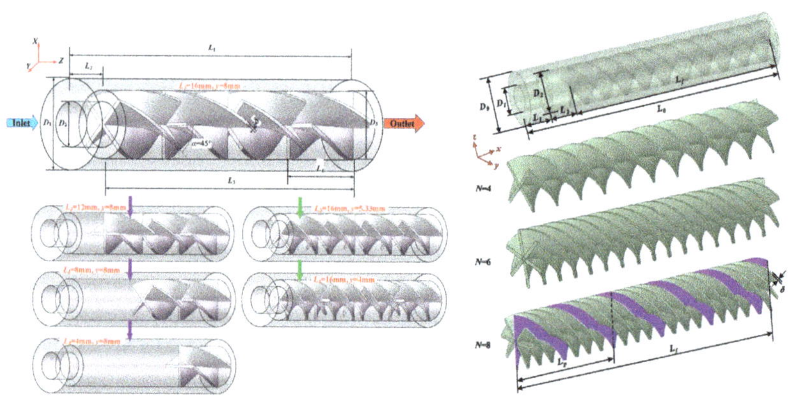

(**a**) (**b**)

Figure 15. Geometrical model of micro-thermophotovoltaic: (**a**) micro-combustor with twisted tape [104]; (**b**) micro-combustor with internal spiral fins [105].

In recent years, studies have explored microburners with nonlinear geometries. Wan et al. [106] numerically studied hydrogen/air mixture combustion in a microburner with wall cavities, as observed in Figure 16, varying cavity depth, and other reaction parameters. As the mixture velocity increased, the flame separated into two distinct zones, leading to significant fuel leakage from the reaction zone and a sharp decrease in the hydrogen conversion rate. They concluded that excessively deep cavities are detrimental when the burner is used as a heat source for thermoelectric or thermophotovoltaic devices, as flame separation results in more temperature peaks along the burner with lower intensities, reducing flame temperature. Yang et al. [107] conducted a similar study, focusing on burners with double cavities.

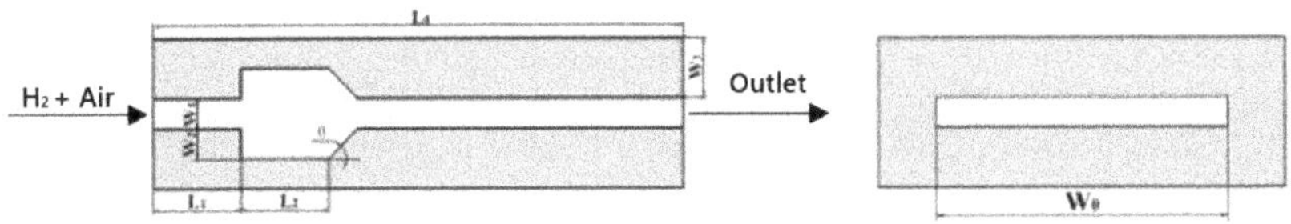

(a) (b)

Figure 16. Microcombustor with cavity schematic [106]: (**a**) longitudinal plane; (**b**) transverse plan.

Various studies have examined the flame dynamics in combustion chambers with diameters close to their extinction diameters. These studies often assume a negligible wall thickness in microburners, creating a significant difference in heat propagation rates between the fluid and walls. A constant temperature profile is typically used as the temperature boundary condition on the wall. Pizza et al. conducted numerous numerical studies on the dynamics of hydrogen/air premixed flames in heated meso channels [108], microchannels [57], and microtubes [56]. Different flame dynamics were observed, including FREI, mild combustion, swirling flames, and stable symmetrical flames. Later, Alipoor and Mazaheri [76] presented flame dynamics maps for hydrogen/air premixed flames based on three parameters: inlet velocity, equivalence ratio, and microchannel width, which are shown below in Figure 17.

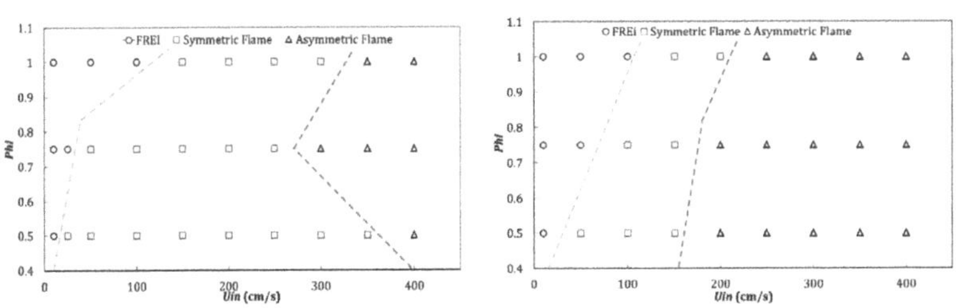

(a) (b)

Figure 17. Flame dynamics diagrams [76]. Effect of U_{in} and ϕ for a channel width of: (**a**) 0.6 mm; (**b**) 0.8 mm.

Boyarko et al. [109] aimed to understand the physics and chemistry of combustion processes in microreactors for microspacecraft propulsion systems. These microspacecraft are designed to have a mass of less than 50 kg, with propulsion systems capable of delivering impulses of 1–10 mN. Micropropulsion systems exhibit distinct behaviors compared to larger-scale systems due to differences in flow rates and heat loss variations to the surroundings. The study analyzed flames in a microtube, chosen for its ease of being manufactured from catalytic materials and its simple geometry, suitable for simulations to validate the case under investigation. Experimentally, they explored the role of catalyst-coated tube surfaces in 0.4 mm and 0.8 mm diameter tubes, with a focus on rich hydrogen and oxygen mixtures.

Investigations in Wavy Wall Geometries

The literature on wavy wall geometries in microcombustion is notably limited. The initial research into this subject was led by Mansouri [110]. This study utilized numerical analysis to evaluate the impact of a newly devised micro-channel characterized by wavy walls on the performance of a micro-thermophotovoltaic system. This investigation proved crucial due to the substantial influence of burner geometry on the system's power output. The study involved a two-dimensional numerical simulation using premixed CH_4/air combustion, maintaining a constant velocity of $U = 1$ [m/s] and an equivalence ratio of $\phi = 1$ across various wavy number configurations. The goal was to gain insights into flame behavior under varying kinetic mechanisms and wave-like designs.

Mansouri [111] later extended the study, examining how the equivalence ratio, inlet velocity, and burner height affect wavy burner geometry's performance. The results showed an impressive 8.3% increase in conversion efficiency compared to flat burners, suggesting the adoption of undulating shapes in future micro-combustor designs and studies.

In their study, Han et al. [112] explored the impact of energy conversion improvement and NOx emissions in the context of thermophotovoltaic applications using ammonia/hydrogen mixtures. Among three distinct wavy combustor designs, it was observed that NOx emissions decreased notably in the wave-shaped design with smooth arcs. Furthermore, this particular design exhibited superior temperature uniformity at velocities exceeding 12 m/s.

In a recent study by Resende et al. [82], flame dynamics within a wave-shaped combustor featuring substantial divergence and convergence variations were investigated. The study employed a hydrogen/air mixture and revealed a pronounced interaction between fluid dynamics and chemical reactions. Specifically, the vortices within the convex regions were identified as acting to resist heat loss flux. Later, the study was extended (Resende et al. [83]) to investigate stable and pulsatory flame burst phenomena in an undulating geometry using premixed hydrogen and air with $\phi = 1.0$. Notably, this work introduced a linear temperature profile along the wall, allowing for an analysis of flow dynamics without imposing fixed flame locations. Two flame configurations were observed (stable for the lowest inlet velocity and PFB). The study emphasized the importance of the interaction between fluid dynamics and combustion, especially due to vortices formed by the burner's nonlinear geometry.

Recently, Abbaspour and Alipoor [113] extended the previous work on convergent–divergent microtubes [102] by using a cosine wave-shaped combustor, and the results indicate that wavy walls facilitate the keeping and propagation of self-sustaining flames within the micro combustion chamber.

7. A Review of Micropropulsion

In this Section, we present a brief review of micropropulsion. The theoretical knowledge presented in the works of [114–116] forms the basis for our discussion here.

7.1. Overview on Micropropulsion Systems

Propulsion can be defined as [117]: "The action or process of drive something forward or onward by or as if by means of a force that imparts motion". In other words, the propulsion system is a mechanical system which turns different forms of energy into kinetic energy.

The limitations of using conventional propulsion systems to deliver small and precise impulses for small satellites have necessitated the development of specific propulsion systems tailored to micro-scale applications [118,119]. In comparison to conventional propulsion systems, micropropulsion systems (MPS) exhibit two main characteristics (though, notably, a propulsion system that solely possesses the first characteristic outlined bellow, is commonly referred to as a micropropulsion system):

(i) They possess the capability to generate relatively minor thrust with magnitudes ranging up to 100 mN and impulse values of μNs up to mNs;
(ii) They display a relatively low volume and weight profile.

MPS based on the use of propellants have traditionally been classified according to the state of the propellant, namely gas, solid, or liquid. However, recent advances in the development of propellantless technologies, such as the Solar Sail, have prompted the need for additional classification schemes. Figure 18 provides a concise summary of the classification schemes utilized.

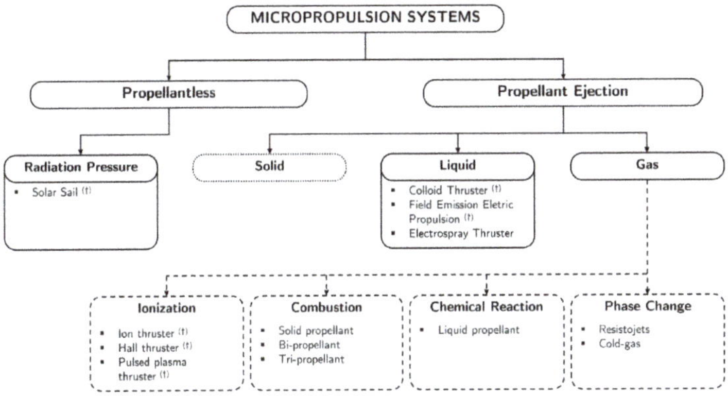

Figure 18. Classification of micropropulsion concepts (based on [120]): the systems highlighted with (†) are manufactured with conventional methods whereas the ones without made using MEMS and silicon technologies.

As shown in Figure 18, another relevant categorization of MPS pertains to the manufacturing techniques employed, with some systems utilizing MEMS and silicon technologies (this research focuses specifically on such systems), while others utilize conventional methods.

7.2. Micropropulsion Systems Performance

MPS performance is commonly assessed under the framework of ideal rocket conditions. Nevertheless, it is crucial to acknowledge that these conditions rely on a set of assumptions that may not perfectly align with the realities of MPS. One such assumption is neglecting friction forces [116]. Hence, a cautious approach is required when applying these idealized models to micropropulsion scenarios. Therefore, the set of equations presented here (the indices correspond to those depicted in Figure 19) serves to provide insights into the ideal performance of micropropulsion systems. When analyzing the performance of the thruster, two key parameters are of major interest: specific impulse and thrust. These equations offer valuable information regarding the performance characteristics of such micropropulsion systems.

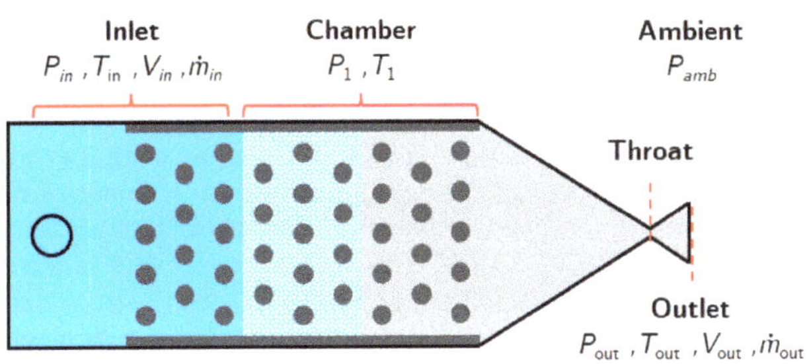

Figure 19. Diagram of a MPS showing the variables and their indexes (adapted from [120]).

7.2.1. Thrust and Specific Impulse

In aircraft, thrust is normally explained through the application of Newton's third law of action and reaction. A gas, or working fluid, is accelerated by the engine, and the reaction to this acceleration produces a force in the engine. The general derivation of the thrust equation shows that the amount of thrust generated depends on the mass flow rate through the engine and the gas exit velocity [121].

From Newton's second law, force can be defined as the change in momentum of an object over time. Momentum is equal to the mass of the object times its velocity. When dealing with a gas, the thrust equation is given as:

$$F = \dot{m}_{\text{out}} \cdot v_{\text{out}} - \dot{m}_{\text{in}} \cdot v_{\text{in}} + (P_{\text{out}} - P_{\text{in}}) \cdot A_{\text{out}} \tag{1}$$

where $\dot{m}_{\text{out}}$ and $\dot{m}_{\text{in}}$ are the mass flow rates at the outlet and inlet of the propulsion system, v_{out} and v_{in} are the respective average fluid velocities at the outlet and inlet of the propulsion system, P_{out} and P_{in} are the outlet and inlet pressures, and A_{out} is the area of the outlet section.

In several cases, the outlet and inlet pressures of the propulsion system are equal to atmospheric pressure. In this case, the term $(P_{\text{out}} - P_{\text{in}})$ becomes zero, and the equation can be simplified as:

$$F = \dot{m}_{\text{out}} \cdot v_{\text{out}} - \dot{m}_{\text{in}} \cdot v_{\text{in}} \tag{2}$$

It has been observed that the mass flow rate at the exit is equal to the sum of the mass flow rate at the inlet and the mass flow rate of fuel. By using the fuel-to-air equivalence ratio, ϕ, the following relation is obtained:

$$\dot{m}_{\text{out}} = (1 + \phi) \cdot \dot{m}_{\text{in}} \tag{3}$$

Let us introduce a new variable termed specific impulse I_{sp}, which serves as a measure of efficiency with respect to propellant consumption.

$$I_{\text{sp}} = \frac{\int_0^t F\mathrm{d}t}{g\int_0^t \dot{m}\mathrm{d}t} \tag{4}$$

where the constant $g = 9.80665\ [\mathrm{m/s^2}]$ represents the gravitational acceleration on Earth at sea level. Despite being expressed in seconds, it does not indicate a measure of time; rather, it denotes the thrust per unit weight of propellant. For optimal propellant consumption efficiency, this value should be as high as possible.

Assuming constant velocity and pressure at the outlet and inlet, it becomes possible to simplify Equation (4). Through algebraic manipulations and by employing Equation (3), we arrive at Equation (5), wherein I_{sp} solely relies on the engine's velocity difference.

$$I_{\text{sp}} = \frac{F}{\dot{m}_{\text{in}}} = (1 + \phi) \cdot v_{\text{out}} - v_{\text{in}} \tag{5}$$

The definition of this new variable brings several advantages to the analysis of propulsion systems. Firstly, the specific impulse is an indicator of the efficiency of the mechanism. Different engines will always have different values of specific impulse. An engine with a higher specific impulse will be more efficient since it produces more thrust for the same airflow. Secondly, as it is a specific property, the inlet mass flow rate can be omitted in all intermediate calculations, greatly simplifying the mathematical procedure. Thirdly, it is a variable that facilitates the design of the propulsion mechanism.

The incoming velocity is typically a known quantity, assuming that the profile is already fully developed and uniform across the inlet area. However, the exit velocity will vary across the outlet area. Therefore, it becomes necessary to calculate the average velocity (referred to as U_{out}) at the system's exit. In mathematical terms, this value is defined as follows:

$$U_{\text{out}} = \frac{1}{A_{\text{out}}} \times \int \|\mathbf{v}_{\text{out}}\| \mathrm{d}A_{\text{out}} \tag{6}$$

in other words, it is the integral of the magnitude of the velocity vector profile over the entire cross-sectional area $(\mathrm{d}A_{\text{out}})$ at the outlet, divided by the cross-sectional outlet area.

7.2.2. Thrust Considering Chamber Pressure

One way to attain more precise impulse values is to simply not disregard the pressure differences between the combustion chamber and ambient pressure. Therefore, while the previous equations are simplified but valid, it is indispensable in the future to grasp the distinctions and advantages of using a more accurate and precise equation, such as [122]:

$$F = \int_{S_{\text{exit}}} \rho u_a (u \cdot n) dA + \int_{S_{\text{exit}}} (P_e - P_\infty) dA \tag{7}$$

where ρ is density, u_{a} is the velocity component in an axial direction, u is a velocity vector, S_{exit} is the area of the fluid outlet, A is control surface, n is a unit vector normal to a plan A, P_e is the pressure at the fluid outlet, and P_∞ is the ambient pressure.

7.2.3. Thrust Key Indicators

In contemporary propulsion systems, multiple components such as turbines, compressors, and combustion chambers are commonly integrated. Typically, several compressors are employed in series to progressively increase the air pressure. Subsequently, the combustion chamber facilitates the reaction between fuel and air, followed by the installation of a turbine ahead of the propeller's exit section. Propulsion is achieved through the difference between the inlet and outlet velocities, generating the desired thrust force.

Two highly relevant indicators in the realm of propulsion mechanisms emerge, although they are not addressed in this review. The first indicator is the Engine Pressure Ratio (EPR), defined as the ratio of the total pressure across the engine. As previously explained, modern engines consist of various components, and in this case, the EPR becomes the product of pressure ratios across all propulsion system components. Consequently, the EPR can be simplified and calculated as follows:

$$\text{EPR} = \frac{P_{\text{tot,out}}}{P_{\text{tot,in}}} \tag{8}$$

The second indicator is the Engine Temperature Ratio (ETR), which, analogously to the previous indicator, represents the ratio of the total temperature across various engine components. Similar to the previous case, given the complexity of modern engines, the ETR becomes the product of temperature ratios across the engine's components. Hence, the ETR is calculated using the following expression:

$$\text{ETR} = \frac{T_{\text{tot,out}}}{T_{\text{tot,in}}} \tag{9}$$

7.3. Conventionally Manufactured Micropropulsion Systems

7.3.1. Pulsed Plasma Thruster

The PPT is an electromagnetic electric thruster that relies on a electromagnetic field to accelerate the plasma to produce thrust. It is based on the principle of the Lorentz force, which creates a force that accelerates the plasma to generate thrust. In other words, a high voltage discharge is generated between the cathode and the anode rod sleeve, which in turns heats the propellant (usually Teflon) making it decompose, which is then ionized to create a plasma. Specifically, the joint action of the magnetic pressure and the gas dynamic pressure in the thin layer of ionized gas accelerates the gas to produce thrust.

PPT operates in a pulsed manner due to the generation of sparks. As a result, continuous operation is not possible, hence the term "pulsed". Figure 20a illustrates the working principle of the PPT, while Figure 20b showcases a working prototype developed by AIS [123].

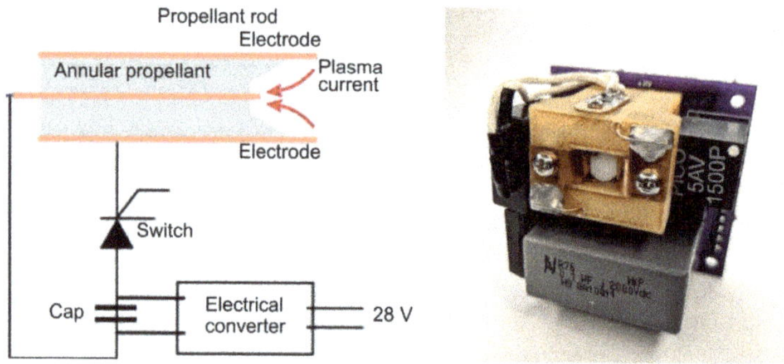

(**a**) (**b**)

Figure 20. Pulsed Plasma Thruster: (**a**) Working principle [124]; (**b**) prototype by AIS [123].

This system has garnered considerable interest among researchers owing to its straightforward design, cost-effectiveness, durability, and capacity for generating low-level outputs. Its track record of successful operation in space dates back to 1964 [125]. Recently, it has been used in CubeSat [126–129] and micro-satellite [130,131] missions successfully.

7.3.2. Field Emission Electric Propulsion

This type of devices uses a liquid propellant ejection system and must employ high-density substances to optimize thruster efficiency. As such, FEEP can be defined as an electrostatic electric propulsion system, that employs a high-voltage electrostatic field to accelerate charged ions for generating thrust. In Figure 21a the fundamental principles of a typical FEEP operation are illustrated, while in Figure 21b a working prototype developed by TU Dresden is presented [132].

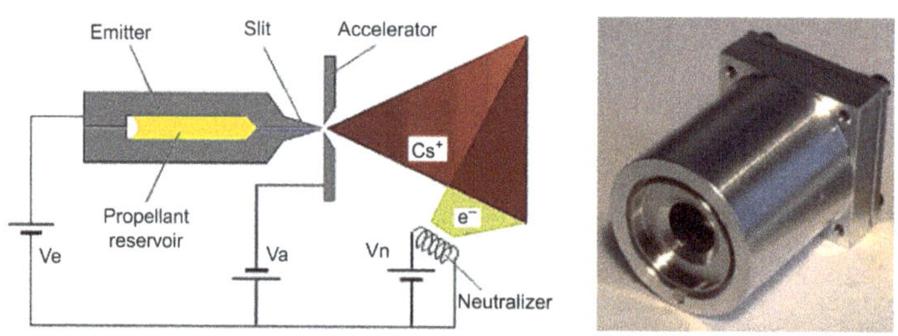

(**a**) (**b**)

Figure 21. Field emission eletric propulsion: (**a**) Working principle [133]; (**b**) prototype by TU Dresden [132].

FEEP consists of a transmitter with a propellant reservoir chamber, an accelerating electrode, and a neutralizer. It employs low-melting-point metals (e.g., cesium, indium) as propellants, stored as solids until liquefied by heating. Capillary action transports the liquid metal to the emitter outlet slit. With 8–15 kV applied between the emitter electrode and the accelerating electric field, metal ions overcome surface tension and gain acceleration, producing thrust [114]. Additionally, a positive voltage on the transmitter and a negative voltage on the accelerator create an electric field, enabling electron beams to discharge metal ions, minimizing pollution plume effects. Zhong et al. [134] studied the effect of the surface morphology of the emitting tip, using dynamic micro-electrochemical etching fabrication techniques to realize the micro-grooves processing of emitting tip surfaces at different depths by regulating roughening time and voltage frequency.

7.3.3. Other Relevant Micropropulsion System Technologies

Besides FEEP and PPT, there are other conventionally manufactured MPS options available, such as:

- Colloid Thruster: An electrostatic electric MPS similar to FEEP that uses high voltage and conductive liquids to accelerate charged particles, producing thrust. The thruster comprises charged particle emitters, extraction electrodes, an accelerating network, and a neutralizer to prevent ionic contamination on satellites. Different electrode potentials generate particle streams of varying polarities. The working principle of the colloid thruster [135] is represented in Figure 22a.
- Hall thruster and Ion Thruster: Devices that generate thrust by ejecting ions at high velocities. They can achieve high specific impulse, but typically produce low thrust levels. There are two main types: Hall thrusters and ion thrusters. Hall thrusters use the Hall effect to trap electrons, ionize the propellant, and then accelerate and exhaust it to create thrust. Ion thrusters, on the other hand, use charged grids to accelerate the ionized propellant [136–138]. This propulsion technology has already seen implementation in space missions and is capable of delivering thrust within the appropriate range for CubeSat missions. The working principle of the Ion thruster [139] and the Hall thruster [135] are represented in Figure 22b,c, respectively. Recently Dong et al. [140] proposed an advanced Watt class of Hall micro thrusters (HMT) for space gravitational wave detection. The research involved creating a circuit for thruster discharge and signal acquisition, using a Faraday probe to measure the plasma plume signal generated by the thruster. The discharge of the HMT thruster [140] using xenon is represented in Figure 22e.
- Solar sail: A spacecraft propulsion method that utilizes sunlight momentum to drive the craft forward [141]. It consists of a lightweight, reflective-coated sail with extreme thinness (down to 4.5 microns) and substantial size (hundreds of meters or kilometers) [142]. Photons from the Sun impart momentum to the sail, propelling the spacecraft and this continuous acceleration allows for high speeds over time. Solar sails have various advantages, including cost effectiveness, sustainability, applicability to small satellites, and the potential for long operating lifetimes [143]. This technology shows great promise for efficient and lightweight space travel. The working principle of the Solar sail [144] is represented in Figure 22).
- Laser Micropropulsion (LMP): The concept of laser propulsion was first introduced by Kantrowitz [145], presenting the advantages of being cost effective, with a high specific impulse, and an excellent thrust power ratio, and has gained significant recognition in the applications in micro/nano satellite propulsion. Nevertheless, the widespread adoption of LMP technology is limited by the proper selection of propellant. Recently, Rao et al. [146,147] presented an innovative LMP technology using metal-organic frameworks (MOFs) and also MOF-derived Carbon-encapsulated-Nano-Metal Composites (CNMC) as the propellant. The characteristic structure in CNMC overcomes the challenge of local hot zones caused by uneven physical mixing encountered in traditional LMP approaches. The ability to manipulate the metal content in MOFs allows for the optimization of laser propulsion performance, paving the way for further enhancements.

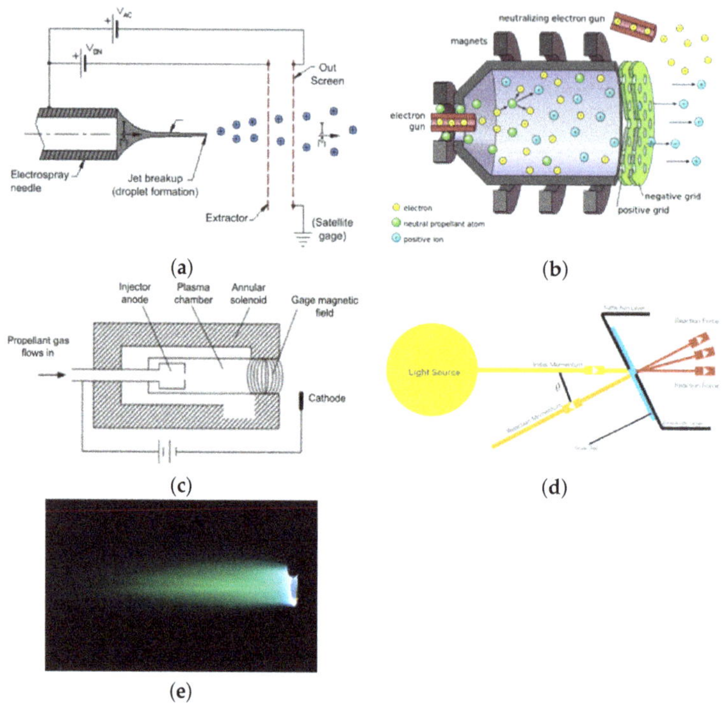

Figure 22. Working principle of: (**a**) colloid thruster [135]; (**b**) ion thruster [139]; (**c**) Hall thruster [135]; (**d**) Solar sail [144]; (**e**) discharge of the HMT thruster [140] using xenon.

7.4. Micro-Electro-Mechanical Systems

The focus of this subsection is a comprehensive investigation of MEMS micropropulsion systems, specifically with regards to their suitability for implementation in nano- and picosatellites. As mentioned previously, the utilization of MEMS holds great potential for the development of highly compact spacecraft, owing to their ability to fabricate extremely small features and seamlessly integrate electronics [148]. More specifically, we delve into systems characterized by the incorporation of MEMS (Micro-Electro-Mechanical Systems) and silicon technologies specifically in the construction of the thruster component. It is noteworthy that some systems may incorporate MEMS elements, such as sensors, but these singular components do not categorically define the entire system as MEMS micropropulsion.

MEMS Manufacturing

Many of these systems operate within the micro to nanometer scales, and they fluidly combine mechanical and electrical components into a unified device. In the context of micropropulsion, the mechanical aspects encompass features like fluidic channels and structural components, while the electrical aspects involve components such as resistive heaters, sensors, and so on [149]. Nevertheless, defining precise boundaries between these mechanical and electrical components within MEMS can be challenging due to the intricate processes occurring within these devices.

The fabrication of these devices, as shown in Figure 23, typically encompasses a sequence of iterative procedures, commencing with a silicon wafer, and then involving these processes [120]:

1. DEPOSITION: This step entails the deposition of minute layers of materials onto the surface of the silicon wafer.

2. PATTERNING: Patterning is employed to imprint a specific design onto the surface, thereby safeguarding certain regions while exposing others to shape the device's features.
3. ETCHING: Etching serves the purpose of material removal from the exposed areas, employing either isotropic or anisotropic techniques.

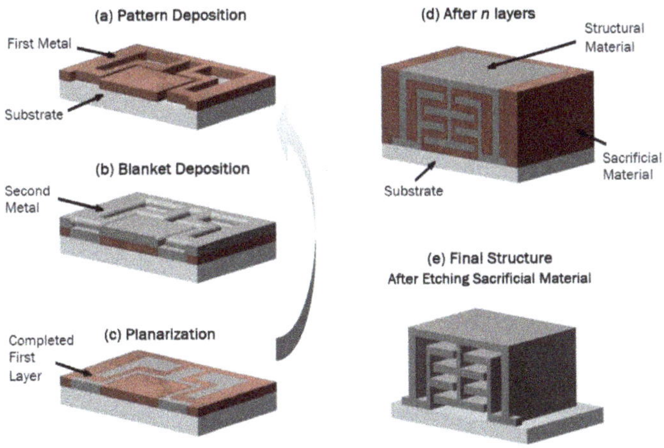

Figure 23. Metal MEMS fabrication process [150].

7.5. MEMS Categories

In summary, there are two major types of micropropulsion systems that utilize MEMS technology, with one of them further dividing into two subgroups. These are MEMS ELECTRICAL THRUSTERS (split into Electrothermal and Electrostatic) and MEMS CHEMICAL THRUSTERS. Additionally, we have a system of microthrusters called COLD-GAS, that are purely mechanical and do not use electrothermal, electrostatic, or chemical methods.

7.5.1. Electrothermal MEMS Electrical Thrusters

These systems operate through a dual process: first, by employing electrical resistance heaters to heat gas within the thrust chamber, and subsequently expelling the heated gas through a nozzle to generate thrust.

Resistojets are among the most prevalent electrothermal MEMS microthrusters. In contrast to certain early designs, like the one introduced by Bayt [151] (GAS-FED RESISTOJETS), which employed propellants in the gaseous phase, some of these systems now utilize propellants in liquid or solid states, resulting in phase changes as the gas undergoes heating.

Based on the specific phase-change characteristics exhibited by these devices, we can categorize them into two types of micro-resistojets, each associated with distinct flow regimes:

1. VAPORIZING LIQUID MICROTHRUSTER (VLM): They have been in development since the 1960s [152–155], and the design comprises an inlet channel for propellant feed, a chamber where a heating element vaporizes the propellant, and a convergent–divergent nozzle to propel the gases to supersonic speeds, as observed in Figure 24a). Several designs demonstrate significant potential, achieving a specific impulse (I_{sp}) exceeding 100 s [156–158];
2. LOW-PRESSURE MICRORESISTOJET (LPM, also referred to as the Free Molecule Micro-Resistojet): They operate in the transitional flow regime ($0.1 < \text{Kn} \leq 10$) [120]. To model and simulate its operation, statistical methods based on gas kinetic theory are required [159,160]. These devices typically consist of an inlet section, a low-pressure plenum (typically below 1000 Pa), and a MEMS-manufactured heater chip with channels to accelerate the gas, as observed in Figure 24b). The design of these

channels is critical for efficient heat transfer and overall thruster efficiency [161,162]. The choice of resistance type and manufacturing approach is also essential for the optimal conversion of electrical to thermal energy. Current designs show I_{sp} of around 7∼39 s [163–165].

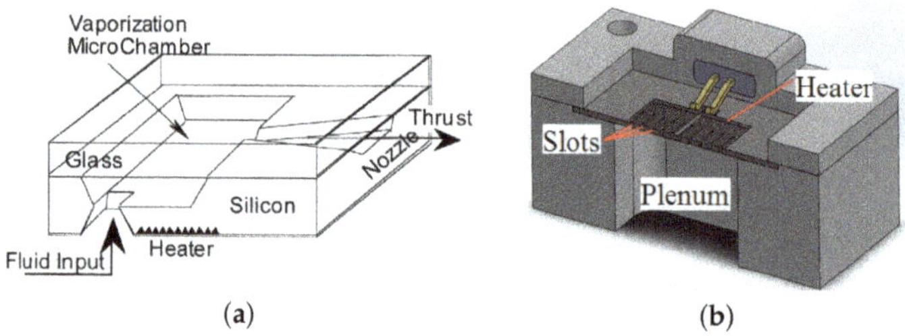

Figure 24. Working principle of: (**a**) VLM [166]; (**b**) LPM [120].

Other electrothermal designs are the SOLID SUBLIMATION MICROTHRUSTER, which operate by sublimating a solid propellant, which means they use electrical energy to heat the solid propellant, causing it to go directly from a solid to a gaseous state, with I_{sp} of around 50∼75 [167].

7.5.2. Electrostatic MEMS Electrical Thrusters

Electrostatic MEMS electrical thrusters function on two separate principles: FIELD-EFFECT ION ENGINES (FET ion engines) and COLLOID THRUSTERS. They operate similarly, with the primary distinction lying in the choice of propellant. While FET ion engines employ low melting point metallic propellants (such as cesium or indium), colloid thrusters utilize ionic liquids or mixtures [167].

In FET ion engines, multiple electrode layers on a silicon substrate create a strong electric field, ionizing liquid metal and expelling positively charged metal atoms for thrust. Electrons are discharged externally. This design offers low power consumption, high specific impulse (up to thousands of seconds), and precise low thrust control.

MEMS microcolloid thrusters operate similarly to field effect particle thrusters but use nonmetallic liquid propellants to accelerate tiny charged droplets. They have a higher operating voltage and thrust density than field effect ion engines but a lower specific impulse compared to electrothermal thrusters (schematic in Figure 25).

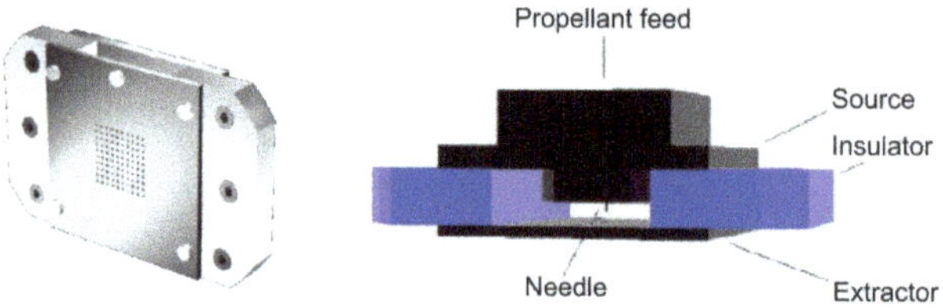

Figure 25. Schematic of advanced microcolloid thruster by Stanford University [168].

A specific type of these systems is the ELECTROSPRAY THRUSTERS, which operate by emitting a particle spray through a Taylor cone effect, which is triggered by applying an electric potential to an ionic liquid in a capillary. This cone sharpens and releases particles, including single ions or droplets, when a threshold voltage is reached. Propellant options include ionic liquids, mixtures, or liquid metals, and an accelerator grid can be added after the extractor to boost particle exit velocity [169,170]. See Figure 26 for a schematic.

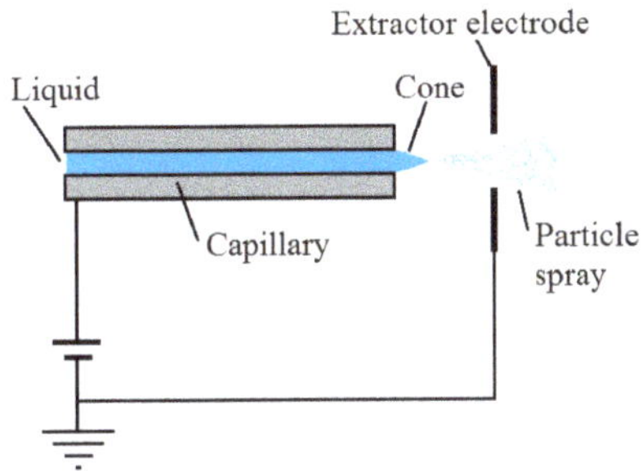

Figure 26. Schematic of an electrospray thruster [120].

7.5.3. MEMS Chemical Thrusters

MEMS microchemical thrusters can employ either SOLID OR LIQUID PROPELLANTS, such as double-base solid nitrocellulose. While they offer higher thrust compared to electrothermal thrusters, their performance lags behind that of electrostatic and electromagnetic thrusters. The typical I_{sp} for MEMS microchemical thrusters falls in the range of 100 to 300 s. A common example of a MEMS microchemical thruster is a digital microarray. Utilizing MEMS (Micro-Electro-Mechanical Systems) technology and SOC (System-on-a-Chip) technology, these thrusters consolidate various components into a single chip. This integration includes the address driving circuit, ignition system, working fluid reservoir, combustion chamber, valves, and microminiature nozzles. A shematics of the working principle of both solid and liquid propellant microthrusters is presented in Figure 27.

SOLID PROPELLANT MICROTHRUSTERS [171–175] comprise a chamber housing a limited propellant quantity, an igniter, often in the form of a heater, and a nozzle for gas acceleration following combustion. These systems are exceptionally compact because they do not require a feeding system or a pressurized container. Recently, Lee et al. [176,177] presented a new PCB micro-ignitor concept without using membranes, using a process based on the lab-on-PCB technique and stencil printing. This PCB micro-ignitor integrates an AP/PDMS solid propellant and stencil-printed heaters. Additionally, Lee et al. [176,177] presented a parametric study to obtain the proper heater dimensions that ensure consistent heat transfer between the heater and the solid propellant. Kan et al. [178] proposed a safety control function for the micro-igniter aiming at improve the output energy. In order to minimize the misaction caused by traditional micro-igniters due to false triggering, Kan et al. [178] proposed a MEMS safety device based on Lorentz force. Kan et al. [179] performed an experimental study of the effect of carbon nanomaterials (CNMs) on the ignition and thrust performance of various types of oxygenated salts/nanothermite mixtures. The authors used CNMs that included graphene–oxide (GO), reduced GO, and carbon nanotubes (CNTs), and demonstrated that non-metallic oxidizers is a good alternative energetic oxidants for thermite mixtures with high combustion performance. The CNMs used by Kan et al. [179] used a simple ultrasonic mixing method to integrate the CNMs into nanothermite compositions.

LIQUID PROPELLANTS MICROTHRUSTERS, using components such as hydrazine and hydrogen peroxide, decompose the propellants into hot gases when catalyzed, which are then accelerated through a nozzle to generate thrust. While hydrazine is a common choice for its medium-level performance, it poses safety challenges due to toxicity and flammability [180]. In contrast, hydrogen peroxide is a safer alternative but can slowly decompose when in contact with organic materials [181,182]. This decomposition risk may lead to propellant loss in CubeSats stored for extended periods before launch.

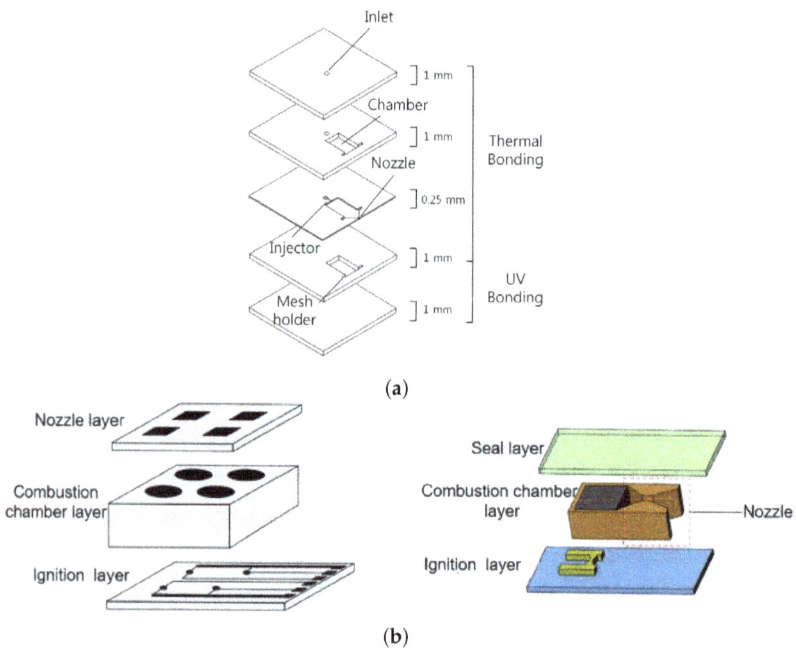

Figure 27. Working principle of: (**a**) Liquid Propellant Microthruster [122]; (**b**) Solid Propellant Microthruster [183].

This review focuses on microcombustion in the field of micropropulsion, a critical component of nanosatellite propulsion systems. Chen et al. [184] developed a hydrogen-based micropropulsion system using catalytic microcombustion to provide power for microsatellites' maneuvers. They experimentally and numerically studied this system's performance, investigating parameters like fuel/air mixture flow rate, equivalence ratio, and nozzle contraction ratio. They used a platinum catalyst-coated tube with a 500 μm diameter and 1 cm length as the reactor, inserted into a quartz tube with various converging-diverging nozzle ratios. It explored different flame behaviors and intrinsic properties, including the temperature and mass concentration of key species (H_2, OH, H_2O).

7.5.4. Cold-Gas Microthrusters

This micropropulsion system employs pressurized gas as propellant, which can be stored in various phases (liquid, gaseous, or solid). They are a type of microthruster that operates based on a simple mechanical principle without the use of electrothermal, electrostatic, or chemical methods. The gas is accelerated through a nozzle to generate thrust, as observed in Figure 28. One challenge is managing leakage, as microscopic particle contamination can affect valve sealing and lead to propellant loss. These systems are well-developed for CubeSats due to their simplicity [120]. Some models even feature integrated control circuits for satellite interfacing [120]. Future advancements may include integrated sensors and control valves. Propellant can be stored as gas, liquid, or solid, with efficient ignition being crucial, and inert gases are commonly used for safety reasons, though using low-boiling-point liquids like butane may improve efficiency [120].

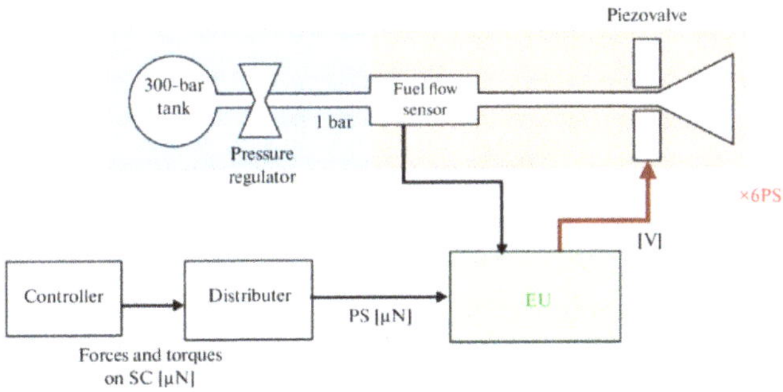

Figure 28. Functional diagram of the circuit of a cold-gas microthruster [185].

8. Conclusions

This work presented a review intended to provide insights into micro-combustion and micro propulsion. Starting with the explorations of the different challenges of micro-combustion and the typical length and time scales involved in micro-combustion. Next we presented a review for the types of micro-flame dynamics, and in the final part of this review, we focused on micro-propulsion systems, their performance metrics, conventional manufacturing methods, and the advancements in Micro-Electro-Mechanical Systems manufacturing. This review is an invaluable resource for researchers and engineers navigating the complexities of micro-scale combustion and propulsion.

Funding: A. M. Afonso acknowledge FCT—Fundação para a Ciência e a Tecnologia for financial support through LA/P/0045/2020 (ALiCE), UIDB/00532/2020 and UIDP/00532/2020 (CEFT), funded by national funds through FCT/MCTES (PIDDAC). P. R. Resende acknowledges the funding by FAPESP through Project No. 2015/26842-3.

Conflicts of Interest: The authors declare no conflicts of interest.

References

1. Ju, Y.; Maruta, K. Microscale combustion: Technology development and fundamental research. *Prog. Energy Combust. Sci.* **2011**, *37*, 669–715. [CrossRef]
2. Nakamura, Y.; Gao, J.; Matsuoka, T. Progress in small-scale combustion. *J. Therm. Sci. Technol.* **2017**, *12*, JTST0001. [CrossRef]
3. Cai, L.; Jiaqiang, E.; Li, J.; Ding, J.; Luo, B. A comprehensive review on combustion stabilization technologies of micro/meso-scale combustors for micro thermophotovoltaic systems: Thermal, emission, and energy conversion. *Fuel* **2023**, *335*, 126660. [CrossRef]
4. Jabar, D.; Al-Fahham, M. A review of combustion parameters affecting length and flame structure. *AIP Conf. Proc.* **2024**, *3051*, 020001.
5. Nauman, M.; Pan, J.; Wang, Y.; Li, F.; Ojo, A.O.; Raza, A. A review of recent advancements in micro combustion techniques to enhance flame stability and fuel residence time. *Int. J. Hydrogen Energy* **2023**, *49*, 1165–1193. [CrossRef]
6. Merriam-Webster. Combustion. Available online: https://www.merriam-webster.com/dictionary/combustion (accessed on 28 March 2023).
7. Fernandez-Pello, A.C. Micropower generation using combustion: Issues and approaches. *Proc. Combust. Inst.* **2002**, *29*, 883–899. [CrossRef]
8. Thomas, G.; Oakley, G.; Bambrey, R. Fundamental studies of explosion arrester mitigation mechanisms. *Process. Saf. Environ. Prot.* **2020**, *137*, 15–33. [CrossRef]
9. Davy, H. VII. Some researches on flame. *Philos. Trans. R. Soc. Lond.* **1817**, *107*, 45–76.
10. Hossain, A.; Nakamura, Y. Thermal and chemical structures formed in the micro burner of miniaturized hydrogen-air jet flames. *Proc. Combust. Inst.* **2015**, *35*, 3413–3420. [CrossRef]
11. Resende, P.R.; Ayoobi, M.; Afonso, A.M. Numerical investigations of micro-scale diffusion combustion: A brief review. *Appl. Sci.* **2019**, *9*, 3356. [CrossRef]
12. Lee, L.N.; Kim, M.J.; Hwang, W.J. Potential of augmented reality and virtual reality technologies to promote wellbeing in older adults. *Appl. Sci.* **2019**, *9*, 3556. [CrossRef]
13. Glassman, I.; Yetter, R. Flame phenomena in premixed combustible gases. In *Combustion*, 4th ed.; Academic Press: Cambridge, MA, USA, 2008; pp. 153–162.

14. Li, Q.; Wang, J.; Meng, L.; Li, J.; Guo, Z. CFD study on stability limits of hydrogen/air premixed flames in planar micro-combustors with catalytic walls. *Appl. Therm. Eng.* **2017**, *121*, 325–335. [CrossRef]
15. Kang, Z.; Shi, Z.; Ye, J.; Tian, X.; Huang, Z.; Wang, H.; Wei, D.; Peng, Q.; Tu, Y. A Review of Micro Power System and Micro Combustion: Present Situation, Techniques and Prospects. *Energies* **2023**, *16*, 3201. [CrossRef]
16. Yun, T.M.; Kottke, P.A.; Anderson, D.M.; Fedorov, A.G. Experimental investigation of hydrogen production by variable volume membrane batch reactors with modulated liquid fuel introduction. *Int. J. Hydrogen Energy* **2015**, *40*, 2601–2612. [CrossRef]
17. Wang, Y. Introduction to Micro-Scale Combustion: University Lecture at Pennsylvania State University. Available online: https://yang.gatech.edu/projects/Meso%20and%20Micro%20Scale%20Propulsion...(Yanxing)/files/micro-combustion%20(for%20webpage%20).pdf (accessed on 19 May 2023).
18. Lee, M.J.; Cho, S.M.; Choi, B.I.; Kim, N.I. Scale and material effects on flame characteristics in small heat recirculation combustors of a counter-current channel type. *Appl. Therm. Eng.* **2010**, *30*, 2227–2235. [CrossRef]
19. Tu, Y.; Liu, H.; Yang, W. Flame characteristics of CH_4/H_2 on a jet-in-hot-coflow burner diluted by N_2, CO_2, and H_2O. *Energy Fuels* **2017**, *31*, 3270–3280. [CrossRef]
20. Mohan, S.; Matalon, M. Diffusion flames and diffusion flame-streets in three dimensional micro-channels. *Combust. Flame* **2017**, *177*, 155–170. [CrossRef]
21. Feldmeier, S.; Schwarz, M.; Wopienka, E.; Pfeifer, C. Categorization of small-scale biomass combustion appliances by characteristic numbers. *Renew. Energy* **2021**, *163*, 2128–2136. [CrossRef]
22. Li, Y.H.; Chen, G.B.; Cheng, T.S.; Yeh, Y.L.; Chao, Y.C. Combustion characteristics of a small-scale combustor with a percolated platinum emitter tube for thermophotovoltaics. *Energy* **2013**, *61*, 150–157. [CrossRef]
23. Kaisare, N.S.; Deshmukh, S.R.; Vlachos, D.G. Stability and performance of catalytic microreactors: Simulations of propane catalytic combustion on Pt. *Chem. Eng. Sci.* **2008**, *63*, 1098–1116. [CrossRef]
24. Bagheri, G.; Hosseini, S.E.; Wahid, M.A. Effects of bluff body shape on the flame stability in premixed micro-combustion of hydrogen–air mixture. *Appl. Therm. Eng.* **2014**, *67*, 266–272. [CrossRef]
25. Akhtar, S.; Khan, M.N.; Kurnia, J.C.; Shamim, T. Investigation of energy conversion and flame stability in a curved micro-combustor for thermo-photovoltaic (TPV) applications. *Appl. Energy* **2017**, *192*, 134–145. [CrossRef]
26. Cao, Z.; Lyu, Y.; Peng, J.; Qiu, P.; Liu, L.; Yang, C.; Yu, Y.; Chang, G.; Yan, B.; Sun, S.; et al. Experimental study of flame evolution, frequency and oscillation characteristics of steam diluted micro-mixing hydrogen flame. *Fuel* **2021**, *301*, 121078. [CrossRef]
27. Aggarwal, S.K. Extinction of laminar partially premixed flames. *Prog. Energy Combust. Sci.* **2009**, *35*, 528–570. [CrossRef]
28. Fu, K.; Knobloch, A.J.; Martinez, F.C.; Walther, D.C.; Fernandez-Pello, C.; Pisano, A.P.; Liepmann, D.; Miyaska, K.; Maruta, K. Design and experimental results of small-scale rotary engines. In Proceedings of the ASME International Mechanical Engineering Congress and Exposition. American Society of Mechanical Engineers, New York, NY, USA, 11–16 November 2001; Volume 35524, pp. 295–301.
29. Shanbhogue, S.; Sanusi, Y.; Taamallah, S.; Habib, M.; Mokheimer, E.; Ghoniem, A. Flame macrostructures, combustion instability and extinction strain scaling in swirl-stabilized premixed CH4/H2 combustion. *Combust. Flame* **2016**, *163*, 494–507. [CrossRef]
30. Eckart, S.; Yu, C.; Maas, U.; Krause, H. Experimental and numerical investigations on extinction strain rates in non-premixed counterflow methane and propane flames in an oxygen reduced environment. *Fuel* **2021**, *298*, 120781. [CrossRef]
31. Spalding, D.B. A theory of inflammability limits and flame-quenching. *Proc. R. Soc. Lond. Ser. A Math. Phys. Sci.* **1957**, *240*, 83–100.
32. Maruta, K.; Yoshida, M.; Ju, Y.; Niioka, T. Experimental study on methane-air premixed flame extinction at small stretch rates in microgravity. In Proceedings of the Symposium (International) on Combustion, Napoli, Italy, 28 July 1996–2 August 1996; Elsevier: Amsterdam, The Netherlands, 1996; Volume 26-1, pp. 1283–1289.
33. Ju, Y.; Maruta, K.; Niioka, T. Combustion Limits. *Appl. Mech. Rev.* **2001**, *54*, 257–277. [CrossRef]
34. Lewis, B.; Von Elbe, G. *Combustion, Flames and Explosions of Gases*; Elsevier: Amsterdam, The Netherlands, 2012.
35. Daou, J.; Matalon, M. Influence of conductive heat-losses on the propagation of premixed flames in channels. *Combust. Flame* **2002**, *128*, 321–339. [CrossRef]
36. Ferguson, C.R.; Keck, J.C. On laminar flame quenching and its application to spark ignition engines. *Combust. Flame* **1977**, *28*, 197–205. [CrossRef]
37. Law, C.K. *Combustion Physics;* Cambridge University Press: Cambridge, UK, 2010.
38. Williams, F.A. *Combustion Theory*; Benjamin-Cummings Publishing Company: San Francisco, CA, USA, 1985.
39. Bockhorn, H.; Habisreuther, P.; Hettel, M. *Numerical Modelling of Technical Combustion: 100 Volumes of 'Notes on Numerical Fluid Mechanics': 40 Years of Numerical Fluid Mechanics and Aerodynamics in Retrospect*; Springer: Berlin/Heidelberg, Germany, 2009; pp. 325–340. [CrossRef]
40. Turns, S.R. *Introduction to Combustion*; McGraw-Hill Companies: New York, NY, USA, 1996; Volume 287.
41. Lieuwen, T.C. *Unsteady Combustor Physics;* Cambridge University Press: Cambridge, UK, 2012. [CrossRef]
42. Jerzak, W.; Kuźnia, M. Experimental study of impact of swirl number as well as oxygen and carbon dioxide content in natural gas combustion air on flame flashback and blow-off. *J. Nat. Gas Sci. Eng.* **2016**, *29*, 46–54. [CrossRef]
43. Thibaut, D.; Candel, S. Numerical Study of Unsteady Turbulent Premixed Combustion: Application to Flashback Simulation. *Combust. Flame* **1998**, *113*, 53–65. [CrossRef]
44. Fruzza, F.; Lamioni, R.; Tognotti, L.; Galletti, C. Flashback of H2-enriched premixed flames in perforated burners: Numerical prediction of critical velocity. *Int. J. Hydrogen Energy* **2023**, *48*, 31790–31801. [CrossRef]

45. Fruzza, F.; Lamioni, R.; Mariotti, A.; Salvetti, M.V.; Galletti, C. Flashback propensity due to hydrogen blending in natural gas: Sensitivity to operating and geometrical parameters. *Fuel* **2024**, *362*, 130838. [CrossRef]
46. Cai, S.; Yang, W.; Wan, J. Combustion and thermal performances of methane-air premixed flame in a novel preheated micro combustor with a flame holder. *Int. J. Therm. Sci.* **2024**, *197*, 108813. [CrossRef]
47. Margolis, S.B. Chaotic Combustion of Solids and High-Density Fluids near Points of Strong Resonance. *Proc. Math. Phys. Sci.* **1991**, *433*, 131–150.
48. Weng, Y.; Unni, V.R.; Sujith, R.; Saha, A. Synchronization framework for modeling transition to thermoacoustic instability in laminar combustors. *Nonlinear Dyn.* **2020**, *100*, 3295–3306. [CrossRef]
49. Kabiraj, L.; Saurabh, A.; Wahi, P.; Sujith, R. Experimental study of thermoacoustic instability in ducted premixed flames: Periodic, quasi-periodic and chaotic oscillations. In Proceedings of the n3l-Int'l Summer School and Workshop on Non-Normal and Nonlinear Effects in Aero-and Thermoacoustics, Munich, Germany, 17–21 May 2010; p. 12.
50. Maruta, K.; Muso, K.; Takeda, K.; Niioka, T. Reaction zone structure in flameless combustion. *Proc. Combust. Inst.* **2000**, *28*, 2117–2123. [CrossRef]
51. Cavaliere, A.; De Joannon, M. Mild combustion. *Prog. Energy Combust. Sci.* **2004**, *30*, 329–366. [CrossRef]
52. Wünning, J.; Wünning, J. Flameless oxidation to reduce thermal NO-formation. *Prog. Energy Combust. Sci.* **1997**, *23*, 81–94. [CrossRef]
53. Katsuki, M.; Hasegawa, T. The science and technology of combustion in highly preheated air. *Symp. Int. Combust.* **1998**, *27*, 3135–3146. [CrossRef]
54. Niioka, T. Fundamentals and applications of high-temperature air combustion. In Proceedings of the 5th ASME/JSME Thermal Engineering Joint Conference, San Diego, CA, USA, 15–19 March 1999; pp. 1–6.
55. Maruta, K.; Kataoka, T.; Kim, N.I.; Minaev, S.; Fursenko, R. Characteristics of combustion in a narrow channel with a temperature gradient. *Proc. Combust. Inst.* **2005**, *30*, 2429–2436. [CrossRef]
56. Pizza, G.; Frouzakis, C.E.; Mantzaras, J.; Tomboulides, A.G.; Boulouchos, K. Three-dimensional simulations of premixed hydrogen/air flames in microtubes. *J. Fluid Mech.* **2010**, *658*, 463–491. [CrossRef]
57. Pizza, G.; Frouzakis, C.E.; Mantzaras, J.; Tomboulides, A.G.; Boulouchos, K. Dynamics of premixed hydrogen/air flames in microchannels. *Combust. Flame* **2008**, *152*, 433–450. [CrossRef]
58. Tsuboi, Y.; Yokomori, T.; Maruta, K. Study on ignition and weak flame in heated meso-scale channel. In Proceedings of the ASME International Mechanical Engineering Congress and Exposition, Proceedings (IMECE), Seattle, WA, USA, 11–15 November 2007; pp. 155–158. [CrossRef]
59. Di Stazio, A.; Chauveau, C.; Dayma, G.; Dagaut, P. Combustion in micro-channels with a controlled temperature gradient. *Exp. Therm. Fluid Sci.* **2016**, *73*, 79–86. [CrossRef]
60. Li, J.; Chou, S.; Yang, W.; Li, Z. A numerical study on premixed micro-combustion of CH4–air mixture: Effects of combustor size, geometry and boundary conditions on flame temperature. *Chem. Eng. J.* **2009**, *150*, 213–222. [CrossRef]
61. Kim, N.I.; Maruta, K. A numerical study on propagation of premixed flames in small tubes. *Combust. Flame* **2006**, *146*, 283–301. [CrossRef]
62. Norton, D.G.; Vlachos, D.G. Combustion characteristics and flame stability at the microscale: A CFD study of premixed methane/air mixtures. *Chem. Eng. Sci.* **2003**, *58*, 4871–4882. [CrossRef]
63. Lamioni, R.; Lapenna, P.E.; Berger, L.; Kleinheinz, K.; Attili, A.; Pitsch, H.; Creta, F. Pressure-induced hydrodynamic instability in premixed methane-air slot flames. *Combust. Sci. Technol.* **2020**, *192*, 1998–2009. [CrossRef]
64. Kurdyumov, V.; Fernandez-Tarrazo, E. Lewis number effect on the propagation of premixed laminar flames in narrow open ducts. *Combust. Flame* **2002**, *128*, 382–394. [CrossRef]
65. Kurdyumov, V.N.; Pizza, G.; Frouzakis, C.E.; Mantzaras, J. Dynamics of premixed flames in a narrow channel with a step-wise wall temperature. *Combust. Flame* **2009**, *156*, 2190–2200. [CrossRef]
66. Dogwiler, U.; Mantzaras, J.; Benz, P.; Kaeppeli, B.; Bombach, R.; Arnold, A. Homogeneous ignition of methane-air mixtures over platinum: Comparison of measurements and detailed numerical predictions. In Proceedings of the Symposium (International) on Combustion, Boulder, CO, USA, 2–7 August 1998; Elsevier: Amsterdam, The Netherlands, 1998; Volume 27-2, pp. 2275–2282.
67. Tsai, C.H. The asymmetric behavior of steady laminar flame propagation in ducts. *Combust. Sci. Technol.* **2008**, *180*, 533–545. [CrossRef]
68. Alipoor, A.; Mazaheri, K. Studying the repetitive extinction-ignition dynamics for lean premixed hydrogen-air combustion in a heated microchannel. *Energy* **2014**, *73*, 367–379. [CrossRef]
69. Lapointe, S.; Druzgalski, C.L.; McNenly, M.J. Numerical study of a micro flow reactor at engine pressures: Flames with repetitive extinction and ignition and simulations with a reduced chemical model. *Combust. Flame* **2018**, *197*, 102–110. [CrossRef]
70. Richecoeur, F.; Kyritsis, D.C. Experimental study of flame stabilization in low Reynolds and Dean number flows in curved mesoscale ducts. *Proc. Combust. Inst.* **2005**, *30*, 2419–2427. [CrossRef]
71. Baigmohammadi, M.; Tabejamaat, S.; Farsiani, Y. Experimental study of the effects of geometrical parameters, Reynolds number, and equivalence ratio on methane–oxygen premixed flame dynamics in non-adiabatic cylinderical meso-scale reactors with the backward facing step. *Chem. Eng. Sci.* **2015**, *132*, 215–233. [CrossRef]
72. Minaev, S.; Maruta, K.; Fursenko, R. Nonlinear dynamics of flame in a narrow channel with a temperature gradient. *Combust. Theory Model.* **2007**, *11*, 187–203. [CrossRef]

73. Jackson, T.; Buckmaster, J.; Lu, Z.; Kyritsis, D.; Massa, L. Flames in narrow circular tubes. *Proc. Combust. Inst.* **2007**, *31*, 955–962. [CrossRef]
74. Miyata, E.; Fukushima, N.; Naka, Y.; Shimura, M.; Tanahashi, M.; Miyauchi, T. Direct numerical simulation of micro combustion in a narrow circular channel with a detailed kinetic mechanism. *Proc. Combust. Inst.* **2015**, *35*, 3421–3427. [CrossRef]
75. Alipoor, A.; Mazaheri, K. Combustion characteristics and flame bifurcation in repetitive extinction-ignition dynamics for premixed hydrogen-air combustion in a heated micro channel. *Energy* **2016**, *109*, 650–663. [CrossRef]
76. Alipoor, A.; Mazaheri, K. Maps of flame dynamics for premixed lean hydrogen-air combustion in a heated microchannel. *Energy* **2020**, *194*, 116852. [CrossRef]
77. Biswas, S. *Physics of Turbulent Jet Ignition: Mechanisms and Dynamics of Ultra-Lean Combustion*; Springer: Berlin/Heidelberg, Germany, 2018.
78. Nakamura, H.; Fan, A.; Minaev, S.; Sereshchenko, E.; Fursenko, R.; Tsuboi, Y.; Maruta, K. Bifurcations and negative propagation speeds of methane/air premixed flames with repetitive extinction and ignition in a heated microchannel. *Combust. Flame* **2012**, *159*, 1631–1643. [CrossRef]
79. Tsuboi, Y.; Yokomori, T.; Maruta, K. Lower limit of weak flame in a heated channel. *Proc. Combust. Inst.* **2009**, *32*, 3075–3081. [CrossRef]
80. Yamamoto, A.; Oshibe, H.; Nakamura, H.; Tezuka, T.; Hasegawa, S.; Maruta, K. Stabilized three-stage oxidation of gaseous n-heptane/air mixture in a micro flow reactor with a controlled temperature profile. *Proc. Combust. Inst.* **2011**, *33*, 3259–3266. [CrossRef]
81. de Macedo, J.M.R.C. Estudos Numéricos em Micropropulsão. Master's Thesis, Universidade do Porto, Porto, Portugal, 2021.
82. Resende, P.R.; Morais, L.C.; Pinho, C.; Afonso, A.M. Combustion characteristics of premixed hydrogen/air in an undulate microchannel. *Energies* **2022**, *15*, 626. [CrossRef]
83. Resende, P.R.; Ferrás, L.L.; Afonso, A.M. Flame dynamics of hydrogen/air mixture in a wavy micro-channel. *Int. J. Hydrog. Energy* **2023**, *48*, 13682–13698. [CrossRef]
84. Li, J.; Chou, S.; Li, Z.; Yang, W. A comparative study of H2-air premixed flame in micro combustors with different physical and boundary conditions. *Combust. Theory Model.* **2008**, *12*, 325–347. [CrossRef]
85. Ronney, P.D. Analysis of non-adiabatic heat-recirculating combustors. *Combust. Flame* **2003**, *135*, 421–439. [CrossRef]
86. Kim, N.I.; Kato, S.; Kataoka, T.; Yokomori, T.; Maruyama, S.; Fujimori, T.; Maruta, K. Flame stabilization and emission of small Swiss-roll combustors as heaters. *Combust. Flame* **2005**, *141*, 229–240. [CrossRef]
87. Kuo, C.; Ronney, P. Numerical modeling of non-adiabatic heat-recirculating combustors. *Proc. Combust. Inst.* **2007**, *31*, 3277–3284. [CrossRef]
88. Kunte, A.; Raghu, A.K.; Kaisare, N.S. A spiral microreactor for improved stability and performance for catalytic combustion of propane. *Chem. Eng. Sci.* **2018**, *187*, 87–97. [CrossRef]
89. Yedala, N.; Raghu, A.K.; Kaisare, N.S. A 3D CFD study of homogeneous-catalytic combustion of hydrogen in a spiral microreactor. *Combust. Flame* **2019**, *206*, 441–450. [CrossRef]
90. Pizza, G.; Mantzaras, J.; Frouzakis, C.E.; Tomboulides, A.G.; Boulouchos, K. Suppression of combustion instabilities of premixed hydrogen/air flames in microchannels using heterogeneous reactions. *Proc. Combust. Inst.* **2009**, *32*, 3051–3058. [CrossRef]
91. Pizza, G.; Mantzaras, J.; Frouzakis, C.E. Flame dynamics in catalytic and non-catalytic mesoscale microreactors. *Catal. Today* **2010**, *155*, 123–130. [CrossRef]
92. Zhou, J.; Wang, Y.; Yang, W.; Liu, J.; Wang, Z.; Cen, K. Combustion of hydrogen–air in catalytic micro-combustors made of different material. *Int. J. Hydrogen Energy* **2009**, *34*, 3535–3545. [CrossRef]
93. Yang, W.M.; Chou, S.; Shu, C.; Li, Z.; Xue, H. Combustion in micro-cylindrical combustors with and without a backward facing step. *Appl. Therm. Eng.* **2002**, *22*, 1777–1787. [CrossRef]
94. Li, J.; Chou, S.; Huang, G.; Yang, W.; Li, Z. Study on premixed combustion in cylindrical micro combustors: Transient flame behavior and wall heat flux. *Exp. Therm. Fluid Sci.* **2009**, *33*, 764–773. [CrossRef]
95. Khandelwal, B.; Deshpande, A.A.; Kumar, S. Experimental studies on flame stabilization in a three step rearward facing configuration based micro channel combustor. *Appl. Therm. Eng.* **2013**, *58*, 363–368. [CrossRef]
96. Faramarzpour, H.; Mazaheri, K.; Alipoor, A. Effect of backward facing step on radiation efficiency in a micro combustor. *Int. J. Therm. Sci.* **2018**, *132*, 129–136. [CrossRef]
97. Fan, A.; Wan, J.; Liu, Y.; Pi, B.; Yao, H.; Liu, W. Effect of bluff body shape on the blow-off limit of hydrogen/air flame in a planar micro-combustor. *Appl. Therm. Eng.* **2014**, *62*, 13–19. [CrossRef]
98. Yang, W.; Deng, C.; Zhou, J.; Liu, J.; Wang, Y.; Cen, K. Experimental and numerical investigations of hydrogen–air premixed combustion in a converging–diverging micro tube. *Int. J. Hydrogen Energy* **2014**, *39*, 3469–3476. [CrossRef]
99. Biswas, S.; Zhang, P.; Wang, H.; Qiao, L. Propagation and extinction behavior of methane/air premixed flames through straight and converging-diverging microchannels. *Appl. Therm. Eng.* **2019**, *148*, 1395–1406. [CrossRef]
100. Rao, S.S.; Raju, V. Effect of Deflector on the Combustion Characteristics of a Micro-Combustor With a Controlled Centrally Slotted Bluff Body. *J. Energy Resour. Technol.* **2024**, *146*, 022303-1.
101. Zhang, C.; Yan, Y.; Shen, K.; Xue, Z.; You, J.; He, Z. Comparative analysis of combustion stability and flow performance in micro combustor based on the synergistic action of slotted blunt body and front-baffle. *Appl. Therm. Eng.* **2024**, *237*, 121802. [CrossRef]

102. Abbaspour, P.; Alipoor, A. Numerical study of combustion characteristics and oscillating behaviors of hydrogen–air combustion in converging–diverging microtubes. *Int. J. Heat Mass Transf.* **2020**, *159*, 120127. [CrossRef]
103. Zhao, H.; Zhao, D.; Becker, S. Thermal performances investigation on an ammonia-fuelled heat-recirculating micro-combustor with reduced chemical mechanism. *Appl. Therm. Eng.* **2024**, *236*, 121685. [CrossRef]
104. Li, J.; Jiaqiang, E.; Ding, J.; Cai, L.; Luo, B. Effect analysis on combustion performance enhancement of the hydrogen-fueled micro-cylindrical combustors with twisted tapes for micro-thermophotovoltaic applications. *Int. J. Hydrogen Energy* **2024**, *49*, 725–743. [CrossRef]
105. He, Z.; Yan, Y.; Zhao, T.; Feng, S.; Li, X.; Zhang, L.; Zhang, Z. Heat transfer enhancement and exergy efficiency improvement of a micro combustor with internal spiral fins for thermophotovoltaic systems. *Appl. Therm. Eng.* **2021**, *189*, 116723. [CrossRef]
106. Wan, J.; Yang, W.; Fan, A.; Liu, Y.; Yao, H.; Liu, W.; Du, Y.; Zhao, D. A numerical investigation on combustion characteristics of H2/air mixture in a micro-combustor with wall cavities. *Int. J. Hydrogen Energy* **2014**, *39*, 8138–8146. [CrossRef]
107. Yang, W.; Xiang, Y.; Fan, A.; Yao, H. Effect of the cavity depth on the combustion efficiency of lean H2/air flames in a micro combustor with dual cavities. *Int. J. Hydrogen Energy* **2017**, *42*, 14312–14320. [CrossRef]
108. Pizza, G.; Frouzakis, C.E.; Mantzaras, J.; Tomboulides, A.G.; Boulouchos, K. Dynamics of premixed hydrogen/air flames in mesoscale channels. *Combust. Flame* **2008**, *155*, 2–20. [CrossRef]
109. Boyarko, G.A.; Sung, C.J.; Schneider, S.J. Catalyzed combustion of hydrogen–oxygen in platinum tubes for micro-propulsion applications. *Proc. Combust. Inst.* **2005**, *30*, 2481–2488. [CrossRef]
110. Mansouri, Z. Combustion in wavy micro-channels for thermo-photovoltaic applications–Part I: Effects of wavy wall geometry, wall temperature profile and reaction mechanism. *Energy Convers. Manag.* **2019**, *198*, 111155. [CrossRef]
111. Mansouri, Z. A novel wavy micro-combustor for micro-thermophotovoltaic applications. *Chem. Eng. Process.-Process Intensif.* **2021**, *163*, 108371. [CrossRef]
112. Han, L.; Li, J.; Zhao, D.; Xi, Y.; Gu, X.; Wang, N. Effect analysis on energy conversion enhancement and NOx emission reduction of ammonia/hydrogen fuelled wavy micro-combustor for micro-thermophotovoltaic application. *Fuel* **2021**, *289*, 119755. [CrossRef]
113. Abbaspour, P.; Alipoor, A. Numerical study of wavy-wall effects on premixed H_2/air flammability limits, propagation modes, and thermal performance of micro combustion chambers. *Appl. Energy* **2024**, *359*, 122727. [CrossRef]
114. You, Z. Chapter 8—Micropropulsion. In *Space Microsystems and Micro/Nano Satellites*; Micro and Nano Technologies; Butterworth-Heinemann: Oxford, UK, 2018; pp. 295–339. [CrossRef]
115. Micci, M.; Ketsdever, A. *Micropropulsion for Small Spacecraft*; American Institute of Aeronautics and Astronautics: Reston, VA, USA, 2000.
116. Sutton, G.; Biblarz, O. *Rocket Propulsion Elements*; A Wiley Interscience Publication, Wiley: Hoboken, NJ, USA, 2001.
117. Merriam-Webster. Propulsion. 2011. Available online: https://www.merriam-webster.com/dictionary/propulsion (accessed on 28 April 2023).
118. Rossi, C.; Rouhani, M.D.; Estève, D. Prediction of the performance of a Si-micromachined microthruster by computing the subsonic gas flow inside the thruster. *Sens. Actuators A Phys.* **2000**, *87*, 96–104. [CrossRef]
119. Youngner, D.; Thai Lu, S.; Choueiri, E.; Neidert, J.; Black, R., III; Graham, K.; Fahey, D.; Lucus, R.; Zhu, X. MEMS mega-pixel micro-thruster arrays for small satellite stationkeeping. In Proceedings of the Small Satellite Conference, Logan, UT, USA, 21–24 August 2000.
120. de Athayde Costa e Silva, M. MEMS Micropropulsion: Design, Modeling and Control of Vaporizing Liquid Microthrusters. Ph.D. Thesis, Delft University of Technology, Delft, The Netherlands, 2018. [CrossRef]
121. NASA Glenn Research Center. Aeronautics: Aircraft Specifications and Performance. Available online: https://www.grc.nasa.gov/www/k-12/airplane/specimp.html (accessed on 21 July 2023).
122. Huh, J.; Park, K.S. Effect of Structural Materials on Monopropellant Thruster Propulsion Performance in Micro Scale. *Aerospace* **2023**, *10*, 362. [CrossRef]
123. Bretti, M. Ais-EPPT1 Pulsed Plasma Thruster. 2020. Available online: https://appliedionsystems.com/portfolio/ais-eppt1-pulsed-plasma-thruster/ (accessed on 10 May 2023).
124. Spanjers, G. *New Satellite Propulsion System Has Mass Below 100 Grams (0.22 Pounds)*; Technical Report; Air Force Research Lab Edwards AFB CA Space and Missile Propulsion Division: Kern County, CA, USA, 2001.
125. Burton, R.L.; Turchi, P. Pulsed plasma thruster. *J. Propuls. Power* **1998**, *14*, 716–735. [CrossRef]
126. Ciaralli, S.; Coletti, M.; Gabriel, S. Performance and lifetime testing of a pulsed plasma thruster for Cubesat applications. *Aerosp. Sci. Technol.* **2015**, *47*, 291–298. [CrossRef]
127. Ciaralli, S.; Coletti, M.; Gabriel, S. Results of the qualification test campaign of a Pulsed Plasma Thruster for Cubesat Propulsion (PPTCUP). *Acta Astronaut.* **2016**, *121*, 314–322. [CrossRef]
128. Coletti, M.; Marques, R.I.; Gabriel, S. Design of a two-stage Pulsed Plasma Thruster for cubesat application. In Proceedings of the 31st IEPC, IEPC-2009-244, Ann Arbor, MI, USA, 20–24 September 2009.
129. Coletti, M.; Guarducci, F.; Gabriel, S. A micro Pulsed Plasma Thruster for Cubesat application: Design and preliminary experimental results. *Acta Astronaut.* **2011**, *69*, 200–208. [CrossRef]
130. Tanaka, M.; Kisaki, S.; Ikeda, T.; Tahara, H. Research and development of pulsed plasma thruster systems for nano-satellites at Osaka Institute of Technology. In Proceedings of the 2012 IEEE Vehicle Power and Propulsion Conference, Seoul, Republic of Korea, 9–12 October 2012; pp. 517–522.

131. Fujita, R.; Tahara, H. Development of electrothermal pulsed plasma thruster systems onboard osaka Institute of technology PROITERES nano-satellites. In Proceedings of the 50th AIAA/ASME/SAE/ASEE Joint Propulsion Conference, Cleveland, OH, USA, 28–30 July 2014; p. 3610.
132. Bock, D.; Tajmar, M. Highly miniaturized FEEP propulsion system (NanoFEEP) for attitude and orbit control of CubeSats. *Acta Astronaut.* **2018**, *144*, 422–428. [CrossRef]
133. Galilei, G. Advanced GG Phase a Study. Available online: http://eotvos.dm.unipi.it/ggweb/phaseA/ (accessed on 22 May 2023).
134. Zhong, Y.; Liu, J.; Li, Z.; Wang, Y.; Li, W.; Zhang, H.; Zou, S.; Cai, B.; Gong, Y.; Tu, L.; et al. Study of microchannel structures for measuring emission conductance of indium field emission electric propulsion micro-thrusters. *Measurement* **2024**, *224*, 113809. [CrossRef]
135. Leach, R.; Neal, K. Discussion of micro-newton thruster requirements for a drag-free control system. In Proceedings of the Small Satellite Conference, Logan, UT, USA, 12–15 August 2002; Session VIII: Advanced Technologies & Subsystems, Components & Sensors II.
136. Mitterauer, J. Micropropulsion for small spacecraft: A new challenge for field effect electric propulsion and microstructured liquid metal ion sources. *Surf. Interface Anal.* **2004**, *36*, 380–386. [CrossRef]
137. Tajmar, M.; Scharlemann, C. Development of electric and chemical microthrusters. *Int. J. Aerosp. Eng.* **2011**, *2011*, 361215. [CrossRef]
138. Rüdenauer, F. Field emission devices for space applications. *Surf. Interface Anal.* **2007**, *39*, 116 – 122. [CrossRef]
139. Räisänen, O. A Diagram of an Electrostatic Ion Thruster. 2012. Available online: https://en.wikipedia.org/wiki/Gridded_ion_thruster#/media/File:Electrostatic_ion_thruster-en.svg (accessed on 6 May 2023).
140. Dong, L.; Lu, S.; Guo, N.; Liang, S.; Long, J.; Luo, W.; Xu, L. Study of the thrust response characteristics of Hall Micro Thruster. *Results Phys.* **2024**, *57*, 107338. [CrossRef]
141. The National Aeronautics and Space Administration. Solar Sail Fact Sheet. 2005. Available online: https://www.nasa.gov/pdf/134645main_solar_sail_fs.pdf (accessed on 27 May 2023).
142. The European Space Agency. Solar Sails. Available online: https://www.esa.int/Education/Solar_sails (accessed on 22 May 2023).
143. Leipold, M.; Eiden, M.; Garner, C.; Herbeck, L.; Kassing, D.; Niederstadt, T.; Krüger, T.; Pagel, G.; Rezazad, M.; Rozemeijer, H.; et al. Solar sail technology development and demonstration. *Acta Astronaut.* **2003**, *52*, 317–326. [CrossRef]
144. Miller, R. Solar Sails: The Math Behind Solar Sails. 2018. Available online: http://ffden-2.phys.uaf.edu/webproj/212_spring_2015/Robert_Miller/physics.html (accessed on 6 May 2023).
145. Kantrowitz, A. Propulsion to orbit by ground based lasers. *Aeronaut. Astronaut.* **1972**, *10*, 74–76.
146. Rao, S.; Xu, Y.; Yuan, J.; Liu, F.; Wang, S.; Jiang, H.; Cheng, G.J. MOFs for Ultrahigh Efficiency Pulsed Laser Micropropulsion. *Adv. Mater.* **2024**, *36*, 2306228. [CrossRef]
147. Rao, S.; Yi, W.; Yuan, J.; Wang, S.; Jiang, H.; Cheng, G.J. Advancing laser micropropulsion: High performance with MOF-derived carbon-encapsulated-nano-metal composites. *Matter* **2024**. [CrossRef]
148. National Research Council. *Microelectromechanical Systems: Advanced Materials and Fabrication Methods*; National Academies Press: Washington, DC, USA, 1998.
149. Müller, J.; Marrese, C.; Polk, J.; Yang, E.H.; Green, A.; White, V.; Bame, D.; Chakraborty, I.; Vargo, S.; Reinicke, R. An overview of MEMS-based micropropulsion developments at JPL. In Proceedings of the 3rd International Symposium of the International Academy of Astronautics (IAA) for Small Satellites for Earth Observation, Berlin, Germany, 2–6 April 2001.
150. Butler, E.; Folk, C.; Cohen, A.; Vasilyev, N.; Chen, R.; del Nido, P.; Dupont, P. Metal MEMS Tools for Beating-heart Tissue Approximation. *IEEE Int. Conf. Robot. Autom.* **2011**, *2011*, 411–416. [CrossRef] [PubMed]
151. Bayt, R.L. *Analysis, Fabrication and Testing of a MEMS-Based Micropropulsion System*; Technical Report; Aerospace Computational Design Laboratory, Department of Aeronautics: Cambridge, MA, USA, 1999.
152. Bennett, S.; Huss, W.; John, R.; Tuchman, A. Experimental propulsion performance of a low power pulsed resistojet. In Proceedings of the 2nd Aerospace Sciences Meeting, New York, NY, USA, 25–27 January 1965; p. 97.
153. Page, R.J.; Halbach, C.R.; Short, R.A. 3-KW concentric tubular resistojet performance. *J. Spacecr. Rocket.* **1966**, *3*, 1669–1674. [CrossRef]
154. Pugmire, T.K.; Shaw, R.; Enos, G.R. Applied resistojet technology. *J. Spacecr. Rocket.* **1971**, *8*, 63–68. [CrossRef]
155. Page, R.; Short, R. Design of high-performance resistojets for advanced spacecraft. In Proceedings of the 9th Aerospace Sciences Meeting, New York, NY, USA, 25–27 January 1971; p. 195.
156. Kundu, P.; Bhattacharyya, T.K.; Das, S. Design, fabrication and performance evaluation of a vaporizing liquid microthruster. *J. Micromech. Microeng.* **2012**, *22*, 025016. [CrossRef]
157. Cen, J.; Xu, J. Performance evaluation and flow visualization of a MEMS based vaporizing liquid micro-thruster. *Acta Astronaut.* **2010**, *67*, 468–482. [CrossRef]
158. Silva, M.A.; Guerrieri, D.C.; van Zeijl, H.; Cervone, A.; Gill, E. Vaporizing liquid microthrusters with integrated heaters and temperature measurement. *Sens. Actuators A Phys.* **2017**, *265*, 261–274. [CrossRef]
159. Ahmed, Z.; Gimelshein, S.F.; Ketsdever, A.D. Numerical analysis of free-molecule microresistojet performance. *J. Propuls. Power* **2006**, *22*, 749–756. [CrossRef]

160. Ketsdever, A.; Wadsworth, D.; Vargo, S.; Muntz, E. The free molecule micro-resistojet-An interesting alternative to nozzle expansion. In Proceedings of the 34th AIAA/ASME/SAE/ASEE Joint Propulsion Conference and Exhibit, Cleveland, OH, USA, 13–15 July 1998; p. 3918.
161. Guerrieri, D.C.; Cervone, A.; Gill, E. Analysis of nonisothermal rarefied gas flow in diverging microchannels for low-pressure microresistojets. *J. Heat Transf.* **2016**, *138*, 112403. [CrossRef]
162. Guerrieri, D.C.; Silva, M.A.; Cervone, A.; Gill, E. Selection and characterization of green propellants for micro-resistojets. *J. Heat Transf.* **2017**, *139*, 102001. [CrossRef]
163. Cervone, A.; Mancas, A.; Zandbergen, B. Conceptual design of a low-pressure micro-resistojet based on a sublimating solid propellant. *Acta Astronaut.* **2015**, *108*, 30–39. [CrossRef]
164. Guerrieri, D.C.; Silva, M.A.; Van Zeijl, H.; Cervone, A.; Gill, E. Fabrication and characterization of low pressure micro-resistojets with integrated heater and temperature measurement. *J. Micromech. Microeng.* **2017**, *27*, 125005. [CrossRef]
165. Guerrieri, D.C.; Silva, M.A.; Cervone, A.; Gill, E. An analytical model for characterizing the thrust performance of a Low-Pressure Micro-Resistojet. *Acta Astronaut.* **2018**, *152*, 719–726. [CrossRef]
166. Mukerjee, E.; Wallace, A.; Yan, K.; Howard, D.; Smith, R.; Collins, S. Vaporizing liquid microthruster. *Sens. Actuators A Phys.* **2000**, *83*, 231–236. [CrossRef]
167. You, Z. *Space Microsystems and Micro/Nano Satellites*; Butterworth-Heinemann: Oxford, UK, 2017.
168. Pranajaya, F. *Progress on Colloid Micro-Thruster Research and Flight Testing*; Space Systems Development Laboratory, Department of Aeronautics and Astronautics, Stanford University: Stanford, CA, USA, 1999 .
169. Berg, S.P.; Rovey, J.; Prince, B.; Miller, S.; Bemish, R. Electrospray of an energetic ionic liquid monopropellant for multi-mode micropropulsion applications. In Proceedings of the 51st AIAA/SAE/ASEE Joint Propulsion Conference, Orlando, FL, USA, 27–29 July 2015; p. 4011.
170. Berg, S.P.; Rovey, J. Decomposition of a double salt ionic liquid monopropellant on heated metallic surfaces. In Proceedings of the 52nd AIAA/SAE/ASEE Joint Propulsion Conference, Salt Lake City, UT, USA, 25–27 July 2016, p. 4578.
171. Lewis, D.H., Jr.; Janson, S.W.; Cohen, R.B.; Antonsson, E.K. Digital micropropulsion. *Sens. Actuators A Phys.* **2000**, *80*, 143–154. [CrossRef]
172. Rossi, C.; Do Conto, T.; Esteve, D.; Larangot, B. Design, fabrication and modelling of MEMS-based microthrusters for space application. *Smart Mater. Struct.* **2001**, *10*, 1156. [CrossRef]
173. Rossi, C.; Orieux, S.; Larangot, B.; Do Conto, T.; Esteve, D. Design, fabrication and modeling of solid propellant microrocket-application to micropropulsion. *Sens. Actuators A Phys.* **2002**, *99*, 125–133. [CrossRef]
174. Rossi, C.; Larangot, B.; Lagrange, D.; Chaalane, A. Final characterizations of MEMS-based pyrotechnical microthrusters. *Sens. Actuators A Phys.* **2005**, *121*, 508–514. [CrossRef]
175. Seo, D.; Lee, J.; Kwon, S. The development of the micro-solid propellant thruster array with improved repeatability. *J. Micromech. Microeng.* **2012**, *22*, 094004. [CrossRef]
176. Lee, J.; Jo, H.; Kim, S.; Lee, E.; Son, Y.; Lee, A. Lab-on-PCB for space propulsion: Integrated membraneless micro-ignitor for MEMS solid propellant thruster. *Sens. Actuators A Phys.* **2023**, *363*, 114696. [CrossRef]
177. Lee, J.; Kim, S.; Jo, H.; Lee, E.; Lee, A. Design and fabrication of a scalable solid-propellant micro-thruster array using lab-on-PCB technology. *Sens. Actuators A Phys.* **2023**, *363*, 114738. [CrossRef]
178. Kan, W.; Liu, W.; Lou, W.; Feng, H.; Chu, E.; Wang, J.; Ren, W. High-Safety Energetic Micro-Igniter for Micro-Thrust System. *Sens. Actuators A Phys.* **2024**, *368*, 115056. [CrossRef]
179. Fahd, A.; Baranovsky, A.; Dubois, C.; Chaouki, J.; Elbasuney, S.; Shokry, S. Thrust characteristics of nano-carbon/Al/oxygenated salt nanothermites for micro-energetic applications. *Def. Technol.* **2023**, *30*, 55–69. [CrossRef]
180. Patel, K.D.; Bartsch, M.S.; McCrink, M.H.; Olsen, J.S.; Mosier, B.P.; Crocker, R.W. Electrokinetic pumping of liquid propellants for small satellite microthruster applications. *Sens. Actuators B Chem.* **2008**, *132*, 461–470. [CrossRef]
181. Hitt, D.L.; Zakrzwski, C.M.; Thomas, M.A. MEMS-based satellite micropropulsion via catalyzed hydrogen peroxide decomposition. *Smart Mater. Struct.* **2001**, *10*, 1163. [CrossRef]
182. Kundu, P.; Sinha, A.K.; Bhattacharyya, T.K.; Das, S. MnO_2 Nanowire Embedded Hydrogen Peroxide Monopropellant MEMS Thruster. *J. Microelectromech. Syst.* **2012**, *22*, 406–417. [CrossRef]
183. Xu, J.; Zhang, J.; Li, F.; Liu, S.; Ye, Y.; Shen, R. A review on solid propellant micro-thruster array based on MEMS technology. *FirePhysChem* **2023**. [CrossRef]
184. Chen, C.P.; Chao, Y.C.; Wu, C.Y.; Lee, J.C.; Chen, G.B. Development of a catalytic hydrogen micro-propulsion system. *Combust. Sci. Technol.* **2006**, *178*, 2039–2060. [CrossRef]
185. Armano, M.; Audley, H.; Baird, J.; Binetruy, P.; Born, M.; Bortoluzzi, D.; Castelli, E.; Cavalleri, A.; Cesarini, A.; Cruise, A.M.; et al. LISA Pathfinder micronewton cold gas thrusters: In-flight characterization. *Phys. Rev. D* **2019**, *99*, 122003. [CrossRef]

energies

MDPI

Article

Comparisons of Different Representative Species Selection Schemes for Reduced-Order Modeling and Chemistry Acceleration of Complex Hydrocarbon Fuels

Kevin M. Gitushi and Tarek Echekki *

Department of Mechanical and Aerospace Engineering, North Carolina State University, Raleigh, NC 27695, USA; kmgitush@ncsu.edu
* Correspondence: techekk@ncsu.edu

Abstract: The simulation of engine combustion processes, such as autoignition, an important process in the co-optimization of fuel-engine design, can be computationally expensive due to the large number of thermo-chemical scalars needed to describe the full chemical system. Yet, the inherent correlations between the different chemical species during oxidation can significantly reduce the complexity of representing this system. One strategy is to select a subset of representative species that accurately captures the combustion process at a fraction of the computational cost of the full system. In this study, we compare the performance of four different techniques to select these species. They include the two-step principal component analysis (PCA) approach, directed relation graphs (DRGs), the global pathway selection (GPS) approach, and the manifold-informed species selection method. A parametric study of the representative species selection is carried out on data from the simulation of homogeneous and perfectly stirred reactors by investigating seven cumulative variances and 47 different cut-off percentages for the two-step PCA, and 65 and 51 thresholds for the DRGs and GPS, respectively. Results show that these selection methods capture key important species that can accurately describe the chemical system and track each stage of oxidation. The two-step PCA is sensitive to the cumulative variance, and DRGs and GPS are sensitive to the choice of target variables. By selecting key representative species and reducing the number of thermo-chemical scalars, these three methods can be used to develop computationally efficient hybrid chemistry schemes.

Keywords: representative species; principal component analysis; directed relation graphs; global pathway selection; manifold-informed method

1. Introduction

The combustion of practical fuels involves the reaction and transport of many chemical species. The multiscale nature of combustion chemistry and the large number of species that must be solved impose significant constraints on the ability to implement high-fidelity simulations of practical combustion devices.

Different approaches have been adopted in the combustion literature to overcome these constraints. Among these approaches, chemistry reduction and acceleration strategies have played a central role in reducing the number of reactions and species and managing the stiffness of the reaction chemistry. Both classes of approaches attempt to reduce the complexity of the chemical mechanisms by identifying a subset of species of the full chemical system, labeled here as representative species [1].

In chemistry reduction, these species are featured in global or skeletal mechanisms, while the contribution of non-representative species and their associated reactions are eliminated through chemistry reduction. The selection of representative species is also a key component in the development of reduced-order or hybrid chemistry (HyChem) models [2–8]. In more recent years, we have implemented the HyChem approach and the choice of representative species with machine learning tools to accelerate the chemistry integration of complex fuel oxidation [9–12]. More recently, Kumar and Echekki [13] proposed a novel chemistry acceleration scheme for use in reacting flow simulations that relies on the transport of representative species, which represent a very small subset of the chemical mechanism. The approach relies on the use of deep operator networks (DeepONets) [14] to map the evolution of the representative species between time increments. The goal of the selection is to retain species that will be adequate markers for the progress of fuel oxidation. Representative species also have played an important role in constructing low-dimensional manifolds for the chemical system [15] as data-driven alternative approaches to traditional turbulent combustion models. Further reduction in this manifold can be achieved through the implementation of principal component analysis (PCA) [16]. Therefore, beyond chemistry reduction, the use of representative species can be extended to different strategies for chemistry acceleration.

Different approaches may serve to select representative species from a full set of thermo-chemical scalars. They include direct relations graphs (DRGs) [17–19], global pathway selection (GPS) [20], the two-step PCA selection process [11] and the more recent manifold-informed reduction method [21]. The first two methods, DRG, and its variants, and GPS, were designed with the primary purpose of chemistry reduction. Regardless, these methods use a range of parameters for their optimum implementation. More strict thresholds for these parameters may be designed to construct skeletal mechanisms for complex fuels. In contrast, less strict thresholds can be used to construct representative species from which other species in a mechanism can be recovered through training data.

In this study, different methods of selecting representative species are compared. The comparison is carried out for n-heptane low-temperature oxidation for DRG, GPS, and the two-step PCA. For the manifold-informed method, the comparison between two-step PCA and this method is implemented on simpler fuels, hydrogen, syngas and ethylene, primarily because of the inherent computational cost of the manifold-informed approach when implemented for complex fuels. One principal goal of the comparisons is to identify whether the various methods yield similar selections of representative species, or whether specific details of their formulations can yield different ones.

2. Methodology

In this section, we briefly discuss the 4 methods considered in this study.

2.1. Two-Step Principal Component Analysis (PCA)

The two-step PCA by Alqahtani and Echekki [11] is used to reduce the total number of species in a detailed mechanism to a lower representative set of the data variance associated with the reaction rates of species in a mechanism. PCA is carried out in two steps: the first step is carried out on the full set of species' reaction rates in the detailed mechanism, excluding the fuel, to select the leading set of species; a second PCA is carried on the species retained from the first PCA.

The species' reaction rates ω_i are centered by subtracting their mean:

$$\omega_i^* = \omega_i - \overline{\omega_i} \quad (1)$$

where i is the ith species index. PCA is carried on the covariance matrix of the centered reaction rates and vector ϕ of the PCs is obtained as:

$$\phi = Q^T \omega^* \quad (2)$$

The obtained PCs are then ordered using as a criterion the magnitude of the eigenvalues from highest to lowest to identify the importance of the associated thermo-chemical scalars. Since the first few PCs represent most of the data variance, we retain the leading

N_{PC} PCs that contain a determined cumulative contribution to the total variance, which is set in the present study at 99%. We express these PCs as follows:

$$\phi_{red} = A^T \omega^* \tag{3}$$

The matrix A contains the leading N_{PC} PCs eigenvectors of Q. To identify the most important species, we implement a cut-off criterion as follows:

$$\frac{\sum_{i=1}^{b} |q_{i.j}|}{\sum_{i=1}^{N} |q_{i.j}|} \times 100 > \text{cut-off\%} \quad j = 1, \ldots, N_{PC} \tag{4}$$

where $q_{i,j}$ are coefficients of the matrix A. A cut-off percentage is used to determine the number b of thermo-chemical scalars with the most contributions to the PCs. Steps 1–4 are repeated on the reaction rates of the retained species from the first PCA to obtain the final set of selected species.

2.2. Manifold-Informed Selection

Recently, Zdybal et al. [21,22] proposed a manifold-informed method for selecting a subset of state variables. This PCA-based method identifies a subset of the state variable that minimizes non-uniqueness and small feature size on low-dimensional manifolds. To obtain the subset state, a backward variable elimination algorithm is used to remove variables that decrease a specified cost function. The cost function $\mathcal{L}$ used by Zdybal et al. [21] is based on the normalized variance derivative $\hat{\mathcal{D}}(\sigma)$ metric [23] and is defined as:

$$\mathcal{L}_{\phi_i} = \int_{\tilde{\sigma}_{min}}^{\tilde{\sigma}_{\max}} P_i(\sigma, \sigma_{p,i}) \cdot \hat{\mathcal{D}}_i(\sigma) d\tilde{\sigma}, \tag{5}$$

where $\sigma = \sigma_{p,i}$ is the largest feature size in the ith manifold ϕ_i, and $P_i(\sigma, \sigma_{p,i})$ is the penalty function defined as

$$P_i(\sigma, \sigma_{p,i}) = |\tilde{\sigma} - \tilde{\sigma}_{p,i}| + \frac{1}{||\tilde{\sigma}_{p,i}||}, \tag{6}$$

where tilde denotes a $\log_{10}$-transformed quantity. The total cost from specified target dependent variables is the cumulative sum $\mathcal{L} = \sum_{i=1}^{n} \mathcal{L}_{\phi_i}$.

2.3. Directed Relation Graphs (DRG)

DRG was developed by Lu and Law [17–19] as a chemistry reduction method that represents kinetic models as graphs. The graph model consists of nodes and directed edges. In the graph model, chemical species are represented by nodes, and their inter-dependence is represented by directed edges. An illustration of a graph of a six-species kinetic model is shown in Figure 1 [17], where line thickness shows the level of dependency between species. Let us consider species B as a target species, which can be the fuel or a major combustion product. The thick directed edge from B to D shows that these two species directly react with each other and are both needed in the model to accurately calculate their production or consumption. B also depends on A and C; however, their dependence is not as strong as the one between B and D. Removing A or C from the kinetic model would not lead to significant errors in the production/consumption of B. Similarly, the dependence between C and both B and D is negligible. E and F are not connected to and do not react with species B, but they strongly depend on one another [24].

In DRG, the direct interaction coefficient is used to measure the dependence between species and to quantify one species's contribution to other species' production or consumption. For our six species graph with target B , the direct interaction coefficient for species B and D is given by:

$$r_{BD} = \frac{\sum_{i=1}^{N_{reaction}} |\nu_{B,i}\omega_i\delta_{D,i}|}{\sum_{i=1}^{N_{reaction}} |\nu_{B,i}\omega_i|} \tag{7}$$

and

$$\delta_{D,i} = \begin{cases} 1, & \text{if } i\text{th reaction involves D} \\ 0, & \text{otherwise} \end{cases}$$

where $\delta_{D,i}$ is an index of species D's presence in the ith reaction, $N_{reaction}$ is the total number of reactions in the kinetic model, $\nu_{B,i}$ is the net stoichiometric coefficient of species B in reaction i, and ω_i is the reaction rate of the ith reaction. To retain important species reacting with the target species, a threshold ε is applied to the graph such that edges where $r_{BD} < \varepsilon$ are removed because the dependence is considered weak and negligible. A deep first search is applied starting from the target species to identify and retain species with strong dependence ($r_{BD} \geq \varepsilon$) to the target species and eliminate the ones with a negligible dependence. Reactions involving species with strong dependence are kept in the model, while reactions involving eliminated species are removed.

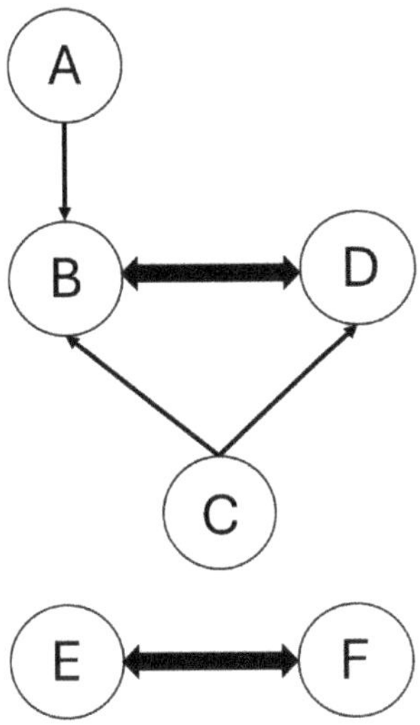

Figure 1. Example of a graph representing a model with six species.

2.4. Global Pathway Selection (GPS)

GPS, first developed by Gao et al. [20], is an algorithm for chemistry reduction that identifies global pathways and removes species and reactions that are not important to these pathways. This is performed by constructing element flux graphs and selecting hub species. The construction of element flux graphs is carried individually on carbon (C), hydrogen (H), and oxygen (O), which are the main elements involved in the combustion of hydrocarbon fuels. For each of these graphs, nodes represent species present in the kinetic mechanism, and directed edges represent element fluxes from one node to another. For GPS, a direct edge is also a measure of the rate of atoms of elements C, H, and O transferred from one species to another from every elementary reaction found in the detailed mechanism. For an element e, the flux or direct edge from ith to j species is calculated as:

$$\boldsymbol{A}_{e,i\rightarrow j} = \sum_r a_{e,r,i\rightarrow j} \tag{8}$$

where $a_{e,r,i\rightarrow j}$, the element flux from the ith to jth species contributed by the rth elementary reaction, is:

$$a_{e,r,i\rightarrow j} = \max(0, C_{e,r,i\rightarrow j}\dot{R}_r) \tag{9}$$

and $\dot{R}_r$ is the net reaction rate of the rth reaction. $C_{e,r,i\rightarrow j}$, the number of e atoms transferred from species i to j in the rth elementary reaction, is given by:

$$C_{e,r,i\rightarrow j} = \begin{cases} n_{e,r,j}\frac{n_{e,r,i}}{n_{e,r}}, & \text{if } v_{e,r,j}v_{e,r,i} < 0 \\ 0, & \text{otherwise} \end{cases} \quad (10)$$

where $n_{e,r,i} = v_{r,i}N_{e,i}$ is the number of eth element atoms transferred from or to species i in reaction r. $v_{r,i}$ is the stoichiometric coefficient, negative if a reactant and positive if a product, of species i in reaction r. $N_{e,i}$ is the number of e atoms in species i. $n_{e,r}$ is the number of e atoms involved in reaction r and can be calculated as $\sum_{(k:n_{e,r,k}>0)} n_{e,r,k}$ or $\sum_{(k:n_{e,r,k}>0)} n_{e,r,k}$.

By defining this directed edge from the ith to jth node for each element, the total element flux between a target species A and other species from all reactions in the detained mechanism is defined as:

$$\alpha_{e,i} = \frac{\max(\sum_k \boldsymbol{A}_{e,i\rightarrow k}, \sum_k \boldsymbol{A}_{e,k\rightarrow i})}{\max_M(\sum_k \boldsymbol{A}_{e,M\rightarrow k}, \sum_k \boldsymbol{A}_{e,k\rightarrow M})} \quad (11)$$

Using Equation (11) and a user-defined threshold value α_{crit}, a set of "hub species" is kept if $\alpha_{crit} > \alpha_{e,i}$ for each element. For each hub species, global pathways are identified. These are chemistry pathways where elemental transfer from initial reactants to final products passes through the hub species. A hub species can have many pathways. The top K fastest pathways can be identified using the graph algorithm by Yen [25] by searching the K shortest pathways from the inverse element graph. Additionally, the top n_r reactions of species found in the global pathways are also selected if $\beta_{e,i\rightarrow j,r=1\sim n_r} > \beta_{crit}$. $\beta_{e,i\rightarrow j,r=1\sim n_r}$, defined in Equation (12), is a measure of the total contribution of the reactions, and β_{crit} is a user-defined threshold value [20,26]. In this work, only the fastest pathways are selected with $K = 1$ and $\beta_{crit} = 0.5$:

$$\beta_{e,i\rightarrow j,r=1\sim n_r} = \frac{\sum_{r=1}^{n_r} a_{e,r,i\rightarrow j}}{A_{e,i\rightarrow j}} \quad (12)$$

3. Results and Discussion

The four methods, which were discussed earlier, are applied to combustion data of perfectly stirred reactors (PSRs) and constant pressure homogeneous reactors to retain representative species, and the results of the number of species retained are shown below. The comparisons between the different methods are implemented for n-heptane. The KAUST lumped mechanism is used as a detailed chemistry mechanism [27], which consists of 538 species and 2824 reactions. For the homogeneous reactor simulations, we consider 5 different equivalence ratios of 0.5, 0.8, 1, 1.2 and 1.5, and initial temperatures varying from 700 K to 850 K at 10-degree increments. For the PSR simulations, 4 residences times of 10^{-7}, 5×10^{-7}, 10^{-6} and 5×10^{-6} s are considered.

The data are generated by solving Equations (13) and (14), describing the PSR and homogeneous reactor, respectively, using Cantera 2.6.0 [28]:

$$\begin{cases} \frac{dY_s}{dt} = \frac{Y_{s,in}-Y_s}{\tau} + \frac{\hat{\omega}_s\hat{W}_s}{\rho}, \; s = 1,\ldots,N \\ \frac{dh}{dt} = \frac{h_{in}-h}{\tau} \end{cases} \quad (13)$$

$$\begin{cases} \frac{dY_s}{dt} = \frac{\hat{\omega}_s\hat{W}_s}{\rho}, \; s = 1,\ldots,N \\ \frac{dT}{dt} = -\frac{\sum_{s=1}^{N} h_s\hat{\omega}_s\hat{W}_s}{\rho c_p} \end{cases} \quad (14)$$

where the subscript in denotes inlet conditions, T is the temperature, h is the enthalpy, ρ is the density, and c_p is the constant pressure specific heat. Y_s, h_s, $\hat{\omega}_s$ and $\hat{W}_s$ are the sth species' mass fraction, enthalpy (sensible and chemical), reaction rate, and molecular weight, respectively.

The selected species are obtained by applying each method above on the data generated by solving Equations (13) and (14), and the final representative species are species common between the selected ones and those that are also part of the list of species found in the smaller C_0–C_4 foundational chemistry mechanism USC Mech-II [29]. After retaining the representative species, we estimate the reduction percentage from the detailed mechanism as $100 \times (1 - N_{rs}/N_{dcs})$, where N_{rs} is the number of representative species, and N_{dcs} is the number of species in the detailed mechanism. A parametric study is conducted for each method by varying (1) the PCA variance and cut-off percentage for the two-step PCA, (2) the threshold ε for DRG, and (3) the threshold value α_{crit} for GPS with constant $\beta_{crit} = 0.5$. For the different methods, we adopt a set of threshold parameters;

1. Two-step PCA: 7 PCA cumulative variances of 85, 90, 95, 99, 99.9, 99.99, and 99.999% are used to retain the first N_{PC} PCs. In total, 47 different cut-off percentages varying from 80 to 99.999% are set as the cut-off criterion to select the number b of thermochemical scalars with the most contributions to the PCs.
2. DRG: 65 thresholds varying from 0.1 to 0.74 by an increment of 0.01 are used as the cut-off threshold ε for the removal of directed edges.
3. GPS: 51 α_{crit} of values of 0.001 and 0.01 to 0.5 by increment of 0.01 are prescribed as the threshold to select hub species.

It is important to note that both DRG and GPS, along with the manifold-informed selection, require the specification of target species. Here, we investigate four different sets of target species. These species are chosen from either reactants, products, or key intermediates for low-temperature and high-temperature oxidation. These sets are summarized as follows:

- Target species set 1: nC_7H_{16}, H_2O, CO_2, CO, CH_2O, H_2O_2, HO_2,
- Target species set 2: nC_7H_{16}, H_2O, CO_2,
- Target species set 3: nC_7H_{16}, H_2O, CO_2, CO,
- Target species set 4: nC_7H_{16}.

3.1. Two-Step PCA

Figure 2 shows the number of selected representative species as a function of the cut-off criterion for the two-step PCA. The PCA is performed on the PSR and homogeneous data from time $t = 0$ s up to equilibrium. The number of selected representative species increases as the PCA cumulative variance and cut-off percentage increase. This trend is expected because a larger number of PCs is needed to retain a higher cumulative variance of the original data. Subsequently, a greater number of b of scalars will be retained to obtain a higher contribution to the PCs. However, for cumulative variances of 85 and 90%, the same number of representative species is obtained as shown by the orange line in Figure 2. This occurs because the same number of PCs is retained to obtain a cumulative variance between 85 and 90%.

Similarly, the number of retained species with a cumulative variance of 95% is very close to 85 and 90%. The number of representative species is also sensitive with regards to PCA variance when they are less than 99.999% and the cut-off percentage is less than 99.9%, but not as sensitive when the variance is greater than 99.999% or the cut-off percentage is greater than 99.9%.

The minimum number of retained species is 8 when the cumulative variance is 85% and cut-off percentage is 80, and the maximum number of retained species is 63 when the cumulative variance and cut-off percentage are both 99.999%. With 8 retained species, a reduction of 99% from the original 538 species is achieved, and we have a reduction of 88% when retaining 63 species. The 8 retained species, which are the common species across all cumulative variances and cut-off percentages analyzed, are n-C_7H_{16}, O_2, H_2O, CO_2, CO, CH_3, C_2H_4 and C_2H_5. The minimum retained species include the fuel, oxidizer, and major products of combustion, H_2O and CO_2, which are great markers of the two stages of ignition for fuel oxidation as shown in Figure 3. CH_3, C_2H_4, and C_2H_5 are great trackers of the second stage of ignition and important markers of the chemistry description in the PSR as shown in the Figure 4.

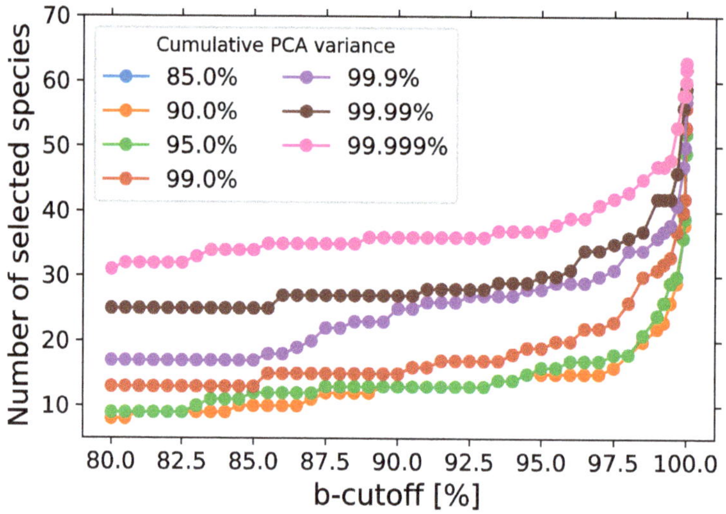

Figure 2. Number of species retained as a function of the cut-off percentage. USC Mech II is used as a foundational chemistry mechanism.

As the cut-off percentage increases, additional species are retained, such as H and O radicals, OH, and H_2. More low-temperature chemistry species that track the first stage of ignition, such as CH_2O and CH_2CO, are retained as the cut-off percentage increases. As the cumulative variance increases, these species are retained at a lower cut-off percentage. For example, CH_2O and CH_2CO are both retained when cumulative variance and cut-off percentage are both, respectively, 90 and 98% with 18 total retained species, 99 and 96.5% with 25 total retained species, and 99.9 and 85.5% with 18 total retained species.

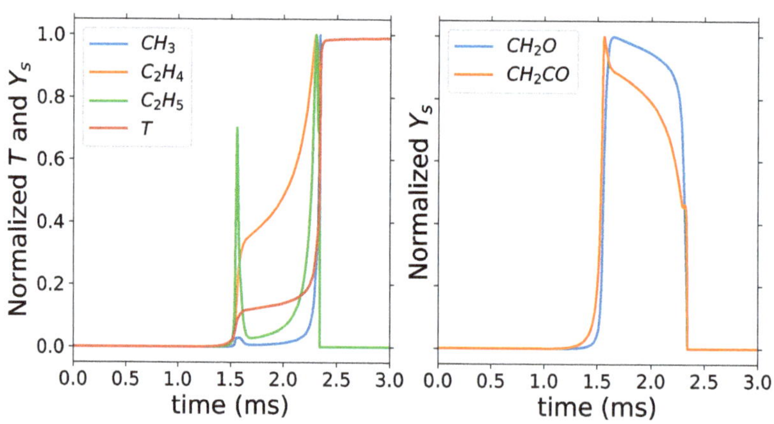

Figure 3. Homogeneous reactor: normalized temperature and mass fractions with T_i = 740 K and equivalence ratio of 1.

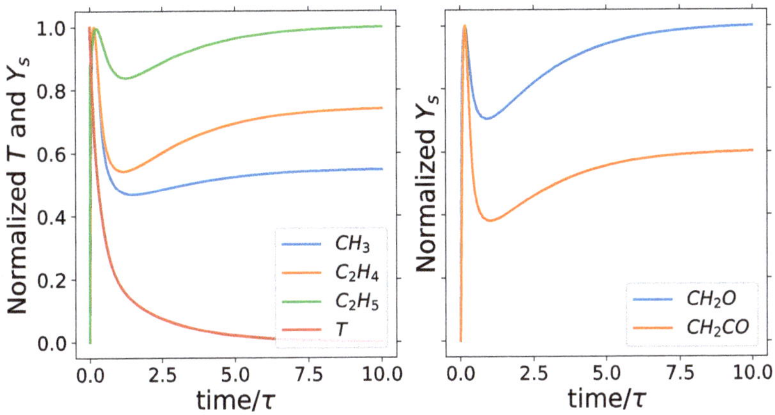

Figure 4. PSR: normalized temperature and mass fractions with T_i = 700 K, $\tau = 4 \times 10^{-5}$ s, and equivalence ratio of 1.

3.2. Manifold-Informed Selection

The manifold-informed approach was applied to our homogeneous reactor and PSR data using PCAfold 2.2.0 [22]; however, no species selection was obtained because of the intensive iterative process and computational cost of the backward variable elimination algorithm, and the size of our data, which have 1.2 million observations and 538 variables. Similarly, when applied to the data provided with the PCAfold 2.2.0 software, which have 50,000 observations and 11 variables, the algorithm still took two days to complete the selection. For reference, the homogeneous reactor and PSR data were generated in less than 12 h, and the two-step PCA for one parameter was completed in under one minute. Data generation and all selection methods were carried out on the Intel Xeon Gold 6234 3.30 GHz CPU. Due to this computational limitation associated with the manifold-informed selection, we applied the two-step PCA using the seven cumulative variances and 47 cut-off percentages listed above to the hydrogen, syngas, and ethylene combustion data provided by Zdybal et al. [21] and compared selected species from the two-step PCA to the manifold-informed selection method. The two-step PCA completed the parametric study of all seven variances and cut-off percentages in less than 3, 2, and 4 s for the hydrogen, syngas, and ethylene data, respectively, on the same CPU used for PCAfold 2.2.0 software. For a very negligible computational cost, our method yielded similar selected species to those reported in Table 2 in [21]. For example, for a cumulative variance of 99.99% and 82.5% cut-off, common species between the two methods are H_2, O, OH, and O_2 for hydrogen combustion, O, OH, CO and CO_2 for syngas combustion, and H_2, O_2, OH, H_2O, CH_3, CO_2, C_2H_2 and C_2H_4. For these parameters, both methods achieved a 58 and 50% reduction from the total number of hydrogen and syngas combustion scalars, respectively. The two-step PCA achieved a reduction of 61% for ethylene combustion, and the manifold-informed reduction achieved 64%. H is a common species among the hydrogen, syngas, and ethylene data that was selected by the manifold-informed selection but not the two-step PCA. This can be attributed to the small magnitude of this species and the lower contribution of its corresponding PC to the overall cumulative variance. Thus, with the two-step PCA, species with small magnitudes close to zero are not retained.

3.3. DRG

The total number of selected representative species resulting from the parametric study of the cut-off threshold ε can be seen in Figure 5. The number of representative species decreases with increasing the threshold. This trend makes sense since increasing the threshold ε would result in the elimination of the directed edges with a small direct interaction coefficient. The selected representative species become very sensitive to the specified target species as the threshold increases, especially between ε values of 0.3 and 0.5, e.g., at $\varepsilon = 0.45$, the number of selected species decreases from 31 for target species set 1, which has 7 target species, to 24 for target species set 4, which has 1 species, the fuel. Increasing the number of target species will increase the number of selected representative species. Target species set 3, which has the same species as target species set 2, in addition to CO, leads to the same number of selected representative species as target species set 2. Changing the additional species CO in the target set 3 to a different species, such as CH_2O, could change the observed trend and result in a different set of selected species depending on the ones involved in the reaction with CH_2O. Target species sets 1, 2, and 3 have the same number of selected species from ε of 0.1 to 0.26, and beyond 0.26, the selected species from target set 1 changes slightly from those of target sets 2 and 3.

At the threshold $\varepsilon = 0.1$, target species sets 1 to 3 lead to the same number of representative species of 51, while target species set 4 has 50, where CH_3COCH_3 is the species not selected for the target set 4. The minimum number of retained species is 19 with target species set 4 and $\varepsilon = 0.66$. This is a 96% reduction from the 538 species in the detailed mechanism. Target species sets 1, 2, and 3 with ε of 0.1 give the maximum number 51 of retained species, which is a reduction of 91% from the total number of species. The minimum retained species include species such as H radical, OH, and HCO, which are great markers of the second stage of ignition, and HO_2, CH_2O, CH_2CO, CH_3CO, CH_2CHO, CH_3CHO, and C_2H_3CHO, which are great markers of the first stage of ignition as shown in Figure 6. CH_3, C_2H_4 and C_2H_5 are also part of the minimum retained species as in the case of the two-step PCA; however, CO and CO_2 are not retained in the minimum list for DRG. While these are important species, they are not retained because reactions involving the fuel do not have a strong dependency, i.e., ε of 0.66, with CO and CO_2. As ε decreases to 0.65 and 0.54, CO_2 and CO are added to the list of retained species for target species set 4.

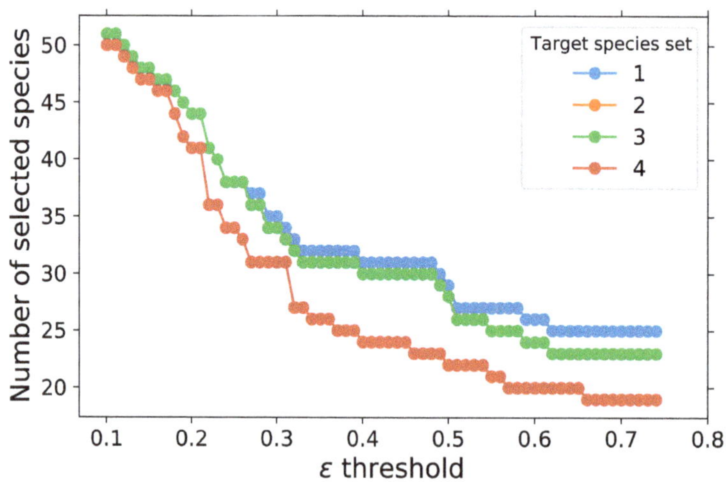

Figure 5. Number of representative species selected as a function of ε threshold with USC Mech II.

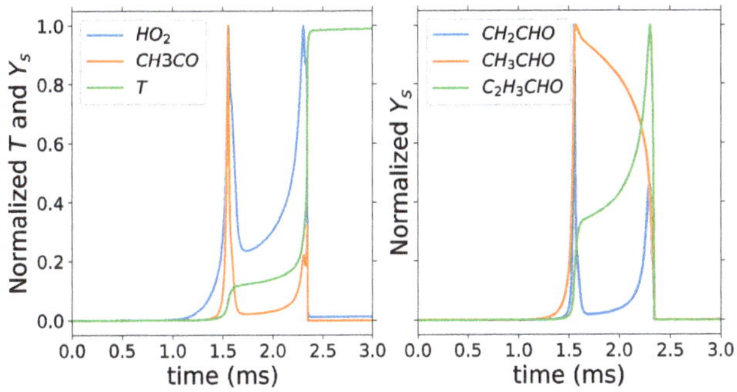

Figure 6. Homogeneous reactor: normalized temperature and mass fractions with T_i = 740 K and equivalence ratio of 1.

3.4. GPS

The number of selected representative species obtained by varying the threshold α_{crit} can be seen in Figure 7 for the GPS method. Similarly to the DRG method, the number of representative species decreases with increasing threshold and increases as the number of target species increases. The selected representative species are very sensitive to the specified target species of the fuel as compared to DRG results shown in the red dotted line in Figure 5. Target species sets 2 and 3 have the same number of representative species for threshold α_{crit} between 0.14 and 0.46, and outside this range, the number of selected species is very close for target species sets 1, 2 and 3. At a small threshold α_{crit} of 0.001, target species sets 1 to 3 have the same number of representative species.

Similar to DRG, the minimum number of selected species is 12 with target species set 4 and α_{crit} of 0.48, and the maximum is 53 with α_{crit} 0.001 and target species sets 1, 2 and 3. With 12 and 53 retained species, reductions of 98 and 90% from the detailed mechanism, respectively, are achieved. At α_{crit}, the target set 4 has 50 retained species, which are also included in the retained species for the first three target sets. N_2, O, and CH_4 are the three additional species retained in the first three target sets. The minimum retained species include H radical, OH, CH_3, C_2H_2, C_2H_3, C_2H_4, CO, and main products of combustion. We can see in Figure 8 that C_2H_4 is a great marker for the first stage of ignition, while the other C_2 species track the first stage of ignition. More low-temperature chemistry-specific species are retained as α_{crit} decreases. For target sets 1 to 3, however, low-temperature specific species such as CH_2O, CH_3O, and CH_3CHO are retained at α_{crit} of 0.48.

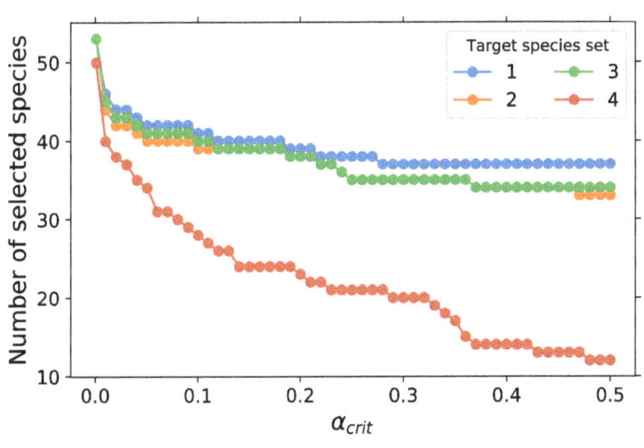

Figure 7. Number of species retained as a function of α_{crit} with USC Mech II.

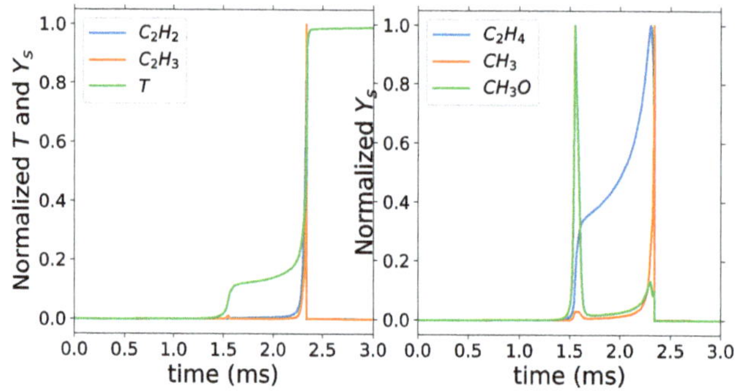

Figure 8. Homogeneous reactor: normalized temperature and mass fractions with T_i = 740 K and equivalence ratio of 1.

3.5. Common Species between Two-Step PCA, DRG and GPS

Table 1 lists the maximum representative species obtained from the three selection methods. For the cut-off conditions selected, we can see that there are common species between the three selection methods. N_2 is selected by DRG and GPS but not two-step PCA. N_2 remains constant and has a reaction rate of zero; therefore, its PC contribution would be minimal and neglected during the two-step PCA. However, O_2 is an oxidizer and has a strong dependence on the fuel because it is involved in fuel consumption. Except for N_2, all remaining species selected by DRG and GPS are also selected by the two-step PCA. DRG yields the smallest number of retained species out of all three methods. The two-step PCA is computationally cheap out of the three methods, and GPS is the most expensive due to the construction element flux graph.

Table 1. List of representative species.

Selection Method	Two-Step PCA	DRG	GPS
Maximum number of representative species	63	51	53
Other Species	C_2H_6, HCCOH, C_4H_{10}, C_4H_4O, $C_4H_6O_{25}$, C_5H_4O, C_5H_4OH, C_5H_5OH, C_6H_2, C_6H_3, C_6H_5, C_6H_6, C_6H_5O, C_6H_5OH	N_2, C_4H_4O	N_2, C_2H_6, HCCOH, C_4H_{10}
Common Species: H, O, OH, HO_2, H_2, H_2O, H_2O_2, O_2, C, CH, CH_2, CH_3, CH_4, HCO, CH_2O, CH_3O, CH_2OH, CH_3OH, CO, CO_2, C_2H, C_2H_2, H_2CC, C_2H_3, C_2H_4, C_2H_5, HCCO, CH_2CO, CH_3CO, CH_2CHO, CH_3CHO, C_3H_3, C_3H_6, C_3H_8, C_2H_3CHO, CH_3COCH_3, C_4H_2, C_4H_4, C_4H_5-2, C_4H_6, C_4H_612, C_4H_6-2, H_2C_4O, $CH_2CHCHCHO$, $C_2H_3CHOCH_2$, $C_4H_6O_{23}$, C_5H_5, C_5H_6, n-C_7H_16			

3.6. Representative Species Coupling with Foundational Chemistry

An application of the representative selection is demonstrated by evaluating a hybrid chemistry model that uses artificial neural networks (ANNs) to model the reaction rates of representative species, and the remaining species that are not selected as representative species are modeled using the foundational chemistry USC Mech II [29]. This validation is performed for the homogeneous chemistry of n-heptane at an initial temperature of 700 K and pressure of 20 atm. From the parametric study conducted on all training data, 20 representative species are selected via two-step PCA for a variance of 85% and 98.5% cut-off, DRG with ε of 0.57 and target species set 4, and GPS with α_{crit} of 0.29 and target species set 4. These parameters are selected because of the number of representative species they lead to, and because the selected species contain the major species observed during n-heptane oxidation. Other parameters and their corresponding representative species can be chosen depending on the accuracy and computational saving that we would like the model to achieve.

ANNs are used to model the reaction rates of the representative species, with a similar implementation to that of our recent work [12]. For this validation, the ANN tabulation is carried only for the reaction rates of DRG-selected representative species. The temperature and mass fractions of the fuel, O_2, H_2O_2, and CH_2, are used as inputs to the ANNs. These species, which are also part of the representative species, are chosen as inputs because they adequately track all stages of ignition, and they are highly correlated with other representative species. Five single-output ANNs, with two hidden layers with 20 neurons per layer, are used to model the reaction rate of each input, while one multiple-output ANN, with three hidden layers with 20 neurons per layer, is used to model the reaction rates of the remaining representative species. Training a separate ANN for each input has shown to improve prediction accuracy [12]. A hyperbolic tangent function is used as the transfer function, and errors between the true and predicted reaction rates are minimized via the mean squared error (MSE) function. The Levenberg–Marquardt backpropagation algorithm is used to optimize the ANN inputs, and each ANN is trained for 5000 epochs. The remaining species' ANN is optimized using PyTorch 1.13.1's Adam optimizer with training carried over 100,000 epochs. Like in [12], the learning rate and weight decay are respectively set to 0.005 and 10^{-6} during the first 20,000 epochs, after which they are decreased to 10^{-5} and 10^{-9}, respectively. The training/test/validation ratio is 70/20/10%, respectively. The ANNs achieved an average accuracy of 2×10^{-6}.

Using USC Mech II, we reconstruct species that are not included in the list of representative ones by solving Equation (14). Figure 9 shows the results of the species reconstructed using foundational chemistry (symbols) and original species from the detailed mechanism (solid lines). CO, CH_4, and C_2H_2 are not part of the DRG-selected representative species, but they are reconstructed well using the foundational chemistry. Similarly, HO_2 and CH_2O, which are included in the representative species, along with the temperature, match well with the detailed chemistry. Between 4 and 4.6 ms in Figure 9, we can observe that the CO, CH_4, and C_2H_2 results from the hybrid chemistry model do not exactly match the detailed mechanism results, although the hybrid model correctly captures the trend and maximum mass fractions. The accuracy of these predictions would be improved if these species were in the list of representative species and modeled by the ANNs. This demonstrates the importance of analyzing a wide range of parameters in the species selection process to achieve better prediction accuracy of important species.

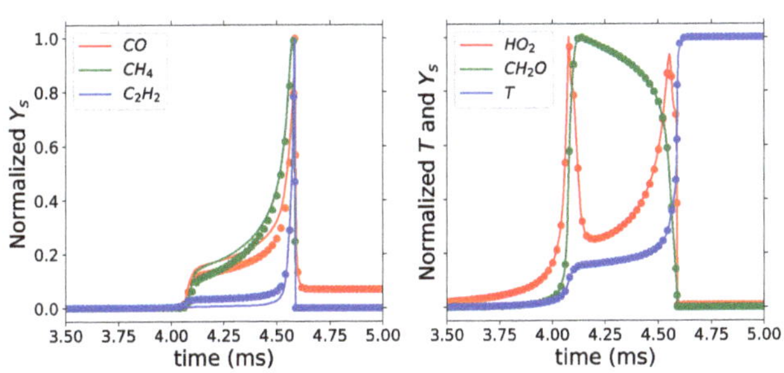

Figure 9. Normalized temperature and mass fractions with T_i = 700 K and equivalence ratio of 1 from detailed mechanism (solid lines) and hybrid chemistry model (symbols).

4. Conclusions

This work compares four methods to select representative species. First, the two-step PCA approach is a PCA-based method selected based on two key parameters, the cumulative variance and the cut-off percentage, which is a measure of the contribution of each species to a PC. The second approach is the manifold-informed reduction, a PCA-based method using the number of PCs as one parameter and selecting species that minimize the non-uniqueness and feature size of the low-dimensional manifold. DRG is the third approach, which selects species based on a threshold parameter ε that measures the dependence strength of species during their production or consumption. GPS is the final method, where species are selected based on a parameter α_{crit} measuring their contribution to the overall combustion.

The selection of representative species via two-step PCA, DRG and GPS is carried out on low temperature, constant pressure homogeneous chemistry and PSR of n-heptane using a detailed mechanism with 538 species. Two-step PCA achieved 88 and 99% reductions from detailed mechanism species, while DRG achieved 91 and 96% reductions, and GPS, 90 and 98%.

The minimum reduction, where most species are retained, is achieved using the highest variance and cut-off percentage for the two-step PCA, and the smallest ε and α_{crit} for DRG and GPS, respectively. Inversely, we can achieve the maximum reduction, a more aggressive reduction where a small number of representative species are selected. In both cases, these three approaches can select key species that track all stages of fuel oxidation.

The manifold-informed reduction is a computationally intensive approach that does not yield results with the homogeneous chemistry and PSR data due to the data size and the expensive iterative optimization of the quality of the low-dimensional manifold. However, a comparison between this reduction method and the two-step PCA yields using hydrogen, syngas, and ethylene data [21] similar selected species, with the two-step PCA carrying a parametric study under 1 min.

DRG and GPS are sensitive to the set of target species, while the two-step PCA is sensitive to the cumulative variance. Additionally, both DRG and GPS require a chemical mechanism to achieve reduction based on reaction paths between species. The two-step PCA does not depend on an existing chemical mechanism. Selected species from the two-step PCA can then be used to develop global reactions, for example, using chemical reaction neural networks [30,31]. Since the latter approach leads to similar species to DRG and GPS, the two-step PCA can successfully be used for data with missing chemical mechanisms or for new fuels with non-existing chemical mechanisms. It can also be used for a faster species' selection given its low computational cost. On the other hand, if a detailed chemical mechanism is available, along with the knowledge of species to target in the selection process, DRG can be used.

Finally, these selection approaches can be used for the development of hybrid chemistry models. Although this study is carried out using n-heptane, the species selection methods in this work can be applied to other hydrocarbons or fuels and yield a smaller set of representative species that correctly describe the full combustion process. Using the selected representative species, a surrogate model or regression of the representative species' reaction rates, such as with artificial or deep neural networks, can be coupled with foundational chemistry [11]. The selected representative species can be adjusted to achieve the desired accuracy of the reduced-order or hybrid chemistry model, and the total number of selected species still achieves a very high reduction rate and has the potential to accelerate chemistry simulation significantly.

Author Contributions: Conceptualization, K.M.G. and T.E.; methodology, K.M.G. and T.E.; software, K.M.G.; validation, K.M.G.; formal analysis, K.M.G. and T.E.; writing—original draft preparation, K.M.G. and T.E.; writing—review and editing, K.M.G. and T.E.; supervision, T.E.; project administration, T.E.; funding acquisition, T.E. All authors have read and agreed to the published version of the manuscript.

Funding: This research was funded by King Abdullah University of Science and Technology (KAUST) through the Competitive Research Grants (CRG) Program, CRG-2020-CRG9-4351.

Data Availability Statement: The data presented in this study are available upon reasonable request from the corresponding author.

Conflicts of Interest: The authors declare no conflicts of interest, and that they have no known competing financial interests or personal relationships that could have appeared to influence the work reported in this paper.

References

1. Pope, S. Small scales, many species and the manifold challenges of turbulent combustion. *Proc. Combust. Inst.* **2013**, *34*, 1–31. [CrossRef]
2. Wang, H.; Xu, R.; Wang, K.; Bowman, C.T.; Hanson, R.K.; Davidson, D.F.; Brezinsky, K.; Egolfopoulos, F.N. A physics-based approach to modeling real-fuel combustion chemistry-I. Evidence from experiments, and thermodynamic, chemical kinetic and statistical considerations. *Combust. Flame* **2018**, *193*, 502–519. [CrossRef]
3. Xu, R.; Wang, K.; Banerjee, S.; Shao, J.; Parise, T.; Zhu, Y.; Wang, S.; Movaghar, A.; Lee, D.J.; Zhao, R.; et al. A physics-based approach to modeling real-fuel combustion chemistry-II. Reaction kinetic models of jet and rocket fuels. *Combust. Flame* **2018**, *193*, 520–537. [CrossRef]
4. Tao, Y.; Xu, R.; Wang, K.; Shao, J.; Johnson, S.E.; Movaghar, A.; Han, X.; Park, J.W.; Lu, T.; Brezinsky, K.; et al. A Physics-based approach to modeling real-fuel combustion chemistry-III. Reaction kinetic model of JP10. *Combust. Flame* **2018**, *198*, 466–476. [CrossRef]
5. Wang, K.; Xu, R.; Parise, T.; Shao, J.; Movaghar, A.; Lee, D.J.; Park, J.W.; Gao, Y.; Lu, T.; Egolfopoulos, F.N.; et al. A physics based approach to modeling real-fuel combustion chemistry-IV. HyChem modeling of combustion kinetics of a bio-derived jet fuel and its blends with a conventional Jet A. *Combust. Flame* **2018**, *198*, 477–489. [CrossRef]
6. Saggese, C.; Wan, K.; Xu, R.; Tao, Y.; Bowman, C.T.; Park, J.W.; Lu, T.; Wang, H. A physics-based approach to modeling real-fuel combustion chemistry-V. NOx formation from a typical Jet A. *Combust. Flame* **2020**, *212*, 270–278. [CrossRef]
7. Xu, R.; Saggese, C.; Lawson, R.; Movaghar, A.; Parise, T.; Shao, J.; Choudhary, R.; Park, J.W.; Lu, T.; Hanson, R.K.; et al. A physics-based approach to modeling real-fuel combustion chemistry-VI. Predictive kinetic models of gasoline fuels. *Combust. Flame* **2020**, *220*, 475–487. [CrossRef]
8. Xu, R.; Wang, H. A physics-based approach to modeling real-fuel combustion chemistry-VII. Relationship between speciation measurement and reaction model accuracy. *Combust. Flame* **2021**, *224*, 126–135. [CrossRef]
9. Ranade, R.; Alqahtani, S.; Farooq, A.; Echekki, T. An ANN based hybrid chemistry framework for complex fuels. *Fuel* **2019**, *241*, 625–636. [CrossRef]
10. Ranade, R.; Alqahtani, S.; Farooq, A.; Echekki, T. An extended hybrid chemistry framework for complex hydrocarbon fuels. *Fuel* **2019**, *251*, 276–284. [CrossRef]
11. Alqahtani, S.; Echekki, T. A data-based hybrid model for complex fuel chemistry acceleration at high temperatures. *Combust. Flame* **2021**, *223*, 142–152. [CrossRef]
12. Alqahtani, S.; Gitushi, K.M.; Echekki, T. A Data-Based Hybrid Chemistry Acceleration Framework for the Low-Temperature Oxidation of Complex Fuels. *Energies* **2024**, *17*, 734. [CrossRef]
13. Kumar, A.; Echekki, T. Combustion chemistry acceleration with DeepONets. *Fuel* **2024**, *365*, 131212. [CrossRef]
14. Lu, L.; Jin, P.; Pang, G.; Zhang, Z.; Karniadakis, G.E. Learning nonlinear operators via DeepONet based on the universal approximation theorem of operators. *Nat. Mach. Int.* **2021**, *3*, 218–229. [CrossRef]
15. Ranade, R.; Echekki, T. A framework for data-based turbulent combustion closure: A priori validation. *Combust. Flame* **2019**, *206*, 490–505. [CrossRef]
16. Jolliffe, I. *Principal Component Analysis*, 2nd ed.; Springer: New York, NY, USA, 2002.
17. Lu, T.; Law, C.K. A directed relation graph method for mechanism reduction. *Proc. Combust. Inst.* **2005**, *30*, 1333–1341. [CrossRef]
18. Lu, T.; Law, C.K. Linear time reduction of large kinetic mechanisms with directed relation graph: N-Heptane and iso-octane. *Combust. Flame* **2006**, *144*, 24–36. [CrossRef]
19. Lu, T.; Law, C.K. On the applicability of directed relation graphs to the reduction of reaction mechanisms. *Combust. Flame* **2006**, *146*, 472–483. [CrossRef]
20. Gao, X.; Yang, S.; Sun, W. A global pathway selection algorithm for the reduction of detailed chemical kinetic mechanisms. *Combust. Flame* **2016**, *167*, 238–247. [CrossRef]
21. Zdybał, K.; Sutherland, J.C.; Parente, A. Manifold-informed state vector subset for reduced-order modeling. *Proc. Combust. Inst.* **2023**, *39*, 5145–5154. [CrossRef]
22. Zdybał, K.; Armstrong, E.; Parente, A.; Sutherland, J.C. PCAfold: Python software to generate, analyze and improve PCA-derived low-dimensional manifolds. *SoftwareX* **2020**, *12*, 100630. [CrossRef]
23. Armstrong, E.; Sutherland, J.C. A technique for characterising feature size and quality of manifolds. *Combust. Theo. Model.* **2021**, *25*, 646–668. [CrossRef]
24. Niemeyer, K.E. Theory: Directed Relation Graph (DRG) Method. Available online: https://niemeyer-research-group.github.io/pyMARS/theory.html (accessed on 14 May 2024).
25. Yen, J.Y. Finding the k shortest loopless paths in a network. *Manag. Sci.* **1971**, *17*, 712–716. [CrossRef]
26. Mishra, R.; Nelson, A.; Jarrahbashi, D. Adaptive global pathway selection using artificial neural networks: A-priori study. *Combust. Flame* **2022**, *244*, 112279. [CrossRef]

27. Xie, C.; Lailliau, M.; Issayev, G.; Xu, Q.; Chen, W.; Dagaut, P.; Farooq, A.; Sarathy, S.M.; Wei, L.; Wang, Z. Revisiting low temperature oxidation chemistry of n-heptane. *Combust. Flame* **2022**, *242*, 112177. [CrossRef]
28. Goodwin, D.G.; Moffat, H.K.; Schoegl, I.; Speth, R.L.; Weber, B.W. Cantera: An Object-oriented Software Toolkit for Chemical Kinetics, Thermodynamics, and Transport Processes. Version 2.6.0. 2022. Available online: https://www.cantera.org (accessed on 2 February 2023).
29. Wang, H.; You, X.; Joshi, A.V.; Davis, S.G.; Laskin, A.; Egolfopoulos, F.; Law, C.K. USC Mech Version II. High-Temperature Combustion Reaction Model of $H_2/CO/C_1$-C_4 Compounds. Available online: https://ignis.usc.edu:80/Mechanisms/USC-Mech%20II/USC_Mech%20II.htm (accessed on 15 May 2021).
30. Ji, W.; Deng, S. Autonomous Discovery of Unknown Reaction Pathways from Data by Chemical Reaction Neural Network. *J. Phys. Chem.* **2021**, *125*, 1082–1092. [CrossRef]
31. Ji, W.; Richter, F.; Gollner, M.J.; Deng, S. Autonomous kinetic modeling of biomass pyrolysis using chemical reaction neural networks. *Combust. Flame* **2022**, *240*, 111992. [CrossRef]

energies

MDPI

Article

Numerical Investigation on the Flame Characteristics of Lean Premixed Methane Flame Piloted with Rich Premixed Flame

Lili Zhang [1], Yongzhang Cui [1,*], Pengfei Yin [1], Wenlong Mao [2] and Pengzhao Zhang [1]

[1] School of Thermal Engineering, Shandong Jianzhu University, Jinan 250101, China; zhanglili10@sdjzu.edu.cn (L.Z.); 24138@sdjzu.edu.cn (P.Y.); 202103103050@stu.sdjzu.edu.cn (P.Z.)
[2] Shandong Special Equipment Inspection Institute Group Co., Ltd., Jinan 250101, China; 10082300@sdic.com.cn
* Correspondence: cyz@sdjzu.edu

Abstract: The introduction of the pilot flame can stabilize the lean premixed flame and promote its industrial application. However, the interaction mechanism between the pilot and main flames is complicated. To reveal the effect of the pilot flame on the main flame, a laminar lean premixed flame adjacent to a rich premixed pilot flame on one side and a similar lean premixed flame on the other side was considered. A two-dimensional numerical model was adopted with detailed chemistry and species transport, also with no artificial flame anchoring boundary conditions. The results show that the pilot flame could promote the main flame stabilized in different locations with various shapes, by adjusting the stretch, heat transfer, and preferential diffusion in a complicated manner. The pilot flame improves the local equivalence ratio and transfer more heat to the main flame. The growth of the pilot flame equivalence ratio and inlet velocity enhances the combustion on the rich side of the main flame and helps it anchor closer to the flame wall. Both the curvature and strain rate show a significant effect on the flame root, which contributes to the main flame bending towards the pilot flame.

Keywords: pilot flame; flame structure; preferential diffusion; heat transfer; flame stretch

1. Introduction

Premixed flame [1–3] has the characteristics of low temperature, completely burnt, and low NOx emissions, while its poor stability and narrow combustion load limit the application. By introducing pilot rich flames (the low-load and fuel-rich flames adjacent to the main fuel-lean flames contributing to stabilize the main flames), the lean flame can be anchored and become stable [4,5]. Although there are many studies about the rich-lean combustion, the dynamic stability mechanisms and flame anchoring of lean premixed flame are not clear yet [6]. However, flame stabilization is a critical and basic problem for the application of burners. It is of great significant to reveal the interaction mechanism between flames, especially the effect of pilot rich flames on the lean main flame.

According to the previous studies [7–9], the pilot flame affects the main flame via supplying heat and radicals to the main flame, modifying the stretch rate of main flame and preventing the cold ambient air reaching the main flame. The shielding effect of the pilot flame was reported by Li et al. [10] and Zhou et al. [11] in the Lund University pilot jet flames, who found that the main jet flame can be stabilized at a higher jet velocity (from 150 m/s to 330 m/s) by increasing the diameter of the pilot flame burner (from 22 mm 60 61 mm). Zhou et al. [8,12] conducted a series of visualization experiments to investigate the interaction mechanism between the pilot and main flame. They drew similar results, that the pilot flames provided not only heat but also fuels and radicals (e.g., carbon monoxide and hydrogen) to the main flame, which could help the stable combustion of a fuel/air mixture in a main flame even at an ultra-low equivalence ratio of 0.4. Tomas and Edgar [13] presented an analytical expression to show how lean flames are contaminated by the presence of rich flames as a function of local velocity gradients and equivalence ratio. Meanwhile, the experimental results showed that the interaction between rich and lean flame in the triple region (the nonpremixed zone enclosed by the two premixed zones (one fuel-rich and one fuel-lean)) contributes to the lean flame continuous bending towards the burner.

The effect of the heat transfer from the pilot flame to the main flame is considerable. Yu et al. [7] used a detailed numerical simulation to elucidate the effect of a pilot flame on a main flame via simplifying the pilot flame as hot gas mixing with an unburned fuel/air mixture. They concluded that the heat transfer from the pilot flame to the main flame had a more significant effect on the propagation and structures of the main flame than the mass transfer. Guo et al. [14] numerically investigated the characteristics of a triple flame (diffusion flame) formed between the rich-lean flames. The conduction heat transfer and radical exchange between the premixed and diffusion flame mainly affected the local burning velocity, while the effect of radiation on local burning properties was negligible.

The different mass diffusivities of species and radicals directly affect the distribution of reactants. Azzoni [15] found that the H_2 and CO, as important intermediate species formed in the inner rich premixed reaction zone, reached a maximum value at the tip of this zone. Thereafter, they diffused toward the nonpremixed reaction zone where the radicals oxidized CO and H_2 into CO_2 and H_2O. Barlow et al. [16] highlighted the significant effect of differential diffusion of heat and mass on the local flame structure and global flame parameters in premixed flames. They found that the H_2 and H_2O diffused preferentially ahead of the CO and CO_2 through the preheated zone and subsequently convected downstream, which was apparently amplified by the recirculation zone accumulating excess CO_2. The accumulation of CO_2 near the flame root was also detected by Wan et al. [17]. They concluded that the preferential diffusion increased the local equivalence ratio around the anchoring location of the flame root (the segments of the flame front near the flame holder), which promoted the stabilization of the flame root. Further studies were conducted by Vance [18] on the heat transfer, preferential diffusion, and flame stretch of a premixed flame. The effects of preferential diffusion on the flame speed were stronger at the flame base due to higher stretch rates. However, with the flame moving closer to the flame wall, the heat loss to the flame wall will counter the enhancement in flame speed due to preferential diffusion.

The flame stretch, related to the flow field, nonuniformity, and flame curvature, may change the flame position and flame speed. Choi and Puri [19,20] emphasized that the flame curvature was the dominating effect on the premixed flames instead of the flow nonuniformity effects. Their experimental research considered the rich and lean premixed methane–air flames in both a slot burner and an axisymmetric cannular burner, where the flame speed in the reaction zone increased significantly due to the curvature effects and reached a maximum value as the curvature change declined near the tip of the reaction zone. Michales and Ghoniem [21] investigated the stability of a laminar lean premixed flame with a heat-conducting bluff body. They emphasized the effect of flame stretch rate on the flame structure and location. Their numerical results showed that the flame stretch rate at the leading edge reached a peak, which is a major factor dictating the location of the flame leading edge.

In summary, the flame characteristics can be affected by the heat transfer, preferential diffusion, and stretch in a complicated manner. However, research about the interaction between the lean main flame and rich pilot flame is rare. The present work aims to reveal the stability mechanism of the rich pilot flame on the lean main flame, considering the heat transfer, diffusion, and fluid dynamics.

2. Model and Numerical Methodology

2.1. Computational Method

Due to the low Reynolds number, the model is assumed to be laminar flow. The mixture follows the ideal gas law. The radiation between the flame and walls and between the flame and environment are ignored [14], considering that the effect of radiative heat on results is unlikely qualitative. However, quantitative predictions of radiation will be attempted in future research. Accordingly, the continuity, momentum, and energy conservation equations are, respectively, expressed as follows.

Continuity

$$\nabla \cdot (\rho \boldsymbol{v}) = 0 \tag{1}$$

Momentum

$$\nabla (\rho \boldsymbol{v} \boldsymbol{v}) = -\nabla p + \nabla \widetilde{\tau} \tag{2}$$

Energy

$$\rho C_p \boldsymbol{v} \cdot \nabla T = \nabla \cdot (\lambda \nabla T) - \sum_{i=1}^{i} C_{pi} \left(\vec{J_i} \cdot \nabla T \right) - \sum_{i=1}^{i} r_i \dot{v}_i W_i \tag{3}$$

where ρ is the density of mixture, C_{pi} and C_p are, respectively, the specific heat of species i and mixture, λ is the mixture thermal conductivity, $\boldsymbol{v} = (u, v)$ is the velocity vector, p is the pressure, $\widetilde{\tau} = \mu\left[(\nabla \boldsymbol{v} + \nabla \boldsymbol{v}^{\mathrm{T}}) - \frac{2}{3}\nabla \cdot \boldsymbol{v} I\right]$ is the viscous stress term, I is the unit vector; $\vec{J_i} = -\rho(W_i/\overline{W}) D_{i,m} \nabla X_i$ is the diffusive mass flux vector, W_i is the molar mass of species i, $\vec{W} = 1/(\sum_{i=1}^{n} Y_i/W_i)$ is the molar mass of mixture, $D_{\mathrm{i,m}}$ is the diffusion coefficient of species i relative to the mixture, X_i is the molar fraction, Y_{i} is the mass fraction, r_i is the enthalpy of species i, and $\dot{v}_i$ is the molar production rate of species i.

The local mass fraction of each species i (i = 1, ..., n − 1) is solved through the following conservation equation. The nth species (nitrogen) with the overall largest mass fraction in the mixture is calculated according to that the mass fraction of all species must sum to unity.

$$\nabla \cdot (\rho \boldsymbol{v} Y_i) = -\nabla \cdot \vec{J}_i + \dot{v}_i W_i \tag{4}$$

$$\sum_{i=1}^{n} Y_i \equiv 1 \tag{5}$$

where the first term on the left of Equation (4) is the convection term, the first term on the right is diffusion term, and the second term on the right is source term.

2.2. Computational Domain and Mesh Independence

The slit burner with rectangular channels and a two-dimensional (2D) computational domain adopted in this work are shown in Figure 1. The computational domain contains one main flame port and a half pilot flame port. The width of the main and pilot flame port is, respectively, 2 mm (d_1) and 1 mm (d_2). The clear distance between the main and pilot flame port is 2.5 mm, and so is the distance between the main and main flames. The computational region extends 20 d_1 in the axial direction downstream from the burner exit plane. To reduce the computational effort, the center between two main flames and center of the pilot flame port are set as symmetry boundary conditions.

The mixture flow in the control volume remains at a constant temperature of T_{in} = 300 K and a uniform velocity of v_{main} = 1 m/s and v_{pilot} = 0.5 m/s. Meanwhile, the mixture at the inlet is steady and is calculated based on the specified equivalence ratio of the reactant mixture. The equivalence ratio of the pilot flame (ϕ_{pilot}) varies from 1.1 to 1.5, meanwhile the equivalence ratio of the main flame (ϕ_{main}) is kept constant at 0.8. It is assumed that there is no heat transfer between the gas mixture and port wall, and the flame wall is at a constant temperature of T_{Wall} = 450 K. The boundary conditions of no-slip and zero species diffusive flux normal to the wall surface are applied on the port and flame wall.

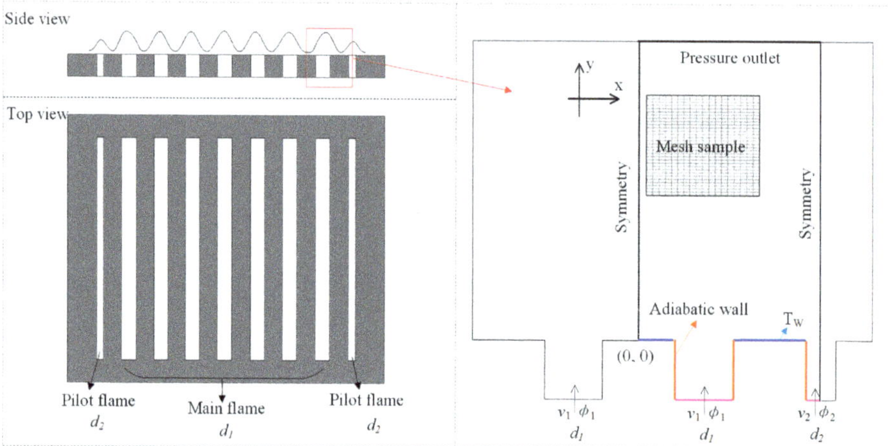

Figure 1. Schematic view of burner and the schematic diagram of the numerical model.

The grid size greatly influences the flame position, which becomes stable only with adequate refinement [22]. Thus, the mesh was refined in the reaction region with a scale factor to make sure the flame was caught accurately. A mesh independence test was conducted to obtain an optimal grid size, which can ensure the accuracy of the results and reduce the calculation time. The temperature profiles along the center line of flame with different grid size are shown in Figure 2. The profile of flame temperature remains nearly unchanged when the grid number reaches 2.02×10^5. Thus, the computational domain is discretized using a grid count of 2.02×10^5 with the grid length of 0.02 mm and width of 0.04 mm.

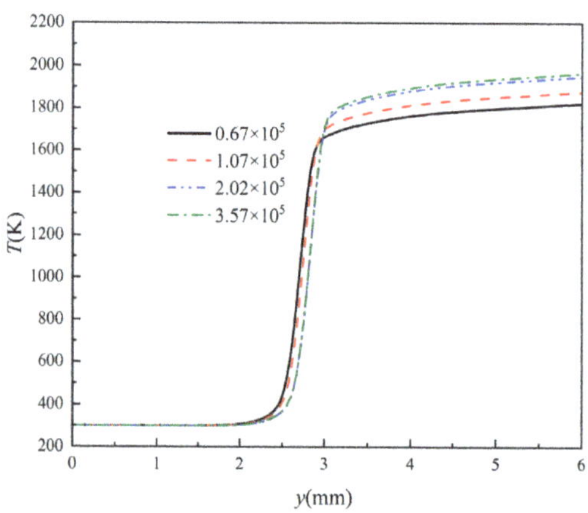

Figure 2. Profiles of flame temperature with different grid quantity.

2.3. Numerical Setup

The numerical calculation was executed in ANSYS FLUENT 16.2. The laminar-finite-rate model was selected to solve the interaction between flow and combustion. A reduction methane–air kinetics mechanism involving 16 species and 41 reactions was imported from ANSYS Chemkin 17.0 which was proved to be accurate in describing the flame characteristics [23,24]. The kinetic theory was detected to solve the mass diffusivity. The density was calculated using incompressible ideal gas assumption, and the specific heat, viscosity, and thermal conductivity were calculated with mixing law.

The second-order upwind scheme was used for spatial discretization, and the SIMPLE algorithm was employed for pressure velocity coupling. The evolution of gradients is developed by the least squares cell-based method. The under-relaxation factors are adjusted with a proper value to achieve convergence considering the finite rate model and species transport model. The convergent criteria of residuals are set as 10^{-6} for the conservation equations.

3. Validation of the Numerical Model

To verify the validity of the numerical model, an experimental study was conducted on a single premixed flame. As shown in Figure 3, the methane and air flow through the mass flow controller and thoroughly mix in the buffer tank. The methane/air mixture flows through the flame arrestor and into the burner with rectangular port. A water-cooling system was installed under the flame wall to adjust the temperature of the surface and keep it constant, which was monitored by a K-type thermocouple and data acquisition system.

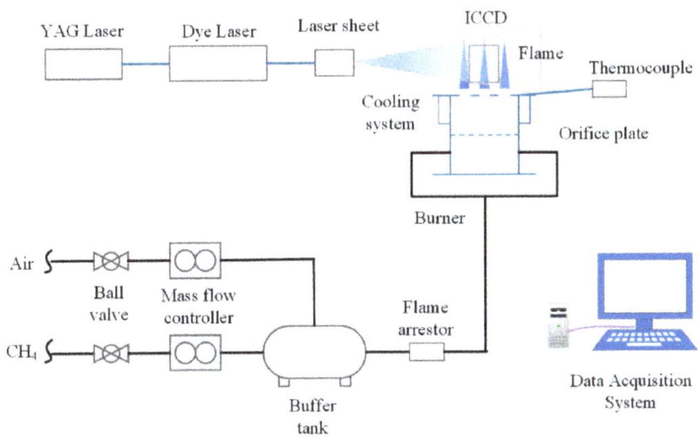

Figure 3. Schematic diagram of experimental system.

The planar-laser-induced fluorescence (PLIF) (La Vision, Göttingen, Germany) of hydroxyl radical (OH) was utilized for the probing of the flame structure. The laser generated in the Nd:YAG Laser at 532 nm was used as the pumping light source of the dye laser. The lase out from the dye laser was frequency-doubled to 283 nm by frequency conversion unit (FCU). Then, a laser sheet was used to shape the light to 50 mm height and 0.1 mm thickness for the flame diagnosis. The OH at the center section of flame was excited by the laser and emitted 308 nm fluorescence, which was captured by the ICCD (intensified charge coupled device) camera.

The spatial distribution of OH under the condition of v = 1 m/s and ϕ = 0.8 is shown in Figure 4. The OH mass concentration from the simulation and the signal intensity detected from PLIF were both normalized ($Y_{OH(n)}$). The OH concentration of numerical simulation is highest at the upper half of the flame and lowest at the root of the flame, which is consistent with the experiment results. The signal intensity of OH at the flame tip is weaker than the numerical results, which is due to the weaker laser intensity near the boundary of laser sheet. Similar results were detected in previous studies [7,25]; the stretch rate at the flame tip is negative, which could strengthen the flame, resulting in higher OH concentration. In this paper, we focus on the characteristics of the flame front (reaction zone) and the preheat zone. Accordingly, the numerical model is convincing to analyze the characteristics of premixed flame.

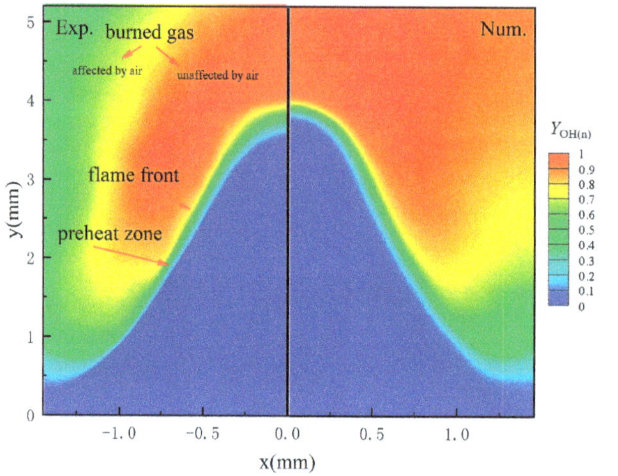

Figure 4. Spatial distribution of normalized OH PLIF signal intensity and numerical OH mass fraction.

4. Result and Discussion

4.1. Flame Characteristic

This paper is mainly focused on the effect of a rich pilot flame on the lean main flame. Thus, the rich side (affected by the rich pilot flame) and lean side (adjacent to the lean main flame) of the main flame are highlighted in this section to analyze the flame characteristics. The flame reaction zone is defined as the domain enclosed by the line of 10% maximum HCO mass concentration. The boundary of the preheat zone is marked by the line of $T = 1.01\ T_{inlet}$. The line of T = 700 K is singled out as a reference surface to analyze the mass transportation in the preheat zone. The flame surface is selected as the line where the mass fraction of OH is equal to 0.05%, which locates in the reaction zone and coincides with the contour of maximum heat release rate.

Figure 5 shows the contours of temperature, T, and heat release rate, ω, for a stable laminar flame at $\phi_{pilot} = 1.2$, $v_{pilot} = 0.5$ m/s, $\phi_{main} = 0.8$, and $v_{main} = 1$ m/s. It is observed that the main flame is a typical bell shape with height (H) of 3.77 mm. The flame tip and base are, respectively, negatively curved and positively curved, and the flame wing between them is weakly curved. The main flame is asymmetrical, that the flame tip inclines toward the pilot flame with a distance of 0.13 mm from the port centerline. Meanwhile, the main flame is anchored with a finite distance from the flame wall. The main flame stand-off distances (the distance between the y-position of the flame root and the flame holder) of the rich side and lean side are, respectively, 0.34 and 0.44 mm. The pilot flame is adjacent to the main flame root, which affects the heat and mass transport.

To investigate the effect of the rich pilot flame on the lean main flame, the flame structure and stand-off distance with different pilot flame equivalence ratio and inlet velocity on condition of constant $\phi_{main} = 0.8$ and $v_{main} = 1$ m/s are shown in Figures 6 and 7. Figure 6a illustrates that as the ϕ_{pilot} increases from 1.1 to 1.2, the flame grows from 3.66 mm to 3.77 mm. Then, as the ϕ_{pilot} increases from 1.2 to 1.5, the flame decreases from 3.77 mm to 3.29 mm. At the same time, the deflection of the main flame towards the pilot flame is reduced gradually. And from Figure 6b, the stand-off distance of the rich side is significantly lower than that of the lean side in the main flame, which declines with the increase in ϕ_{pilot}. Hence, the pilot flame could help the main flame anchor closer to the flame wall, and this influence is more remarkable as the ϕ_{pilot} increases (as the ϕ_{pilot} from 1.1 to 1.5, the h from 0.38 mm to 0.28 mm). The increasing inlet velocity of the pilot flame from 0.5 m/s to 0.9 m/s, shown in Figure 7, leads to the main flame height declining from 3.77 mm to 3.32 mm. Meanwhile, the deviation of the main flame to the pilot flame is reduced gradually. The increase in the pilot flame inlet velocity also plays an active role in reducing the stand-off distance on the rich side of the main flame, in that h_{rich} declines from 0.34 mm to 0.28 mm as v_{pilot} increases from 0.5 m/s to 0.9 m/s.

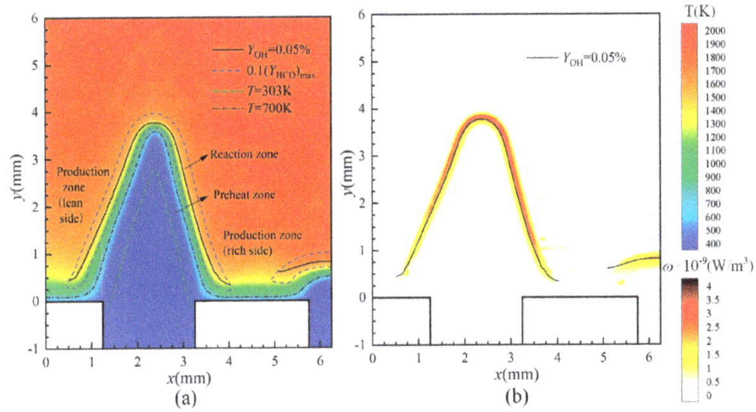

Figure 5. Contours of T and ω at ϕ_{main} = 0.8, v_{main} = 1 m/s, ϕ_{pilot} = 1.2, and v_{pilot} = 0.5 m/s. (**a**) Temperature; (**b**) volumetric heat release rate.

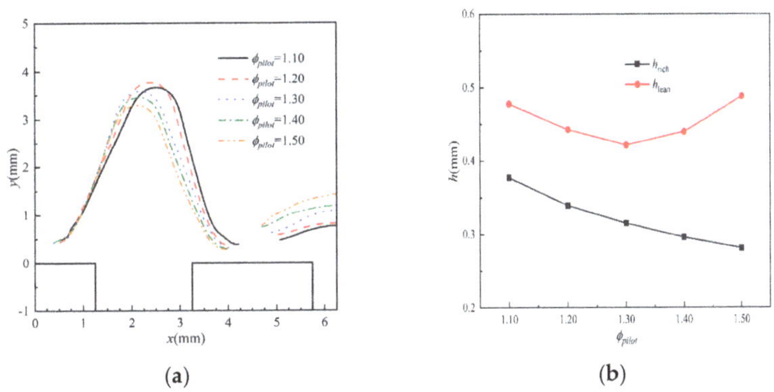

Figure 6. Profile of flame position and stand-off distance on the rich and lean side of the main flame with different pilot flame equivalence ratio at ϕ_{main} = 0.8 and v_{main} = 1 m/s. (**a**) Flame position; (**b**) stand-off distance.

4.2. Consumption Speed

The consumption speed (S_c) describes the rate that the fuel (in lean mixture) or oxidizer (in rich mixture) is consumed. The consumption speed [2] is calculated via integrating the energy equation across the flame.

$$S_c = \frac{\int_0^\infty \omega / c_p dn}{\rho_u (T_b - T_u)} \tag{6}$$

where ρ_b and ρ_u are the density of the burned and unburned mixture, ω is the volumetric heat release rate, n is the normal to the flame or the normal to the constant temperature contours, c_p is specific heat of the mixture, and T_b and T_u are the burned and unburned temperature.

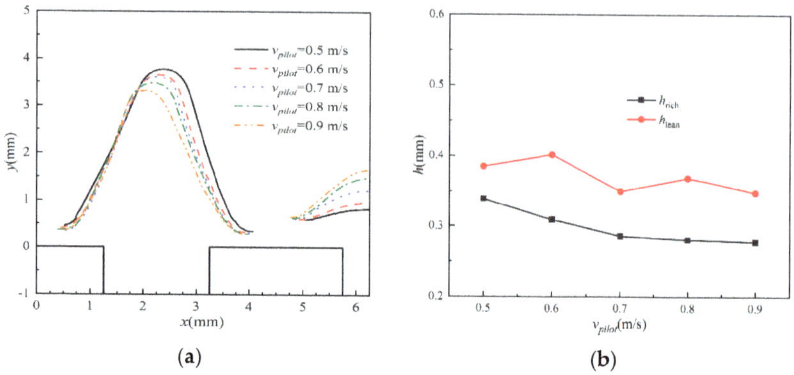

Figure 7. Profile of flame position and stand-off distance on the rich and lean side of the main flame with different pilot flame inlet velocity at $\phi_{\text{main}} = 0.8$, $v_{\text{main}} = 1$ m/s, and $\phi_{\text{pilot}} = 1.2$. (**a**) Flame position; (**b**) stand-off distance.

The distributions of the consumption speed on the flame surface with intervals of 0.5 mm in the y-direction are presented in Figure 8. It can be observed that the consumption speed increases gradually from the flame root to the flame tip. It is worth noting that the combustion speeds on the rich side are generally greater than those on the lean side, while this discrepancy becomes weaker from the flame root to the tip. The flame speed is balanced with the normal flow velocity on the flame front, which contributes to the flame deflection. Considering that the pilot flame affects the main flame majorly on the flame root, the S_c on both rich and lean sides of the main flame at y = 0.5 mm is shown in Figure 9. As we can see, the increasing ϕ_{pilot} and v_{pilot} promote the combustion on the rich side of the main flame root, and this effect becomes weaker gradually. However, the various pilot flame conditions show inconspicuous influence on the S_c of the lean side. Accordingly, the rich side of the main flame was anchored closer to the flame wall, as shown in Figure 7.

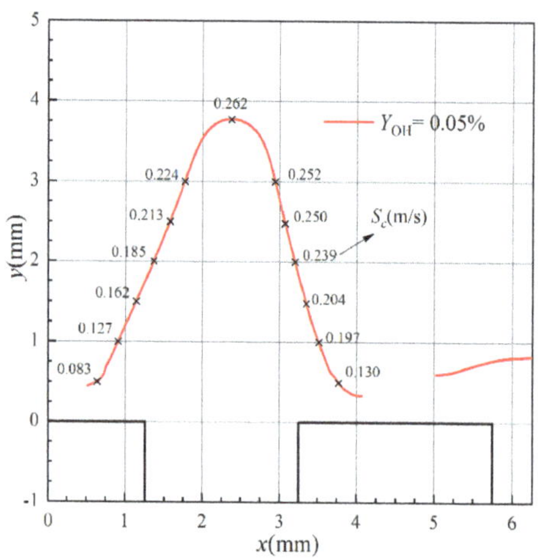

Figure 8. Distribution of the consumption speed on the flame surface, at $\phi_{\text{pilot}} = 1.2$, $v_{\text{pilot}} = 0.5$ m/s, $\phi_{\text{main}} = 0.8$, and $v_{\text{main}} = 1$ m/s.

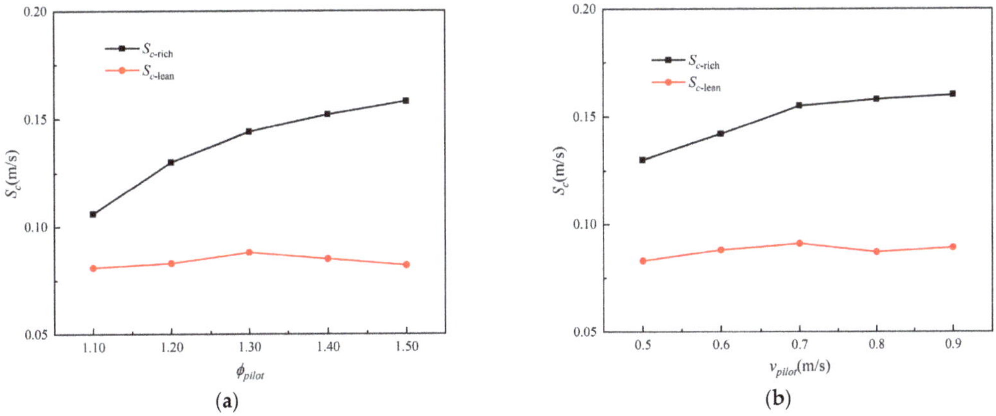

Figure 9. The Sc on the lean and rich sides of the main flame root (y = 0.5 mm) with different pilot flame equivalence ratio and inlet velocity at ϕ_{main} = 0.8 and v_{main} = 1 m/s. (**a**) With different ϕ_{pilot}; (**b**) with different v_{pilot}.

4.2.1. Effect of Preferential Diffusion

In this study, as the fuel of CH_4 and operating pressure of atmosphere are settled, the consumption speed is mainly related to the fuel concentration and the temperature of the gas mixture. Hence, the reactant concentration and heat transfer in the preheat zone are discussed in Figures 10–12. To quantify the effect of difference in a two-dimensional flow and mass diffusivities of species and radicals (described as preferential diffusion) on the distribution of the reactants, the local equivalence ratio is adopted as Equation (8). Barlow et al. [16] proposed it via calculating the local fuel/oxygen atom balance for major species (CH_4, O_2, H_2O, CO_2, CO, and H_2).

$$\phi_{\text{local}} = \frac{0.5(X_{H_2} + X_{H_2O}) + X_{CO_2} + X_{CO} + 2X_{CH_4}}{0.5(X_{CO} + X_{H_2O}) + X_{O_2} + X_{CO_2}} \tag{7}$$

where X_i is the molar fraction of species *i*.

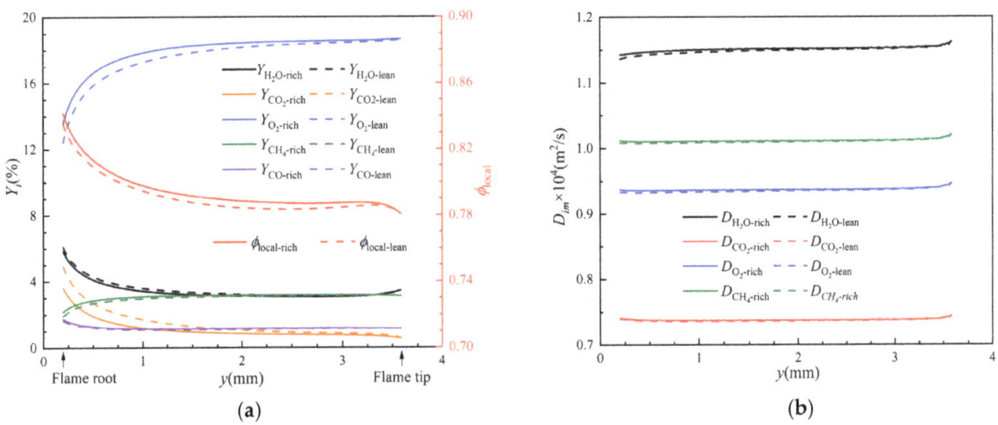

Figure 10. Distribution of local equivalence ratio, mass fraction, and diffusivities of major species on the surface of T = 700 K at ϕ_{main} = 0.8, v_{main} = 1 m/s, ϕ_{pilot} = 1.2, and v_{pilot} = 0.5 m/s. (**a**) Local equivalence ratio and mass fraction; (**b**) diffusivity.

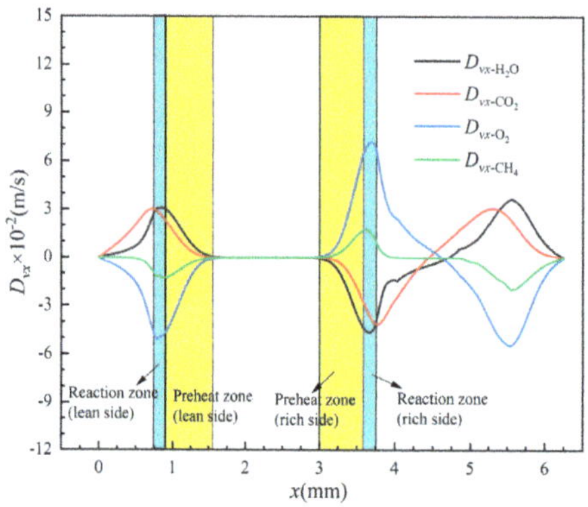

Figure 11. Diffusion velocity in x-direction at y = 0.5 mm.

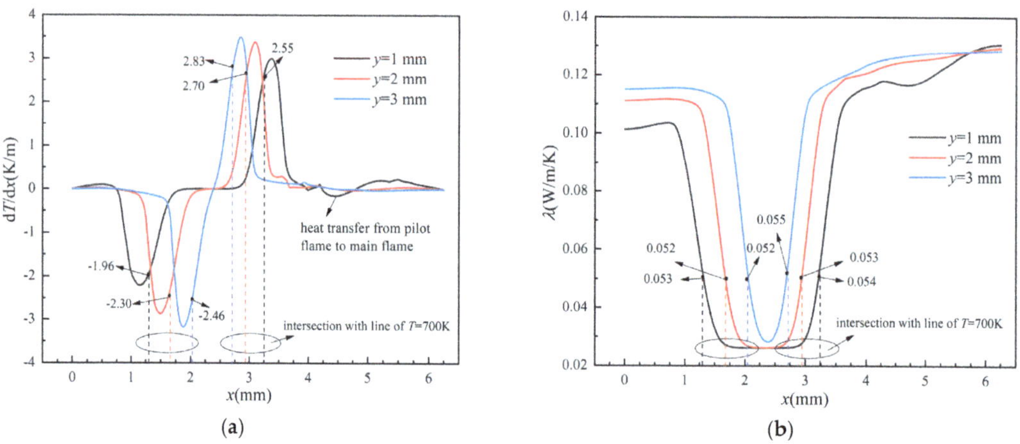

Figure 12. Distribution of temperature gradient in x-direction and thermal conductivity at y = 1 mm, 2 mm, and 3 mm. (**a**) Temperature gradient; (**b**) thermal conductivity.

The local equivalence ratio, mass fraction, and diffusivities of major species in the preheat zone of the main flame are presented in Figure 10. The reference surface of T = 700 K is divided into rich side and lean side by the peak of flame. The concentration of the major species and ϕ_{local} is steady from the flame wing to flame tip, illustrating that the increase in S_c (in Figure 8) from the flame wing to tip exhibits little correlation with species concentration. However, the Y_{H2O}, Y_{CO2} decrease and Y_{O2}, Y_{CH4} increase rapidly from flame root to flame wing; moreover, the growing ϕ_{local} contributes to the increase in S_c. The mass fractions of O_2 and CH_4 on the rich side are higher than those of the lean side, while the mass fractions of H_2O and CO_2 on the rich side are lower than those of the lean side. The mass fraction of CO on the rich side is nearly equal to the lean side. The difference between the rich and lean sides becomes weaker from the flame root to the tip, which is in accordance with the distribution of S_c present in Figure 8. Consider the variation of mass fraction of CH_4, O_2, H_2O, and CO_2 from the flame root to the tip; the diffusivities are discussed in Figure 10b. The diffusivity of H_2O is dramatically larger than that of the CO_2, which is produced in the reaction zone and diffused to the preheat zone. Moreover, the diffusivity of CH_4 is greater than that of O_2, which diffuses from the preheat zone to the reaction zone.

To reveal the effects of preferential diffusion on the main flame through the flame root, the diffusion velocity in the x-direction on section of y = 0.5 mm is presented in Figure 11.

As can be seen, the H_2O and CO_2 diffuse from the production zone to the preheat zone, while CH_4 and O_2 diffuse from the preheat zone to the production zone both on the lean and the rich sides of the main flame. It is important to note that the D_{vx} of the considered species on the rich side is obviously higher than that on the lean side, and reaches peak value in the reaction zone. The diffusion difference of species between the rich and the lean sides of the main flame results in the species concentration distribution in the preheat zone, shown in Figure 10a.

4.2.2. Effect of Heat Transfer

As can be seen in the contour of temperature shown in Figure 5a, the temperature of the rich pilot flame tip and the lean main flame tip are, respectively, 1900 K and 1800 K. Accordingly, the temperature of production zone on the rich side is also higher than that of the production zone on the lean side. The higher reactant concentrate of the rich pilot flame causes more heat release.

The temperature gradient in the x-direction and the thermal conductivity can express the heat transfer within and between the flames. To analyze the effect of heat transfer on the main flame, Figure 12 shows the distribution of dT/dx and λ on sections of y = 1, 2, and 3 mm. The maximum value of dT/dx is located near the inner boundary of the flame reaction zone. The maximum value of dT/dx on the rich side is larger than that of the lean side, while this difference decreases as the y-coordinate increases. The values of dT/dx and λ on the intersection points of T = 700 K reference line and three y-cross section lines are marked in the figures. Significantly, these values on the rich side are greater than those on the lean side. In conclusion, the conductive heat flux in the preheat and reaction zones of the rich side is greater than that of the lean side, and the difference decreases gradually from the flame root to the flame tip.

4.3. Flame Stretch

The flame stretch is mainly related to the flow motion, flame curvature, and local nonuniformity. According to Figure 8, the entire laminar premixed flame surface is smooth and planar. Hence, the weak stretch effect on the local flame speed can be expressed as

$$S_{\rm u} = S_{\rm u}^0 - L\kappa \tag{8}$$

where $S_u{}^0$ is the unstretched flame speed; L is the Markstein length, which is related to the order of the flame thickness and the properties of gas mixtures [26]; κ is the flame stretch rate given by

$$\kappa = \kappa_{\rm s} + \kappa_{\rm c} \tag{9}$$

where κ_s and κ_c are, respectively, strain rate and curvature rate. The strain rate results from the flow strain which is related to the velocity field, and it consists of the shear strain rate $\kappa_{s,s}$ and normal strain rate $\kappa_{s,n}$,

$$\kappa_{\rm s,s} = -n_{\rm y}n_{\rm x}\left(\frac{\partial u}{\partial x} + \frac{\partial v}{\partial y}\right) \tag{10}$$

$$\kappa_{\rm s,n} = -n_{\rm y}^2\frac{\partial u}{\partial y} - n_{\rm x}^2\frac{\partial v}{\partial x} \tag{11}$$

where u and v are y-component and x-component velocity, respectively, and n is the normal of the flame front towards the reactants. Meanwhile, the curvature rate, κ_c, caused by the flame curvature, can be calculated as [27]

$$\kappa_{\rm c} = -\left(un_{\rm y} + vn_{\rm x}\right)\left(\frac{dn_{\rm y}}{dy} + \frac{dn_{\rm x}}{dx}\right) \tag{12}$$

Considering the flow nonuniformity along the normal direction of the flame surface, a reference surface is chosen by 99% of maximum methane mass fraction for the calculation of the stretch rate [28]. The profiles of κ, κ_c, κ_s, $\kappa_{s,s}$, and $\kappa_{s,n}$ (Equations (10)–(13)) along the reference surface in the x-direction are presented in Figure 13. The positive peak values of the stretch rate are observed near the flame base, with $\kappa = 0.81 \times 10^4$ 1/s on the rich side and $\kappa = 0.57 \times 10^4$ 1/s on the lean side, illustrating that the stretch rate on the flame base of the rich side is greater than that of the lean side. Both the values of κ_c and κ_s on the rich side are greater than those of the lean side, while the κ_c plays a more important role. The curves of $\kappa_{s,s}$ and κ_s almost coincide, proving that the κ_s is dominated by the shear strain rate and the normal strain rate $\kappa_{s,n}$ is negligible. Moreover, the negative peak values of the κ and κ_c are near the flame tip, where $|\kappa_s| << |\kappa_c|$. Accordingly, the stretch rate on the flame tip is mainly dominated by the curvature rate. Thus, the quantity of the stretch rate is negative near the flame tip, while it is positive near the flame root. The negative stretch rate will increase the reaction rate and heat generation, which contributes to increasing the local flame speed. The positive stretch rate has the opposite effects.

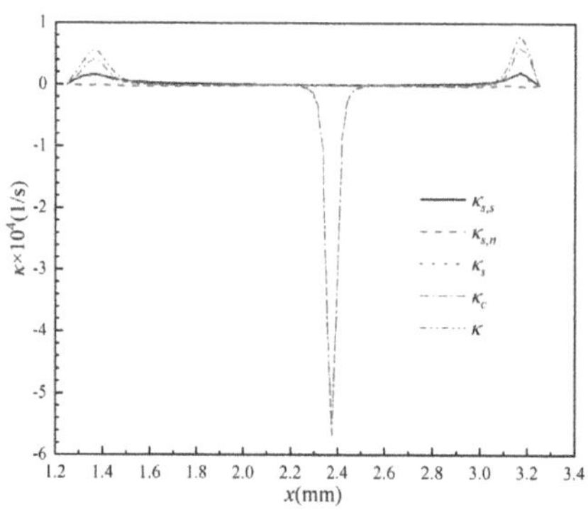

Figure 13. Profile of flame stretch rate at $\phi_{main} = 0.8$, $v_{main} = 1.0$ m/s, $\phi_{pilot} = 1.2$, and $v_{pilot} = 0.5$ m/s.

The effects of ϕ_{pilot} on the flame stretch are shown in Figure 14. According to the above analysis, the stretch rate near the flame tip is mainly dominated by the curvature rate, while it is related to both the curvature rate and strain rate near the flame root. From Figure 14a, as the ϕ_{pilot} increases from 1.10 to 1.50, the peak value of $|\kappa_c|$ (near the flame tip) increases by 21%, and its location moves towards the lean main flame. The increase in the pilot flame equivalence ratio promotes the curvature of the flame tip; meanwhile, the displacement of the peak stretch rate contributes to the reduction in the flame tip deflection, shown in Figure 6a. And from the enlarged curves in Figure 14a,b, the values of $|\kappa_c|$ and $|\kappa_s|$ near the flame root on the rich side are generally greater than those on the lean side under conditions of various ϕ_{pilot}. Hence, the curvature rate and strain rate show more significant effects on the rich side of the main flame root. As the ϕ_{pilot} increases from 1.10 to 1.40, and then to 1.5, the κ_c (near the flame root) first decreases dramatically and then increases slightly; meanwhile, the location of the peak value moves towards the center of the main flame, whereas the κ_s increases and then decreases as the ϕ_{pilot} increases from 1.10 to 1.40, and then to 1.5. The displacement of the peak κ_c and κ_s towards the main flame near the flame root on the rich sides coincides with the profile of the flame root with different ϕ_{pilot}, shown in Figure 6a.

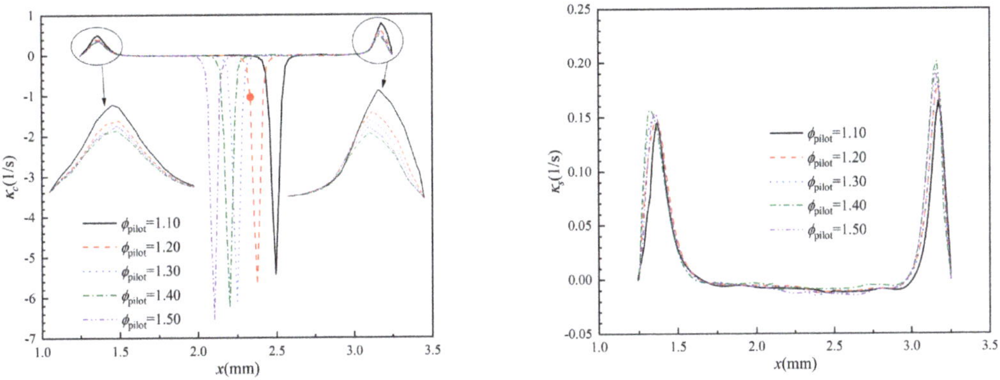

(a) (b)

Figure 14. Profile of flame stretch rate with different pilot flame equivalence ratios at ϕ_{main} = 0.8, v_{main} = 1.0 m/s, and v_{pilot} = 0.5 m/s. (**a**) κ_c; (**b**) κ_s.

The effect of v_{pilot} on the flame stretch is shown in Figure 15. As the v_{pilot} increases from 1.10 to 1.50, the peak value of $|\kappa_c|$ (near the flame tip) increases by 34%, and its location moves towards the lean main flame. Similarly, the increase in pilot flame inlet velocity promotes the curvature of the flame tip and contributes to the reduction in the flame tip deflection. Considering the stretch rate near the flame root, the κ_c and κ_s increase in a reducing rate as the v_{pilot} increases from 0.5 m/s to 0.9 m/s. Meanwhile, both the κ_c and κ_s near the flame root on the rich side are generally greater than those on the lean side under conditions of various v_{pilot}. The locations of peak κ_c and κ_s (near the flame root) on the rich side move towards the center of the main flame, which contributes to the profile of the flame root, shown in Figure 7a.

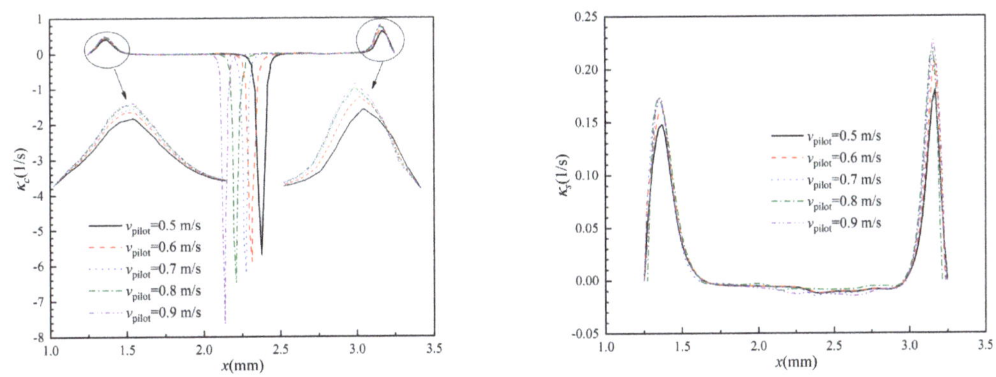

(a) (b)

Figure 15. Profile of flame stretch rate with different pilot flame inlet velocity at ϕ_{main} = 0.8, v_{main} = 1.0 m/s, and v_{pilot} = 0.5 m/s. (**a**) κ_c; (**b**) κ_s.

5. Conclusions

A numerical investigation was conducted in this paper to analyze the effect of a laminar rich pilot methane/air jet flame on the flame structure and burning speed of a lean main flame. Accordingly, the flame characteristics on the rich and lean sides of the main flame were highlighted, considering the influence of preferential diffusion, heat transfer, and flame stretch. The main conclusions are as follows.

(1) The main flame is asymmetrical, with the flame tip inclining toward the pilot flame. Meanwhile, the main flame stand-off distances of the lean side are larger than those of the rich side. The pilot flame could help the main flame anchor closer to the flame wall, and this influence is more remarkable as the ϕ_{pilot} and v_{pilot} increase.
(2) Affected by the pilot flame, the combustion speeds on the rich side are generally greater than those on the lean side. The increasing ϕ_{pilot} and v_{pilot} promote the combustion on the rich side of the main flame root, and this effect becomes weaker gradually as ϕ_{pilot} increases from 1.1 to 1.5 and v_{pilot} increases from 0.5 m/s to 0.9 m/s.
(3) The local equivalence ratio on the main flame preheat zone of the rich side is generally greater than that of the lean side. Meanwhile, the conductive heat flux on the preheat and reaction zones of the rich side is greater than that of the lean side. The distribution of the local equivalence ratio and heat transfer contribute to the increase in combustion speeds on the rich side of the main flame.
(4) The flame stretch rate near the flame tip is negative and dominated by the curvature rate, while it is positive and related to both the curvature rate and strain rate near the flame root. Both the curvature and strain rate show more significant effects on the rich side of the main flame root, which contributes to the main flame bending towards the pilot flame.

Author Contributions: Conceptualization, L.Z. and Y.C.; methodology, Y.C. and L.Z.; software, L.Z. and P.Y.; validation, L.Z. and P.Y.; data curation, L.Z. and W.M.; writing—original draft preparation, L.Z.; writing—review and editing, L.Z., W.M., and P.Z.; funding acquisition, L.Z. All authors have read and agreed to the published version of the manuscript.

Funding: This research was funded by the Shandong Provincial Natural Science Foundation of China, grant numbers ZR2021QE273.

Data Availability Statement: Datasets are available upon reasonable request.

Conflicts of Interest: Author Wenlong Mao was employed by the company Shandong Special Equipment Inspection Institute Group Co., Ltd. The remaining authors declare that the research was conducted in the absence of any commercial or financial relationships that could be construed as a potential conflict of interest.

Nomenclature

C_p	Specific heat capacity (J/(kg·K))
d	Width of flame port (mm)
D	Diffusion coefficient (m^2/s)
D_v	Diffusion velocity (m/s)
H	Flame height (mm)
h	Stand-off distance (mm)
p	Pressure (Pa)
r_i	Enthalpy of species i
S_c	Consumption speed (m/s)
T	Temperature (K)
u	Velocity in x-direction (m/s)
v	Velocity in y-direction (m/s)
W	Molar mass (g/mol)
x	x-coordinate (m)
y	y-coordinate (m)
X_i	Molar fraction of the *i*th species
Y_i	Mass fraction of the *i*th species

Greek symbols	
ρ	Density (kg/m^3)
ϕ	Equivalence ratio
$\dot{\nu}_i$	Molar production rate of species i
ω	Volumetric heat release rate (W/m^3)
λ	Thermal conductivity (W/(m·K))
κ	Flame stretch rate (1/s)
Subscripts	
main	Main flame
pilot	Pilot flame
lean	Lean side of main flame
rich	Rich side of main flame
b	Burned mixture
u	Unburned mixture

References

1. Altay, H.M.; Kedia, K.S.; Speth, R.L.; Ghoniem, A.F. Two-dimensional simulations of steady perforated-plate stabilized premixed flames. *Combust. Theory Model.* **2010**, *14*, 125–154. [CrossRef]
2. Kedia, K.S.; Ghoniem, A.F. Mechanisms of stabilization and blowoff of a premixed flame downstream of a heat-conducting perforated plate. *Combust. Flame* **2012**, *159*, 1055–1069. [CrossRef]
3. Qin, F.; Shah, A.; Huang, Z.W.; Peng, L.N.; Tunestal, P.; Bai, X.S. Detailed numerical simulation of transient mixing and combustion of premixed methane/air mixtures in a pre-chamber/main-chamber system relevant to internal combustion engines. *Combust. Flame* **2018**, *188*, 357–366. [CrossRef]
4. Yasuda, H.; Asato, K.; Miyasaka, T.; Miyashita, T.; Sakakibara, D.; Kurachi, S.; Hagi, S.; Umeda, Y. Characteristics of Combustion of Rich-Lean Burner Controlled Boundary Region between Rich and Lean Flames. *Trans. Jpn. Soc. Mech. Eng. B* **2006**, *73*, 3143–3150.
5. Aoki, S.; Yamazaki, H. Combustion Mechanism of Rich-Lean Flame Burner Controlled Boundary Zone. In Proceedings of the ASME/JSME Thermal Engineering Heat Transfer Summer Conference collocated with the ASME 2007 InterPACK Conference, Vancouver, BC, Canada, 8–12 July 2007.
6. Vance, F.H.; Shoshin, Y.; Goey, L.; Oijen, J. An investigation into flashback and blow-off for premixed flames stabilized without a recirculation vortex. *Combust. Flame* **2021**, *235*, 111690. [CrossRef]
7. Yu, S.; Bai, X.S.; Zhou, B.; Wang, Z.; Aldén, M. Numerical Studies of the Pilot Flame Effect on a Piloted Jet Flame. *Combust. Sci. Technol.* **2019**, *194*, 351–364. [CrossRef]
8. Zhou, B.; Brackmann, C.; Li, Z.; Aldén, M.; Bai, X.S. Simultaneous multi-species and temperature visualization of premixed flames in the distributed reaction zone regime. *Proc. Combust. Inst.* **2015**, *35*, 1409–1416. [CrossRef]
9. Pires, J.M.; Fernandes, E.C. Combined effect of equivalence ratio and velocity gradients on flame stability and emission formation. *Fuel* **2018**, *222*, 800–809. [CrossRef]
10. Li, Z.S.; Li, B.; Sun, Z.W.; Bai, X.S.; Aldén, M. Turbulence and combustion interaction: High resolution local flame front structure visualization using simultaneous single-shot PLIF imaging of CH, OH, and CH_2O in a piloted premixed jet flame. *Combust. Flame* **2010**, *157*, 1087–1096. [CrossRef]
11. Zhou, B.; Costa, M.; Li, Z.; Aldén, M.; Bai, X.S. Characterization of the reaction zone structures in a laboratory combustor using optical diagnostics: From flame to flameless combustion. *Proc. Combust. Inst.* **2016**, *36*, 4305–4312. [CrossRef]
12. Zhou, B.; Brackmann, C.; Wang, Z.; Li, Z.; Richter, M.; Alden, M.; Bai, X.S. Thin reaction zone and distributed reaction zone regimes in turbulent premixed methane/air flames: Scalar distributions and correlations. *Combust. Flame* **2017**, *175*, 220–236. [CrossRef]
13. Lúcio, T.; Fernandes, E. Rich–Lean Flame Interaction in a Lamella-Type Burner. *Combust. Sci. Technol.* **2015**, *188*, 416–438. [CrossRef]
14. Guo, H.; Liu, F.; Smallwood, G.J. A numerical study of laminar methane/air triple flames in two-dimensional mixing layers. *Int. J. Therm. Sci.* **2006**, *45*, 586–594. [CrossRef]
15. Azzoni, R.; Ratti, S.; Puri, I.K.; Aggarwal, S.K. The structure of triple flames stabilized on slot burner. *Combust. Flame* **1999**, *119*, 23–40. [CrossRef]
16. Barlow, R.S.; Dunn, M.J.; Sweeney, M.S.; Hochgreb, S. Effects of preferential transport in turbulent bluff-body-stabilized lean premixed CH_4/air flames. *Combust. Flame* **2012**, *159*, 2563–2575. [CrossRef]
17. Wan, J.; Zhao, H.; Akkerman, V.Y. Anchoring mechanisms of a holder-stabilized premixed flame in a preheated mesoscale combustor. *Phys. Fluids* **2020**, *32*, 97103. [CrossRef]
18. Vance, F.H.; Shoshin, Y.; de Goey, L.P.H.; van Oijen, J.A. Quantifying the impact of heat loss, stretch and preferential diffusion effects to the anchoring of bluff body stabilized premixed flames. *Combust. Flame* **2022**, *237*, 111729. [CrossRef]
19. Choi, C.W.; Puri, I.K. Flame stretch effects on partially premixed flames. *Combust. Flame* **2000**, *123*, 119–139. [CrossRef]

20. Choi, C.W.; Puri, I.K. Contribution of curvature to flame-stretch effects on premixed flames. *Combust. Flame* **2001**, *126*, 1640–1654. [CrossRef]
21. Michaels, D.; Ghoniem, A.F. Impact of the bluff-body material on the flame leading edge structure and flame–flow interaction of premixed CH_4/air flames. *Combust. Flame* **2016**, *172*, 62–78. [CrossRef]
22. Bennett, B.A.V.; Fielding, J.; Mauro, R.J.; Long, M.B.; Smooke, M.D. A comparison of the structures of lean and rich axisymmetric laminar Bunsen flames: Application of local rectangular refinement solution-adaptive gridding. *Combust. Theory Model.* **1999**, *3*, 657–687. [CrossRef]
23. Smooke, M.D.; Puri, I.K.; Seshadri, K. A comparison between numerical calculations and experimental measurements of the structure of a counterflow diffusion flame burning diluted methane in diluted air. *Symp. (Int.) Combust.* **1988**, *21*, 1783–1792. [CrossRef]
24. Kedia, K.S. Development of a Multi-Scale Projection Method with Immersed Boundaries for Chemically Reactive Flows and Its Application to Examine Flame Stabilization and Blow-Off Mechanisms. Ph.D. Thesis, Massachusetts Institute of Technology, Cambridge, MA, USA, 2013.
25. Hu, S.; Gao, J.; Gong, C.; Zhou, Y.; Bai, X.S.; Li, Z.S.; Alden, M. Assessment of uncertainties of laminar flame speed of premixed flames as determined using a Bunsen burner at varying pressures. *Appl. Energy* **2017**, *227*, 149–158. [CrossRef]
26. Bechtold, J.K.; Matalon, M. The dependence of the Markstein length on stoichiometry. *Combust. Flame* **2001**, *127*, 1906–1913. [CrossRef]
27. Kedia, K.S.; Ghoniem, A.F. The blow-off mechanism of a bluff-body stabilized laminar premixed flame. *Combust. Flame* **2015**, *162*, 1304–1315. [CrossRef]
28. Chen, T.; Yu, S.; Liu, Y.C. Effects of pressure on propagation characteristics of methane-air edge flames within two-dimensional mixing layers: A numerical study. *Fuel* **2021**, *301*, 120857. [CrossRef]

aerospace

MDPI

Article

Numerical Study of High-g Combustion Characteristics in a Channel with Backward-Facing Steps

Zhen Gong [1] and Hao Tang [2,*]

[1] Aero-Engine Thermal Environment and Structure Key Laboratory, Ministry of Industry and Information Technology, Nanjing University of Aeronautics and Astronautics, Nanjing 210016, China; gongzhen@nuaa.edu.cn

[2] Jiangsu Province Key Laboratory of Aerospace Power System, Nanjing University of Aeronautics and Astronautics, Nanjing 210016, China

* Correspondence: hao.tang@nuaa.edu.cn

Abstract: High gravity (high-g) combustion can significantly increase flame propagation speed, thereby potentially shortening the axial length of aero-engines and increasing their thrust-to-weight ratio. In this study, we utilized the large eddy simulation model to investigate the combustion characteristics and flame morphology evolution of premixed propane–air flames in a channel with a backward-facing step. The study reveals that both the increase in centrifugal force and flow velocity can enhance pressure fluctuations during combustion and increase the turbulence intensity. The presence of centrifugal force promotes the occurrence of Rayleigh–Taylor instability (RTI) between hot and cold fluids. The combined effects of RTI and Kelvin–Helmholtz instability (KHI) enhance the disturbance between hot and cold fluids, shorten the fuel combustion time, and intensify the dissipation of large-scale vortices. The increase in fluid flow velocity can raise the flame front's hydrodynamic stretch rate, thereby enhancing the turbulence level during combustion to a certain extent and increasing the fuel consumption rate. When a strong centrifugal force is applied, the global flame propagation speed can be more than doubled. Within a certain range, the increase in high-g field strength can enhance the intensity of RTI and accelerate the transition of RTI to the nonlinear stage.

Keywords: high-g combustion; Rayleigh–Taylor instability; curved channel; centrifugal force

1. Introduction

Modern aero gas turbines aim to improve performance, enhance durability, and reduce emissions. Ultra-compact combustion technology adopts the rich burn–quick quench–lean burn (RQL) combustion organization method, which significantly shortens the fuel combustion length, thereby reducing the length of the combustion chamber and increasing the thrust-to-weight ratio of the gas turbine [1,2]. Figure 1 presents a comparative schematic of the ultra-compact combustor and the conventional combustor. As shown, compared to the conventional combustor, the ultra-compact combustor features a cavity near the engine's outer diameter, where fuel and air mix and burn in a circumferential rotation within the cavity. This design, to some extent, transforms traditional axial combustion into circumferential combustion, greatly reducing the axial length of the combustion chamber. The key to ultra-compact combustion technology is the creation of a high-g combustion environment within the combustor components, achieved by strong swirling or rotating the combustor to generate high centrifugal forces. The intensity of these centrifugal forces is typically represented by a dimensionless ratio relative to gravity, and this mode of combustion is also known as high-g combustion.

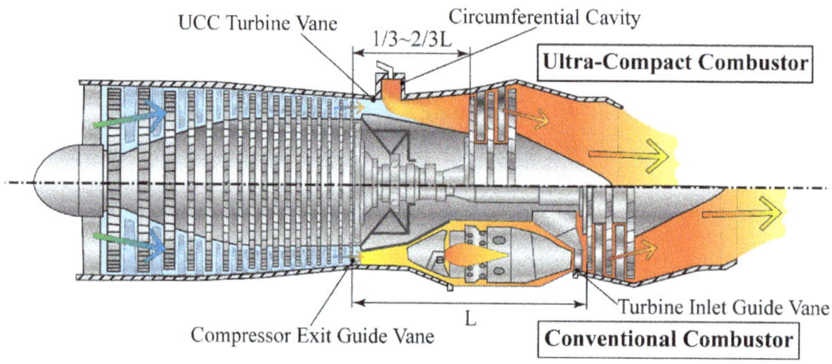

Figure 1. Schematic comparison of ultra-compact combustion chamber and traditional combustion chamber [3].

In 1971, Lewis [4] conducted combustion experiments with premixed propane air within a rotating tube, demonstrating that within a certain range of high-g combustion environments, the flame propagation speed during the combustion process increases with the strength of the high-g field. The experiments also revealed that the accelerated flame propagation phenomenon only occurs within the high-g range of 600 g to 3500 g. Within this range, the flame propagation speed Sf is approximately proportional to the square root of the high-g factor g_0, as expressed by the equation $S_f = C_1 g_0^{0.5}$. Subsequently, Lewis proposed the "bubble transport principle" to explain this phenomenon and noted that buoyancy forces within a specific high-g range significantly impact the flame propagation speed [5,6]. Later, Sakai [7] conducted combustion experiments on flame propagation in a rigid-body rotating flow field using an open-ended tube. The experiments indicated that as the rotational speed increased, the flame propagation speed at specific points in the tube exhibited an approximately linear increase. Additionally, it was found that the presence of centrifugal forces could expand the flammability limits of the fuel.

Katta [8] simplified Lewis's experiment into a two-dimensional model extending infinitely in the direction of flame propagation. By applying a force parallel to the flow direction in the momentum equation to simulate the body force generated by rotation, the phenomenon of buoyancy accelerating the flame propagation speed was observed when the body force increased to 500 g. This further validated the rationality of the "bubble transport principle". Briones, A.M. et al. [9] used a closed, X-axis symmetric two-dimensional model to perform numerical simulations of propane–air premixed gas in a rotating reference frame. They pointed out that the centrifugal force promotes the occurrence of Rayleigh–Taylor instability (RTI), increases the contact area between the flame front and unburned gas, accelerates the flame combustion rate, and enhances flame propagation speed. Liuyu et al. [10] considered the effects of Coriolis force and centrifugal force on flame propagation speed in a closed duct. They pointed out that in a closed rotating combustor, the Coriolis force also has a certain impact on flame propagation speed and cannot be ignored.

As high-speed fluids flow through a curved channel, a significant centrifugal force field can form within the channel. When fuel combusts in this environment, high-g combustion phenomena may occur. Based on this, Lapsa [11,12] conducted combustion experiments with high-speed fuel flow in a curved channel featuring a backward-facing step. The study revealed that when the step is located on the outer side, the centrifugal force can induce Rayleigh–Taylor Instability (RTI) between the burned and unburned gases, increasing their contact area, enhancing heat transfer, and thereby increasing the flame propagation rate. Building on Lapsa's experiments, Erdmann et al. [13] further investigated the combustion of propane–air premixed gas in a curved channel. Their research indicated that due to the presence of the centrifugal force field, high-density reactants near the channel's inner radius are driven outward by RTI, while low-density combustion products near the outer radius are driven inward, creating intense turbulent structures that greatly promote mixing in the combustion zone. Sykes et al. [14] optimized Lapsa's backward-facing step curvature model into a curved channel model with a splitter plate. They studied the development of RTI and pressure fluctuation characteristics within the curved channel. The research revealed that the model with the partition plate reduced the pressure variation amplitude by an order of magnitude compared to Lapsa's original model.

It is evident that current research on high-g combustion mechanisms is primarily divided into two directions: combustion in rotational systems based on Lewis's experiments and combustion of high-speed fluids in curved channels based on Lapsa's experiments. High-g combustion is mainly applied in the field of aero-engine turbojet engines, but utilizing rotational combustion poses significant challenges to the overall structure, fuel supply system, and ignition system of existing engines. Current ultra-compact combustors are mainly designed using specific combustion organization methods that generate high-g fields through high-speed fluid flow. Therefore, Lapsa's experiments offer better guidance for the design of ultra-compact combustors, which is the main reason for choosing Lapsa's experimental model for the study in this paper.

Although considerable research has been conducted based on Lapsa's experiments, there is still a lack of studies on the potential factors that enhance flame propagation speed in high-g combustion. Therefore, this paper, based on the experimental model of Lapsa, employs large eddy simulation (LES) to investigate the mechanism of high-g combustion. By analyzing flame morphology evolution, flame surface area changes, and the distribution of flame surface stretch rate, this study explores the mechanisms by which high-g combustion enhances flame propagation speed. The findings of this study can provide relevant references for the selection of high-g intensity and combustion organization methods in the design of ultra-compact combustors.

2. Computational Model and Method

2.1. Computational Model

As shown in Figure 2, the computational model used in this paper is based on the experimental model with a backward-facing step by Lapsa [11,12]. Model 1 is illustrated in Figure 2a. Due to the high computational resource requirements of LES simulations, and given the limited computational resources, we focused on the core combustion region downstream of the backward-facing step. Therefore, the model was further simplified to the core combustion region spanning -46° to 45°, totaling 91° (the shaded area in Figure 2). Model 2, shown in Figure 2b (only the core region is displayed), serves as a comparative study in a non-high-g environment against Model 1. The length and circumference at the curved channel center correspond, and various boundary conditions are identical to those of the curved channel. To minimize the influence of the front and rear walls on the flame morphology, all model front and rear walls were simplified to translational periodic boundaries. Additionally, the green-shaded area—Region A—in the corresponding positions of each model in Figure 2 is designated as the primary region for later analysis, and a monitoring point, Point P0, was set up during the solution process to analyze its pressure fluctuations later.

In the calculations, it was ensured that the mesh resolution for turbulent kinetic energy was above 85% in all regions, with $Y^+ < 1$. A boundary layer with a total thickness of 1 mm was set near the wall, comprising 20 layers, with the first layer height being 0.005 mm and a growth rate of 1.2, to better capture the significant variations in parameters close to the wall. After initial calculations, multiple modifications, and repeated computations, the final mesh used had a primary mesh size of 0.3 mm, with approximately 2.8 million mesh cells. The yellow area in Figure 2a shows half of the final mesh at the inlet of Model 1.

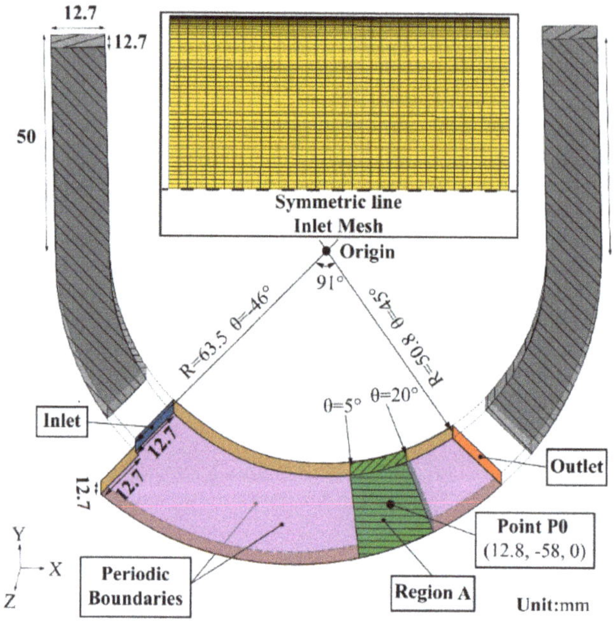

(a) Model 1

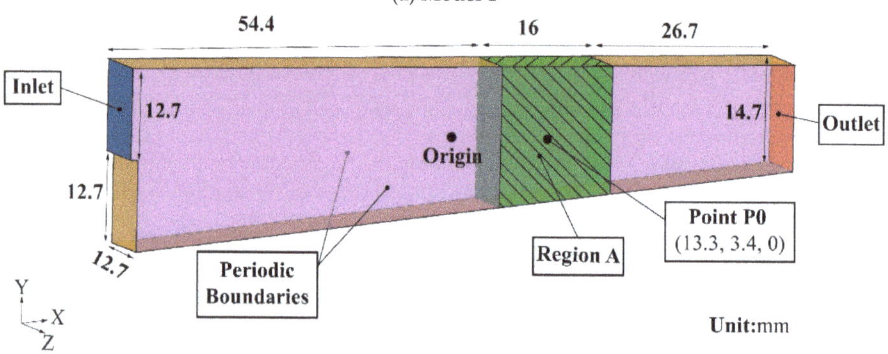

(b) Model 2

Figure 2. Schematic diagram of the computational model.

2.2. Computational Method

In this study, ANSYS FLUENT 2022 R2 software was used for the numerical simulation of the model. This software is widely applied in computational fluid dynamics (CFD) and is known for its stability and robustness. To better simulate the turbulent characteristics within the channel, the large eddy simulation (LES) model was chosen for the turbulence model, and the wall-adapting local eddy viscosity (WALE) model was selected for the subgrid–scale model. The species transport equation utilized the eddy dissipation concept (EDC) model, and the combustion mechanism employed a global two-step reaction mechanism for propane air.

A second-order implicit difference scheme was used to discretize the time term. The pressure–velocity coupling scheme employed the Coupled method. For spatial discretization, the PRESTO! scheme was used for pressure, while bounded central differencing schemes with low dissipation were selected for density, momentum, turbulent kinetic energy, species, and energy. The gradient spatial discretization used the least squares cell-based scheme. The time step was determined using the Courant–Friedrichs–Lewys condition (CFL) method, with multiple calculations monitoring the maximum Courant number (C_{max}) to ensure the global C_{max} remained below 1, resulting in a final time step of $\Delta t = 10^{-6}$ s.

To eliminate the influence of steady-state initial conditions on the LES calculations, an initial calculation was conducted for 1.5 to 2 times the fluid residence time to serve as the initial condition for steady-state calculations. Subsequently, the statistical function was activated for formal statistical calculations, with the statistical time also being 1.5 to 2 times the residence time. Since LES calculations require substantial computational resources, to ensure accuracy and save computation time, the initial and statistical times were set to 1.5 times the residence time at low speed (inlet velocity of 20 m/s) and 2 times at high speed (inlet velocity of 40 m/s).

2.3. Boundary Conditions

During the simulation process, the Reynolds-averaged Navier–Stokes (RANS) method was first used for unsteady solution calculations of the complete model. After obtaining the steady-state solution, the temperature, velocity, pressure distribution, and turbulence information at the corresponding interface of the core region inlet (the blue areas in Figure 2a,b) were extracted as the boundary conditions for the core region inlet. The inlet composition for each case was a stoichiometric propane–air premixed gas, and the other inlet boundary conditions for each case are shown in Table 1, where the prefix BFS represents the channel with a backward-facing step (BFS).

Table 1. Inlet boundary conditions for each case.

Case	Model	Velocity [m/s]	Temperature [K]	High-g Field [g]
BFS A	Model1	20	300	642
BFS B	Model1	40	300	2568
BFS C	Model2	40	300	0

2.4. Mesh Scale Selection and Independence Verification

The requirement for mesh resolution in LES simulations is primarily evaluated by the percentage of resolved turbulent kinetic energy k_{res} to the total turbulent kinetic energy k_t, with a general requirement of exceeding 80%. The total turbulent kinetic energy is the sum of the resolved turbulent kinetic energy k_{res} and the subgrid–scale turbulent kinetic energy k_{sgs}. Prior to determining the final mesh scale, multiple preliminary calculations were conducted. Initially, the RANS method was used to calculate the integral length scale l0, and the initial mesh Δ was set to one-fifth of the integral length scale. The calculation formula for the initial mesh scale Δ is provided in Equation (1):

$$\Delta = \frac{l_0}{5} = \frac{\sqrt{k}}{5C_\mu \omega} \quad (1)$$

where k is the turbulent kinetic energy, C_μ is the Smagorinsky constant (0.325), and ω is the specific dissipation rate, with units of S^{-1}.

After determining the initial mesh scale, the mesh was adjusted multiple times based on the distribution of resolved turbulent kinetic energy and Y^+, ensuring that all Y^+ values were less than 1. A boundary region with a total thickness of 1 mm was set near the wall, containing 20 boundary layers. The height of the first layer was 0.005 mm, with a growth rate of 1.2, to better capture the significant variations in parameters close to the wall. After preliminary calculations, multiple modifications, and recalculations, the final mesh used had a main mesh size of 0.3 mm, with approximately 2.8 million mesh nodes. The yellow region in Figure 2a shows half of the final mesh at the inlet for model 1. Figure 3 provides the proportion of resolved turbulent kinetic energy for BFS B (left) and BFS C (right) with the final mesh. As shown in the figure, the adopted mesh resolves more than 96% of the turbulent kinetic energy in most regions, with only a few small areas, such as near the inlet or close to the upper wall, exhibiting relatively lower resolution. However, even in these areas, the resolution exceeds 85%, indicating that the chosen mesh scale adequately meets the mesh independence requirements for LES simulations.

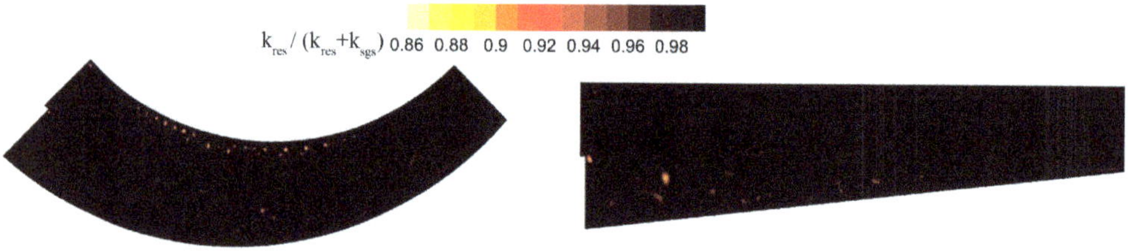

Figure 3. Proportion of Resolved Turbulent Kinetic Energy for BFS B (**left**) and BFS C (**right**) with the Final Mesh.

3. Results and Discussion

To better study the combustion characteristics of fuel in a combustion channel with a backward-facing step, the following sections will discuss various aspects of the combustion process, including pressure fluctuations, temperature field distribution, flame surface morphology and area, flame stretch rate, and flame propagation speed.

3.1. Analysis of Pressure Fluctuation Characteristics in the Channel

Figure 4 (top) shows the variation curves of pressure fluctuations at monitoring point P0 over time for each case. Comparing BFS A and BFS B, it can be seen that BFS B exhibits more intense pressure fluctuations with a higher overall amplitude. The standard deviation (SD) of BFS B is approximately four times that of BFS A, indicating that an increase in flow rate enhances pressure fluctuations during the flow process. The possible reasons for this phenomenon are as follows: On the one hand, the increase in flow rate leads to a higher Reynolds number, directly enhancing the turbulence intensity in the flow process. On the other hand, the increase in flow rate in the curved channel results in greater centrifugal force during the flow process, promoting the mixing of fluids with different densities, which also increases turbulence intensity and thereby enhances pressure fluctuations.

Comparing BFS B and BFS C, it is observed that their overall pressure fluctuation trends are quite similar, with maximum pressure fluctuations around 1000 Pa. However, BFS B shows significantly stronger local pressure fluctuation amplitudes and a higher standard deviation, nearly 1.5 times that of BFS C. This indicates that, under the same conditions, pressure fluctuations in a curved channel with a backward-facing step are stronger than in a straight channel. This further demonstrates that the presence of centrifugal force can enhance the turbulence intensity within the channel, leading to pressure fluctuations during the flow process.

Figure 4 (bottom) shows the power spectrums of pressure fluctuations for each case, with the dashed line representing the classical turbulent dissipation curve with a slope of $-5/3$. From the power spectrum, it is evident that all cases conform to the Kolmogorov universal equilibrium theory in the inertial range, further validating the rationality of the computational models. All cases exhibit typical turbulent decay; however, it can be seen that BFS B and BFS C have stronger power densities in the low-frequency range, indicating that their energy density is more concentrated in the low-frequency region.

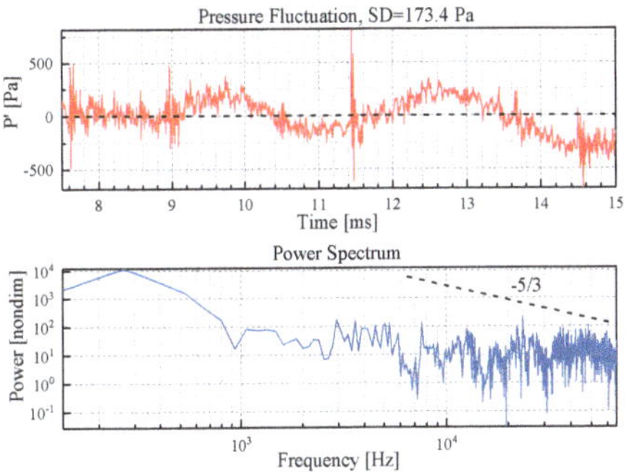

(**a**) BFS A

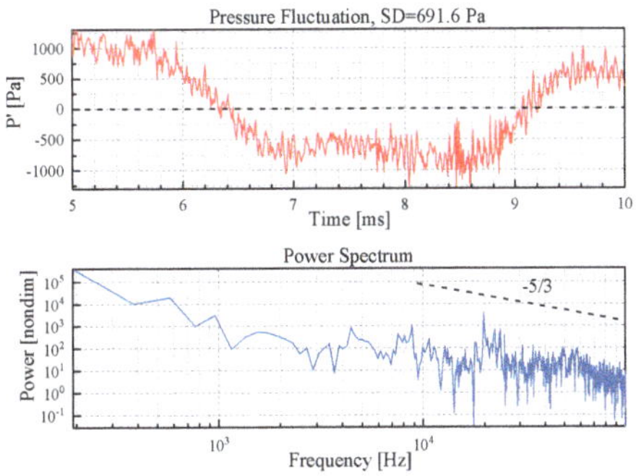

(**b**) BFS B

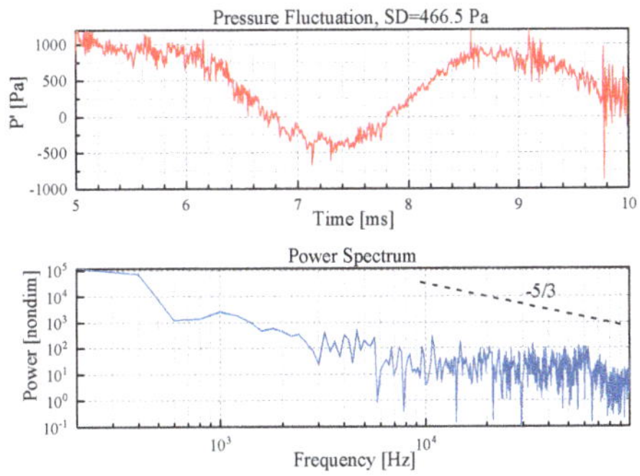

(**c**) BFS C

Figure 4. Pressure fluctuation curves (**top**) and power spectrum curves (**bottom**) for each case.

3.2. Analysis of High-g Field Distribution in the Channel

High-speed fluid flow in a curved channel generates centrifugal force, creating a high-g combustion environment that affects the combustion process and flame propagation speed. Figure 5 shows the instantaneous high-g field contour during the combustion process for BFS A and BFS B. The high-g field strength g_0 is expressed as a dimensionless factor, representing the ratio of the resulting centrifugal acceleration to gravitational acceleration. Its value can be calculated using Equation (2).

$$g_0 = \frac{{V_{\tan}}^2}{gr} \tag{2}$$

In the equation, V_{tan} is the tangential velocity of the fluid, g is the gravitational acceleration, and r is the radius at the corresponding position.

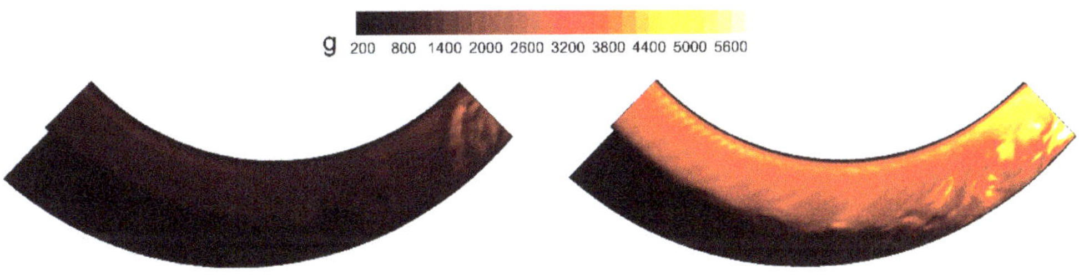

Figure 5. High-g field contours for BFS A (**left**) and BFS B (**right**).

From the figure, it can be observed that the high-g field is stronger near the inner radius for both cases because the high-g field is inversely proportional to the channel radius. Additionally, strong high-g fields are formed at certain local points near the exit. This phenomenon is likely due to the intense mixing between the burnt and unburnt gases at these locations, which disrupts the shear layer formed between the hot and cold fluids, enhancing the turbulence level and consequently increasing the tangential velocity of the local fluid. This results in a stronger high-g field at these local points.

Both cases exhibit large areas of low-intensity high-g fields (around 200 g) near the backward-facing step region. The gas velocity in these regions is relatively low, and the high-temperature combustion products mix with the low-temperature unburnt premixed fuel. This region is referred to as the "recirculation zone" in some literature [15]. It can be observed that the low-g field region in BFS B is smaller than that in BFS A, mainly due to the difference in fluid velocity. When the flow velocity increases from 20 m/s to 40 m/s, the high-g field at the leading edge of the backward-facing step increases from around 800 g to 2600 g, indicating that increasing the inlet velocity enhances the high-g intensity within the flow field.

The high-g field at the flame front significantly impacts the fuel consumption rate, which in turn alters the combustion characteristics of the turbulent flame. Further analysis will be conducted by examining other parameters during the combustion process.

3.3. *Analysis of Flame Combustion Characteristics in the Channel*

To study the flame combustion characteristics during the combustion process for each case, time-averaged temperature contour plots were obtained for the Z = 0 mm section (Z0) and X = 0 mm section (X0), as shown in Figure 6. The blue dashed lines in the figure indicate the position of the X0 section.

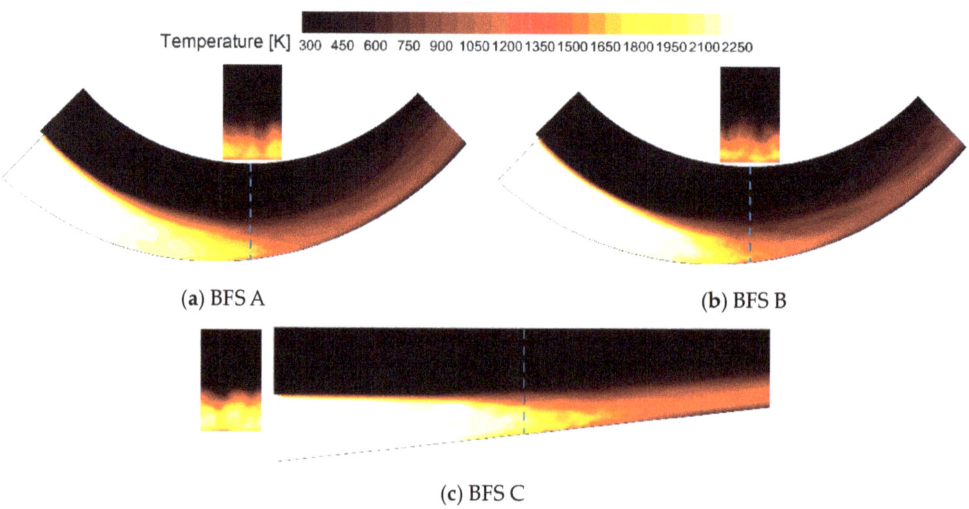

Figure 6. Time-averaged temperature contours for each case.

Figure 6 shows that all cases exhibit a large high-temperature area near the backward-facing step, corresponding to the recirculation zone mentioned earlier. This further demonstrates that the recirculation zone in the backward-facing step combustion channel maintains a high temperature. Comparing the downstream extension distance of the high-temperature regions for each case, BFS C has the longest extension distance, followed by BFS A, and BFS B the shortest. The downstream extension distance can reflect the intensity of the combustion process to some extent. A shorter extension distance indicates that the fuel completes combustion within a short distance, leading to a more intense initial combustion process. Conversely, a longer extension distance implies that more time is required for combustion at the same speed, resulting in a relatively mild combustion process.

This observation suggests that combustion in a high-g environment can shorten the fuel combustion distance, and within a certain range, the combustion distance is inversely related to the intensity of the high-g field.

Comparing the shear layers between the hot and cold fluids in each case reveals that in the non-high-g environment (BFS C), the shear layer changes more gradually, with a more apparent transition between the hot and cold fluids and a thinner shear layer. In contrast, in high-g environments (BFS A and BFS B), the shear layer between the hot and cold fluids changes more dramatically and is relatively thicker. In cases of higher high-g fields, the changes in the shear layer are even more intense. The thickness of the shear layer can reflect the intensity of the mixing process between the fluids to some extent. Under the same conditions, a thicker shear layer indicates stronger mixing, resulting in a shorter mixing time. This suggests that the presence of a high-g field can enhance the mixing degree between the hot and cold fluids and shorten the mixing time, which might be one of the factors contributing to the increased flame propagation speed in high-g fields.

Figure 7 presents the instantaneous temperature contours for each case during the combustion process. The combustion time increases sequentially from top to bottom, with the combustion time indicated in the top left corner of each image. The 0 ms time frame (the topmost image) for each case represents the steady-state results calculated using the RANS turbulence model, while the bottommost image represents the final time frame of the calculation for each case. Following the method of Sykes [16], the instantaneous flame front is characterized by the T = 1275 K temperature isosurface, depicted with green lines in the figures.

From the figure, it can be observed that at the initial moment (0 ms), the distribution of high- and low-temperature regions is quite distinct. The high-temperature regions are predominantly located downstream of the backward-facing step. Compared to BFS A, the high-temperature region in BFS B is smaller. This phenomenon can be explained as follows: On the one hand, the increased flow velocity can quickly blow the flame front downstream; while on the other hand, the increased flow velocity in the curved channel accompanies an increase in centrifugal force, resulting in a stronger high-g field. This intensifies the strain rate of the flame front, leading to local flame quenching and thereby reducing the high-temperature region. At the initial moment, although the RANS turbulence model can roughly calculate the temperature distribution within the channel, it averages the global turbulent kinetic energy, lacking the details of local turbulent kinetic energy, and thus cannot fully capture the combustion process accurately.

By comparing the temperature contours at different times, it can be observed that the flame fronts in each case generally appear on the right and upper parts of the recirculation zone, consistent with the steady flame stabilization mechanism behind the backward-facing step studied by Rasmussen [17]. Additionally, in the straight channel, the high-temperature regions are mainly located on the right side of the backward-facing step, whereas in the curved channel, the flame gradually expands toward the inner wall downstream, significantly enlarging the combustion region.

Comparing BFS B and BFS C, it is noted that in the straight channel, there is a segment near the backward-facing step where the boundary between high and low temperatures remains undisturbed. Only as it approaches the middle section does minor perturbation appear. This can be explained by the Kelvin–Helmholtz instability (KHI) affecting the disturbance between hot and cold fluids primarily due to differing flow velocities. Near the backward-facing step, the short-wavelength perturbations formed by KHI are counteracted by the surface tension of the fluid, hence minimal changes occur at the boundary. As KHI intensifies, the induced disturbances gradually increase, breaking the surface tension and forming small amplitude disturbances at the boundary. However, the perturbation caused by KHI over short distances is limited. In contrast, BFS B shows disturbances at the high- and low-temperature boundary over a shorter distance from the backward-facing step, with much stronger amplitudes compared to BFS C. This phenomenon can be explained as follows: During the flow, the density of the low-temperature free stream gas is approximately 1.19 m^3/kg, while the high-temperature recirculation zone density is about 0.15 m^3/kg, an eightfold difference. The rapid flow in the curved channel generates high body forces due to centripetal acceleration, creating a centripetal buoyancy effect. Under the influence of this density difference and force, the hot products in the recirculation zone are radially forced inward into the reactive flow, while the cooler reactants are radially forced outward into the recirculation zone, resulting in an intense mixing region between reactants and combustion products, promoting the formation and development of RTI. Therefore, the fluid in BFS B is affected by both RTI and KHI, forming strong disturbances at the high- and low-temperature interface.

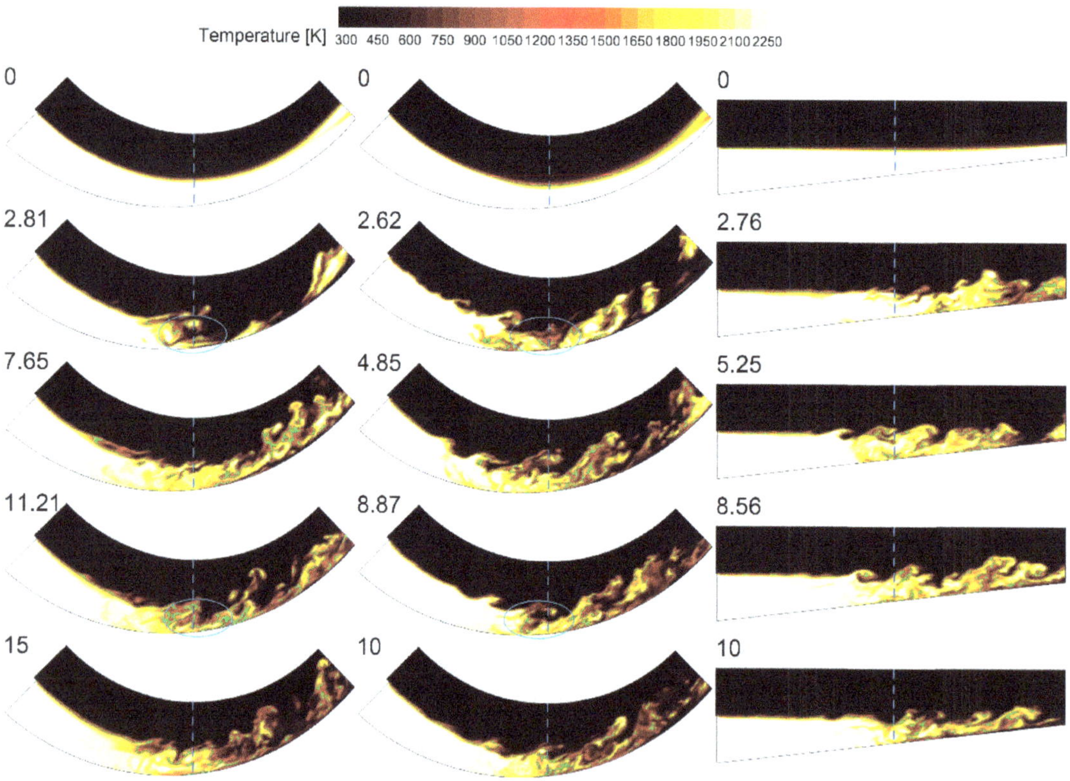

(**a**) BFS A (**b**) BFS B (**c**) BFSC

Figure 7. Instantaneous temperature contours at different time instances for each case (Time unit: ms).

Although the high- and low-temperature interface in the straight channel gradually exhibits curling and wrinkling downstream, the extent is far less than in BFS A and BFS B. Comparing BFS A and BFS B, it is evident that at higher velocities, the disturbances at the high- and low-temperature interface are stronger, and the distance over which disturbances transition to wrinkled flames is shorter. This phenomenon likely occurs because a stronger high-g field is formed at higher velocities, which more rapidly promotes RTI to reach a nonlinear stage, resulting in stronger disturbance variations.

Additionally, comparing the instantaneous temperature contours of each case in Figure 7, it can be observed that large-scale vortices periodically appear in the curved channel, as indicated by the blue ellipses, whereas such vortices are not observed in the straight channel. This phenomenon can be explained as follows: In the curved channel, due to the influence of RTI, the perturbations between the high- and low-temperature interfaces gradually increase, eventually forming large-scale vortices. However, as these large-scale vortices flow downstream, according to the energy cascade hypothesis in turbulence theory, the larger vortices gradually break down into smaller vortices, which then further decompose into even smaller vortices, ultimately dissipating as internal energy. Consequently, large-scale vortices periodically appear in the curved channel. Additionally, the combustion process within the channel generates a certain degree of acoustic instability, which periodically influences the vortices formed within the channel.

Based on the above discussion, it can be concluded that the fluid in the curved channel promotes the occurrence of RTI between the hot and cold fluids. The combined effect of RTI and KHI enhances the perturbations between the hot and cold fluids, shortens the combustion time of the fuel, and increases the dissipation of large-scale vortices. These effects, in turn, have a significant impact on the flame propagation speed. In the design of actual ultra-compact combustors, it is crucial to adopt appropriate combustion organization methods to reduce the generation of periodic large vortices, thereby ensuring stability during the combustion process.

3.4. Analysis of Flame Stretch Rate and Instantaneous Flame Morphology in the Channel

In nonuniform flow processes, the flame front is influenced by both strain and curvature effects. The combined impact of these two factors leads to changes in the flame front. The flame stretch rate κ can be used to quantitatively reflect these changes. The total flame stretch rate κ_t can be represented by the sum of the curvature-induced stretch rate (κ_c) and the hydrodynamic stretch rate (κ_h) [18,19]. The κ_c primarily reflects the stretching caused by the local curvature effects on the flame front, while the κ_h mainly reflects the stretching due to strain caused by local flow nonuniformity.

In this section, the instantaneous isotherms at T = 1275 K are used to represent the flame surface. Figure 8 shows the instantaneous flame surfaces for two translation periods for each case, with the flame surfaces colored by the curvature-induced stretch rate and the hydrodynamic stretch rate, respectively.

From the figure, it can be seen that for all three cases, the flame front in the upstream region near the rear inlet is primarily unaffected by curvature-induced stretch. The total stretch rate is mainly controlled by hydrodynamic stretch. This stretch rate is primarily due to the flow imbalance caused by the velocity and density differences between the free stream and the recirculation zone. Although this phase can induce some undulations on the flame surface, the influence on the flame surface area is limited due to the surface tension of the fluid itself. As the flow progresses downstream, the free stream gradually approaches the flame stabilization region, where intense chemical reactions begin to occur. Under the influence of chemical reaction instability, KHI, and RTI, the curvature-induced stretch rate of the flame front increases rapidly, causing the flame front to fold and wrinkle. This indicates that in the backward-facing step channel studied in this paper, the primary cause of flame front wrinkling and folding is curvature-induced stretch.

Comparing BFS A and BFS B, it can be observed that in the upstream region, BFS B exhibits a higher hydrodynamic stretch rate on the flame front. Additionally, the location where the flame front starts to show ripples, folds, and wrinkles is closer to the inlet in BFS B. This phenomenon can be explained as follows: The free stream in BFS B has a higher velocity compared to BFS A, resulting in a greater velocity difference between the free stream and the recirculation zone, which leads to a larger flow imbalance between the two regions. Consequently, BFS B initially exhibits a higher hydrodynamic stretch rate. Simultaneously, a stronger high-g field is formed in the BFS B flow, causing the distance for RTI to develop to the nonlinear stage between the free stream and the recirculation zone to be shorter. Therefore, the flame front in BFS B starts to fold and wrinkle earlier. Overall, comparing the flame front morphology in the mid-downstream regions of BFS A and BFS B, it is evident that BFS B has more wrinkles and a higher degree of flame front fragmentation. However, in the flame stabilization region, the wrinkles growing toward the inner wall in BFS A are more pronounced. The reasons for these observed flame front characteristics can be explained as follows: The higher velocity of the free stream in BFS B can partially blow away the wrinkled flame front growing toward the inner wall, resulting in less pronounced wrinkles toward the inner wall. Additionally, BFS B is subjected to a stronger high-g field, leading to more intense RTI at the shear interface between the free stream and the recirculation zone, causing large-scale vortices to transition into small-scale vortices more rapidly. Furthermore, the stronger high-g field can cause local flame front fragmentation and quenching to some extent. Therefore, BFS B exhibits a higher degree of flame front fragmentation.

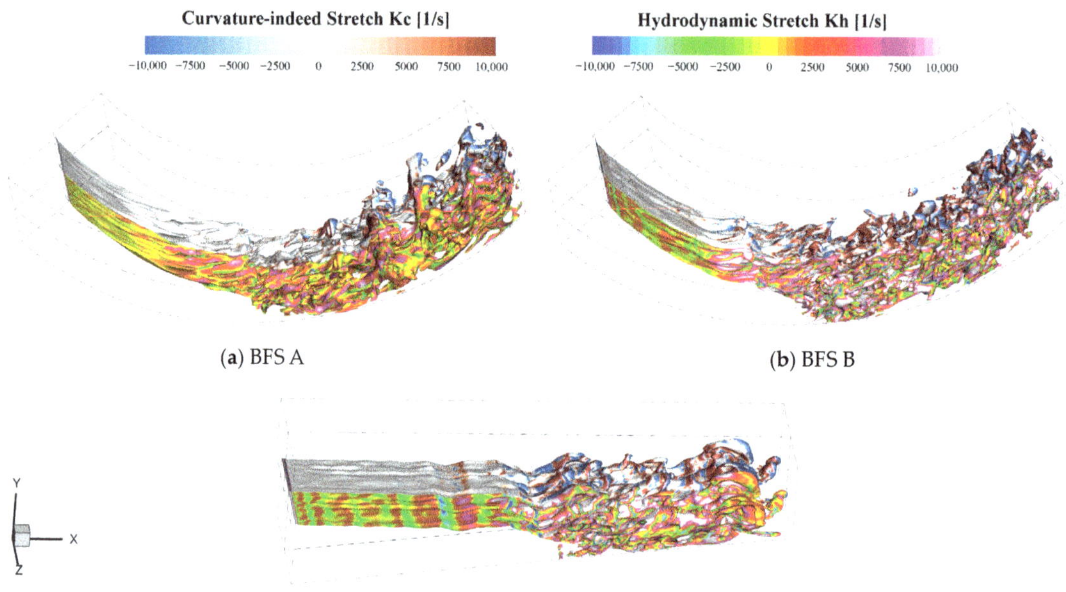

(**a**) BFS A

(**b**) BFS B

(**c**) BFS C

Figure 8. Instantaneous flame front and stretch rate distribution contours for each case.

Comparing BFS B with BFS C, it can be observed that the hydrodynamic stretch rates near the inlet upstream are numerically similar. In contrast to BFS A at the same location, it further indicates that the initial hydrodynamic stretch rate is mainly due to the velocity difference between the free stream and the recirculation zone fluids. Additionally, the greater the velocity difference, the higher the hydrodynamic stretch rate to some extent. Compared to BFS C, BFS B exhibits flame front folding and wrinkling earlier, and the magnitude of these folds and wrinkles is significantly greater.

3.5. Analysis of Flame Area Evolution and Flame Propagation Speed in the Channel

To quantitatively study the flame evolution characteristics during the combustion process of each case, Figure 9 presents the curve of flame surface area over time. As shown in the figure, the flame surface area in the curved channel under both flow velocity conditions is larger than that in the straight channel. This indicates that the presence of a high-g field can increase the flame surface area to some extent. Comparing BFS A and BFS B, although their average flame areas are numerically close, BFS B exhibits larger maximum values and smaller minimum values for the flame area. The overall standard deviation (SD) of BFS B is 2.4 times that of BFS A, indicating that BFS B has stronger fluctuations in flame area. The possible reasons for this phenomenon are as follows: In BFS B, the free stream has a higher flow velocity and can generate a stronger high-g field. Under the combined effect of high flow velocity and strong high-g field, stronger turbulence is induced in the middle and downstream regions. Additionally, due to the stronger RTI, the flame surface breaks up more quickly and even extinguishes locally in some areas. Consequently, BFS B shows greater fluctuations in flame surface area during combustion.

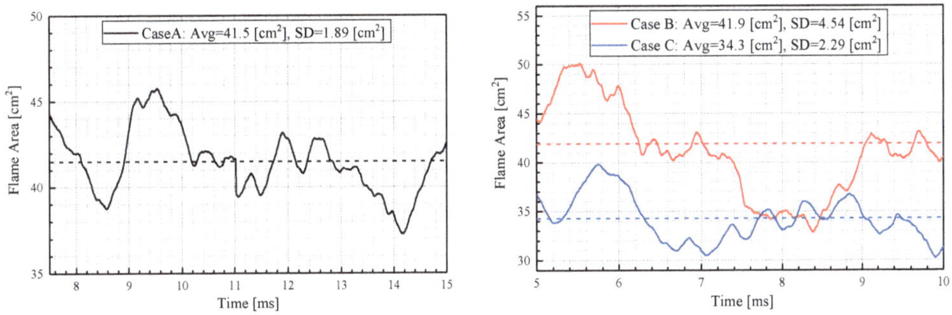

(**a**) 20 m/s (**b**) 40 m/s

Figure 9. The variation curve of flame surface area with time for each calculation.

To further study the combustion characteristics of each case, this paper uses the method of global consumption speed [20] to calculate the flame propagation speed during the combustion process. Equation (3) provides the calculation formula. To eliminate the influence of the initial stage on the flame propagation speed, Region A (as shown by the shaded area in Figure 2) was selected as the calculation region for the flame propagation speed. This region is located in the stabilized flame zone behind the step for all three cases and can accurately reflect the flame propagation speed during the rapid combustion stage.

$$S_f = \frac{\dot{\omega}_R}{\rho_R A_{\overline{C}=0.5}} = \frac{\iiint \dot{\omega}_R dV}{\rho_R A_{flame}} \tag{3}$$

In the above equation, $\dot{\omega}_R$ represents the fuel consumption rate; ρ_R represents the density of the reactants; $A_{\overline{C}=0.5}$ represents the area of the isosurface where the reaction progress variable is 0.5; and A_{flame} represents the flame surface area during the combustion process.

The flame propagation speed for BFS A, as calculated using the aforementioned definitions, is 0.47 m/s, whereas BFS B achieves a speed of 0.97 m/s, approximately double that of BFS A. This indicates that increasing the inlet flow velocity in the curved channel used in this study can significantly enhance the global propagation speed. For BFS C, the calculated flame propagation speed is 0.53 m/s, slightly higher than BFS A but significantly lower than BFS B, suggesting that at the same inlet velocity, the curved channel can significantly improve the global flame propagation speed during combustion.

The observed results for the flame propagation speed can be explained as follows: In the curved channels (BFS A and BFS B), BFS B's higher flow velocity forms a stronger KHI with the recirculation zone. The turbulence level in the stabilization zone is more chaotic, and the RTI developed during BFS B's flow is stronger. The combined effect of strong turbulence and strong RTI enhances the mixing between the low-temperature fuel and high-temperature recirculation zone gases. Additionally, it increases the flame front's twisting and stretching (as seen in Figure 8), thereby increasing the fuel consumption rate and enhancing the global flame propagation speed. At the same flow velocity, BFS B and BFS C have similar hydrodynamic stretch rates and KHI intensity, but BFS B has the presence of RTI in the channel and a higher curvature-induced stretch rate of the flame front (as seen in Figure 8), which accelerates the fuel consumption rate, resulting in a faster global flame propagation speed.

Additionally, the combustion efficiency of the three cases during the combustion process was calculated based on the relevant standards of the Society of Automotive Engineers (SAE) [21]. The average combustion efficiencies for BFS A, BFS B, and BFS C were 25.32%, 22.18%, and 19.58%, respectively. BFS A achieved the highest combustion efficiency primarily due to its lower inlet velocity, which resulted in a longer residence time for the fuel within the channel, leading to relatively higher combustion efficiency. Under the same inlet velocity, BFS B, which experienced centrifugal force, exhibited higher combustion efficiency compared to the BFS C case. This indicates that high-g combustion also has the potential to improve combustion efficiency to a certain extent.

A comprehensive comparison of the flame propagation speeds of the three cases shows that within a certain range, increasing the fluid velocity and enhancing the strength of the high-g field can both increase the fuel consumption rate during combustion, thereby improving the flame propagation speed. Based on the above research results, the following conclusions can be drawn: In channels with a backward-facing step, increasing the inlet velocity can increase the hydrodynamic stretch rate of the flame front, and to a certain extent, enhance the turbulence level during combustion, thereby increasing the fuel consumption rate. In curved channels, increasing the inlet velocity is accompanied by an increase in centrifugal force, which can enhance the strength of RTI, accelerate the transition of RTI to the nonlinear stage, increase the curvature-induced stretch rate of the flame front, promote the mixing of the free stream and the recirculation zone, thereby increasing the fuel consumption rate and flame propagation speed. However, the presence of centrifugal force also increases the flame front area. However, it is worth noting that the presence of centrifugal force also increases the flame front area. When the relative increase in overall combustion consumption rate is less than the relative increase in the flame front area, it can potentially reduce the overall flame propagation speed.

4. Conclusions

In this study, a LES model was used to perform numerical simulations of propane -air premixed fuel in channels with backward-facing steps at different flow velocities. The combustion characteristics in high-g and normal gravity environments were investigated in curved and straight channels. The influence of flame expansion characteristics and flame morphology within the channels was qualitatively analyzed, and the flame propagation speed and flame front area were quantitatively compared. The following conclusions were obtained:

(1) The increase in centrifugal force and flow velocity both enhance pressure fluctuations during combustion and increase the turbulence intensity. The presence of centrifugal force promotes RTI between hot and cold fluids. The combination of RTI and KHI enhances the disturbance between hot and cold fluids, shortens the combustion time of the fuel, and enhances the dissipation of large-scale vortices.

(2) Increasing the flow velocity of the fluid can increase the hydrodynamic stretch rate of the flame front, enhance the turbulence intensity during the combustion process to a certain extent, and increase the fuel consumption rate. Within a certain range, increasing the intensity of the high-g field can enhance the strength of RTI, accelerate the transition of RTI to the nonlinear stage, increase the curvature stretch rate of the flame front, enhance the mixing of the free stream and the recirculation zone, increase the fuel consumption rate, and increase the flame propagation speed. When a strong centrifugal force is applied, the global flame propagation speed can be more than doubled.

(3) In the curved channel, the increase in flow velocity accompanied by the increase in centrifugal force triggers stronger RTI, resulting in increased hydrodynamic and curvature stretch rates of the flame front. This leads to enhanced flame front curvature and wrinkling, improving the mixing of unburned premixed gas with high-temperature combustion products, and strengthening heat and mass transfer. Under these conditions, the overall fuel consumption rate is increased, potentially leading to increased

combustion efficiency and power density, thus improving the overall flame propagation speed. However, the presence of centrifugal force also increases the flame front area.

(4) It is worth noting that the presence of centrifugal force also increases the flame front area. When the relative increase in overall combustion consumption rate is less than the relative increase in the flame front area, it can potentially reduce the overall flame propagation speed.

(5) In the design of ultra-compact combustors, structures that maintain a high-g environment should be incorporated to achieve faster combustion rates and higher flame propagation speeds. This will reduce the fuel residence time within the combustion chamber and consequently shorten the chamber's axial length.

Although this study qualitatively explains the mechanisms by which high-g environments accelerate flame propagation, it is limited by the computational resources required for large-eddy simulations of the combustion process. As a result, only three cases under two models were computed and analyzed, and a sensitivity analysis of boundary conditions was not performed. Therefore, the generalization of high-g characteristics may have certain limitations, which is a shortcoming of this study. Related research and analysis are ongoing and will be published in the future. Additionally, while high-g combustion has potential for reducing the axial length of combustion chambers and enhancing overall engine performance, our research has identified that the stability of the ignition process in a high-g environment is relatively poor. In the design of ultra-compact combustors, additional measures such as flame stabilization structures or multi-point ignition strategies may be required to ensure engine start-up and stable operation. We are also investigating the variation characteristics and potential impact mechanisms of fuel ignition in high-g environments, with results to be published in future studies.

Author Contributions: Conceptualization, Z.G.; methodology, Z.G.; software, Z.G.; validation, H.T.; formal analysis, Z.G.; investigation, Z.G.; resources, H.T.; data curation, H.T.; writing—original draft preparation, Z.G.; writing—review and editing, Z.G.; visualization, Z.G.; supervision, H.T.; project administration, H.T.; funding acquisition, H.T. All authors have read and agreed to the published version of the manuscript.

Funding: This research was supported by the National Natural Science Foundation of China (No. 51076064 and No. 91641131), and the Fundamental Research Funds for the Central Universities (No. 56XBA22001).

Data Availability Statement: Some data, models, or code that support the findings of this study are available from the corresponding author upon reasonable request.

Conflicts of Interest: The authors declare no conflicts of interest.

Nomenclature

BFS	Backward-Facing Step
CFD	Computational Fluid Dynamics
CFL	Courant–Friedrichs–Lewys Condition
C_{max}	Maximum Courant number
EDC	Eddy-Dissipation-Concept
high-g	high gravity
K_c	Curvature-induced Stretch
K_h	Hydrodynamic Stretch
KHI	Kelvin–Helmholtz Instability
LES	Large Eddy Simulation
RANS	Reynolds-averaged Navier–Stokes
RQL	Rich burn-Quick quench-Lean burn
RTI	Rayleigh–Taylor Instability
SAE	Society of Automotive Engineers

SD	Standard Deviation
WALE	Wall-Adapting Local Eddy-Viscosity
X0	X = 0 mm section
Z0	Z = 0 mm section

References

1. Straub, D.L.; Casleton, K.H.; Lewis, R.E.; Sidwell, T.G.; Maloney, D.J.; Richards, G.A. Assessment of rich-burn, quick-mix, lean-burn trapped vortex combustor for stationary gas turbines. *J. Eng. Gas Turbines Power* **2005**, *127*, 36–41. [CrossRef]
2. Peterson, C.O.; Sowa, W.A.; Samuelsen, G. *Performance of a Model Rich Burn-Quick Mix-Lean Burn Combustor at Elevated Temperature and Pressure*; NASA: Washington, DC, USA, 2002.
3. Bohan, B.T. Analysis of Flow Migration in an Ultra-Compact Combustor. Master's Thesis, Air Force Institute of Technology, Dayton, OH, USA, 2011.
4. Lewis, G.D. Combustion in a centrifugal-force field. *Symp. (Int.) Combust.* **1971**, *13*, 625–629. [CrossRef]
5. Lewis, G.D. Centrifugal-force effects on combustion. *Symp. (Int.) Combust.* **1973**, *14*, 413–419. [CrossRef]
6. Lewis, G.D.; Shadowen, J.H.; Thayer, E.B. Swirling Flow Combustion. *J. Energy* **1977**, *1*, 201–205. [CrossRef]
7. Sakai, Y.; Ishiuka, S. The phenomena of flame propagation in a rotating tube. *Symp. (Int.) Combust.* **1996**, *26*, 847–853. [CrossRef]
8. Katta, V.; Blunck, D.; Roquemore, M. Effect of Centrifugal Forces on Flame Stability in an Ultra-Compact Combustor. In Proceedings of the 51st AIAA Aerospace Sciences Meeting Including the New Horizons Forum and Aerospace Exposition, Grapevine, TX, USA, 7–10 January 2013.
9. Briones, A.M.; Burrus, D.L.; Erdmann, T.J.; Shouse, D.T. Effect of Centrifugal Force on the Performance of High-G Ultra Compact Combustor. In Proceedings of the Asme Turbo Expo: Turbine Technical Conference & Exposition, Montréal, QC, Canada, 15–19 June 2015.
10. Liu, Y.; Wang, Z.; Tang, H. Numerical Investigation of Turbulent Premixed Combustion in a High Acceleration Field. *J. Therm. Sci. Eng. Appl.* **2020**, *12*, 041010. [CrossRef]
11. Lapsa, A.; Dahm, W. Experimental study on the effects of large centrifugal forces on step-stabilized flames. In Proceedings of the 5th US Combustion Meeting, San Diego, CA, USA, 25–28 March 2007; pp. 3388–3399.
12. Lapsa, A.P.; Dahm, W.J.A. Hyperacceleration effects on turbulent combustion in premixed step-stabilized flames. *Proc. Combust. Inst.* **2009**, *32*, 1731–1738. [CrossRef]
13. Erdmann, T.J.; Caswell, A.W.; Gutmark, E. Experimental Study of the Impact of High Centrifugal Body Forces on Constant Pressure, Propane-Air Flames. In Proceedings of the AIAA Scitech 2019 Forum, San Diego, CA, USA, 7–11 January 2019.
14. Sykes, J.; Gallagher, T.; Rankin, B.A. Numerical Design of an Experiment to Investigate Rayleigh-Taylor Instabilities in Reacting Flows. In Proceedings of the AIAA Scitech 2019 Forum, San Diego, CA, USA, 7–11 January 2019.
15. Driscoll, J.F.; Rasmussen, C.C. Correlation and analysis of blowout limits of flames in high-speed airflows. *J. Propuls. Power* **2005**, *21*, 1035–1044. [CrossRef]
16. Sykes, J.P.; Gallagher, T.P.; Rankin, B.A. Effects of Rayleigh-Taylor instabilities on turbulent premixed flames in a curved rectangular duct. *Proc. Combust. Inst.* **2021**, *38*, 6059–6066. [CrossRef]
17. Rasmussen, C.C.; Driscoll, J.F.; Carter, C.D.; Hsu, K.-Y. Characteristics of cavity-stabilized flames in a supersonic flow. *J. Propuls. Power* **2005**, *21*, 765–768. [CrossRef]
18. Karlovitz, B. Open turbulent flames. *Symp. (Int.) Combust.* **1953**, *4*, 60–67. [CrossRef]
19. Poinsot, T.; Candel, S.; Trouvé, A. Applications of direct numerical simulation to premixed turbulent combustion. *Prog. Energy Combust. Sci.* **1995**, *21*, 531–576. [CrossRef]
20. Poinsot, T.; Echekki, T.; Mungal, M.G. A Study of the Laminar Flame Tip and Implications for Premixed Turbulent Combustion. *Combust. Sci. Technol.* **1992**, *81*, 45–73. [CrossRef]
21. *SAE ARP1533B*; Procedure for the Analysis and Evaluation of Gaseous Emissions from Aircraft Engines. SAE: Warrendale, PA, USA, 2013.

applied sciences

MDPI

Article

Heat Release Rates of Straight Soybean and Diesel Oil Blends in a Compression Ignition Engine

Nury A. Nieto Garzón *, Amir A. Martins Oliveira and Edson Bazzo

Laboratory of Combustion and Thermal Systems Engineering—LabCET, Mechanical Engineering Department, Federal University of Santa Catarina, Florianópolis 88040-900, SC, Brazil
* Correspondence: nury@labcet.ufsc.br

Abstract: Straight soybean and diesel oil blends are proposed as alternatives for electricity generation in isolated regions. The compression ignition engine is considered the prime mover and has the attractive potential to distribute electricity generation for supply in isolated regions, as well as for small applications demanded by the agro-industrial sector. The heat release rate evaluation of straight soybean blends is the main focus of this paper. A single-cylinder compression ignition engine with a nominal power of 14.7 kW/2200 rpm fueled with blends of 50% and 80% v/v straight soybean oil with commercial diesel oil was tested on a dynamometer bench. The heat release rate and ignition delay were determined from in-cylinder pressure measurements using zero-dimensional modeling. The experimental results showed a promising performance and coherent behavior with the physicochemical fuel properties and load conditions tested. The highest fraction of vegetable oil led to a combustion delay, characterized by high diffusive and residual combustion phases, although the fuel oxygen content favored the combustion. Finally, this work allowed observation of the development of the heat release rate of straight soybean blends in a diesel engine, understanding the influence of the fuel properties and in-cylinder gas properties on the combustion process.

Keywords: straight soybean oil; biofuels; heat release rate; physicochemical fuel properties; distributed electricity generation

1. Introduction

Global climate change and worldwide dependency on fossil fuels have encouraged the study of alternative energy sources. Biofuels are an interesting alternative for energy production in countries with an agricultural basis, such as Brazil, who presents the availability of land for energy crops, favorable climatic conditions, and well-established infrastructure for crop production, transport, and processing. In this context, straight vegetable oils offer opportunities for the distributed electricity generation in the energy, agriculture, and agro-industrial sectors. They are the preferred choice for replacing diesel oil in isolated regions, where the perils of transporting and storing diesel oil are an environmental and social hazard [1,2].

Currently, vegetable oil is converted into biodiesel through the transesterification process and is used in defined proportions in drop-in fuels for transportation. The diesel oil used in Brazil carries 13% (vol.) of biodiesel in its formulation [3]. However, in distributed energy production, straight vegetable oil (SVO) can be used directly in compression ignition engines, thus reducing costs and providing fuel availability in remote areas, such as the Brazilian rainforest. Worldwide, different vegetable oils have been used, either directly or after conversion to biodiesel, as compression engine fuel. Examples include soybean oil in the USA, rapeseed and sunflower oils in Europe, and palm, jatropha, and coconut oils in Asia and Africa [4–7]. In Brazil, soybean and palm oils have been widely used in the production of biodiesel.

Most studies on the use of SVO in compression ignition engines have focused on dynamometer-based bench and thermodynamic analyses of compression ignition engines operating with different vegetable oils. The main technological challenge addressed is associated with the differences in the physicochemical properties of vegetable oils in comparison with diesel oil. The larger molecular structure, higher viscosity, and lower volatility of SVOs compared to diesel oil significantly influence the atomization and combustion processes [8]. Heating and blending are normally used to approximate the properties of SVOs to those of diesel oil [9,10]. The comparable performance of diesel oil in relation to thermal efficiency and other performance parameters has been reported in the literature [7,10–15]. Additionally, studies based on pressure readings to calculate the heat release rate have increased in the last decade for several pure vegetable oils and blends [16–33]. Results regarding the performance, emissions, and combustion parameters are compared in relation to diesel oil.

This work focuses on the heat release rate of straight soybean and diesel oil blends as fuels in compression ignition engines. Soybean oil was chosen for this study because of its availability in the Brazilian market. Tests were carried out measuring the dynamic pressure in the combustion chamber of a single-cylinder engine. The heat release rate was estimated from the transient pressure measurements using a well-known zero-dimensional approach [34,35]. The main objective of this work was to understand the influence of the physicochemical fuel and in-cylinder gas properties on the combustion process in order to contribute to further studies concerning the development of a combustion model for the analysis and design of electricity generation systems using straight vegetable oils in compression ignition engines.

2. Experimental Procedure

2.1. Fuels

The fuels tested were diesel oil and two blends of soybean and diesel oil. The blends tested were 50/50% v/v soybean and diesel oils and 80/20% v/v soybean and diesel oils. Table 1 shows the fuels tested with their respective label and temperature for testing. The blend of 50/50% v/v was tested at two temperature levels. The high level (85 °C) was determined in order to obtain a similar break-up regime to diesel oil at 25 °C. This condition was evaluated following the criterion for the onset of jet atomization reported by Reitz and Bracco [36]. Density, dynamic viscosity, and surface tension are significant physical properties in the atomization process. These properties were tested for diesel and soybean oils and their blends at different temperatures ranging from 25 °C to 85 °C. Subsequently, correlations were proposed as a function of the temperature and volume fraction of soybean oil, allowing the calculation of the fuel properties at different injection conditions. These correlations were previously reported by Garzón et al. [37], and some measurements of the physical properties tested are presented in Table 2, including the respective expanded uncertainties. One can observe the increase in the density, dynamic viscosity, and surface tension with an increase in the percentage of vegetable oil in the blend, as well as the effect of heating on the reduction in these properties.

Table 1. Identification of the fuels tested and the respective injection temperatures.

Fuel	Injection Temperature (°C)	Label
Diesel oil	25	100D(25)
50/50% v/v soybean/diesel	25	50S/50D(25)
	85	50S/50D(85)
80/20% v/v soybean/diesel	85	80S/20D(85)

Soybean oil tested suffered no refining or transesterification, only filtering. Vegetable oils are composed of fatty acid triglycerides, which determine the different physical and chemical properties of each vegetable oil. The diesel oil tested was commercial Brazilian diesel oil (known as diesel S10). The commercial Brazilian diesel oil used had a volumetric addition of 8% of biodiesel in accordance with national regulations. Currently, the commercial Brazilian diesel oil carries 13% (vol.) of biodiesel in its formulation [3]. The higher heating value (*HHV*), lower heating value (*LHV*), and elemental chemical composition were measured for soybean and diesel oils, and the respective results are presented in Table 3. In relation to *LHV*, the highest content of carbon and hydrogen in diesel oil influences its *LHV*, which is about 13% higher than the *LHV* of soybean oil. An empirical formula of soybean oil was calculated from the content of carbon, hydrogen, and oxygen. The molecular formula of soybean oil was estimated from its empirical formula and the assumption of soybean oil as a triglyceride (a molecule with six atoms of oxygen). Diesel oil was modeled as dodecane.

Table 2. Measurements of physical properties of the fuels tested.

Fuel	Temperature (°C)	Density (kg/m^3)	Dynamic Viscosity (mPa·s)	Surface Tension (mN/m)
Diesel oil	25	828.7 ± 3.0	3.51 ± 0.22	27.76 ± 0.40
50/50% *v*/*v* soybean/diesel	25	871.7 ± 2.2	13.19 ± 0.43	30.12 ± 0.28
50/50% *v*/*v* soybean/diesel	85	832.9 ± 2.2	3.62 ± 0.18	25.54 ± 0.28
80/20% *v*/*v* soybean/diesel	85	858.8 ± 2.7	6.47 ± 0.23	26.83 ± 0.32
Soybean oil	85	876.1 ± 3.3	9.53 ± 0.37	27.72 ± 0.39

Table 3. Values of *HHV*, *LHV*, C, H, S, N, and O of soybean and diesel oils.

Property	Soybean Oil	Diesel Oil
HHV (kJ/kg)	39,440	45,345
LHV (kJ/kg)	36,950	42,435
Carbon, C (%)	77.67	85.79
Hydrogen, H (%)	11.52	13.52
Sulfur, S (%)	<0.04	<0.04
Nitrogen, N (%)	0.02	0.05
Oxygen, O (%)	10.76	0.61
Molecular formula	$C_{57.4}H_{101.5}O_6$	$C_{12}H_{26}$

2.2. Experimental Setup

The experimental setup consists of a 14.7 kW, 4-stroke single-cylinder, direct injection diesel engine (Yanmar, model YT22E, Yanmar Co., Ltd., Osaka, Japan) coupled to an electromagnetic dynamometer (Schenk, model W70, Schenck RoTec GmbH, Darmstadt, Germany). The dynamometric bench is equipped with a fuel supply system, a measurement system of the in-cylinder pressure, an air measuring system, and a control and data acquisition system. Figure 1 shows a schematic diagram of the experimental setup. Technical specifications of the direct injection diesel engine are shown in Table 4.

An electric heater comprising an aluminum tube (12.7 mm diameter), electrical resistance of 119 Ω, and ceramic insulation was manufactured to heat the vegetable oil blend before it entered the engine. An electromagnetic 3/2 valve was installed to switch to the fuel employed. This enables the passage of diesel oil from the original tank or the fuel under test. The fuel consumption was measured using an electronic balance (Marte, model AD5000) with serial communication. The instant reading of the data allowed the calculation of the fuel flow. The engine torque was measured with an extensometer-type load cell installed on the dynamometer arm. The speed was measured with an incremental encoder 360 pulses/revolution (Autonics, model E40S, Seoul, Republic of Korea).

The inducted air mass was measured with a hot-film mass air-flow meter (Bosch, model HFM 2, Gerlingen, Germany). An air reservoir was coupled between the intake manifold and the air-mass meter to ensure uniformity of the flow and the measurement of the air-mass rate. Temperature and relative humidity of the air inducted were measured through a humidity sensor and a temperature sensor installed before the mass air-flow meter. A thermocouple was installed in the intake manifold, close to the intake valve, to register the air temperature inducted. Additionally, a set of thermocouples was installed at the engine cooling water outlet, the exhaust manifold, the heater outlet, and the injection pump inlet. An electronic control and data acquisition system was developed with LabVIEW software 2011 [38] using two acquisition boards to register the different variables.

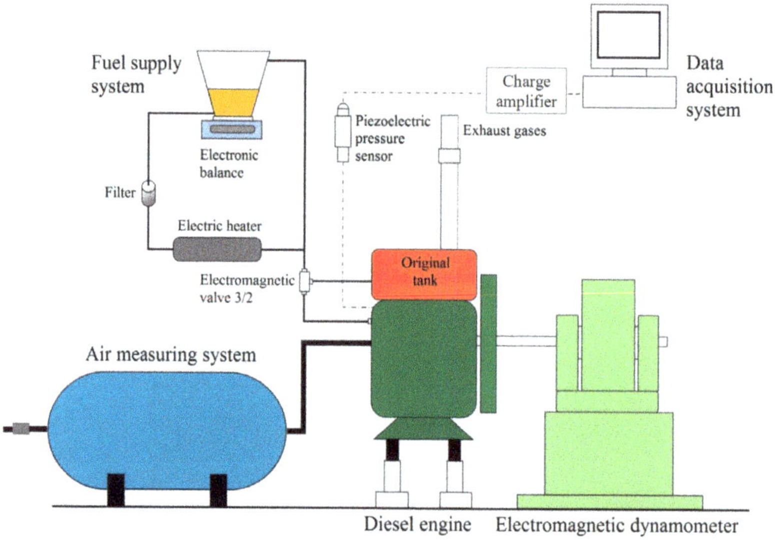

Figure 1. Schematic diagram of the experimental setup.

Table 4. Technical specifications of the diesel engine tested.

Item	Specifications
Type	Diesel, 1 horizontal cylinder, 4-stroke
Injection system	Direct
Injection pressure	200 bar
Injection timing	15° BTC
Cylinder diameter	115 mm
Stroke	115 mm
Cylinder capacity	1194 cm^3
Compression ratio	17.3
IVO-IVC	654° CA–278° CA
EVO-EVC	444° CA–64° CA
Power	14.7 kW (20 CV)/2200 rpm
Specific fuel consumption	238 g/kWh (175 g/CVh)
Cooling	Water
Starting system	Manual or electric
Net weight	195 kg–180 kg
Dimensions	965 × 450 × 699 mm
Lubrication system	Forced, trochoid pump

Experimental Testing

The tests were performed at the maximum flow rate of the injection pump. In each test, the engine operation was started with diesel fuel until the heating period was completed, i.e., when the cooling water temperature reached 60 °C. The diesel oil was then replaced by the fuel under test. After achieving stable operation with the new fuel (steady readings), the brake process was started with the dynamometer. The load was applied to accomplish 1800 and 2100 rpm, which correspond to the conditions of high load and maximum power, respectively. Measurements of torque, speed, power output, fuel consumption, and air flow mass were recorded at the steady-state condition. The measurements were registered at intervals of five seconds.

Additionally, a piezoelectric pressure sensor (Kistler, model 6041B, Winterthur, Switzerland) was used to measure the in-cylinder pressure. The sensor was installed in the cylinder head, according to technical specifications, and its signal was synchronized with the incremental encoder to register the pressure values each 0.5° CA (crank angle). The pressure sensor was connected to a charge amplifier (Kistler, model 5018A), and its signal was registered through a data acquisition system (National Instrument, model SCB-68, Austin, TX, USA). The in-cylinder pressure readings were recorded at the frequency of 10,000 S/s.

For each test condition, the in-cylinder pressure readings were processed to obtain a representative pressure curve as a function of the crank angle. It was verified that a sample of 50 cycles is a representative sample to calculate the pressure curve without disturbing the accuracy of the results. This data processing consisted of the spurious data elimination, cycles averaging, filtering, and fitting of average pressure data. The complete description of the in-cylinder pressure data processing is presented in Garzón et al. [39].

2.3. Determination of Ignition Delay

The time interval between the start of the injection and the start of the combustion, related to the ignition delay, was here determined using the method proposed by Reddy et al. [40].

The method is based on the analysis of the pressure curve as a function of the crank angle, considering that the behavior of the pressure is influenced by the injection and the burning of the fuel. As soon as the fuel is injected, the rate of pressure rise inside the combustion chamber decreases due to evaporation of the fuel, which reduces the in-cylinder temperature and slows down the rate of pressure rise. This phenomenon represents an inflection point in the pressure curve and a maximum point on the corresponding first derivative curve. At the beginning of the combustion, an inverse phenomenon occurs. The release of energy by combustion produces a sudden rise in temperature and pressure, representing a new inflection point on the pressure curve and a minimum point on the first derivative curve. The inflection points represent maximum or minimum points on the first derivative curve and, therefore, zero values on the corresponding second derivative curve. Following this analysis, the injection and ignition angles were determined as the crank angles where the second derivative of the pressure with respect to the crank angle was zero in the time interval analyzed. The complete description of the determining of ignition delay is presented in Garzón et al. [37].

2.4. Experimental Heat Release Rate

The post-processing of pressure measurements is known as the heat release rate analysis, and it is performed during the compression and expansion strokes, while the valves remain closed. The combustion chamber is modeled according to zero-dimensional modeling proposed by Krieger and Borman [34] applied to the control volume defined by the engine cylinder. The only mass flow across the boundary is the fuel injected, assuming no leaks through possible gaps in the cylinder. The energy and the mass conservation equations for this control volume are

$$\frac{d(mu)}{dt} = \frac{dQ}{dt} - p\frac{dV}{dt} + h_f\frac{dm_f}{dt} \quad (1)$$

$$\frac{dm}{dt} = \frac{dm_f}{dt} \quad (2)$$

where dQ/dt represents the heat transfer rate through system boundaries, pdV/dt is the work rate performed by the system due to system boundary displacement, dm/dt is the mass flow rate across the system boundary, u is the specific internal energy of the gas inside the system boundary, h_f is the fuel enthalpy, and dm_f/dt is the mass fuel flow rate.

Following the hypothesis that fuel injection is the only mass changing in the cylinder and that the single combustion model assumes a complete burning instantaneously as it enters the combustion chamber, and also that the thermodynamics states of the working fluid are affected by the chemical energy release, the term dm_f/dt can be considered as the mass burning rate of the fuel.

The equivalence ratio ϕ at each instant is

$$\frac{d\phi}{dt} = \frac{1}{m_a(FA)_s}\frac{dm_f}{dt} \tag{3}$$

where $(FA)_s$ corresponds to the stoichiometric fuel/air ratio and m_a to the mass air. Since the properties of the gases in the cylinder are a function of T, p, and ϕ, the internal energy u and the gas constant R_g can be expressed as

$$\frac{du}{dt} = \frac{\partial u}{\partial T}\frac{dT}{dt} + \frac{\partial u}{\partial p}\frac{dp}{dt} + \frac{\partial u}{\partial \phi}\frac{d\phi}{dt} \tag{4}$$

$$\frac{dR_g}{dt} = \frac{\partial R_g}{\partial T}\frac{dT}{dt} + \frac{\partial R_g}{\partial p}\frac{dp}{dt} + \frac{\partial R_g}{\partial \phi}\frac{d\phi}{dt} \tag{5}$$

The equation of state is also considered in the analysis as an ideal gas mixture. Thus, in a differential form,

$$\frac{1}{V}\frac{dV}{dt} + \frac{1}{p}\frac{dp}{dt} = \frac{1}{T}\frac{dT}{dt} + \frac{1}{R_g}\frac{dR_g}{dt} + \frac{1}{m}\frac{dm}{dt} \tag{6}$$

The temperature variation as a function of time and the fuel burning rate is obtained from Equations (1)–(6). Thus,

$$\frac{dT}{dt} = \frac{\left[\frac{1}{V}\frac{dV}{dt} + \left(\frac{1}{p} - \frac{1}{R_g}\frac{\partial R_g}{\partial p}\right)\frac{dp}{dt} - \left(\frac{1}{R_g m_a(FA)_s}\frac{\partial R_g}{\partial \phi} + \frac{1}{m}\right)\frac{dm_f}{dt}\right]}{\frac{1}{T} + \frac{1}{R_g}\frac{\partial R_g}{\partial T}} \tag{7}$$

$$\frac{dm_f}{dt} = \frac{\frac{dQ}{dt} - \left(\frac{m}{V}D + p\right)\frac{dV}{dt} - \left[mD\left(\frac{1}{p} - \frac{1}{R_g}\frac{\partial R_g}{\partial p}\right) + m\frac{\partial u}{\partial p}\right]\frac{dp}{dt}}{u - h_f + \frac{m}{m_a(FA)_s}\frac{\partial u}{\partial \phi} - D - \frac{mD}{R_g m_a(FA)_s}\frac{\partial R_g}{\partial \phi}} \tag{8}$$

where

$$D = \frac{\partial u}{\partial T}\left(\frac{1}{T} + \frac{1}{R_g}\frac{\partial R_g}{\partial T}\right)^{-1} \tag{9}$$

The volume variation is calculated from the geometrical relations of the engine. The pressure values and the corresponding first derivative were obtained from experimental data collected from the in-cylinder pressure readings. The thermodynamic properties and the corresponding partial derivatives related to the temperature, pressure, and equivalence ratio are determined by using the routines proposed by Olikara and Borman [41].

The heat transfer to the walls was modeled considering heat exchange by convection and radiation, using the Hohenberg correlation for calculating the heat transfer coefficient [42] and the gas global emissivity as reported by Watson and Janota [43].

Equations (7) and (8) are solved numerically to obtain the mean gas temperature and the fuel mass burned. It is a nonlinear ordinary differential equations system, in this work, solved by the Runge–Kutta method using MATLAB software 2013 [44].

The heat release rate is calculated from the fuel burning rate and its lower heating value (*LHV*), so that

$$\frac{dQ_f}{dt} = LHV\frac{dm_f}{dt} \tag{10}$$

3. Results and Discussion

In order to fulfill the proposed objectives, the main results are presented here, concerning the in-cylinder pressure, ignition delay, heat release rate, and atomization analysis of fuel spray. Regarding performance parameters, the main results were comparable for all fuel tested in correspondence with other works reported in the literature. These results are not presented in order to emphasize the analysis of the combustion process.

3.1. In-Cylinder Pressure

The in-cylinder pressure data as a function of crank angle for the fuels tested at 1800 and 2100 rpm are shown in Figure 2. The highest pressures are observed at the lowest speed as a consequence of the high load and, therefore, the highest fuel amount injected per cycle. High-pressure values are related to the fuel *LHV*, the ignition delay, and the development of the heat release rate in the premixed phase during the combustion of each sample.

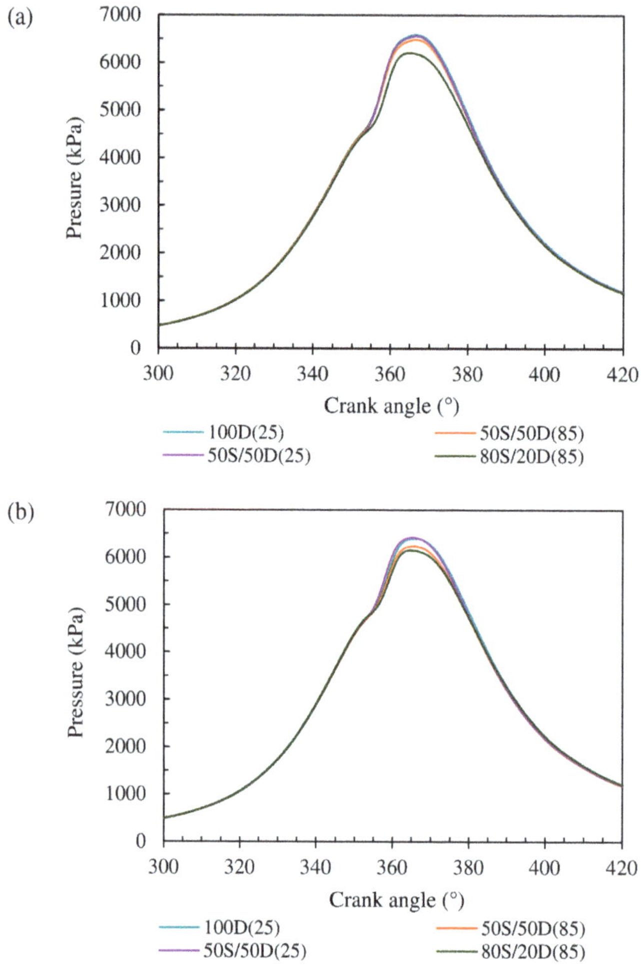

Figure 2. Experimental in-cylinder pressure data as a function of crank angle for the fuels tested at (**a**) 1800 rpm and (**b**) 2100 rpm.

The start of the ignition produces a slope change of the pressure curves, close to 360° CA. In this context, one observes an ignition more delayed for the blend 80S/20D(85) for both engine speeds. The maximum pressure value presented for each fuel at each speed and the corresponding crank angle are shown in Table 5, including the percentage difference relative to diesel oil. At 1800 rpm, the highest maximum pressure was presented for diesel oil 100D(25) and the blend 50S/50D(25). The lowest maximum pressure was presented by the blend 80S/20D(85), being about 6% lower than the maximum pressure of the diesel oil. Similar behavior was observed at 2100 rpm, decreasing to 4% of the difference between the maximum pressure of the diesel oil and the blend 80S/20D(85).

Table 5. Maximum pressure and percentage difference relative to diesel oil at 1800 and 2100 rpm.

Engine Speed (rpm)	Fuel	Maximum Pressure (kPa)	Difference (%)	Angle at Maximum Pressure (° CA)
1800	100D(25)	6584.7 ± 22.4		366.5
	50S/50D(25)	6557.7 ± 19.3	−0.41	366.5
	50S/50D(85)	6484.8 ± 23.1	−1.52	366.5
	80S/20D(85)	6204.8 ± 36.0	−5.77	365.0
2100	100D(25)	6395.1 ± 25.6		365.5
	50S/50D(25)	6413.3 ± 19.1	0.28	365.5
	50S/50D(85)	6234.6 ± 26.1	−2.51	365.5
	80S/20D(85)	6152.3 ± 16.2	−3.80	364.5

3.2. Ignition Delay

The experimental results regarding ignition delay are presented for all fuels tested at 1800 and 2100 rpm in Table 6. The expanded uncertainty for ignition delay was calculated as ±0.71° CA. The results confirm the highest ignition delay for the 80S/20D(85), as previously mentioned, regarding the slope change of the pressure curves.

Table 6. Ignition delay of the fuel tested at 1800 rpm.

Engine Speed (rpm)	Fuel	Start of Injection (° CA)	Start of Ignition (° CA)	Ignition Delay (° CA)	Ignition Delay (ms)
1800	100D(25)	345.0	352.5	7.5	0.694
	50S/50D(25)	345.0	352.5	7.5	0.694
	50S/50D(85)	345.5	353.0	7.5	0.694
	80S/20D(85)	345.5	354.0	8.5	0.787
2100	100D(25)	345.0	353.5	8.5	0.675
	50S/50D(25)	344.5	353.0	8.5	0.675
	50S/50D(85)	345.0	353.5	8.5	0.675
	80S/20D(85)	345.0	354.5	9.5	0.754

3.3. Heat Release Rate

The heat release rate and the corresponding cumulative energy release obtained for the fuels at 1800 and 2100 are presented in Figures 3 and 4, respectively. The heat release rate curves evidence the premixed and nonpremixed (diffusive) combustion phases. Additionally, a significant residual combustion phase is also observed.

The heat release rate increases with the load, i.e., with the speed reduction, where the mass fuel injected per cycle increases. Observing the curves of cumulative energy release, the combustion performance presented a similar behavior in the premixed combustion phase for all fuels tested, changing in the nonpremixed combustion phase, where we observed a significant difference among the fuels.

The heat release rate in the premixed combustion phase is influenced by the ignition delay. Consequently, for both engine speeds, one observes that all fuels present a similar premixed combustion phase, corresponding to a similar ignition delay, as shown in Table 6. In the case of the blend 80S/20D(85), the heat release curve begins slightly delayed as a result of its ignition delay that is slightly higher with respect to the other fuels. At 1800 rpm, the higher premixed combustion rate is related to the load increase and the time increase per each crank angle degree corresponding to the speed reduction, which favors the air–fuel mixing and the burning.

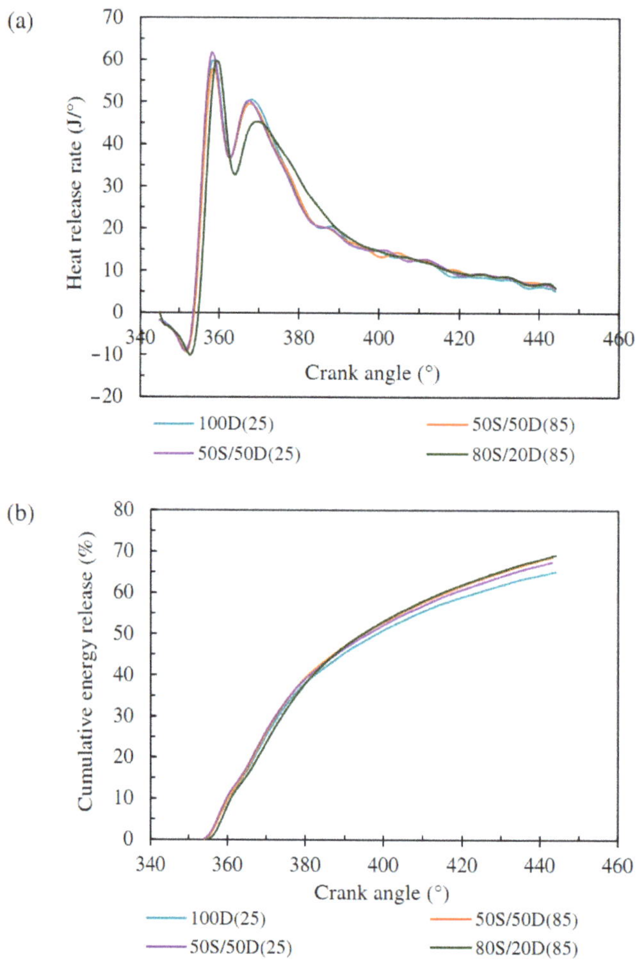

Figure 3. (**a**) Heat release rate and (**b**) cumulative energy release as a function of crank angle for the fuels tested at 1800 rpm and full load condition.

At 1800 rpm, the diesel oil 100D(25) and the blends 50S/50D(25) and 50S/50D(85) presented the highest peak of the heat release rate in the diffusive phase. From 380° CA, approximately, the heat release rate increased for the blend 80S/20D(85) in comparison with the other fuels (residual combustion phase), which shows the combustion delay of this blend and the effect of the highest oxygen content in the fuel that favors the burning, such that the cumulative energy release is high for the blends 80S/20D(85) and 50S/50D(85). The 100D(25) oil presented the lowest cumulative energy release. Similar results were reported by Kannan et al. [45] in their work with oxygenated fuels.

At 2100 rpm, the diesel oil 100D(25) and the blend 80S/20D(85) presented the highest peak of the heat release rate in the diffusive phase. In the residual combustion phase, the blend 80S/20D(85) presented the highest heat release rate. Once more, the blend 80S/20D(85) presented the highest cumulative energy release, increasing the overall reactive rate in the diffusive and residual combustion phases. The burning of the blends 50S/50D(25) and 50S/50D(85) presented similar cumulative energy release, being slightly higher for the blend 50S/50D(85). The diesel oil presented the lowest cumulative energy release. For the blends, the cumulative release energy was higher at 2100 rpm than at 1800 rpm.

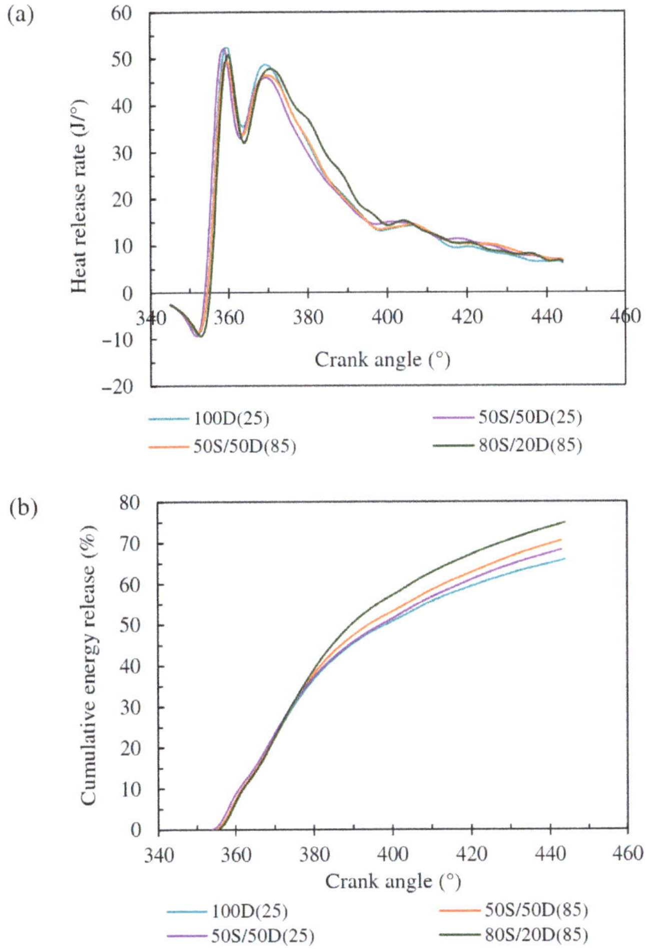

Figure 4. (**a**) Heat release rate and (**b**) cumulative energy release as a function of crank angle for the fuels tested at 2100 rpm and full load condition.

In general, for fuels 100D(25), 50S/50D(25), and 50S/50D(85), the highest diffusive combustion rate was obtained at 1800 rpm (higher mass fuel injected), decreasing in the residual combustion phase as a consequence of high heat transfer (higher time per crank angle and higher gas temperature). At 2100 rpm, we observed the highest residual combustion rate, favored by the tumble and swirl effects, possibly present in the combustion chamber at that engine speed. In the case of the blend 80S/20D(85), the highest diffusive and residual combustion rates were observed at 2100 rpm as a consequence of the combustion delay and the effect of the turbulence and blend oxygen content that favor the burning at this speed.

The curves of the heat release rate were developed over compression and expansion strokes of the engine operating cycle when valves were closed. Consequently, the final point of the curves was 444° CA, where the exhaust valve started to open. Based on the curves of cumulative energy release, one can observe that the total mass fuel injected did not burn completely, continuing the burning during the exhaust stroke. Table 7 presents the mass fraction burned at 444° CA for all fuels at the two speeds tested. The values of the mass fraction burned are lower than 100%. One can consider that the low values calculated for the total mass fraction burned are a consequence of the model employed for the heat transfer coefficient, which probably underestimates the heat transfer for the walls; therefore, the heat release rate is also underestimated, signifying low mass fraction burned.

Table 7. Mass fraction burned at 444° CA for all fuels at the two speeds tested.

Fuel	Mass Fraction Burned (mg Burned/mg Injected)	
	1800 rpm	2100 rpm
100D(25)	0.650	0.657
50S/50D(25)	0.675	0.683
50S/50D(85)	0.687	0.705
80S/20D(85)	0.689	0.746

As shown in Table 7, the mass fraction burned increased with the speed increase or load reduction where the mass fuel injected per cycle and the global equivalence ratio were less (see Table 8). The highest mass fraction burned was presented for the blend 80S/20D(85) at both speeds, with the difference in comparison to 100D(25) at 2100 rpm being more significant. Additionally, one can see that the lowest global equivalence ratio and the turbulence produced by high speeds favor the air–fuel mixing and, consequently, the mass fraction burned.

Table 8. Stoichiometric air/fuel ratio and global equivalence ratio for all fuels at the two speeds tested.

Fuel	Stoichiometric Air/Fuel Ratio (AF)s	Global Equivalence Ratio ϕ_{gl}	
		1800 rpm	2100 rpm
100D(25)	14.92	0.83	0.81
50S/50D(25)	14.23	0.77	0.76
50S/50D(85)	14.23	0.74	0.74
80S/20D(85)	12.82	0.70	0.69

3.4. Atomization Analysis

The jet pattern is a significant factor in the combustion process. It is influenced by the injection parameters and the physical fuel properties. The primary disintegration or break-up regime of the fuel jet in diesel engines must be the atomization regime. The Ohnesorge diagram permits the identification of the break-up regime, relating the Ohnesorge number with the Reynolds number of the spray. The dimensionless numbers required to build the Ohnesorge diagram are

$$Re_{sp} = \frac{\rho_f u_{sp} d_o}{\mu_f} \tag{11}$$

$$We_{sp} = \frac{\rho_f u_{sp}^2 d_o}{\sigma_f} \tag{12}$$

$$Oh = \frac{\sqrt{We_{sp}}}{Re_{sp}} \tag{13}$$

where Re_{sp}, We_{sp}, and Oh are Reynolds, Weber, and Ohnesorge dimensionless numbers of the spray, respectively; u_{sp} is the spray velocity; d_o is the injector nozzle hole diameter; and ρ_f, μ_f, and σ_f are the density, dynamic viscosity, and surface tension of the fuel, respectively. The injector nozzle hole diameter was 0.3 mm, and the fuel properties were calculated at respective injection temperatures using the correlations previously reported in Garzón et al. [37].

In order to help understand the combustion process, the break-up regime (Ohnesorge diagram), *SMD* (Sauter mean diameter), droplet Reynold number, and oxygen content were estimated for all fuels tested at 1800 rpm. The results are shown in Figure 5 and

Table 9. The *SMD* was calculated using the correlation proposed by Hiroyasu and Arai [46]. Thus,

$$\frac{SMD}{d_o} = 0.38\mathrm{Re}_{sp}^{0.25} We_{sp}^{-0.32} \left(\frac{\mu_f}{\mu_g}\right)^{0.37} \left(\frac{\rho_f}{\rho_g}\right)^{-0.47} \quad (14)$$

where μ_g and ρ_g are the dynamic viscosity and density of the gas, respectively. The gas dynamic viscosity was calculated as $\mu_g = 4.57 \times 10^{-7} T^{0.645}$, according to Annand [47]. The droplet Reynold number was calculated as follows:

$$\mathrm{Re}_D = \frac{\rho_g (u_{sp} - S_p) SMD}{\mu_g} \quad (15)$$

where S_p is the piston velocity (gas velocity).

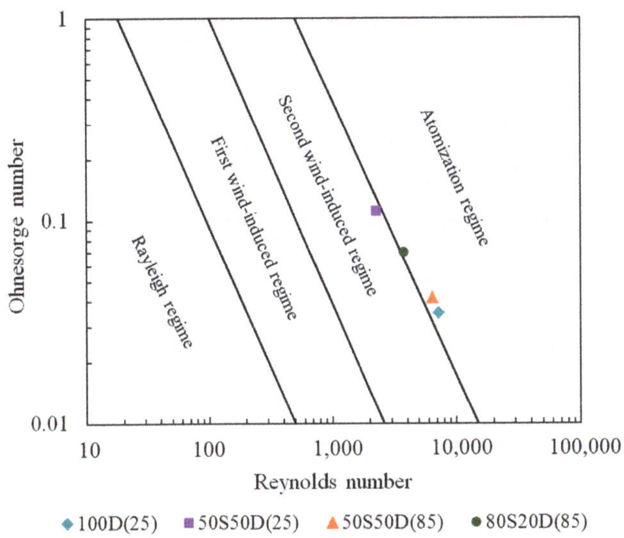

Figure 5. Ohnesorge diagram for all fuels tested at 1800 rpm and full load condition.

Table 9. *SMD*, droplet Reynold number, and oxygen content for all fuels tested at 1800 rpm and full load condition.

Fuel	*SMD* (μm)	*Re*$_D$	O_2 (%)
100D(25)	22.20	737.24	-
50S50D(25)	25.65	851.84	5.66
50S50D(85)	21.91	727.58	5.66
80S20D(85)	23.40	776.87	8.81

According to Ohnesorge diagram, the blend 50S/50D(25) does not present the atomization regime required for diesel engine combustion, and the break-up regime of the blend 80S/20D(85) is between second-wind-induced and atomization regimes. In the case of the blend 50S/50D(85), the heating favored the break-up, reaching the atomization regime.

Concerning the results shown in Table 9, the blend 50S/50D(25) presents the highest *SMD*, being 15.5% higher than the *SMD* of 100D(25). The highest droplet Reynold number is also presented by blend 50S/50D(25), and the blend 80S/20D(85) has the maximum oxygen content.

Although blends 80S/20D(85) and 50S/50D(25) are not in an atomization regime and present the highest *SMD*, the diesel engine operated satisfactorily with these blends and the combustion results were comparable with the other two fuels. In the case of the blend 50S/50D(25), one can suggest that the highest *SMD* favored the droplet Reynold number and, therefore, its evaporation constant or average area reduction rate of the droplet. For the blend 80S/20D(85), the high *SMD* and the highest fraction of vegetable oil led to a combustion delay, characterized by high diffusive and residual combustion phases; however, the oxygen content favored its combustion.

Additionally, in relation to the combustion of a binary liquid blend droplet (diesel + straight vegetable oil), products of oil degradation can help or can inhibit the ignition; therefore, it is not clear what occurs in the premixed combustion phase when comparing the blends 50S/50D(25) and 50S/50D(85).

The results analyzed confirm that the jet pattern is influenced by the injection parameters and the physical fuel properties. The combustion of soybean oil blends will be enhanced with high fuel temperature, high engine speed, low injection timing, high injection pressure, and high compression ratio (high in-cylinder temperature and pressure at injection timing).

4. Conclusions

This work was aimed at the applicability analysis of soybean oil blends in compression ignition engines, studying the heat release rate obtained from in-cylinder pressure measurements.

Tests on the dynamometric bench were carried out for diesel oil and two blends, 50/50% v/v and 80/20% v/v of soybean and diesel oils. The engine was tested at 1800 and 2100 rpm. In-cylinder pressure readings were recorded and processed to calculate the heat release rate using a zero-dimensional approach. The maximum in-cylinder pressure, ignition delay, and spray parameters were also determined. The results evidenced the premixed, nonpremixed (diffusive), and residual combustion phases. The influence of the physicochemical fuel properties was also evidenced in the experimental results.

The premixed combustion phase was similar for all fuels tested at 1800 and 2100 rpm. For the blend of 80% soybean oil, the high *SMD* (Sauter mean diameter) and the highest fraction of vegetable oil led to a combustion delay, characterized by high diffusive and residual combustion phases. However, the oxygen content, heating, and high engine speeds favored its combustion. Products of vegetable oil degradation may increase or inhibit the ignition, depending on their chemical nature. In the case of the blend of 50% soybean oil, it is suggested that the highest *SMD* increased the droplet Reynold number and, therefore, the evaporation constant.

Finally, the results allowed us to observe the feasibility of using straight soybean oil and diesel oil blends as alternatives for electricity generation for replacing diesel oil in isolated regions and also small applications demanded by the agro-industrial sector. Additionally, the results will support the development of a specific combustion model for straight vegetable oils in order to model and optimize the diesel engine operation with these biofuels.

Author Contributions: Conceptualization, N.A.N.G., A.A.M.O. and E.B.; methodology, N.A.N.G., A.A.M.O. and E.B.; software, N.A.N.G.; validation, N.A.N.G.; data curation, N.A.N.G.; formal analysis, N.A.N.G., A.A.M.O. and E.B.; writing—original draft preparation, N.A.N.G.; writing—review and editing, N.A.N.G., A.A.M.O. and E.B.; supervision, A.A.M.O. and E.B. All authors have read and agreed to the published version of the manuscript.

Funding: This research was funded by CNPq (National Council for Scientific and Technological Development, Grant number: 201205451), and by Mechanical Engineering Graduate Program at Federal University of Santa Catarina.

Institutional Review Board Statement: Not applicable.

Informed Consent Statement: Not applicable.

Data Availability Statement: The original contributions presented in the study are included in the article, and university library collection at https://pergamum.ufsc.br/acervo/352068 (accessed on 23 September 2024). Further inquiries can be directed to the corresponding author.

Acknowledgments: The authors acknowledge CNPq (National Council for Scientific and Technological Development) for the scholarship for N.A. Nieto Garzón, and Mechanical Engineering Graduate Program at Federal University of Santa Catarina.

Conflicts of Interest: The authors declare no conflicts of interest.

Nomenclature

Latin Symbols

d_o	Diameter of the injector nozzle	[m]
$(FA)_s$	Stoichiometric fuel/air ratio	[kg/kg]
h_f	Enthalpy of fuel flow	[kJ/kg]
HHV	Higher heating value	[kJ/kg]
LHV	Lower heating value	[kJ/kg]
m	Total mass	[kg]
m_a	Mass of air inducted into the cylinder	[kg]
m_f	Mass fuel or fuel burned	[kg]
Oh	Ohnesorge number	[-]
p	Pressure	[kPa]
Q	Heat transfer through of system boundaries	[kJ]
Q_f	Heat release for the fuel	[kJ]
Re_D	Droplet Reynold number	[-]
Re_{sp}	Spray Reynold number	[-]
R_g	Gas constant	[kJ/(kg·K)]
S_p	Instantaneous piston velocity	[m/s]
SMD	Sauter mean diameter	[m]
t	Time	[s]
T	Temperature	[K]
u	Specific internal energy	[kJ/kg]
V	Volume	[m^3]
u_{sp}	Spray velocity	[m/s]
We_{sp}	Spray Weber number	[-]

Greek Symbols

μ_f	Fuel dynamic viscosity	[Pa·s]
μ_g	Gas dynamic viscosity	[Pa·s]
ρ_f	Fuel density	[kg/m^3]
ρ_g	Gas density	[kg/m^3]
σ_f	Fuel surface tension	[N/m]
ϕ	Equivalence ratio at any instant	[-]
ϕ_{gl}	Global equivalence ratio	[-]

Abbreviations

BTC	Before top-center crank position
CA	Crank angle
EVC	Exhaust valve closing
EVO	Exhaust valve opening
IVC	Intake valve closing
IVO	Intake valve opening
SVO	Straight vegetable oil

References

1. de Almeida, S.C.A.; Belchior, C.R.; Nascimento, M.V.G.; dos Vieira, L.S.R.; Fleury, G. Performance of a diesel generator fuelled with palm oil. *Fuel* **2002**, *81*, 2097–2102. [CrossRef]
2. Duarte, A.R.; Bezerra, U.H.; de Lima Tostes, M.E.; Duarte, A.M.; da Rocha Filho, G.N. A proposal of electrical power supply to Brazilian Amazon remote communities. *Biomass Bioenergy* **2010**, *34*, 1314–1320. [CrossRef]

3. Ministry of Mines and Energy—Brazil. Resolution N° 3, March 20, 2023. Available online: https://www.gov.br/mme/pt-br/assuntos/conselhos-e-comites/cnpe/resolucoes-do-cnpe/2023/ResCNPE32023.pdf (accessed on 8 October 2024). (In Portuguese)
4. Graboski, M.S.; McCormick, R.L. Combustion of fat and vegetable oil derived fuels in diesel engines. *Prog. Energy Combust. Sci.* **1998**, *24*, 125–164. [CrossRef]
5. Agarwal, A.K. Biofuels (alcohols and biodiesel) applications as fuels for internal combustion engines. *Prog. Energy Combust. Sci.* **2007**, *33*, 233–271. [CrossRef]
6. Sidibé, S.S.; Blin, J.; Vaitilingom, G.; Azoumah, Y. Use of crude filtered vegetable oil as a fuel in diesel engines state of the art: Literature review. *Renew. Sustain. Energy Rev.* **2010**, *14*, 2748–2759. [CrossRef]
7. Misra, R.D.; Murthy, M.S. Straight vegetable oils usage in a compression ignition engine—A review. *Renew. Sustain. Energy Rev.* **2010**, *14*, 3005–3013. [CrossRef]
8. Agarwal, D.; Kumar, L.; Agarwal, A.K. Performance evaluation of a vegetable oil fuelled compression ignition engine. *Renew. Energy* **2008**, *33*, 1147–1156. [CrossRef]
9. Hartmann, R.M.; Garzón, N.N.; Hartmann, E.M.; Oliveira, A.A.M.; Bazzo, E. Vegetable Oils of Soybean, Sunflower and Tung as Alternative Fuels for Compression Ignition Engines. *Int. J. Thermodyn.* **2013**, *16*, 87–96. [CrossRef]
10. Garzón, N.A.N.; Oliveira, A.A.M.; Hartmann, R.M.; Bazzo, E. Experimental and thermodynamic analysis of a compression ignition engine operating with straight soybean oil. *J. Braz. Soc. Mech. Sci. Eng.* **2015**, *37*, 1467–1478. [CrossRef]
11. Nwafor, O.M.I. The effect of elevated fuel inlet temperature on performance of diesel engine running on neat vegetable oil at constant speed conditions. *Renew. Energy* **2003**, *28*, 171–181. [CrossRef]
12. Belagur, V.K.; Reddy, V.; Wadawadagi, S.B. Effect of Injection Pressure on Performance, Emission and Combustion Characteristics of Direct Injection Diesel Engine Running on Blends of Pongamia Pinnata Linn (Honge oil) Oil and Diesel Fuel. *Agric. Eng. Int. CIGR J.* **2009**, *11/1316*, 1–17.
13. Sarada, S.N.; Shailaja, M.; Raju, A.S.R.; Radha, K.K. Optimization of injection pressure for a compression ignition engine with cotton seed oil as an alternate fuel. *Int. J. Eng. Sci. Technol.* **2010**, *2*, 142–149. [CrossRef]
14. Rakopoulos, D.C.; Rakopoulos, C.D.; Giakoumis, E.G.; Dimaratos, A.M.; Founti, M.A. Comparative environmental behavior of bus engine operating on blends of diesel fuel with four straight vegetable oils of Greek origin: Sunflower, cottonseed, corn and olive. *Fuel* **2011**, *90*, 3439–3446. [CrossRef]
15. D'Alessandro, B.; Bidini, G.; Zampilli, M.; Laranci, P.; Bartocci, P.; Fantozzi, F. Straight and waste vegetable oil in engines: Review and experimental measurement of emissions, fuel consumption and injector fouling on a turbocharged commercial engine. *Fuel* **2016**, *182*, 198–209. [CrossRef]
16. Kumar, M.S.; Ramesh, A.; Nagalingam, B. An experimental comparison of methods to use methanol and Jatropha oil in a compression ignition engine. *Biomass Bioenergy* **2003**, *25*, 309–318. [CrossRef]
17. Reddy, J.N.; Ramesh, A. Parametric studies for improving the performance of a Jatropha oil-fuelled compression ignition engine. *Renew. Energy* **2006**, *31*, 1994–2016. [CrossRef]
18. Geo, E.V.; Nagarajan, G.; Nagalingam, B. Studies on dual fuel operation of rubber seed oil and its bio-diesel with hydrogen as the inducted fuel. *Int. J. Hydrog. Energy* **2008**, *33*, 6357–6367. [CrossRef]
19. Agarwal, A.K.; Dhar, A. Karanja oil utilization in a direct-injection engine by preheating. Part 1: Experimental investigations of engine performance, emissions, and combustion characteristics. *Proc. Inst. Mech. Eng. Part D J. Automob. Eng.* **2010**, *224*, 73–84. [CrossRef]
20. Geo, V.E.; Nagarajan, G.; Nagalingam, B. Studies on improving the performance of rubber seed oil fuel for diesel engine with DEE port injection. *Fuel* **2010**, *89*, 3559–3567. [CrossRef]
21. Rakopoulos, C.D.; Rakopoulos, D.C.; Giakoumis, E.G.; Dimaratos, A.M. Investigation of the combustion of neat cottonseed oil or its neat bio-diesel in a HSDI diesel engine by experimental heat release and statistical analyses. *Fuel* **2010**, *89*, 3814–3826. [CrossRef]
22. Leenus, J.M.; Varuvel, E.G.; Prithviraj, D. Effect of diesel addition on the performance of cottonseed oil fuelled DI diesel engine. *Int. J. Energy Environ.* **2011**, *2*, 321–330.
23. Kasiraman, G.; Nagalingam, B.; Balakrishnan, M. Performance, emission and combustion improvements in a direct injection diesel engine using cashew nut shell oil as fuel with camphor oil blending. *Energy* **2012**, *47*, 116–124. [CrossRef]
24. Leenus, M.; Edwin Geo, V.; Kingsly Jeba Singh, D.; Nagalingam, B. A comparative analysis of different methods to improve the performance of cotton seed oil fuelled diesel engine. *Fuel* **2012**, *102*, 372–378. [CrossRef]
25. Rakopoulos, D.C. Heat release analysis of combustion in heavy-duty turbocharged diesel engine operating on blends of diesel fuel with cottonseed or sunflower oils and their bio-diesel. *Fuel* **2012**, *96*, 524–534. [CrossRef]
26. Agarwal, A.K.; Dhar, A. Experimental investigations of performance, emission and combustion characteristics of Karanja oil blends fuelled DICI engine. *Renew. Energy* **2013**, *52*, 283–291. [CrossRef]
27. Daho, T.; Vaitilingom, G.; Ouiminga, S.K.; Piriou, B.; Zongo, A.S.; Ouoba, S.; Koulidiati, J. Influence of engine load and fuel droplet size on performance of a CI engine fueled with cottonseed oil and its blends with diesel fuel. *Appl. Energy* **2013**, *111*, 1046–1053. [CrossRef]
28. Li, H.; Biller, P.; Hadavi, S.A.; Andrews, G.E.; Przybyla, G.; Lea-Langton, A. Assessing combustion and emission performance of direct use of SVO in a diesel engine by oxygen enrichment of intake air method. *Biomass Bioenergy* **2013**, *51*, 43–52. [CrossRef]

29. Rakopoulos, D.C. Combustion and emissions of cottonseed oil and its bio-diesel in blends with either n-butanol or diethyl ether in HSDI diesel engine. *Fuel* **2013**, *105*, 603–613. [CrossRef]
30. Sharon, H.; Jai Shiva Ram, P.; Jenis Fernando, K.; Murali, S.; Muthusamy, R. Fueling a stationary direct injection diesel engine with diesel-used palm oil–butanol blends—An experimental study. *Energy Convers. Manag.* **2013**, *73*, 95–105. [CrossRef]
31. Vallinayagam, R.; Vedharaj, S.; Yang, W.M.; Lee, P.S.; Chua, K.J.E.; Chou, S.K. Combustion performance and emission characteristics study of pine oil in a diesel engine. *Energy* **2013**, *57*, 344–351. [CrossRef]
32. Qi, D.H.; Lee, C.F.; Jia, C.C.; Wang, P.P.; Wu, S.T. Experimental investigations of combustion and emission characteristics of rapeseed oil–diesel blends in a two cylinder agricultural diesel engine. *Energy Convers. Manag.* **2014**, *77*, 227–232. [CrossRef]
33. Rakopoulos, D.; Rakopoulos, C.; Giakoumis, E.; Dimaratos, A.; Kakaras, E. Comparative Evaluation of Two Straight Vegetable Oils and Their Methyl Ester Biodiesels as Fuel Extenders in HDDI Diesel Engines: Performance and Emissions. *J. Energy Eng.* **2014**, *140*, A4014001. [CrossRef]
34. Krieger, R.B.; Borman, G.L. The computation of apparent heat release for internal combustion engines. *ASME* **1966**, *66-WA/DGP-4*, 1–16.
35. Heywood, J.B. *Internal Combustion Engine Fundamentals*, 1st ed.; McGraw-Hill: New York, NY, USA, 1988.
36. Reitz, R.D.; Bracco, F.B. On the dependence of spray angle and other spray parameters on nozzle design and operating conditions. *SAE* **1979**, 1–19. [CrossRef]
37. Nieto Garzón, N.A.; Oliveira, A.A.M.; Bazzo, E. An ignition delay correlation for compression ignition engines fueled with straight soybean oil and diesel oil blends. *Fuel* **2019**, *257*, 116050. [CrossRef]
38. *LabVIEW Software 2011*; National Instruments: Austin, TX, USA, 2016.
39. Garzón, N.A.N.; Zarza, H.M.; Schroeder, F.; Santos, O.A.A.M.; Bazzo, E. *Application of Statistical Procedures for the Analysis of the Combustion Chamber Pressure Curve of Compression-Ignition Engines*; Actas de MTL 2016 Jornadas Iberoamericanas de Motores Térmicos y Lubricación; Universidad Nacional de La Plata: La Plata, Argentina, 2016; pp. 279–294. (In Portuguese)
40. Reddy, P.; Krishna, D.; Mallan, K.R.; Ganesan, V. Evaluation of combustion parameters in direct injection diesel engines—An easy and reliable method. *SAE* **1993**, 159–165. [CrossRef]
41. Olikara, C.; Borman, G.L. A computer program for calculating properties of equilibrium combustion products with some applications to I.C. engines. *SAE* **1975**, 1–21.
42. Hohenberg, G.F. Advanced approaches for heat transfer calculations. *SAE* **1979**, 88.
43. Watson, N.; Janota, M.S. *Turbocharging the Internal Combustion Engine*, 1st ed.; John Wiley & Sons, Inc.: New York, NY, USA, 1982.
44. MATLAB Natick, MA: The MathWorks, Inc.; 2013. Available online: https://la.mathworks.com/products/matlab.html (accessed on 9 October 2024).
45. Kannan, D.; Pachamuthu, S.; Nabi, M.N.; Hustad, J.E.; Løvås, T. Theoretical and experimental investigation of diesel engine performance, combustion and emissions analysis fuelled with the blends of ethanol, diesel and jatropha methyl ester. *Energy Convers. Manag.* **2012**, *53*, 322–331. [CrossRef]
46. Hiroyasu, H.; Arai, M. Structures of fuel sprays in diesel engines. *SAE* **1990**, 1–15. [CrossRef]
47. Annand, W.J.D. Heat transfer in the cylinders of reciprocating internal combustion engines. *Proc. Inst. Mech. Eng.* **1963**, *177*, 973–990.

aerospace

MDPI

Article

Numerical Investigation of the Effect of Equivalent Ratio on Detonation Characteristics and Performance of CH_4/O_2 Rotating Detonation Rocket Engine

Xiao Xu [1], Qixiang Han [1,*] and Yining Zhang [2]

[1] College of Energy and Power Engineering, Nanjing University of Aeronautics and Astronautics, Nanjing 210016, China; bz2202002@nuaa.edu.cn
[2] Beijing Power Machinery Institute, Beijing 100074, China; zhangyining@casic.com.cn
* Correspondence: hqx205@nuaa.edu.cn

Abstract: Equivalent ratio (ER) is an important factor affecting detonation characteristics and propulsion performance of rotating detonation rocket engine (RDRE). In this paper, the effects of different equivalent ratios detonation characteristics and thrust performance of methane-oxygen RDRE were studied by 2D numerical simulation. The premixed reactants were injected through the injection holes to simulate the discrete injection of reactants on the injection panel in actual RDRE, the number of injection holes was 60 and 120. The results show that there is hybrid detonation mode (HDM), co-direction multi-wave detonation mode (CMM) and unstable detonation mode (UDM) in detonation combustion due to the influence of equivalent ratio and the number of injection holes, and the co-directional multi-wave detonation mode is beneficial to the thrust stability of RDRE. At the last, the number of detonation waves in RDRE decreases with the increase in the equivalent ratio, and the specific impulse (I_{sp}) increases with the increase of the equivalent ratio.

Keywords: equivalent ratio; detonation characteristics; propulsion performance

1. Introduction

Detonation is a combustion process characterized by the strong coupling of shock waves and chemical reactions. As the detonation wave propagates through the reactants, its high temperature and pressure cause spontaneous ignition. Theoretically, detonation offers an exceptionally efficient method for combustion [1]. It has the advantages of high thermal cycle efficiency and fast heat release speed [2–6]. Compared to deflagration, detonation can achieve more intense and thermodynamically favorable fuel combustion within a smaller combustion chamber [7]. Therefore, the utilization of detonation as a novel approach to enhance the performance of ramjet and rocket engines has garnered significant attention [5], leading to the development of numerous prototypes for rotating detonation engines (RDEs) and rotating detonation rocket engines (RDREs).

The RDE combustion chamber typically adopts a ring structure, with fresh reactants being sprayed from the head of the chamber. After ignition, the rotating detonation wave (RDW) propagates around the combustion chamber. The detonation product is ejected at high speed from the bottom of the combustion chamber to produce continuous and stable thrust. However, due to the instability of RDE [8,9] and propellant equivalent ratio (ER), many problems still need to be studied.

In experimental investigations, Bykovskii et al. [10] validated the rotating detonation wave in liquid rocket and ramjet engines using air and liquid oxygen as oxidants, and ascertained the control parameters for attaining continuous rotating detonation. Fotia et al. [11] carried out experimental studies on the performance of RDEs with nozzles and measured the equivalence ratio to explore the relationship between specific impulse and thrust. Russo et al. [12] analyzed the pressure rise characteristics of RDW and discovered that the propagation speed of RDW declines as the equivalence ratio decreases. Anand et al. [13] studied the propagation characteristics of hydrogen/air RDE under different injection modes, and the results showed that the ratio of reactant filling height to cell width of detonation gradually decreased under lean-fuel and rich-fuel conditions, and the rise of static pressure generated in RDE depends on the equivalent ratio of flow rate and reactant. Li et al. [14] studied the RDW propagation characteristics of hydrogen/air mixtures under various equivalence ratios and found that the average propagation frequency and speed of the RDW initially increase and subsequently decrease as the equivalence ratio rises. Deng et al. [15] found that RDE has the ability to adapt to the change in the equivalent ratio. The decrease of equivalent ratio leads to the deterioration of the reactivity of the mixture, which reduces the wave velocity in single-wave mode and leads to the conversion of the two-wave mode to single-wave mode. Xie et al. [16] examined the combustion modes of the H2/Air RDE with air mass flow rates ranging from 25 to 225 g/s and equivalence ratios from 0.6 to 1.0, identified four combustion mode subregions, including fast deflagration, unstable detonation, quasi-stable detonation, and stable detonation.

In numerical simulations, Wang et al. [17] conducted numerical studies on three-dimensional Hydrogen/Air RDE and found that the RDW propagation speed initially increases and then decreases as the equivalence ratio increases. In 2020, Zheng et al. [18] conducted numerical studies on hydrogen/air non-premixed RDE under different equivalent ratios, and the results showed that with the increase of equivalent ratio, the mode of RDW changed from single wave mode to double wave mode, and the change of equivalent ratio had an impact on the specific impulse and thrust of RDE. Zhao et al. [19,20] conducted a two-dimensional numerical study on the RDE of hydrogen/air premixed mixtures. It was found that the detonation wave number and propagation direction change with the increase in injection pressure, and the equivalent ratio is also one of the factors causing RDW chaotic propagation. Lietz et al. [21] used the large eddy simulation method to study the effects of equivalent ratio and mass flow rate on combustion performance of methane-oxygen RDRE. The results showed that increasing mass flow rate would increase combustion chamber pressure, thrust, and specific impulse, while increasing equivalent ratio would cause detonation waves change into bidirectional or flatting mode, which could not produce significant performance gain. Huang et al. [22,23] studied the expansion characteristics of nozzle and the effect of equivalent ratio on the thrust of RDE, and the results showed that the addition of a nozzle could improve RDE thrust, and the specific impulse of RDE increased with the increase of equivalent ratio. Yao et al. [24] studied the spontaneous formation of multiple detonation wavefronts in RDE through three-dimensional numerical simulation, and the results showed that the explosion caused by the collision of detonation wavefronts could induce new detonation waves and possibly change the propagation direction. The numerical study of Wu et al. [25] shows that the double wave mode improves the stability of RDE combustor, and the average outlet pressure in the double wave mode is higher than that in the single wave mode. Zhou et al. [26] found that oblique shock waves have little effect on RDE performance through two-dimensional numerical simulation, and the detonation thermodynamic cycle model in numerical simulation is consistent with the ideal ZND model. Subramanian et al. [27] found that detonation and deflagration existed simultaneously in RDE through two-dimensional numerical study, and stratification of fuel and product would cause the detonation wave velocity to decrease. Chen et al. [28] found that with the change of equivalent ratio, the propagation mode of RDW changes from single wave to double wave in the same direction and double wave collision. The velocity deficit of RDW in double wave mode is larger than that in single wave mode, specific impulse decreases with the increase of equivalent ratio and specific thrust increases with the increase of equivalent ratio in single wave mode.

According to the above research, it was found that there are many factors affecting RDW propagation characteristics and propulsion performance, and the equivalent ratio is one of the important influencing factors. The chemical characteristics of reactants, reaction heat release rate, detonation wave velocity, and propagation stability are all affected by the equivalent ratio. Therefore, the effect of equivalent ratio on RDW propagation mode and propulsive performance has important research value. Two-dimensional numerical simulations of premixed CH_4/O_2 RDRE with 60 and 120 injection holes were carried out under different equivalent ratios to solve these problems. Different from the ramjet RDE, RDRE operates in the fuel-rich range and it is necessary to study the detonation combustion characteristics and propulsion performance of RDRE under different equivalent ratios and high inlet total pressure. The purpose of this paper is to investigate the influence of equivalent ratio on detonation combustion characteristics and RDW propagation modes of RDRE with varying numbers of holes, as well as to examine the impact of different RDW propagation modes on the propulsion performance of RDRE.

2. Numerical Methods and Computational Models

2.1. Numerical Method

Fluent is used for numerical simulation in this study, the simulation involves the high-speed flow process and chemical reaction process of fluid in rotating detonation, and the transport phenomena such as viscosity, thermal diffusion, and mass diffusion are considered. Therefore, two-dimensional compressible Navier-Stokes equations in cartesian coordinates are used in numerical simulation:

$$\frac{\partial U}{\partial t}+\frac{\partial F_1}{\partial x}+\frac{\partial F_2}{\partial y}=\frac{\partial G_1}{\partial x}+\frac{\partial G_2}{\partial y}+S \tag{1}$$

$$U=(\rho_1,\cdots,\rho_N,\rho u,\rho v,E)^T \tag{2}$$

$$F_1=[\rho_1 u,\cdots,\rho_N u,\rho u^2+p,\rho uv,u(E+p)]^T \tag{3}$$

$$F_2=[\rho_1 v,\cdots,\rho_N v,\rho uv,\rho v^2+p,v(E+p)]^T \tag{4}$$

$$G_1=\left[\rho D_1\frac{\partial Y_1}{\partial x},\cdots,\rho D_N\frac{\partial Y_N}{\partial x},\tau_{xx},\tau_{xy},q_x+u\tau_{xx}+v\tau_{xy}\right]^T \tag{5}$$

$$G_2=\left[\rho D_1\frac{\partial Y_1}{\partial y},\cdots,\rho D_N\frac{\partial Y_N}{\partial y},\tau_{yx},\tau_{yy},q_y+u\tau_{yx}+v\tau_{yy}\right]^T \tag{6}$$

$$S=[0,\cdots,0,\rho f_x,\rho f_y,\rho(uf_x+vf_y)+\rho\dot{q}]^T \tag{7}$$

In the above equations, F_1 and F_2 are the convection vectors, G_1 and G_2 are the viscous convection fluxes, p, u, and v represent pressure, velocity in the x direction and velocity in the y direction, respectively, D is the Diffusion coefficient. The total energy E and the density ρ are calculated by:

$$\rho=\sum_{i=1}^{N}\rho_i, i=1,2,\cdots,N \tag{8}$$

$$E = \sum_{i=1}^{N} \rho_i h_i + \frac{1}{2}\rho(u^2 + v^2) \quad (9)$$

where h is the specific enthalpy.

The turbulence model adopted the SST k-ω model, which is detailed as follows, and the specific symbols and parameters should be seen in relevant literature [29]:

$$\frac{\partial}{\partial t}(\rho k) + \frac{\partial}{\partial x_i}(\rho k u_i) = \frac{\partial}{\partial x_j}(\Gamma_k \frac{\partial k}{\partial x_j}) + G_k - Y_k + S_k \quad (10)$$

$$\frac{\partial}{\partial t}(\rho \omega) + \frac{\partial}{\partial x_i}(\rho \omega u_i) = \frac{\partial}{\partial x_j}(\Gamma_\omega \frac{\partial \omega}{\partial x_j}) + G_\omega - Y_\omega + S_\omega \quad (11)$$

A coupled algorithm was used in this paper, the stiff chemical solver based on Arrhenius equation was used for the chemical reaction, and the FFCMy-12 [30] chemical mechanism was used to model the chemical reactions. This chemical mechanism has been successfully applied to the three-dimensional large eddy simulation (LES) numerical calculation of RDRE [21,31]. The second order upwind scheme was used for pressure discretization and the third order QUICK scheme was used for discretization of other spatial convection terms to achieve accurate solutions for the flow field. The second order implicit scheme was used for time discretization. The specific heat, thermal conductivity, and viscosity of the material are calculated by ideal gas mixture model based on kinetic theory.

2.2. Computational Model

The annular RDRE combustion chamber can be approximated as a two-dimensional computational domain for numerical analysis due to its small height relative to its diameter and limited radial variation imposed by the flow field, which helps save calculation resources. Although this approach neglects three-dimensional effects such as wall boundary layer, channel curvature, and detonation wave structure [32,33], numerous two-dimensional RDE numerical studies [23,34–41] have demonstrated that it effectively captures the key characteristics and flow field structure of rotating detonation propagation in RDRE, including the presence of detonation waves, oblique shock waves, contact surfaces between fresh reactants and combustion products, and slip lines.

The two-dimensional calculation model of RDRE is shown in Figure 1. The total length in x direction is 300 mm, the total height in y direction is 95 mm, the height of the plenum is 20 mm, the height of the injection holes is 5 mm, and the height of the combustion zone is 70 mm. Since the reactants in the experiment are usually injected into the combustion chamber through small holes or slits, discrete injection holes were used to simulate the discrete distribution of the reactants in practice in this paper. The injection hole of the calculation domain is shown in Figure 1. The inlet boundary is alternatively distributed by both the inlet and the wall. The wall is the adiabatic slip boundary. The total number of injection holes is 60 and 120, with a consistent injection area. The ratio between the injection holes and the wall is $\frac{1}{4}$. The patch method is adopted for detonation ignition, the specific location is shown in Figure 1. The size of the ignition region is 10 mm $\times$ 10 mm square, the ignition temperature is 2000 K, and the pressure is 2 MPa. At the same time, the ignition region is given at a speed of 2000 m/s in the x direction. The red and dark blue regions are filled with the premix CH_4/O_2 gas from the plenum, which is maintained at the pressure of 1 MPa and at the temperature of 300 K. The equivalent ratio is set according to the requirements of different cases. The gas composition in the yellow region is modified from CH_4/O_2 premixed gas to CO_2/H_2O premixed gas with a mass ratio of 1:1 by patch method. A non-reflecting pressure outlet boundary is applied at the outlet, where the conditions are set to the ambient temperature (300 K) and pressure (101,325 Pa). It should be noted that chemical reactions in the regions of injection holes and plenum are shut off to prevent flame flash back into the plenum.

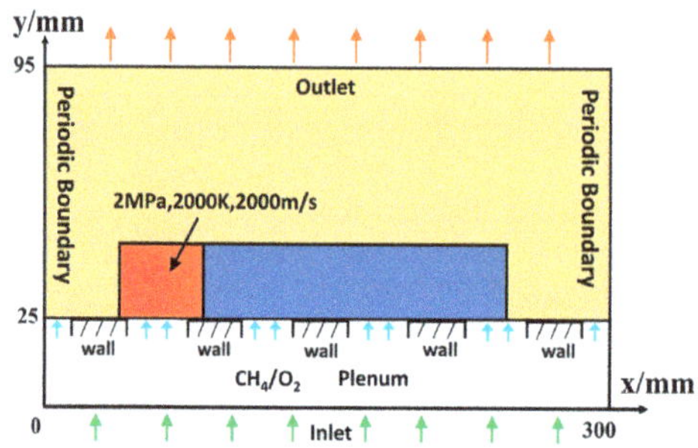

Figure 1. Two-dimensional RDRE calculation domain.

3. Verification of Numerical Simulation Method

A 40 cm long two-dimensional detonation tube is numerically simulated and compared with the Chapman-Jouguet (C-J) theoretical velocity and theoretical temperature calculated by NASA CEA [42] to verify the reliability of the numerical method and the mesh size. Three mesh sizes of 0.2 mm, 0.4 mm, and 0.5 mm were used to verify the mesh resolution. The detonation tube is filled with methane-oxygen premixed gas with an equivalent ratio of 1, a gas pressure of 101,325 Pa and a gas temperature of 300 K. Figure 2 depicts the distribution of detonation pressure with three different grid resolutions. The distribution of pressure drop in the three grids is basically the same, which means that the grid resolution has been reasonably converged. Therefore, a mesh size of 0.4 mm is used in numerical simulation to ensure the calculation accuracy and save the calculation resources. Figure 3 shows the distribution of detonation wave pressure and velocity at different times with 0.4 mm grid. In this study, the propagation velocity and the temperature of detonation wave are 2659 m/s and 3958 K, the C-J theoretical velocity is 2390 m/s with an error of 11% and the C-J theoretical temperature is 3721 K with an error of 6.4%. The results show that the simulation results are in good agreement with the C-J theoretical value calculated by NASA CEA. Due to the high activity of oxygen as an oxidizer, the simulation results are acceptable in this paper and the reliability of the numerical method in this study can be verified from C-J velocity and C-J temperature. The method proposed in this paper is suitable for the two-dimensional numerical study of methane-oxygen detonation in compressible reactive fluids.

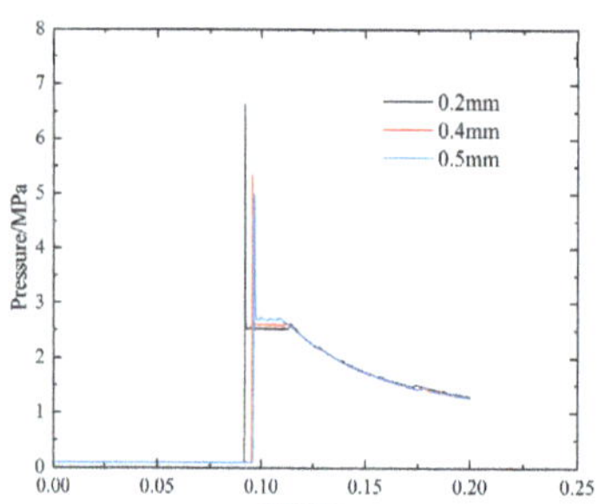

Figure 2. Pressure distribution of different grid numbers.

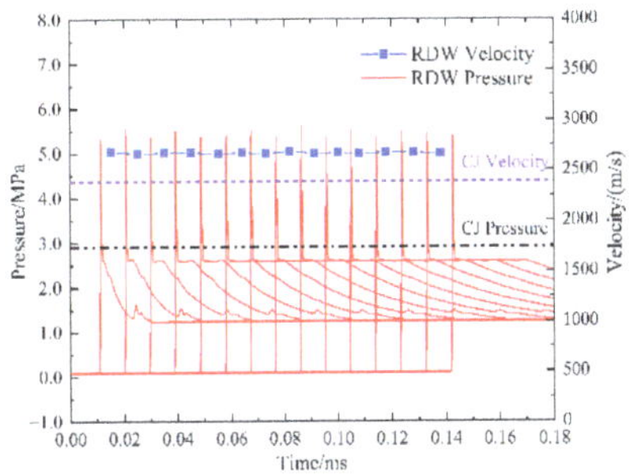

Figure 3. Pressure and velocity distribution in 0.4 mm grid at different times.

Figure 4 shows the comparison between the numerical simulation flow field in this study and the experimental results of Bykovskii et al. [10]. The numerical simulation of methane-oxygen RDRE with a total inlet temperature of 300 K, a total pressure of 1 MPa and an equivalent ratio of 1.4. The result shows that the numerical flow field structure in this study is consistent with the propane-oxygen RDE flow field structure obtained by Bykovskii et al. [10], which further verifies the reliability of the numerical calculation methods in this paper.

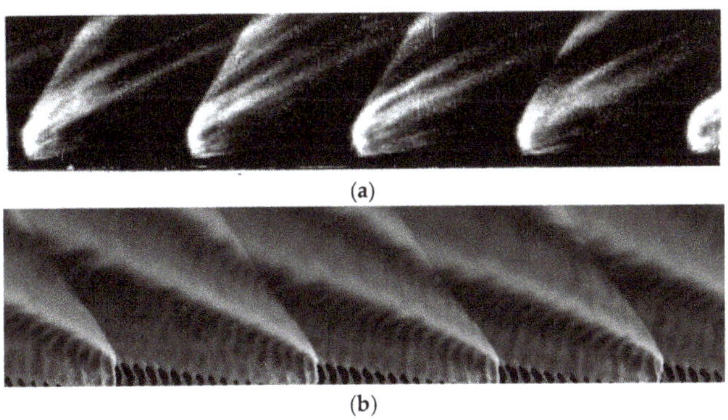

Figure 4. Comparison of numerical and experimental flow field. (**a**) Flow field structure of propane-oxygen RDE captured by Bykovskii et al. [10]. (**b**) Numerical flow field structure of RDRE in this study.

4. Results and Discussion

4.1. Influence of Equivalent Ratio on RDRE Flow Field Structure

This section discusses the effect of equivalent ratio on the structure of RDRE detonation flow field. RDRE simulation results under different equivalent ratios are shown in Table 1. P_{inlet} and T_{inlet} are the inlet total pressure and the inlet total temperature of premixed CH_4/O_2 in plenum. According to the results, there are three modes of detonation propagation. They are hybrid detonation mode (HDM), co-directional multi-wave mode (CMM), and unstable detonation mode (UDM). Detonation ignition failed in case B5, while the other cases are in detonation combustion mode.

Table 1. Numerical simulation results.

ER	P_{inlet}	T_{inlet}	60 Injection Holes			120 Injection Holes		
	MPa	K	Case	Wave Number	Mode	Case	Wave Number	Mode
1.0	1.0	300	A1	Unstable	HDM	B1	10	CMM
1.2	1.0	300	A2	6	CMM	B2	8	CMM
1.4	1.0	300	A3	4	CMM	B3	8	CMM
1.6	1.0	300	A4	3	CMM	B4	Unstable	UDM
1.8	1.0	300	A5	2	CMM	B5	Failure	Failure

When the number of injection holes is 60, two RDRE detonation flow field structures appear: the hybrid detonation mode with a large amount of deflagration in case A1 and the co-directional multi-wave mode in the other cases. As shown in Figure 5, the temperature and static pressure contours of RDRE flow field from top to bottom are case A1~A5. Since the flow field of the plenum and injection holes is not the focus of research, it is not discussed in this study.

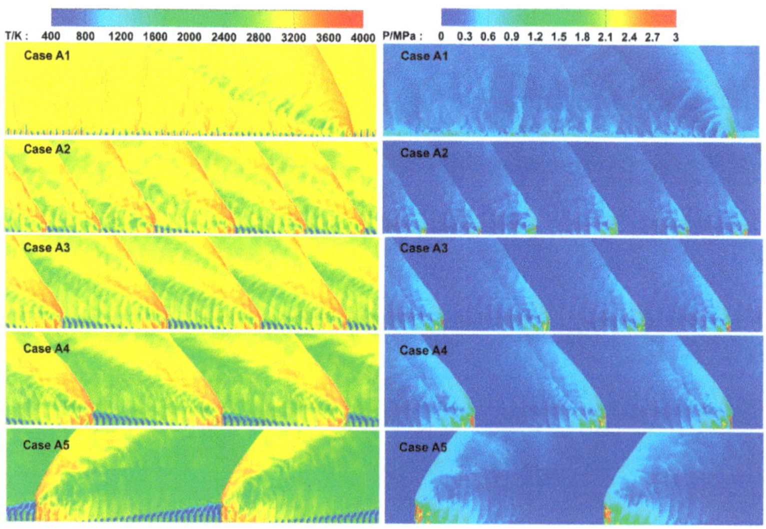

Figure 5. Contours of temperature and static pressure in case A1~A5.

As shown in Figure 6, the flow field of case A4 is a typical two-dimensional multi-wave detonation flow structure, where A is the detonation wave, B is the oblique shock wave, C is the slip line, D is the contact surface, and E is the triangular reactant filling zone. The detonation wave and oblique shock wave are uniformly distributed, which is a standard multi-wave stable propagation structure of the detonation in the same direction, it is also consistent with the flow field structure simulated by Huang et al. [23]. The reactants enter the RDRE through the injection holes. After the detonation wave passes through, deflagration regions are distributed in strips. Under this condition, the proportion of deflagration in the reactant filling zone is small, because the high equivalent ratio reduces reactant activity, resulting in less heat accumulation in the deflagration zone, and large-scale deflagration could not be formed.

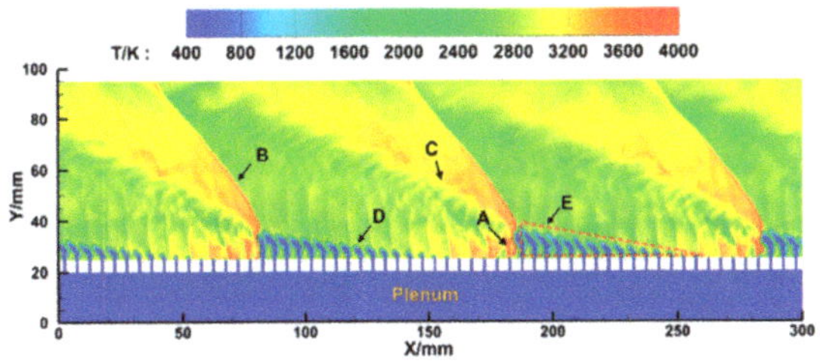

Figure 6. Contours of temperature in case A4.

The RDRE flow field structure of case A1 is relatively special. There are both detonation waves and a large amount of deflagration in the flow field, which belongs to the detonation-deflagration mixed combustion. The flow field structure of case A1 is similar to the tangential instability in conventional rocket engines. When ER = 1, the reactant activity is the highest, and the reactant starts to burn when it enters the combustion chamber, and a reactant layer of sufficient height cannot be formed. It also can be noted that the height of the reactant filling zone under this condition is the lowest. A large amount of fuel is consumed by deflagration, resulting in insufficient energy to support the propagation of detonation waves. Meanwhile, the accumulated heat of deflagration also induces the formation of new detonation waves. These two factors together lead to the instability of detonation wave propagation. Detonation-deflagration mixed combustion and any non-ideal, non-detonation processes are harmful to the performance of RDRE and will consume chemical energy and reduce the proportion of energy used by detonation [43,44]. The optimal combustion method in RDRE is to maximize the proportion of fuel used for detonation in the combustion chamber. Figure 7 shows the propagation process of detonation waves in case A1, and the number and direction of RDWs vary at different times.

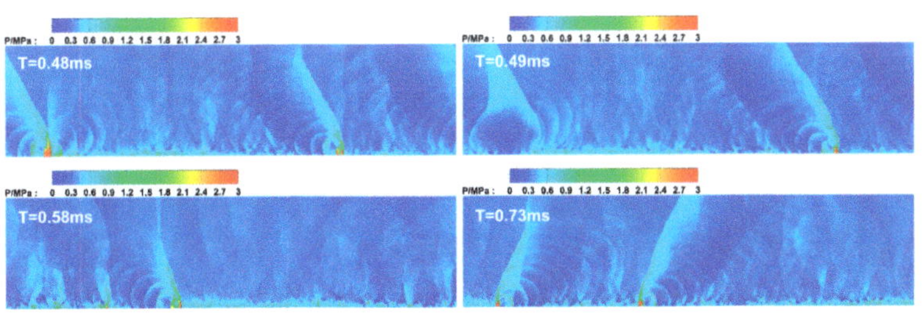

Figure 7. Contours of static pressure of case A1 at different time.

Figure 8 shows the local OH distribution contour of Case A1~A5. There is a qualitative positive correlation between the concentration of OH and the heat release rate, and the mass fraction of OH represents the magnitude of reaction intensity and heat release rate [45,46]. It can be clearly found that a large amount of deflagration occurs after fuel injection under condition A1. A large amount of OH was distributed around the injection holes and in the whole chamber. For the remaining cases, the OH mass fraction in the detonation wave front and post-wave regions decreases with the increase of the equivalent ratio, and the OH distribution is concentrated in the detonation wave and oblique shock wave regions, indicating that the decrease of deflagration ratio in RDRE and post-detonation wave region. The decrease in reactant activity leads to the decrease in heat release rate and intensity of detonation combustion, reduces the probability of inducing new detonation wave formation, reduces the number of detonation waves, and increases the proportion of reactant used for detonation. In general, the decrease in reactant activity will lead to a decrease in the number of detonation waves in the RDRE and an increase in the height of the reactant filling zone, thus affecting the flow field structure in the RDRE.

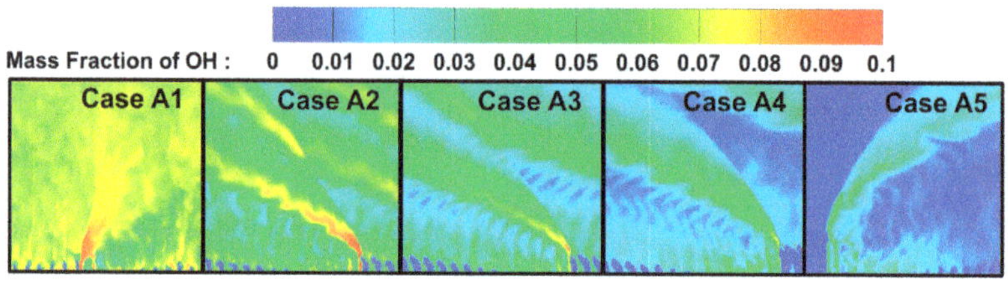

Figure 8. OH distribution contours of detonation wave front in case A1~A5.

When the number of injection holes is 120, CMM and UDM appear as shown in Figure 9. Figure 9 is the temperature and pressure contours of Cases B1~B4. The flow field structure in CMM will not be discussed. For UDM, take case B4 as an example, the flow field is in the state of unstable detonation propagation. As shown in Figure 10, in the unstable propagation mode, the number of detonation waves in the flow field is lower, the height of the reactant filling zone is higher, the distribution of detonation waves in the flow field is irregular, and linear oblique shock waves are not formed due to the impact of detonation wave collisions. When the detonation wave collides at t = 1.1 ms, the weaker detonation wave is decoupling and reburning after the collision, repeating the process of detonation wave collision, initiation, and annihilation, resulting in the unstable propagation mode of RDRE, Yao et al. [24] and Xia et al. [47] also observed a similar reinitiation phenomenon. The heat accumulated in small deflagration region is insufficient to induce the formation of RDW in the condition of low reactant activity, and even the deflagration in some areas disappears, such as the filling zone of the detonation wave front at t = 1.04 ms. The unsteady propagation mode eventually leads to the uneven distribution of temperature and pressure in the flow field and the uneven expansion of combustion products.

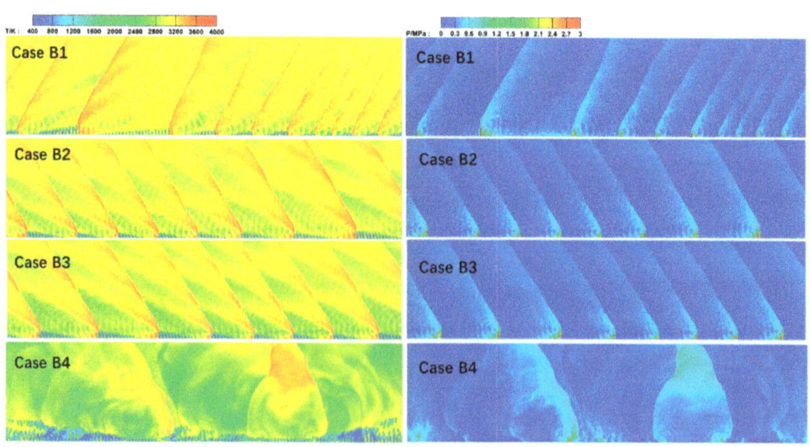

Figure 9. Contours of temperature and static pressure in cases B1~B4.

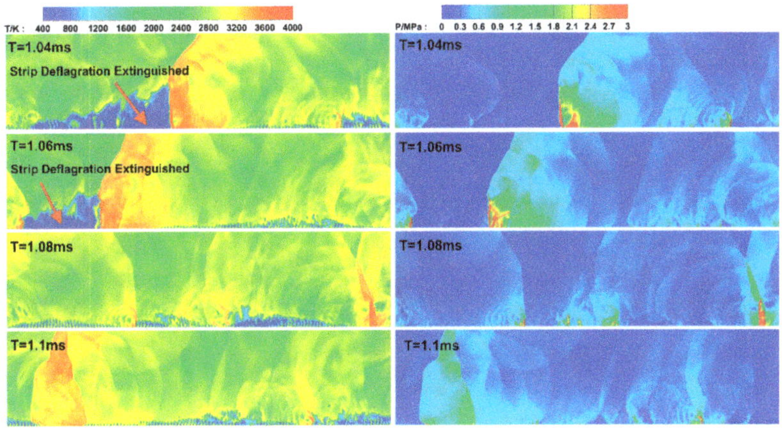

Figure 10. Detonation wave propagation process in case B4.

Different from case A1, case B1 does not operate in hybrid detonation mode. Figure 11 shows the contours of reaction heat in Case A1 and B1. It can be observed that a large amount of deflagration consumes reactants under the condition of high reactant activity, resulting in the accumulation of a large amount of reaction heat. It is known that hotspots easily form detonation wave under conditions of high reactant activity. However, excessive consumption of reactant during deflagration results in insufficient energy for the stable propagation of detonation waves [43,44], ultimately leading to unstable propagation.

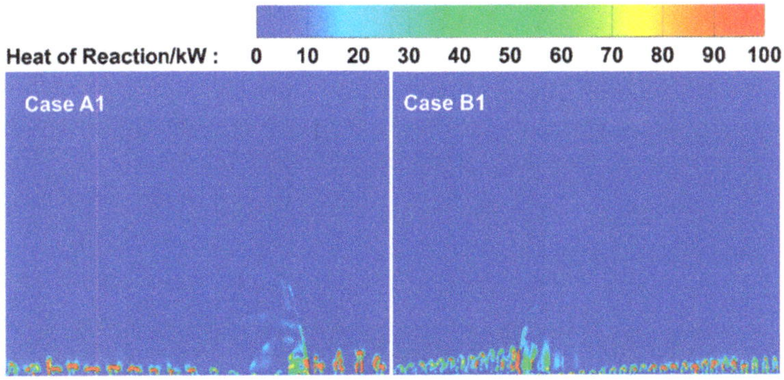

Figure 11. Heat of reaction counters in case A1 and case B1.

In conclusion, under the same inlet conditions, the increase in the number of injection holes improves the dispersion degree of the reactants, induces the formation of multi-wave propagation modes under the condition of high reactant activity (ER $\leq$ 1.4), and leads to the formation of more RDWs but reduces the accumulated heat in the deflagration region of the reactant filling zone. When the reactant activity is low (ER = 1.6) and the number of injection holes is 120, an unstable detonation propagation mode was formed, and the detonation failed at higher equivalent ratio. However, stable detonation is achieved with 60 injection holes under low reactant activity, because more heat accumulated in large strip deflagration regions increases the reactant activity. In general, the increase in the equivalent ratio leads to the decrease in reactant activity, resulting in a decrease of the probability of multiple detonation wave formation in the initiation stage, a decrease in the number of

RDWs and a decrease of the ratio of deflagration in RDRE. These also lead to an increase in the height of the reactant filling zone and RDWs. For optimal performance of RDRE, the ratio of detonation combustion should be maximized, and the detonation propagation mode should be in CMM.

4.2. Effect of Equivalent Ratio on Propagation Characteristics of RDW

The above research indicates that the equivalent ratio has a significant influence on the flow field of RDRE, thereby exerting a substantial impact on the propagation characteristics of RDW. Figures 12–14 illustrate the temporal evolution of static pressure and temperature at the sampling point of the RDRE (x = 90 mm, y = 28 mm). Figure 12 shows the temperature and static pressure history diagram of case A1. The static pressure and temperature curve shows that the propagation mode of condition A1 is in the single wave/multi-wave conversion mode. The arrows in the figure indicate the multi-wave region, where the single wave/multi-wave conversion is continuously carried out in the flow field. The reason for this chaotic pattern [48] is that hotspots around the reactants will lead to the formation of new detonation waves, while there are not enough reactants to maintain the intensity and propagation of the detonation waves, resulting in constant decoupling and re-initiation of the detonation waves. In addition, discrete injection holes will generate multiple reflected waves after the detonation wave, and stronger reflected waves may also be transformed into detonation waves [35,49]. According to previous studies, when the equivalent ratio is 1, the reactant activity is the largest, and a large number of hotspots near the injection holes can induce the formation of new detonation waves, but there are not enough reactants to maintain the stable propagation of detonation waves, and it is impossible to convert from chaotic mode to stable detonation propagation mode. Therefore, the HDM is an unstable propagation mode. Under the condition of 60 injection holes, a stable multi-wave propagation mode is formed. Except for case A1, the number of detonation waves in other stable propagation conditions shows a downward trend, and the time required to form a stable detonation mode is relatively similar. Meanwhile, the peak temperature and pressure of detonation wave are less affected by the changes in equivalent ratio. The peak temperature of detonation wave is around 3900 K, and the peak static pressure of the RDW is around 2 MPa.

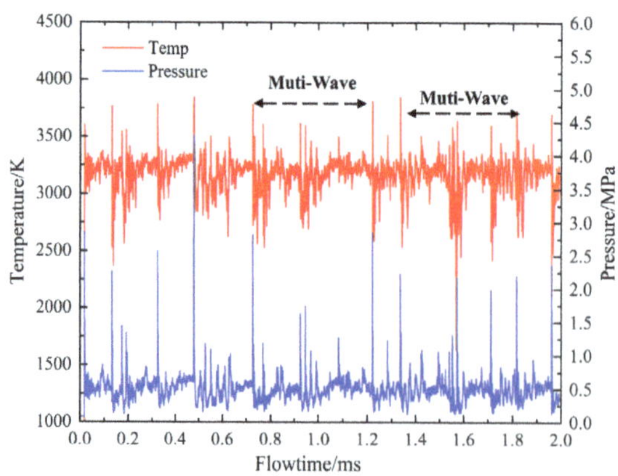

Figure 12. Pressure and temperature history diagram of case A1.

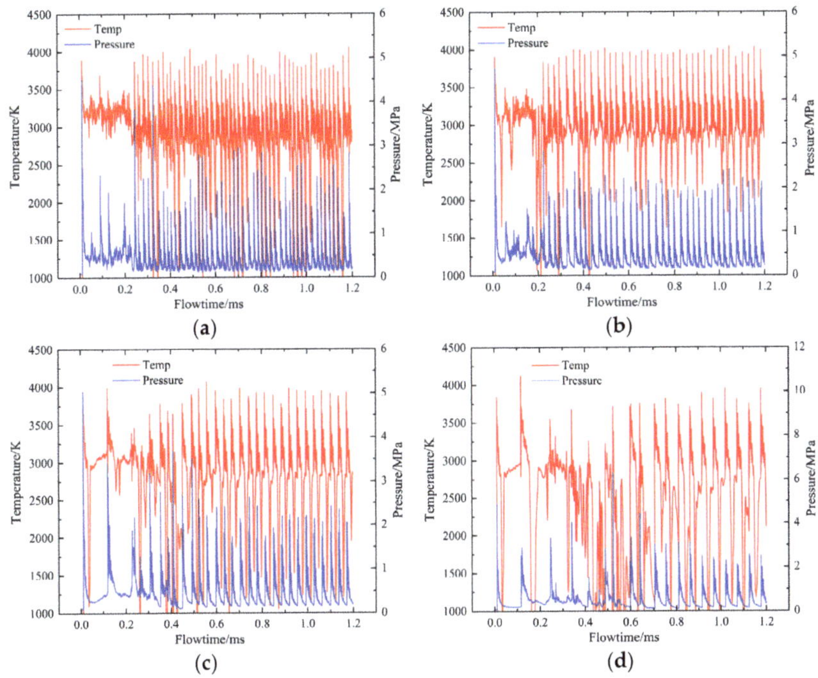

Figure 13. Pressure and temperature history diagram of case A2~A5. (**a**) Case A2. (**b**) Case A3. (**c**) Case A4. (**d**) Case A5.

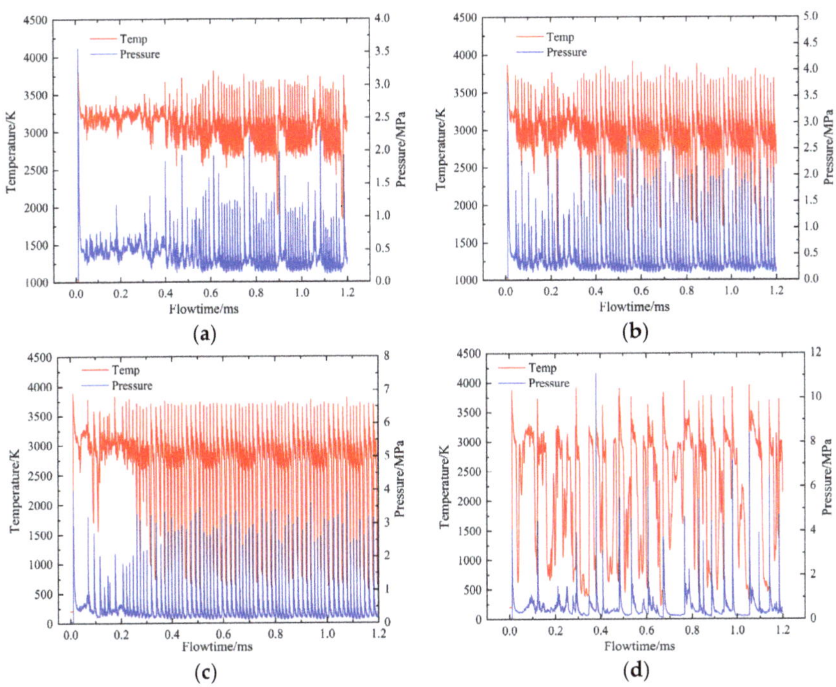

Figure 14. Pressure and temperature history diagram of cases B1~B4. (**a**) Case B1. (**b**) Case B2. (**c**) Case B3. (**d**) Case B4.

Figure 14 shows the static pressure and temperature history diagram of case B1, B2, B3, and B4 when the number of injection holes is 120. RDRE is in a stable multi-wave detonation propagation mode with equivalent ratios of 1, 1.2, and 1.4, the temperature and pressure peaks are coupled, and the flow field of RDRE is in a stable multi-wave propagation state. In case B1, with an equivalent ratio of 1, the peak temperature and pressure of RDW are significantly lower than those in other cases. This is primarily attributed to a higher proportion of reactants being consumed through deflagration rather than detonation, resulting in a lower intensity and pressure of RDW. The quantity of reactants consumed by deflagration in case B1, however, is insufficient to generate a detonation-deflagration mixing mode. When the equivalent ratio is 1.6, RDRE is in the UDM state, and it can be clearly observed that the peak pressure and temperature do not coincide, exhibiting an irregular distribution. It indicates that the detonation wave in the flow field is constantly colliding, annihilating, and reigniting. In general, as the equivalent ratio increases, there is a decreasing trend in the number of detonation waves with a closer formation time for stable detonation conditions. Simultaneously, changes in the equivalent ratio have less impact on the temperature and pressure of the detonation wave, the peak temperature of RDWs in cases B1~B3 reaches approximately 3700 K, the peak temperature is marginally lower compared to other cases with 60 injection holes under the same equivalent ratio and propagation mode. Meanwhile, the peak pressure of cases B2~B3 reaches around 2 MPa and the peak pressure of RDWs in case B1 is marginally lower compared to other cases due to a relatively reduced proportion of reactants used for detonation.

Table 2 shows the propagation velocity V_{DW} of RDWs at t = 1.2 ms and Δt_1 is the time for forming a stable co-directional multi-wave propagation mode under different equivalent ratios. The results show that with the increase of equivalent ratio, the propagation speed of RDW in stable propagation mode increases, and Δt_1 required to form stable co-directional propagation mode is within 1 ms. At the same equivalent ratio, the detonation wave propagation velocity of RDRE with fewer injection holes is faster, because more injection streams and strip deflagration regions hinder the propagation of detonation wave and resulting in the loss of detonation wave propagation velocity. On the other hand, the decrease in detonation wave velocity will also lead to the increase in the height of the filling zone, but the main factor affecting the height of the reactant filling zone is the number of detonation waves. Considering the different number of injection holes and fuel-rich conditions, stable co-directional multi-wave detonation propagation modes are more easily formed under equivalent ratio 1.2 and 1.4. However, when equivalent ratio is 1, due to the highest reactant activity, the increase in the deflagration fuel consumption ratio can form an HDM, which is not conducive to the detonation combustion of RDRE.

Table 2. V_{DW} and Δt_1 under different equivalent ratios.

ER	P_{inlet}	T_{inlet}	60 Injection Holes			120 Injection Holes		
	MPa	K	Case	Δt_1/ms	V_{DW}/(m/s)	Case	Δt_1/ms	V_{DW}/(m/s)
1.0	1.0	300	A1	-	2275	B1	0.6	2165
1.2	1.0	300	A2	0.3	2488	B2	0.39	2402
1.4	1.0	300	A3	0.26	2743	B3	0.26	2442
1.6	1.0	300	A4	0.48	2769	B4	-	2383
1.8	1.0	300	A5	0.72	2836	B5	-	-

4.3. Influence of Equivalent Ratio on RDRE Propulsion Performance

This section mainly studies the effect of equivalent ratio on RDRE propulsion performance. Specific impulse (I_{sp})is an important parameter for evaluating propulsive performance, so it is necessary to calculate I_{sp}. I_{sp} can be calculated by the following formula:

$$F = m_e v_e + (p_e - p_0)A_e \tag{12}$$

$$I_{sp} = \frac{F}{m_t g} \tag{13}$$

where F is the thrust of RDRE, m_t is the mass flow rate at the inlet of the plenum, m_e is the mass flow rate at the outlet of the RDR, v_e is the axial velocity at outlet of the RDRE, p_e is the static pressure at the outlet, p_0 is the environmental pressure at the outlet, A_e is the outlet area of RDRE, and g is the standard gravitational acceleration.

In order to eliminate the influence of the ignition stage, the data between 0.8 ms and 1.2 ms after the ignition were averaged in this paper. Figure 15 shows the average specific impulse (I_{spavg}) under different ER. When the number of injection holes is 60, the equivalent ratio increases from 1.0 to 1.8, and I_{spavg} increases from 192 s to 220 s by 14.6%. When the number of injection holes is 120, the equivalent ratio increases from 1.0 to 1.4 and from 192 s to 205 s, and the I_{spavg} of Case B4 under the unsteady detonation propagation mode is 237 s, but it has no reference value. The overall results show that when the equivalent ratio is between 1.0 and 1.8, the I_{sp} of RDRE increases with the increase of the equivalent ratio. We know that rocket engines can increase the specific impulse by increasing the gas temperature or decreasing the propellant molecular weight [50]. RDRE is consistent with conventional rocket engines, and the operating range of RDRE is different from ramjet RDE which is under lean fuel equivalent ratio. Figure 16 shows the average exit total temperature (T_{exit}) and molecular weight of the reactants (M_{molar}) of RDRE. The results demonstrate a decrease in T_{exit} and M_{molar} with an increase in the equivalent ratio.

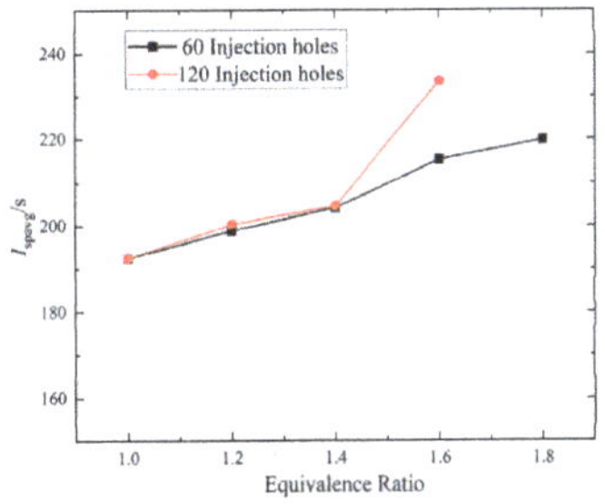

Figure 15. I_{spavg} under different equivalent ratios.

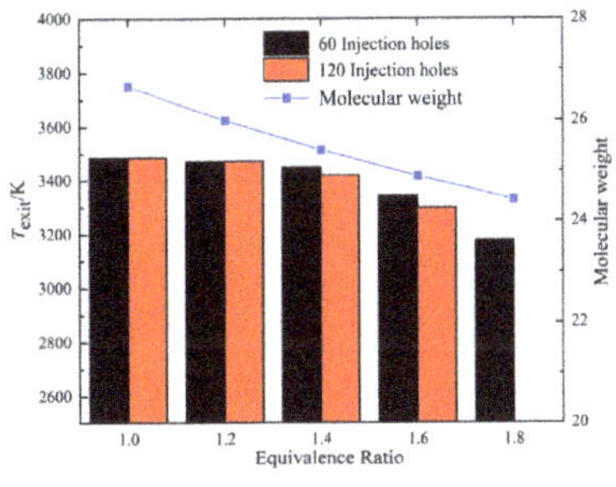

Figure 16. T_{exit} and M_{molar} under different equivalent ratios.

The curves of I_{sp} under different equivalent ratios are presented in Figure 17, exhibiting a significant oscillation during the initiation stage, which is caused by the blockage of part of the injection holes by irregular detonation waves that lead to the instability of inlet mass flow rate in the initiation stage. However, once the multi-wave stable propagation mode is reached, co-directional and regular RDWs have little effect on the blockage of injection holes and the specific impulse curve becomes stable. Under the condition of high equivalent ratio, case B4 experiences an unstable detonation mode leading to substantial oscillations in specific impulse. This instability arises due to variations in quantity, direction, and intensity of detonation waves and oblique shock waves within the RDRE flow field at different times during unstable detonation propagation mode, resulting in irregular and uneven gas expansion. Similarly, case A1 with a detonation-deflagration mixing mode also exhibits an unstable propagation mode with greater oscillation amplitude compared to other cases with 60 injection holes.

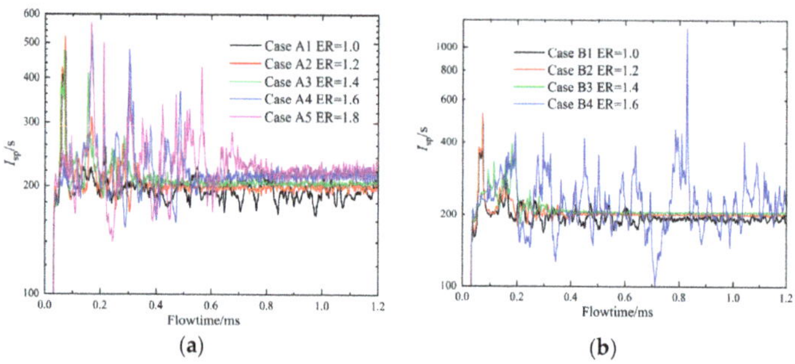

Figure 17. Specific impulse of different cases. (**a**) Specific impulse of cases with 60 injections holes. (**b**) Specific impulse of cases with 120 injections holes.

Figure 18 shows the standard deviation of thrust under different equivalent ratios between 0.8 ms and 1.2 ms after ignition. Standard deviation of thrust reflects the deviation degree between instantaneous thrust and average thrust. The thrust standard deviation of RDRE under unstable detonation propagation is greater than that under stable detonation propagation. In general, the thrust of RDRE is more stable in CMM, detonation waves are evenly distributed in RDRE and the spacing between oblique shock waves is uniform. The expansion of combustion products after oblique shock waves is more uniform and regular, thus generating more stable thrust. Combining the specific impulse and the stability of thrust, RDRE should be maintained in a state of co-directional multi-wave stable propagation to achieve more stable thrust within the range of the equivalent ratio where maximum specific impulse can be attained.

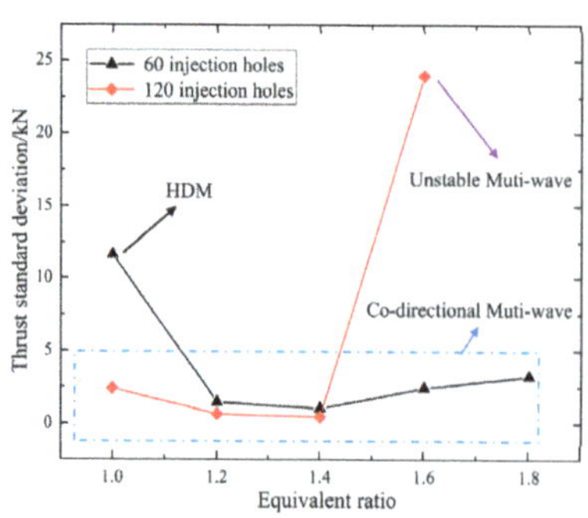

Figure 18. Standard deviation of thrust under different equivalent ratios.

5. Conclusions

In this study, two-dimensional numerical simulation of methane-oxygen RDRE was conducted. Two types of discrete injection holes were used to study the effects of different equivalent ratios (ER ranging from 1.0 to 1.8) on detonation characteristics and propulsion performance of RDRE. The conclusions are as follows:

(1) The numerical simulation results of this paper show that three distinct modes of detonation propagation are observed due to the influence of equivalent ratio and the number of injection holes. These modes include hybrid detonation mode, co-directional multi-wave detonation mode, and unstable detonation mode. The optimal propagation mode for RDRE is found to be the co-directional multi-wave detonation mode.
(2) The reactant activity has a significant influence on the flow field structure of RDRE. With the increase of equivalent ratio, the reactivity of the reactant decreases and the accumulated reaction heat in the strip deflagration regions decreases, which makes it difficult to induce more detonation waves. This leads to a decrease in the number of detonation waves and an increase in the height of both the reactant filling zone and the detonation waves.
(3) Owing to the presence of strip deflagration regions in the reactant filling zone, the accumulation of reaction heat under conditions approaching the stoichiometric ratio can readily induce the formation of new detonation waves, resulting in the unsteady propagation of detonation waves during the initiation stage. When the reactant activity is low, if the strip deflagration regions are large and concentrated, such as in cases with 60 injection holes, the accumulated heat can improve the activity of reactant and eventually lead to the formation of a stable co-directional multi-wave detonation mode; conversely, if the strip deflagration regions are small and dispersed, the accumulated reaction heat is insufficient to increase the reactant activity sufficiently to sustain a stable detonation propagation mode.
(4) Under fuel-rich conditions, the average specific impulse of RDRE increases with the increase of the equivalent ratio, which is consistent with the conventional rocket engines. Meanwhile, when RDRE operates in co-directional multi-wave detonation mode, it enhances the stability of both specific impulse and thrust, minimizing the oscillation amplitude of the specific impulse.

Author Contributions: Conceptualization, X.X.; methodology, X.X.; validation, X.X.; formal analysis, X.X.; investigation, X.X.; resources, Y.Z. and Q.H.; data curation, X.X.; writing—original draft preparation, X.X.; writing—review and editing, Q.H. and Y.Z.; supervision, Y.Z.; project administration, Y.Z. All authors have read and agreed to the published version of the manuscript.

Funding: This research received no external funding.

Data Availability Statement: The original contributions presented in this study are included in the article. Further inquiries can be directed to the corresponding author.

Conflicts of Interest: The authors declare no conflicts of interest.

References

1. Kailasanath, K. Review of propulsion applications of detonation waves. *AIAA J.* **2000**, *38*, 1698–1708. [CrossRef]
2. Zhao, X.; Wang, J.; Gao, L.; Wang, X.; Zhu, Y. Flame acceleration and onset of detonation in inhomogeneous mixture of hydrogen-air in an obstructed channel. *Aerosp. Sci. Technol.* **2022**, *130*, 107944. [CrossRef]
3. Zhao, X.; Wang, J.; Gao, L.; Pan, J.; Zhu, Y. Effect of hydrogen concentration distribution on flame acceleration and deflagration-to-detonation transition in staggered obstacle-laden channel. *Phys. Fluids* **2023**, *35*, 016124. [CrossRef]
4. Lu, F.K.; Braun, E.M. Rotating detonation wave propulsion: Experimental challenges, modeling, and engine concepts. *J. Propuls. Power* **2014**, *30*, 1125–1142. [CrossRef]

5. Zhou, R.; Wu, D.; Wang, J. Progress of continuously rotating detonation engines. *Chin. J. Aeronaut.* **2016**, *29*, 15–29. [CrossRef]
6. Zhou, S.; Ma, H.; Ma, Y.; Zhou, C.; Hu, N. Investigation on C_2H_4-Air combustion mode in a non-premixed rotating detonation combustor. *Propuls. Power Res.* **2022**, *11*, 85–96. [CrossRef]
7. Roy, G.; Frolov, S.; Borisov, A.; Netzer, D. Pulse detonation propulsion: Challenges, current status, and future perspective. *Prog. Energy Combust. Sci.* **2004**, *30*, 545–672. [CrossRef]
8. Anand, V.; George, A.S.; Driscoll, R.; Gutmark, E. Characterization of instabilities in a rotating detonation combustor. *Int. J. Hydrog. Energy* **2015**, *40*, 16649–16659. [CrossRef]
9. Anand, V.; George, A.S.; Gutmark, E. Amplitude modulated instability in reactants plenum of a rotating detonation combustor. *Int. J. Hydrog. Energy* **2017**, *42*, 12629–12644. [CrossRef]
10. Bykovskii, F.A.; Zhdan, S.A.; Vedernikov, E.F. Continuous spin detonations. *J. Propuls. Power* **2006**, *22*, 1204–1216. [CrossRef]
11. Fotia, M.L.; Schauer, F.; Kaemming, T.; Hoke, J. Experimental study of the performance of a rotating detonation engine with nozzle. *J. Propuls. Power* **2016**, *32*, 674–681. [CrossRef]
12. Russo, R.; King, P.; Schauer, F.; Thomas, L. Characterization of pressure rise across a continuous detonation engine. In Proceedings of the 47th AI-AA/ASME/SAE/ASEE Joint Propulsion Conference & Exhibit, San Diego, CA, USA, 31 July–3 August 2011; p. 6046. [CrossRef]
13. Anand, V.; George, A.S.; Driscoll, R.; Gutmark, E. Investigation of rotating detonation combustor operation with H_2-Air mixtures. *Int. J. Hydrog. Energy* **2016**, *41*, 1281–1292. [CrossRef]
14. Li, B.; Wu, Y.; Weng, C.; Zheng, Q.; Wei, W. Influence of equivalence ratio on the propagation characteristics of rotating detonation wave. *Exp. Therm. Fluid Sci.* **2018**, *93*, 366–378. [CrossRef]
15. Deng, L.; Ma, H.; Xu, C.; Liu, X.; Zhou, C. The feasibility of mode control in rotating detonation engine. *Appl. Therm. Eng.* **2018**, *129*, 1538–1550. [CrossRef]
16. Xie, Q.; Wen, H.; Li, W.; Ji, Z.; Wang, B.; Wolanski, P. Analysis of operating diagram for H2/Air rotating detonation combustors under lean fuel condition. *Energy* **2018**, *151*, 408–419. [CrossRef]
17. Wang, Y.; Wang, J. Effect of equivalence ratio on the velocity of rotating detonation. *Int. J. Hydrog. Energy* **2015**, *40*, 7949–7955. [CrossRef]
18. Zheng, H.; Meng, Q.; Zhao, N.; Li, Z.; Deng, F. Numerical investigation on H2/Air non-premixed rotating detonation engine under different equivalence ratios. *Int. J. Hydrog. Energy* **2020**, *45*, 2289–2307. [CrossRef]
19. Zhao, M.; Zhang, H. Origin and chaotic propagation of multiple rotating detonation waves in hydrogen/air mixtures. *Fuel* **2020**, *275*, 117986. [CrossRef]
20. Zhao, M.; Cleary, M.J.; Zhang, H. Combustion mode and wave multiplicity in rotating detonative combustion with separate reactant injection. *Combust. Flame* **2021**, *225*, 291–304. [CrossRef]
21. Lietz, C.; Ross, M.; Desai, Y.; Hargus, W.A.; Kailasanath, K.; Wilhite, J.; Driscoll, R.B.; George, A.C.S.; Anand, V.; Gutmark, E.J.; et al. Numerical investigation of operational performance in a methane-oxygen rotating detonation rocket engine. In Proceedings of the AIAA Scitech 2020 Forum, Orlando, FL, USA, 6–10 January 2020. [CrossRef]
22. Huang, Y.; Xia, H.; Chen, X.; Luan, Z.; You, Y. Shock dynamics and expansion characteristics of an aerospike nozzle and its interaction with the rotating detonation combustor. *Aerosp. Sci. Technol.* **2021**, *117*, 106969. [CrossRef]
23. Luan, Z.; Huang, Y.; Gao, S.; You, Y. Formation of multiple detonation waves in rotating detonation engines with inhomogeneous methane/oxygen mixtures under different equivalence ratios. *Combust. Flame* **2022**, *241*, 112091. [CrossRef]
24. Yao, S.; Liu, M.; Wang, J. Numerical investigation of spontaneous formation of multiple detonation wave fronts in rotating detonation engine. *Combust. Sci. Technol.* **2015**, *187*, 1867–1878. [CrossRef]
25. Wu, D.; Zhou, R.; Liu, M.; Wang, J. Numerical investigation of the stability of rotating detonation engines. *Combust. Sci. Technol.* **2014**, *186*, 1699–1715. [CrossRef]
26. Zhou, R.; Wang, J.-P. Numerical investigation of flow particle paths and thermodynamic performance of continuously rotating detonation engines. *Combust. Flame* **2012**, *159*, 3632–3645. [CrossRef]
27. Subramanian, S.; Meadows, J. Novel approach for computational modeling of a non-premixed rotating detonation engine. *J. Propuls. Power* **2020**, *36*, 617–631. [CrossRef]
28. Chen, H.; Si, C.; Wu, Y.; Hu, H.; Zhu, Y. Numerical investigation of the effect of equivalence ratio on the propagation characteristics and performance of rotating detonation engine. *Int. J. Hydrog. Energy* **2023**, *48*, 24074–24088. [CrossRef]
29. Menter, F.R. Two-equation eddy-viscosity turbulence models for engineering applications. *AIAA J.* **1994**, *32*, 1598–1605. [CrossRef]
30. Xu, R.; Dammati, S.S.; Shi, X.; Genter, E.S.; Jozefik, Z.; Harvazinski, M.E.; Lu, T.; Poludnenko, A.Y.; Sankaran, V.; Kerstein, A.R.; et al. Modeling of high-speed, methane-air, turbulent combustion, Part II: Reduced methane oxidation chemistry. *Combust. Flame* **2024**, *263*, 113380. [CrossRef]
31. Schau, K.A.; Oefelein, J.C. Numerical analysis of wave characteristics in a methane-oxygen rotating detonation engine. *AIAA J.* **2023**, *61*, 97–111. [CrossRef]

32. Smirnov, N.; Nikitin, V.; Stamov, L.; Mikhalchenko, E.; Tyurenkova, V. Three-dimensional modeling of rotating detonation in a ramjet engine. *Acta Astronaut.* **2019**, *163*, 168–176. [CrossRef]
33. Smirnov, N.; Nikitin, V.; Stamov, L.; Mikhalchenko, E.; Tyurenkova, V. Rotating detonation in a ramjet engine three-dimensional modeling. *Aerosp. Sci. Technol.* **2018**, *81*, 213–224. [CrossRef]
34. Liu, P.; Li, Q.; Huang, Z.; Zhang, H. Interpretation of wake instability at slip line in rotating detonation. *Int. J. Comput. Fluid Dyn.* **2018**, *32*, 379–394. [CrossRef]
35. Zhao, M.; Li, J.-M.; Teo, C.J.; Khoo, B.C.; Zhang, H. Effects of variable total pressures on instability and extinction of rotating detonation combustion. *Flow Turbul. Combust.* **2020**, *104*, 261–290. [CrossRef]
36. Schwer, D.; Kailasanath, K. Feedback into mixture plenums in rotating detonation engines. In Proceedings of the 50th AIAA Aerospace Sciences Meeting Including the New Horizons Forum and Aerospace Exposition, Nashville, TN, USA, 9–12 January 2012; p. 617. [CrossRef]
37. Liu, X.-Y.; Luan, M.-Y.; Chen, Y.-L.; Wang, J.-P. Propagation behavior of rotating detonation waves with premixed kerosene/air mixtures. *Fuel* **2021**, *294*, 120253. [CrossRef]
38. Wen, H.; Fan, W.; Xu, S.; Wang, B. Numerical study on droplet evaporation and propagation stability in normal-temperature two-phase rotating detonation system. *Aerosp. Sci. Technol.* **2023**, *138*, 108324. [CrossRef]
39. Li, Q.; Liu, P.; Zhang, H. Further investigations on the interface instability between fresh injections and burnt products in 2-D rotating detonation. *Comput. Fluids* **2018**, *170*, 261–272. [CrossRef]
40. Schwer, D.; Kailasanath, K. Numerical investigation of the physics of rotating-detonation-engines. *Proc. Combust. Inst.* **2011**, *33*, 2195–2202. [CrossRef]
41. Zhang, S.; Yao, S.; Luan, M.; Zhang, L.; Wang, J. Effects of injection conditions on the stability of rotating detonation waves. *Shock. Waves* **2018**, *28*, 1079–1087. [CrossRef]
42. NASA Glenn Research Center. *CEAgui-Win, Chemical Equilibrium with Applications GUI for Windows, Software Package, Version 28*; NASA Glenn Research Center: Cleveland, OH, USA, 2005.
43. Chacon, F.; Gamba, M. Study of parasitic combustion in an optically accessible continuous wave rotating detonation engine. In Proceedings of the AIAA Scitech 2019 Forum, San Diego, CA, USA, 7–11 January 2019. [CrossRef]
44. Burr, J.R.; Yu, K. Detonation wave propagation in cross-flow of discretely spaced reactant jets. In Proceedings of the 53rd AIAA/SAE/ASEE Joint Propulsion Conference, Atlanta, GA, USA, 10–12 July 2017; p. 4908. [CrossRef]
45. Hardalupas, Y.; Orain, M. Local measurements of the time-dependent heat release rate and equivalence ratio using chemiluminescent emission from a flame. *Combust. Flame* **2004**, *139*, 188–207. [CrossRef]
46. Röder, M.; Dreier, T.; Schulz, C. Simultaneous measurement of localized heat-release with OH/CH2O–LIF imaging and spatially integrated OH∗ chemiluminescence in turbulent swirl flames. *Proc. Combust. Inst.* **2013**, *34*, 3549–3556. [CrossRef]
47. Xia, Z.-J.; Luan, M.-Y.; Liu, X.-Y.; Wang, J.-P. Numerical simulation of wave mode transition in rotating detonation engine with OpenFOAM. *Int. J. Hydrog. Energy* **2020**, *45*, 19989–19995. [CrossRef]
48. Chen, H.; Si, C.; Hu, H.; Jin, Y.; Zhu, Y. Effects of the perturbation inlet on the evolution and oscillation characteristics of multiple rotating detonation waves. *Aerosp. Sci. Technol.* **2023**, *141*, 108586. [CrossRef]
49. Meng, H.; Weng, C.; Xiao, Q.; Zheng, Q.; Wu, Y.; Wang, F. Numerical analysis on evolution process of multiple rotating detonation waves with ethylene–oxygen–nitrogen mixture. *Proc. Inst. Mech. Eng. Part G J. Aerosp. Eng.* **2022**, *236*, 1304–1317. [CrossRef]
50. George, P.S.; Oscar, B. *Rocket Propulsion Elements*, 7th ed.; Science Press: Beijing, China, 2003; p. 38.

aerospace

MDPI

Article

Expanding Known Performance Capabilities of Geared Turbofan Engine When Powered by LNG and Methanol

Sergios Villette *, Alexios Alexiou, Nikolaos Aretakis and Konstantinos Mathioudakis

Laboratory of Thermal Turbomachines, School of Mechanical Engineering, National Technical University of Athens, 15780 Athens, Greece; alexioua@mail.ntua.gr (A.A.); naret@mail.ntua.gr (N.A.); kmathiou@mail.ntua.gr (K.M.)

* Correspondence: ser_villet@mail.ntua.gr

Abstract: As aviation demand rises, fossil jet fuel consumption follows, thus increasing focus on sustainable aviation fuels to reduce aircraft greenhouse gas emissions. While advanced technologies and optimized operations play a role, alternative fuels, especially non-drop-in options like Liquefied Natural Gas (LNG) and methanol, offer promising potential for significant emission reductions if used in current aero-engines. LNG, a candidate near-term replacement of fossil jet fuel and methanol, even though a less conventional option in aviation, present advantages. Both fuels showcase the ability to generate the same thrust output by also achieving lower post-combustion temperatures, thereby enhancing component life and reducing emissions. Inversely, requesting equal post-combustion temperature as the baseline kerosene operation of the engine can produce greater thrust output, a much needed result for such fuels with low volumetric energy density, which causes greater take-off thrust demand mainly due to their larger tank requirements. This study uses advanced 0-D engine models coupled with detailed chemistry 1-D burner models and mission analysis tools to assess the aforementioned trends of LNG and methanol used to power a current geared turbofan engine. The aim of this work is to provide insights into the advantages, the limitations and the overall viability of the fuels in question as less polluting aviation fuels, addressing both environmental impact and operational feasibility in future aviation applications. According to findings of this article, when compared with Jet-A, LNG can reduce post-combustion temperature by an average of 1% or increase net-thrust by 3% while lowering CO_2, NO_x and CO emissions by 20%, 46% and 39%, respectively. Adversely, methanol is capable of lessening post-combustion temperature by 3% or enhancing thrust output by 10% while also reducing CO_2, NO_x and CO emissions by an average of 6%, 60% and 38%, respectively.

Keywords: aero-engine model; geared turbofan; methanol; LNG; chemical reactor network; PROOSIS

1. Introduction

The aviation sector contributes to greenhouse gas (GHG) emissions, producing 2–3% of global carbon dioxide (CO_2) [1] and 2–5% of total anthropogenic GHGs [2–4], including CO_2, nitrous oxides and methane (CH_4) [5,6]. These emissions are apparently linked with global warming through radiative forcing, aviation-induced cloudiness, such as contrails, etc. [1,7,8]. While aviation's global impact is relatively small, its regional effects on air pollution can rival those of ground transportation, affecting public health [9,10].

Aviation's heavy reliance on fossil fuels challenges the sector's sustainability. Rising demand, geopolitical conflicts, such as the 2022 Ukraine War, and supply shortages have driven up fuel prices, creating an energy crisis [11]. Despite modern aircraft being 75–80% more fuel-efficient than those of 50 years ago [3,12], projected air traffic growth of 4.3% annually over the next 20 years [13] will lead to increased fuel consumption and a corresponding 4% rise in emissions [4].

There is thus need to explore new, alternative fuels. International agreements, like the Paris Agreement, and industrial strategies, such as the European Green Deal, aim to achieve carbon neutrality by 2050, requiring annual CO_2 emission reductions of 4–8% [12,14]. These initiatives call for significant investments in alternative aviation fuels, fuel-efficient technologies and infrastructures to meet these goals [12,15].

Out of the above pathways, this study focuses on alternative aviation fuels and specifically non drop-in fuels, namely fuels that, due to their dissimilar properties (heating value, atomization, density, etc.) compared to conventional jet fuel, require alterations on the fuel delivery systems, fuel tanks, injectors and other infrastructure. According to the road-map towards carbon neutrality [1,12], non-drop in fuels, such as Liquefied Natural Gas (LNG), are placed as a mid-term solution, while liquefied hydrogen and battery-assisted flight finally dominate in the long term [6,16]. The near-term solution of bioenergy-derived kerosene (also known as bio-kerosene) is credited with the ability to decrease net CO_2 emissions up to 80% within its life-cycle [1–3,11] and is already certified for use on commercial flights when blended with fossil kerosene [11,15,17,18]. A comparison between the more technologically mature bio-kerosene and non drop-in fuels would hold more interest from a techno-economic or life-cycle perspective. However, this falls outside the scope of the current research, which focuses on evaluating non-drop-in fuels in current aero-engines and advancing the corresponding simulation methods required to achieve this.

A recent study [19] tackles the integration of LNG in aviation by evaluating the mission of retrofitted aircraft, so as assess the trade-offs between energy consumption and CO_2 emission per range unit by taking into account only its different heating value, density and pre-calculated Emission Indexes (EIs). Another approach [20] extends even further by re-designing the engine cycle according to aircraft thrust requirements and running an aircraft mission analysis for a plethora of alternative fuels, representing each main fuel family (i.e., methane for alkanes, butanol for alcohols, bio-kerosene for drop-in SAFs, etc.). A different approach [21] suggests, via simulation, a novel propulsion system of a turbofan coupled with a fuel cell to power the aircraft's auxiliary power unit in order to evaluate performance enhancement and CO_2 reductions, with respect to only the design operating conditions when utilizing non-drop-in fuels including hydrogen, methane, methanol, ethanol, dimethyl ether and their blends.

LNG offers several advantages and drawbacks as an aviation fuel. One of its key benefits is its ability to generate about 20% less CO_2 emissions for the same heat output compared to conventional jet fuel. It is also economically feasible when derived from fossil sources due to the vast natural gas reserves and low production costs, making it a viable alternative aviation fuel [1,19]. Additionally, LNG has a higher specific energy density, leading to reduced fuel consumption. However, LNG has notable drawbacks. Its low volumetric energy density necessitates larger storage tanks, which is challenging in the limited space available on aircraft. Significant infrastructure investments are required for airports, including LNG storage tanks, liquefaction facilities, and refueling pipelines. LNG's reactive nature also introduces safety concerns, particularly in crash scenarios. Furthermore, using LNG increases the aircraft's tank weight and drag, leading to a 7–10% rise in energy consumption according to respective literature [19].

MeOH, on the other hand, is one of the most commonly produced chemicals with a broad range of applications [22]. While it is not widely used directly as aviation fuel, it mostly serves as feedstock to produce synthetic kerosene [1]. MeOH can be synthesized from fossil sources or biogenic materials, offering potential life-cycle CO_2 reduction, depending on the methods used for hydrogen production and CO_2 sourcing [1,22,23]. Combustion of MeOH emits 5.5% less CO_2 than kerosene for the same energy output while offering significant emission reduction potential such as NO_x by 75–90% and particulate matter emissions by 50% in certain applications [10,24,25]. Its high vaporization heat also lowers the post-combustion temperature of a gas turbine powerplant, reducing thermal NO_x formation and potentially enhancing turbine lifespan [24,25]. Despite these advantages, MeOH has several drawbacks. Its low specific energy density requires approximately twice more fuel mass flow compared to kerosene for the same net thrust which, when coupled with its low volumetric energy density, causes increased size and weight requirements for the aircraft, engine and wings, making the aircraft mission much more energy consuming [1]. Retrofitting propulsion systems for MeOH use requires specialized injectors and more energy-intensive pumps to handle the increased fuel workload and compensate for methanol's low lubricating properties [22,23,25]. Therefore, is a highly toxic and corrosive fuel, which necessitates the use of specialized storage containers and strict safety measures for its handling. Furthermore, methanol is only poorly miscible with kerosene, making it unsuitable for blending [25]. While MeOH shows promise in reducing emissions, its usefulness in aviation remains under-explored, and further research is needed to fully assess its potential.

While the aforementioned computational studies provide valuable insights into the integration of non drop-in fuels, they assume that the combustion process remains unaffected by changes in engine design or fuel type. To evaluate emissions within the design space, these studies rely on either fixed Emission Indices (EIs) or simple (one-reaction) stoichiometric calculations, neglecting potential fluctuations in combustion efficiency. As a result, they are unable to fully quantify key aviation pollutants such as NO_x, CO and UHC, leaving gaps in understanding the environmental impacts of such fuels. Other studies [26–30] focus on accurately predicting combustion and post-combustion properties (e.g., emissions, ignition delay time, flame speed) of alternative fuels by employing state-of-the-art physics-based tools using detailed chemical and thermodynamic calculations. Addressing this gap, the current study proposes integrating advanced 0-D turbomachinery component models [31] with a physics-based Chemical Reactor Network (CRN) model for isobaric gas turbine combustors [30] within a unified simulation framework. This framework ensures transparency, seamless intercommunication between tools, and ease of use. The proposed model is applied to evaluate the performance and emission characteristics of a modern Geared Turbofan (GTF) engine powered by two liquid non-drop-in fuels: LNG and methanol (MeOH).

The engine model developed in this study is first applied and adapted to the layout of a current generation GTF. Once validated, the adapted model is used to simulate the performance of the aforementioned non-drop-in fuels. Initial simulations assume the same output thrust conditions as those of the engine's baseline Jet-A operating conditions. Subsequently, simulations are performed under set turbine inlet temperature (TIT) conditions matching the baseline, thereby expanding our understanding of the engine's operating limits and its associated emissions. Finally, a real mission scenario within the operational scope of the studied engine is simulated to demonstrate that the advantages of these fuels persist under actual flight objectives and constraints.

2. Methodology

In this section, the assumptions made, the numerical methods used, the proposed model concept, virtual structure and configuration are discussed.

2.1. Engine Performance Model

To analyze the impact of alternative fuels on current aeroengines, a general model of a high-bypass, dual-spool, axial-flow, geared turbofan (GTF) engine [32] is generated in the PROOSIS v6.2.0 object-oriented gas-turbine simulation environment [33]. The engine consists of the following components and each of them is separately modeled and coupled accordingly within the PROOSIS framework as illustrated in Figure 1:

- Inlet Duct (InEng);
- Fan (CmpFan)— A single-stage fan;
- Gearbox (GearFan)—Transmits torque from the low-pressure compressor to the fan with a gear ratio of approximately 1:3 and an assumed mechanical efficiency of 99.5%;
- Low-Pressure Compressor (CmpL)—A three-stage compressor;
- High-Pressure Compressor (CmpH)—An eight-stage compressor;
- Combustion Chamber (Brn)—An isobaric TALON X burner [34];
- High-Pressure Turbine (TrbH)—A two-stage cooled turbine that drives the high-pressure compressor;
- Low-Pressure Turbine (TrbL)—A three-stage turbine that drives the fan and the low-pressure compressor;
- Primary Nozzle (NozPri)—A fixed-geometry convergent nozzle that directs the main exhaust stream;
- Bypass Nozzle (NozSec)—A fixed-geometry convergent nozzle that directs the bypass air stream.

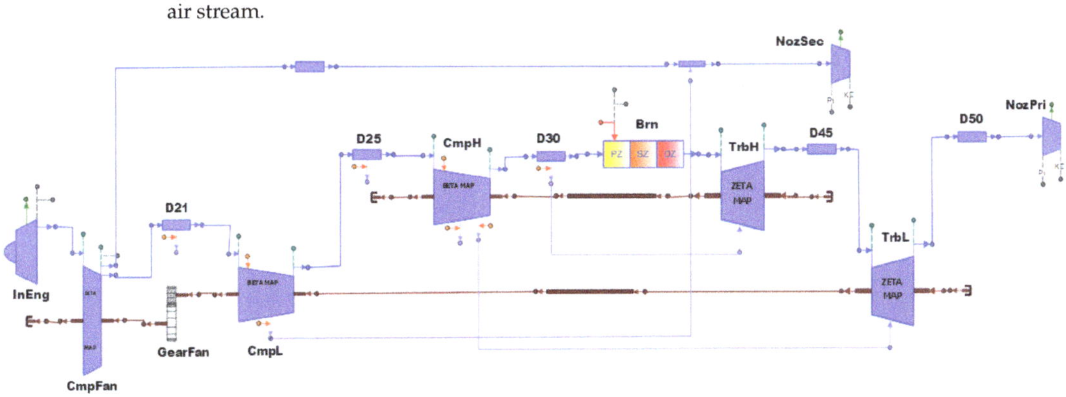

Figure 1. GTF engine model schematic. The interconnecting ducts (D) are labeled according to the engine station numbering, which is consistently followed from this point onward.

The novelty of the proposed engine model compared to previous respective attempts which utilize empirical algebraic models to estimate emissions and combustion efficiency [31] is that it incorporates a physics-based 1-D combustor model. According to the literature [26,27,30], such models are essential to provide deeper insights into the combustion characteristics and pollutant formation mechanisms of less-studied fuels compared to conventional Jet-A, such as MeOH and LNG. All the other components of the engine seen in Figure 1 are sourced from the TURBO library [33] of the environment.

2.2. Combustor Model

The model used is a three sub-volume generalist model of an isobaric combustor [30] which is developed and tested by LTT/NTUA in the CANTERA environment [35]. It is introduced for the first time in the PROOSIS framework, to work in unison with the other component models. To accomplish this, the common C++ interface between the two software platforms is employed, enabling the burner's main C++ function to be defined externally while remaining accessible within the PROOSIS environment. This shared function serves as foundation for the construction of the CRN burner component in PROOSIS, represented as "Brn" in Figure 1.

The division of the combustor volume into zones, along with the corresponding interconnections between these zones, creates a Chemical Reactor Network (CRN), illustrated in Figure 2, which is also used to reference the model. The latest version of the utilized CRN model, though following the same principal modeling strategy of Primary, Secondary and Dilution zone partitioning of the burner, has a slightly altered structure, giving it the ability to include water vapor inflow, which is modeled separately than the dry air intake. Also, computational time is reduced to less than a third of the time needed by the initial version, as discussed in [30]. This is achieved by introducing a dynamic initialization mixture within each zone, which is forced to react at the start of the simulation, accelerating the convergence of the integral solution.

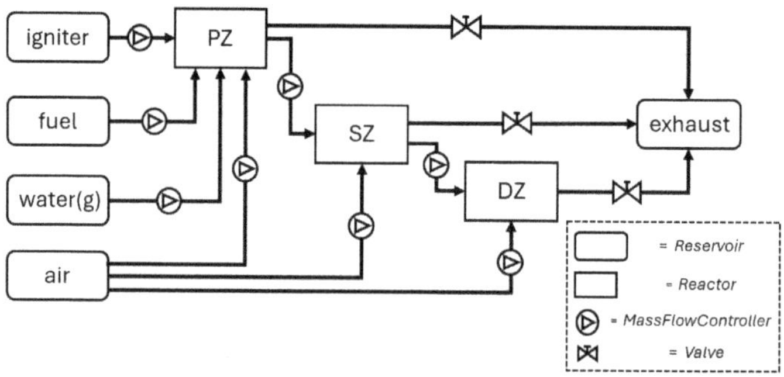

Figure 2. Virtual architecture of the CRN isobaric combustion chamber model.

The fuel information required by the model to simulate a specific fuel comes in the form of a combustion kinetic mechanism (defining a range of possible reactions among a specific set of potential species) as well as the assumed composition of the fuel. For the applications discussed in this work, any information about the fuel-combustion mechanisms and the assumed fuel compositions are disclosed in Table A1 of Appendix A and further explained.

Additionally, typical 0-D engine PROOSIS models rely on precomputed 3-D interpolation tables of thermodynamic data to simulate the effects of different fuels on exhaust gases. These tables are generated using NASA-CEA 2 software [36] and integrated into the model as working medium model. The current engine model follows this approach for all components except the CRN burner, which operates independently of this working medium model.

2.3. *Combustor Model Parameter Adaptation*

Separately, yet following the same adaptation principles, the CRN model of the burner has a number of geometric and flow parameters which are user-defined so as to produce more accurate emission estimations of the engine in question. The geometric parameters, specifically the length L and the reference area A_{ref} of the combustor, together with the length ratios of each zone (LR_{PZ}, LR_{SZ} and LR_{DZ}) are either known or easily estimated.

Meanwhile, the ratios of the core airflow (ARs) entering each zone or equivalently the equivalence ratio (ϕ) in each zone (given that they are explicitly linked) are often a priori unknown and thus usually serve as Design Variables (DVs) within the adaptation process in order to minimize the discrepancy between the model emission estimations and their known values. These parameters determine the quality of combustion in each zone of the burner, highly influencing the post-deflagration temperature and the formation of pollutants [30]. Yet when using the CRN model in question, two distinct assumptions for these parameters can be made. The first, which is also used in the introductory study concerning the model [30], states that the *AR* of each zone is considered unaffected by the operating conditions and thus the equivalence ratio of each zone accordingly altered with respect to changes in Fuel-to-Air Ratio (FAR). The second assumes that the different ϕ values of each zone are kept constant within the whole operating range of the engine, thus modifying the *ARs* accordingly to achieve this. Which of the two assumptions better reflects the operation of the current burner is examined in more detail in Section 3.2.

The implemented adaptation flow chart is illustrated in Figure 3, while the Objective Function (OF) is formulated as follows:

$$OF = \sum_i \omega \left(1 - \frac{EI_{NOx}^{(i)}}{EI_{NOx}^{(ref,i)}}\right)^2 + (1-\omega)\left(1 - \frac{EI_{CO}^{(i)}}{EI_{CO}^{(ref,i)}}\right)^2, \quad i = \{TO, Cl, App, Id\} \tag{1}$$

where a choice for $\omega = 85\%$ is made in order to place more emphasis on NO_x accuracy.

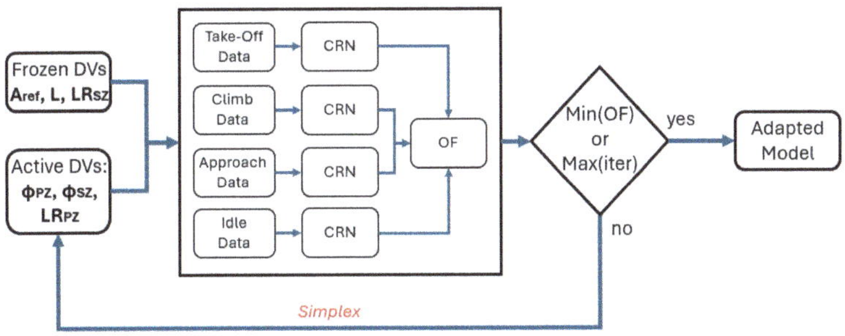

Figure 3. CRN model adaptation loop.

3. Application Test Case

A contemporary GTF engine, the Pratt and Whitney PW1127G-JM engine, is selected to be simulated using the general GTF model. Enough engine-specific experimental and geometric data are available to enable the model's adaptation. A blueprint of the engine's cross-section can be found in literature [32], while its fan diameter and length are known to be 2.10 m and 3.08 m, respectively. This engine powers the Airbus A320neo and was first introduced in January 2016. It is a high-bypass, dual-spool, axial-flow geared turbofan engine.

Fuel and exhaust emission data are procured from the ICAO Aircraft Engine Emissions Databank [37] as well as the Overall Pressure Ratio (OPR) and the Bypass Ratio (BPR) of the engine at ICAO Take-Off (TO) conditions. These measurements pertain to an engine mounted on a test rig configuration and tested under static atmospheric conditions at sea level. All operational parameters of the GTF under consideration constitute the assumed reference operating conditions of this study and are presented in Table A2, Appendix B.

The general twin-spool geared turbofan achitecture of Figure 1 is adjusted to simulate the specific PW1127G engine by a process of minimizing deviations between the model-computed net thrust (F_n) and fuel mass flow ($\dot{m}_f$) and their reference values of Table A2 by modifying design point parameters and locally adapting the maps of the turbo-machinery components.

3.1. Engine Cycle

In Figure 4, the ICAO operating line, also known as the net thrust and Thrust Specific Fuel Consumption (TSFC) correlation, is shown as computed by PROOSIS, with thrust varying by approximately 2.5% in each calculation. The four discrete reference operating points are displayed along with their deviations from their respective calculated values. Figure 4 also includes the deviation from the reference performance parameters (OPR, BPR, $\dot{m}_f$) corresponding to the TO thrust of 120.43 KN (see Table A2), which serves as the design point of the PROOSIS model.

The maximum deviation at the design point is around 0.4%, while across the entire range of operating conditions, it remains below 3%. The model is thus considered certified in accordance to the values provided by ICAO.

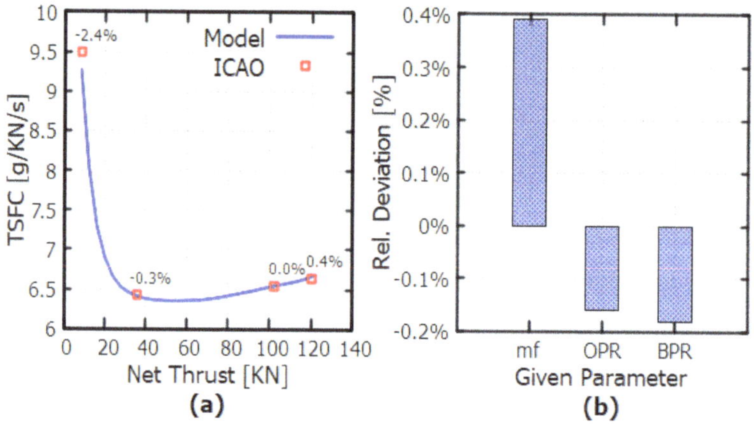

Figure 4. Engine model verification: (**a**) comparison of model-estimated operating line and ICAO-given operating points along with relative variation of the line with respect to each point, (**b**) design point model-estimated parameter variation with respect to ICAO-given values.

Total pressure, total temperature and core airflows of the simulated and verified GTF/Jet-A engine cycle are presented in Figure 5 per engine station in order to serve as a baseline cycle reference for the alternative fuel evaluations of Section 4.

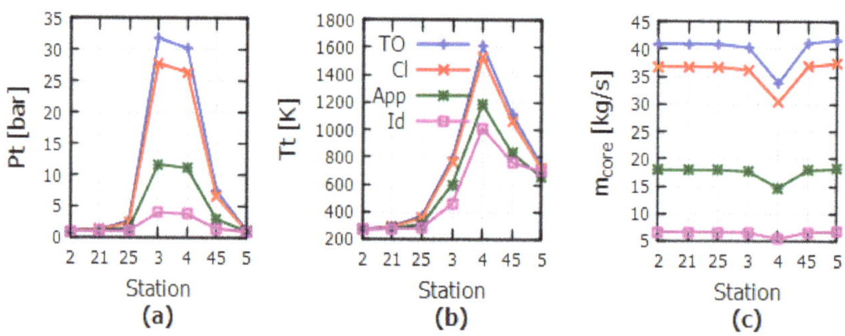

Figure 5. Total pressure (**a**), temperature (**b**) and core mass flow (**c**) of Jet-A-powered GTF engine (baseline operation) for 4 operating modes (station numbering according to Figure 1).

3.2. Combustor and Emissions

First, the constant CRN parameters (L, A_{ref}, LR_{SZ} and LR_{DZ}) are deduced by knowing the GTF's length and fan diameter and assuming its cross-section (found in [32]) is drawn in scale. Second, for the adaptation process to better represent the physics of the combustor design, the correct assumption has to be made concerning the combustion process.

According to the documentation of engine's burner, referred as the TalonX combustor, it cannot be modeled with the first approach mentioned in Section 2.3, given that its low

NO_x design uses special High Shear swirlers that uniformly create a fuel-rich mixture ($\phi_{PZ} > 1$) within the PZ by changing the swirl angles of the two swirling air passages and the air flow split between them [34]. The family of burners working with this general principle is referred to as Rich-Quench-Lean, a.k.a. RQL for short, and are represented in Figure 6a. Thus, the initial combustion occurring in the PZ is conducted within the low-NO_x-formation-rich region of ϕ, while downstream of the PZ, the flue gasses are quenched and diluted by the SZ and DZ air flows, decreasing the global ϕ towards its lean region along with their mean-adiabatic temperature T_{ad}. This trend is qualitatively recreated using the CRN model for the current burner by varying the ϕ_{PZ} through adjustments to the ARs entering each zone (T_{t3}, p_{t3} & $FAR = const.$). The effects on EI_{NOx} and T_{ad} are depicted in Figure 6b to further ensure the model's validity.

Figure 6b, consistent with related literature [6,38], shows that the peak of T_{ad} slightly differs from the peak of NOx with respect to ϕ, as they appear on opposite sides of the $\phi = 1$ line. This trend is expected to be within the predictive ability of the utilized Jet-A combustion mechanism (see Appendix A), which accounts not only for thermal NO_x but also for the prompt NO_x and N_2O pathways. These pathways play a significant role in NO_x formation in the $\phi < 1$ region, leading to the observed optimum in Figure 6b.

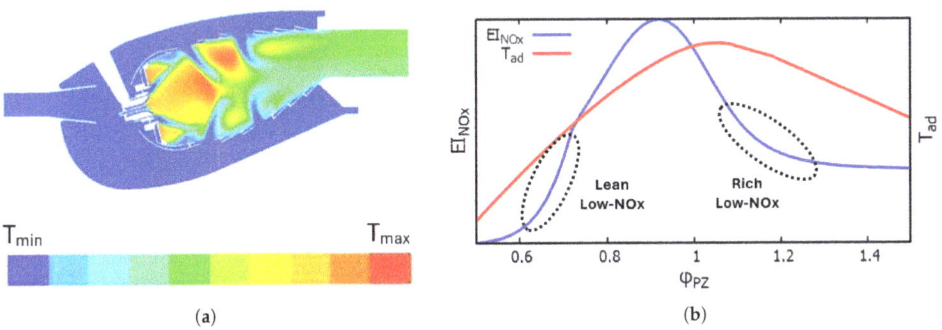

Figure 6. TalonX characteristics. (**a**) RQL burner-family cross-section with qualitative temperature contours [39]. (**b**) Qualitative EI_{NOx} & T_{ad} vs ϕ_{PZ} map for the TalonX at TO conditions.

Assuming that the TalonX burner can maintain constant ϕ values per zone throughout its operational range, the second approach is followed. Therefore, the ϕ_{PZ} and ϕ_{SZ} along with the LR_{PZ} are considered active DVs of the adaptation/optimization process visualized in Figure 3, while L, LR_{SZ} and A_{ref} are deemed frozen DVs.

The adaptation results of the CRN model are shown in Figure 7. Specifically, the initial and adapted CRN-derived results for EI_{NOx}, EI_{CO}, EI_{UHC}, and EI_{CO_2} profiles, relative to the engine's net thrust, are compared against the corresponding four reference operating points (TO, Cl, App, Id). Both initially and finally selected DVs of the adaptation process of Figure 3 are included in Appendix C and remain unchanged after the adaptation for the following calculations.

Figure 7 demonstrates that the engine model effectively captures the true emission trends across the entire operational range, akin to the validation of the working line accuracy discussed previously. The adaptation successfully achieved to improve the estimation of EI_{NOx} by making trade-offs with the EI_{CO} predictions, which is expected given that NO_x formation is inversely proportional to unburnt particle formation such as CO [38]. It is worth noting that the relatively low differences between the final and initial DVs (see Table A3 of Appendix C) indicates that the adaptation effectively preserves the physical characteristics of the combustor in question. Also, for the current adjusted configuration, the model seems unable to yield accurate EI_{UHC} estimations, neither qualitative nor quantitative and therefore such predictions are not included in the later parts of this work.

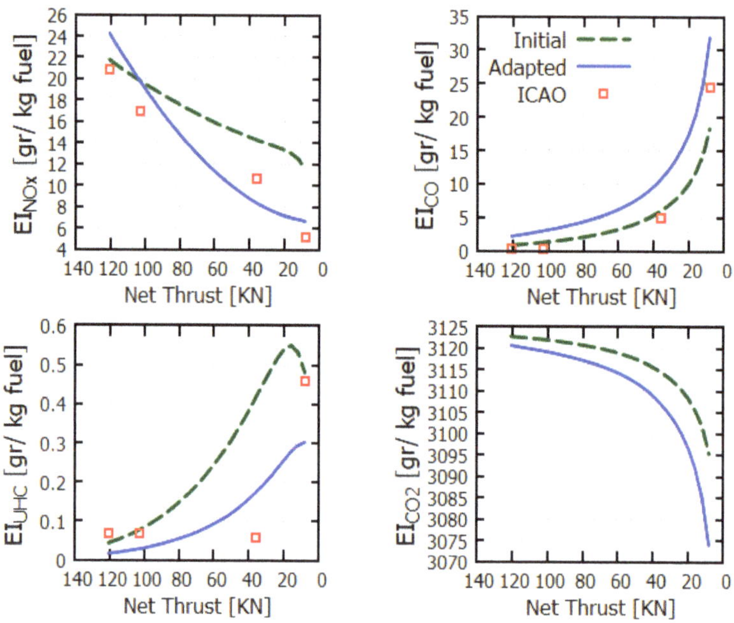

Figure 7. GTF model results: comparison between initial, adapted and ICAO (if available) NO_x-CO-UHC-CO_2 emission values with respect to engine thrust.

Furthermore, LTO emission comparisons are meant to highlight the joint accuracy of cycle and emission predictions, given that they are calculated using $\dot{m}_f$ and EI estimations as in Equation (2):

$$LTO_k\,[g] = \sum_i 60\,\dot{m}_{f,i}\,t_i\,EI_{k,i}, \quad i = \{TO, Cl, App, Id\} \;\;\&\;\; k = \{NO_x, CO\} \tag{2}$$

where $t = \{0.7, 2.0, 4.0, 26.0\}$ min is the time spent by the aircraft in each of the TO, Cl, App and Id operating modes, respectively.

The LTO NO_x and CO emissions of the GTF model are shown in Figure 8. Both prediction evaluations clearly demonstrate the qualitative reliability of the presented model in estimating NO_x and CO emissions, with discrepancies of 11% and 33%, respectively. This capability was previously demonstrated by the authors for the standalone CRN model [30], but it has now been extended for use as part of a complete aero-engine model, fully integrated within the PROOSIS framework. It should be noted that the quantitative predictive accuracy for NO_x is significantly better compared to CO. Nonetheless, the PROOSIS/CRN model can still be utilized to assess emission performance by slight overestimation given that its overall fidelity is deemed sufficient for a 0-D engine model.

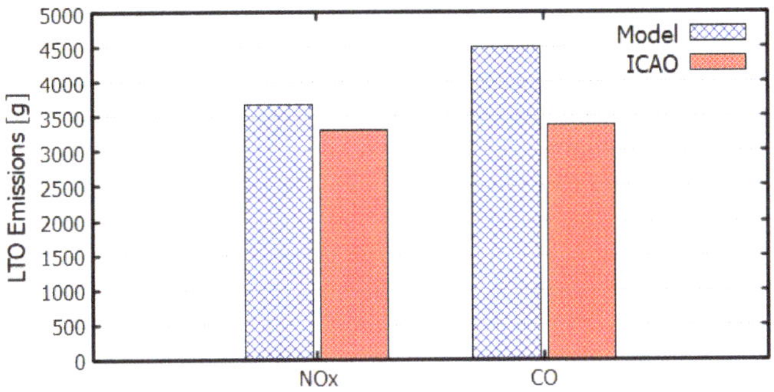

Figure 8. Calculated (model) and reference (ICAO) LTO NO_x and CO emitted mass comparison.

4. LNG and MeOH Evaluation

The calibrated GTF model is employed to evaluate the use of MeOH and LNG on the engine in question. Each fuel has physical and chemical properties that differentiate it significantly from typical jet fuel as shown in Table 1, those most concerning this work being Lower Heating Value (LHV) and latent vaporization heat.

Table 1. Properties of simulated fuels.

Properties	Jet-A	LNG	MeOH
Chemical formula	$C_{11}H_{21}$ *	CH_4, C_2H_6, C_3H_8	CH_3OH
Density [kg/m^3]	775–840 (at 15 °C)	420–470 (at −160 °C)	798 (at 15 °C)
Lower heating value [MJ/kg]	42.8–43.2	48–55	20.09
Latent heat of vaporization [kJ/kg]	250	510	1100
Freezing point [°C]	−47	−182.2	−97.6
Autoignition Temperature [°C]	255	535	465
Min. Autoignition Energy [MJ]	0.25–0.4	0.28–0.30	0.21

* an approximate formula [38], Jet-A is a varying HC mixture consisting mostly of paraffins. (n- and iso-alkanes), napthalenes and aromatic HCs [38,40] (more in Appendix A).

Given that both alternative fuels have lower carbon content and higher vaporization heats than Jet-A, they are anticipated to reduce the cycle maximum temperature and the overall engine emissions for the same thrust level attained with Jet-A; set-F_N. Also, compared to Jet-A, LNG's higher LHV implores a decrease in fuel flow, while MeOH's lower LHV indicates the opposite. This initial evaluation involves comparisons of fuel consumption, Turbine Inlet Temperature (TIT), rotational speed (N) and emission results (including both EI and LTO calculations). Furthermore, variations of the cycle temperature, pressure and core airflow per station for each fuel are analyzed to better understand their overall impact on engine performance. The set-thrust approach evaluates the ability of the studied fuels to meet the propulsive requirements of the current aircraft design, thus assuming their conceptual application in the current aircraft fleet.

Subsequently, the LNG and MeOH are evaluated and compared in the same manner, but this time under the assumption that the same TIT as the baseline Jet-A operation is maintained. This analysis, referred as set-TIT approach, serves to enhance our understanding of the expansion of current technological capabilities of aero-engines when fueled by LNG or MeOH. It assumes that retrofitting requirements for such fuels affects the propulsive needs of the existing fleet [1,19]. Thus, this analysis opts to provide insights into the adaptability of current engine technology to alternative fuels.

In both evaluations, it is assumed that the CRN burner model configuration remains the same according to the adaptation process followed in Section 2.3, meaning that the zones of the model are expected to operate at the adapted ϕ conditions, regardless of the fuel used. This assumption is based on the premise that the TalonX operation remains consistent and adapts to the new fuels in a similar manner as it does to Jet-A, given that no specific information is provided for the burner for LNG or MeOH-powered operation.

4.1. Set-Thrust Operation

The set-F_N condition denotes the possibility of operating current engines using LNG or MeOH with minimized changes in the engine cycle and aircraft infrastructure, assuming alterations be limited to the aircraft's fuel storage and supply systems. Thereby, the set-F_N approach constitutes a more conservative alternative fuel evaluation scenario. The main thermodynamic and emission parameters yielded from this analysis are shown in Table 2.

Table 2. LNG and MeOH evaluation in GTF requesting the same net thrust as the baseline Jet-A operation of the GTF.

	Take-Off	**Climb**	**Approach**	**Idle**
F_N [N]	120,430.0	102,365.5	36,129.0	8430.1
		Jet-A (baseline)		
$\dot{m}_f$ [kg/s]	0.803	0.670	0.232	0.078
TIT [K]	1614.44	1535.39	1188.18	1013.96
N_H [rpm]	20,649	19,948	15,776	11,343
N_L [rpm]	8848	8286	5193	2620
EI_{NOx} [g/kg]	24.22	19.68	8.36	6.66
EI_{CO} [g/kg]	2.02	2.82	10.02	30.02
EI_{CO_2} [g/kg]	3120.65	3119.33	3107.27	3073.95
		LNG		
$\dot{m}_f$ [kg/s]	0.707	0.590	0.204	0.069
TIT [K]	1592.25	1515.58	1177.64	1006.24
N_H [rpm]	20,660	19,960	15,781	11,344
N_L [rpm]	8855	8293	5195	2621
EI_{NOx} [g/kg]	11.18	9.31	5.60	5.86
EI_{CO} [g/kg]	1.35	1.88	7.09	20.50
EI_{CO_2} [g/kg]	2781.47	2780.64	2772.42	2751.34
		MeOH		
$\dot{m}_f$ [kg/s]	1.644	1.373	0.477	0.161
TIT [K]	1547.28	1475.73	1155.01	988.37
N_H [rpm]	20,660	19,976	15,785	11,338
N_L [rpm]	8855	8293	5196	2621
EI_{NOx} [g/kg]	4.28	3.48	1.50	1.09
EI_{CO} [g/kg]	0.75	0.99	3.17	8.69
EI_{CO_2} [g/kg]	1370.52	1370.13	1366.68	1357.99

For the same thrust, LNG fuel consumption drops by approximately 12%, while MeOH produces a fuel flow increase by 105%, a finding also observed in [19,24,25]. TIT is lowered for both fuels, yet the greatest decrease of 2.5–4% is observed for MeOH. Both aforementioned trends are visualized in Figure 9. As stated above, the heat of vaporization, being greater than Jet-A in both fuels, causes the TIT drop, thus indicating the potential of improving the engine life expectancy. Yet the cooling advantage of MeOH lies also in its lower LHV, which causes more fuel mass to be injected into the burner, which in turn requires even more heat to ignite.

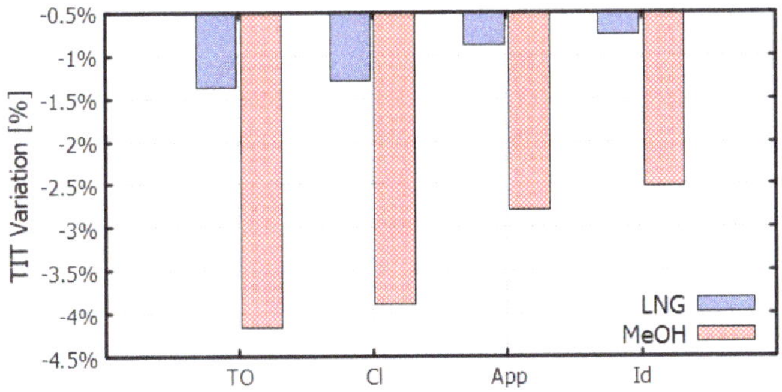

Figure 9. Set-thrust: TIT variation of LNG and MeOH-powered GTF, relative to its baseline Jet-A operation.

The decrease in all post-combustion temperatures (Stations 4, 45, 5) caused by both fuels can be observed also in Figure 10, where the covered cycle parameter variations relative to GTF/Jet-A operation are displayed, yet only for the Take-Off and Idle (Id) modes, given that they constitute the boundaries of possible variation. Figure 10 shows the aforementioned total mass-flow differentiation induced by each fuel due to their heating properties. Since the rotational speeds of both the high and low pressure spools remain virtually unchanged (see Table 2), the relatively minor variations in the engine's baseline total pressure profiles are reasonable.

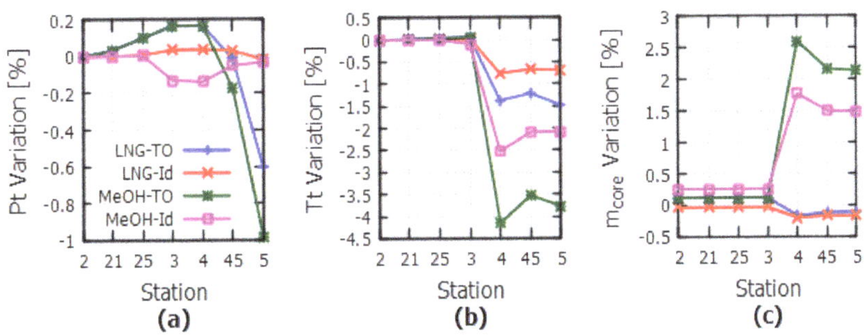

Figure 10. Set-thrust: total pressure (**a**), temperature (**b**) and core mass flow (**c**) relative variation of LNG and MeOH-powered GTF, for TO and Id operating modes.

Regarding the EIs covered in Table 2, both fuels severely reduce emissions for the same thrust setting. Examining only the EIs indicates that MeOH shows a clear advantage over LNG when it comes to curbing of pollutant emissions. Yet when analyzed as part of the engine LTO cycle for each fuel, the results differ, because the variation of $\dot{m}_f$ and the time of each phase play a significant factor as well, according to Equation (2). The increase in fuel consumption of MeOH, presented in Table 2, plays a significant factor in the LTO emitted mass of pollutants shown in Figure 11. Specifically, the CO_2 reduction potential of LNG is about 21%, while MeOH's potential amounts to 10%, always relative to the GTF/JetA results of the model, presented in Figure 8. In contrast, MeOH clearly displays the capability to reduce LTO NO_x by 64%, thus overcoming the LNG's potential of 48%, proving that when it comes to NO_x the EI reduction prevails over the increase in fuel flow. The emitted mass of LTO CO remains practically the same for both fuels.

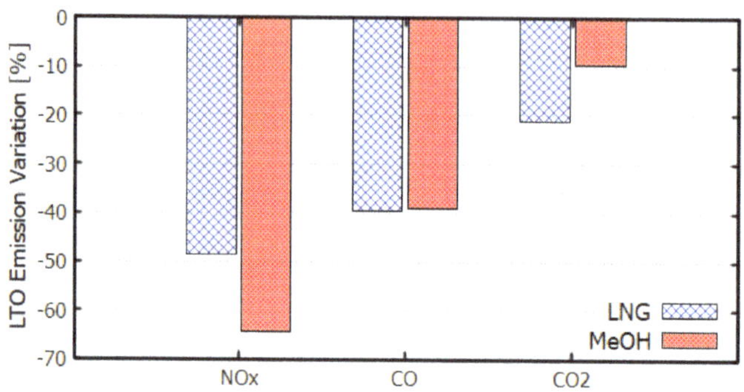

Figure 11. Set-thrust: LTO NO_x, CO and CO_2 variation of LNG and MeOH-powered GTF, relative to its baseline Jet-A operation.

4.2. Set-TIT Operation

Having shown that the GTF can produce the desired thrust using MeOH and LNG while reducing post-combustion temperatures, the potential for producing more thrust than the Jet-A operation comes into question. More importantly, the thrust-increasing potential is expected [24,25] and should be quantified for the current aero-engine, along with assessing its influence on emission and other cycle parameters. The results of the evaluation approach in question for LNG and MeOH are displayed in Table 3 (the baseline Jet-A operation is the one shown in Table 2). Idle is excluded from this analysis, as there is no practical need for additional F_N for this operating regime.

Table 3. LNG and MeOH evaluation in GTF requesting the same TIT as the baseline Jet-A operation of the GTF.

	Take-Off	**Climb**	**Approach**
TIT [K]	1614.44	1535.39	1188.18
		LNG	
$\dot{m}_f$ [kg/s]	0.737	0.621	0.213
F_N [N]	124,606.6	107,319.3	37,639.4
N_H [rpm]	20,857	20,138	15,917
N_L [rpm]	8969	8460	5294
EI_{NOx} [g/kg]	11.73	9.77	5.63
EI_{CO} [g/kg]	1.23	1.73	6.81
EI_{CO_2} [g/kg]	2781.67	2780.88	2772.86
		MeOH	
$\dot{m}_f$ [kg/s]	1.871	1.601	0.542
F_N [N]	132,755.7	117,779.7	41,317.9
N_H [rpm]	21,231	20,545	16,249
N_L [rpm]	9178	8779	5527
EI_{NOx} [g/kg]	4.97	4.15	1.60
EI_{CO} [g/kg]	0.57	0.78	2.80
EI_{CO_2} [g/kg]	1370.81	1370.46	1367.26

The initial research enquiry of this work is answered for both fuels and is visualized in Figure 12. LNG is capable of increasing net thrust by 3.5% for TO to 5% for Climb (Cl) operation, while still consuming 8–7% less fuel mass than Jet-A, respectively. Furthermore, MeOH can boost net thrust by up to 10% during TO and 15% during Cl while also leading to a 132–138% increase in fuel consumption, respectively.

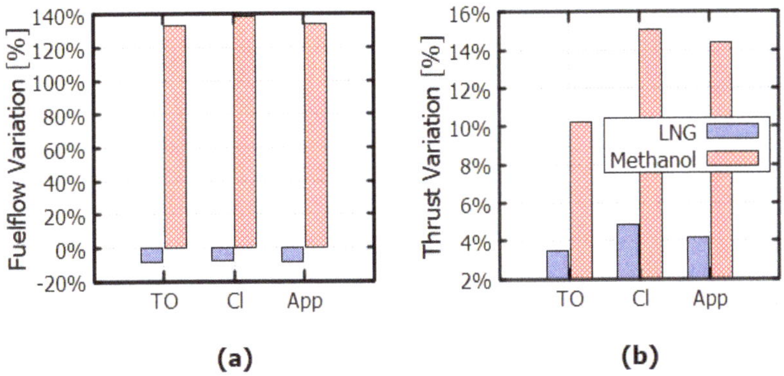

Figure 12. Set-TIT: fuel massflow (**a**) and thrust (**b**) variation of LNG and MeOH-powered GTF, relative to its baseline Jet-A operation.

By further examining the cycle parameter variations in the TO and Cl modes displayed in Figure 13, the cycle parameter trends are similar for both fuels but more emphatic for MeOH. It can be deduced that both fuels lead to a significant increase in core airflow (2–10%) before combustion, indicating that more air is drawn into the engine's intake. This is accompanied by a rise in the overall pressure ratio (OPR) and the expected temperature increases at Stations 25 and 3. The resulting increase in the load on the compressor components, which is primarily responsible for the thrust augmentation, may necessitate modifications to the compressor block, such as a proportional increase in size. This is also expected to result in an increase in engine dry weight, a conclusion similarly highlighted in [1] for ethanol, a closely related but distinct alternative aviation fuel. For both fuel types, the additional fuel required to meet the set-TIT condition results in a 1–6% increase in rotational speed, as shown in Figure 14. This change in power plant operation reasserts the previous statement as higher rotational speeds could negatively impact engine lifespan and should be thoroughly assessed. Also, the increased $\dot{m}_f$ (see Figure 12) causes exhaust gases with greater specific heat capacities to pass through the turbine [20]. This in turn leads more energy to be recuperated from the turbines and transferred to the compressors working at greater speeds, thus causing airflow and compression ratio to increase. This trend applies to both fuels across all studied modes but is more pronounced for MeOH, particularly during Cl due to its significant increase in $\dot{m}_f$.

Finally, the emission assessment shows a slightly smaller reduction potential for all studied pollutants from an EI perspective (see Table 3). Yet, once again, the LTO calculations for NO_x, CO and CO_2 presented in Figure 15 indicate otherwise. LNG overall keeps its LTO decreasing ability, though slightly mitigated. It is clear that in the equal-TIT condition, MeOH loses its GHG reduction potential, given that the CO_2 emitted mass for MeOH rises by 10% compared to Jet-A, predominately due the greatly increased Idle fuel consumption given that the aircraft spends the most time in that phase. Otherwise, for MeOH, the LTO emitted mass of CO remains virtually unchanged, while its NO_x reduction ability remains more potent than LNG.

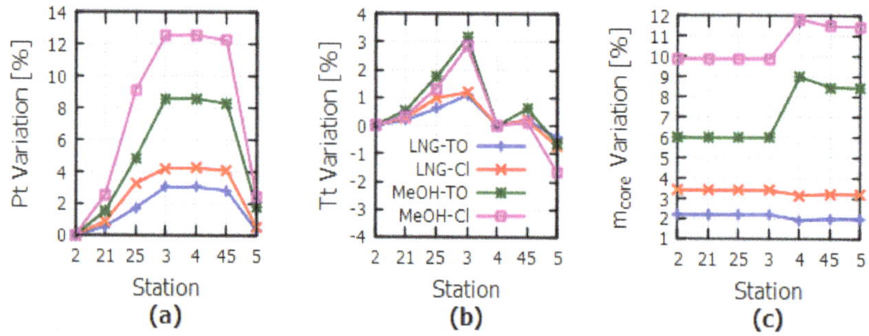

Figure 13. Set-TIT: total pressure (**a**), temperature (**b**) and core mass flow (**c**) relative variation of LNG and MeOH-powered GTF for TO and Cl operating modes.

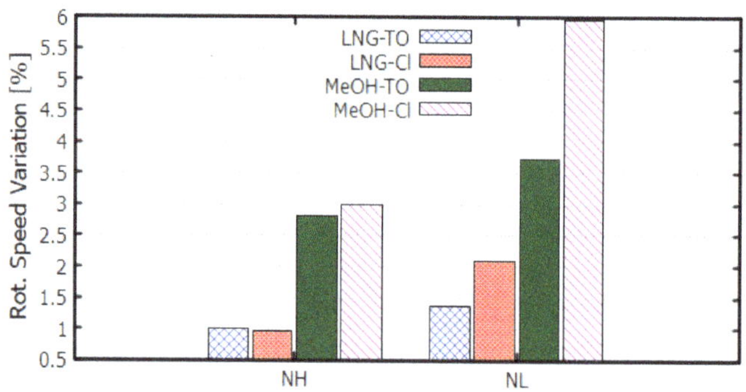

Figure 14. Set-TIT: low and high pressure spool speed variation for LNG and MeOH-powered GTF for TO and App operating modes.

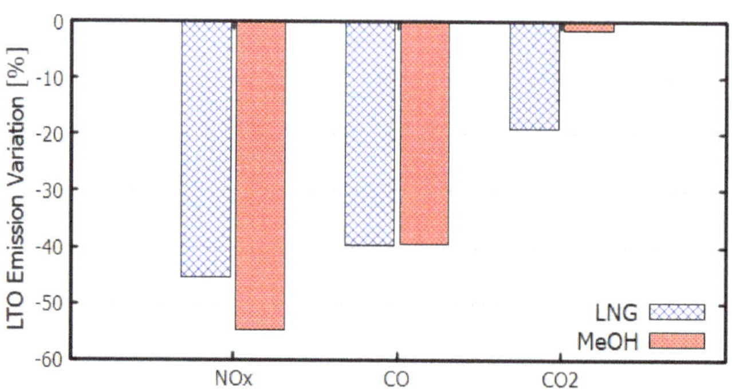

Figure 15. Set-TIT: LTO NO_x, CO and CO_2 variation of LNG and MeOH-powered GTF, relative to its baseline Jet-A operation.

4.3. Mission Analysis

The selected mission is meant to prove the mere feasibility of LNG- and MeOH-powered flight under simple weight restrictions of current aircraft, not to evaluate the mission-wise operational limits of the studied fuels, engine and aircraft. According to data from Eurocontrol, the average flight distance in Europe for 2019 (pre-pandemic) was 1062 km, while over 80% of flights departing from Europe (EU27 + U.K.) involved distances of up to 2000 km [41]. Therefore, different fuel scenarios are evaluated on an average real flight scenario from London to Barcelona, with a range of 1109 km. The flight model used to compute the mission parameters as well as the assumptions made for the aircraft weights and retrofitting are discussed in Appendix D, by explaining the information of Figure A1. The mission profile specifics and flight route are presented in Figure A2 of the same Appendix. Additional aircraft data, such as aerodynamic parameters needed for mission analysis, are sourced from Base of Aircraft Data (BADA) [42].

The proposed GTF model is implemented to predict engine performance during flight by inputting the flight speed, altitude and net-thrust requirements of the mission. The emission evaluation of an aircraft during flight using engine-integrated detailed chemistry models, all executed within the PROOSIS environment, is also an important originality of this work. The mission analysis results are presented in Table 4 and Figures 16 and 17 covering variations in emission, net-thrust, Take-off Weight (TOW), total Fuel Weight (FW) and aircraft weight for each fuel scenario.

Table 4. Mission analysis aircraft and emission weight results.

Fuel	TOW [tn]	FW [kg]	CO_2 [kg]	NO_x [kg]	CO [kg]
Jet-A	62.35	4576.44	12,399.66	73.19	20.71
LNG	62.59	4048.16	9779.91	33.96	11.92
MeOH	67.40	9635.97	11,472.13	27.73	13.60

As established above, the low LHV of MeOH increases FW by 110% and TOW by 8%, since the payload and OEW remain the same for both fuels. In contrast, LNG reduces FW by 11.5% while increasing slightly TOW by 0.4% due to the increased OEW, which is caused by the LNG tank retrofitting assumption. However, the TOW and FW stay within the maximum allowable limit of 77 and 20 tons, respectively (see Appendix D), for both alternative fuels. This signifies that the flight with LNG and MeOH remains within the weight restrictions of the A320 for the mission in question.

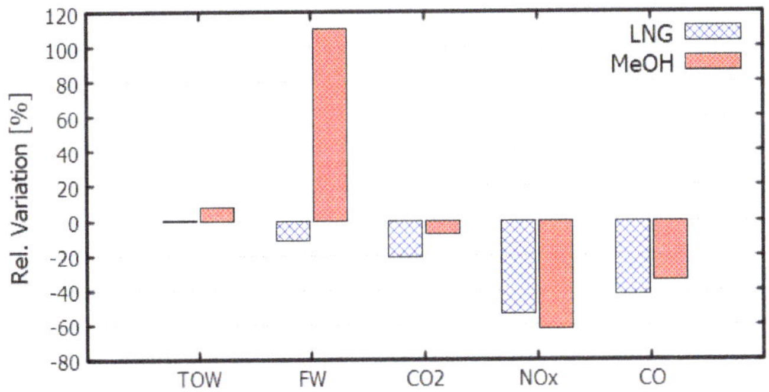

Figure 16. Mission analysis: comparison of aircraft TOW, FW and emission weight variations with respect to Jet-A operation, for LNG and MeOH.

Additionally, this specific flight scenario, under the assumptions made, grossly falls within the set-F_N category regardless of the fuel used, as indicated by the thrust time-series in Figure 17. This demonstrates that, for missions of comparable range, the set-F_N assumption is viable when evaluating MeOH or LNG for current aircraft. Finally, total CO_2, NO_x and CO emissions are all lowered by both fuels as illustrated in Figure 16. The trends of NO_x and CO EIs in Figure 17 once again appear to be inversely correlated. This observation is also supported by theory [38], which predicts a decrease in NO_x and an increase in CO during the progression of a typical mission.

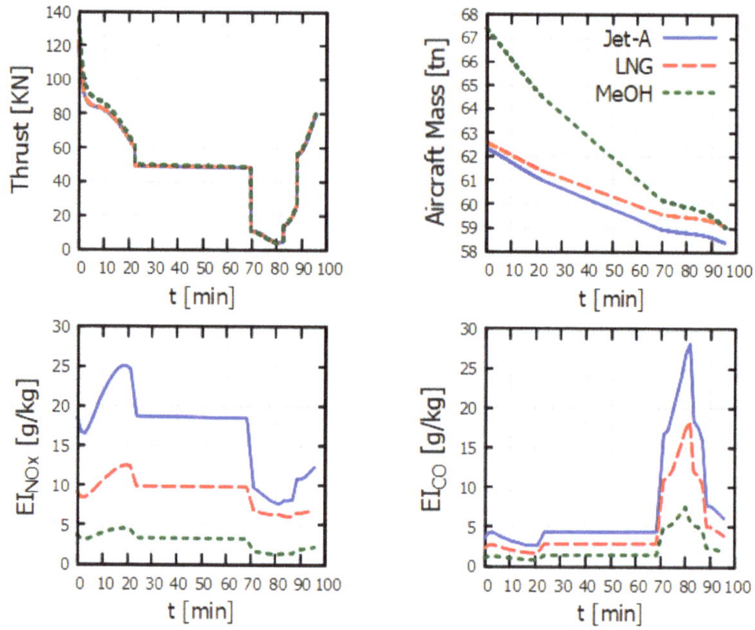

Figure 17. Thrust, aircraft total mass and EI time series for the London–Barcelona mission.

5. Discussion

The integration of a CANTERA-based component into the PROOSIS environment represents the primary novelty of this study, enabling aero-engine models to efficiently evaluate a wide range of fuels, provided a corresponding plug-in combustion mechanism is available. This approach eliminates the need to determine fuel-specific thermal properties (e.g., LHV) and combustion efficiency correlations, typically bare minimum requirements for evaluating alternative fuels on a 0-D level. Instead, if the base chemical composition of the fuel is known or approximated, the evaluation process becomes significantly streamlined. Previous studies using PROOSIS relied on correlation methods to predict emission trends and fuel efficiency variations [31,33]. In contrast, the proposed model introduces added generality for estimating aero-engine emissions and design-sensitive parameters related to the burner. This added capability is particularly valuable for engine and mission design analyses [43].

The evaluation of GTF operation fuelled by LNG and MeOH reasserts their literature-covered thrust-production ability within current engine setups coupled with their overall emission reduction potential. Furthermore, this study aims to facilitate the preliminary re-design of existing engines to optimally accommodate alternative fuels, both in terms of emissions and fuel consumption [30]. This is achieved by employing advanced, generalist burner models rather than the simplistic iterations commonly used in prior studies [20]. In practice, the GTF-adapted model was verified against corresponding ICAO data for Jet-A operation, demonstrating sufficient fidelity in predicting engine thermodynamic cycle and emissions, particularly for NO_x trends.

Considering the evaluation results of LNG and MeOH, focusing solely on the engine performance independent of its mission area of application, two approaches were adopted: one with a set-thrust equal to the baseline and the other with a set-TIT condition.

- Set-thrust evaluation of LNG and MeOH successfully addresses the current propulsive need of the A320neo and produces estimations which are in agreement with previous studies. The results show that both fuels can store enough energy in the working medium to power the turbine and convert the rest into thrust while decreasing its post-combustion temperature and its overall emissions [10,19,20,24,25]. This further confirms that both fuels are potentially capable of powering an A320 aircraft with benefits for engine life and emissions, provided the modifications to the fuel supply and storage systems remain within the aircraft's weight and volume constraints. However, this mission evaluation approach should be further investigated to understand how these fuels are expected to reduce the operational range of the aircraft [1]. Moreover, the proven decrease in TIT when using such fuels suggests a possible reduction in the need for turbine cooling air bleeds. This opens up additional opportunities for optimization when considering an engine redesign with a set-F_N objective, as less cooling air bleed could improve overall engine efficiency and performance (through a snowball effect concerning the beneficial effect of bleed lowering to turbine efficiency reducing TIT even further).

- The set-TIT approach causes new insights to emerge regarding the impact of LNG and MeOH on the engine cycle and emission profiles. The previously discussed TIT-decrease margin can be forfeited to generate even greater thrust to accommodate the highlighted need of a greater TO thrust due to maximum TOW augmentation caused by either the cryogenic tank configuration for LNG [19] or by the absolute increase in $\dot{m}_f$ for MeOH [10,24]. This evaluation strategy achieves extra thrust by increasing fuel consumption, which in turn raises rotational speeds and enhances airflow and the OPR. The additional energy from the increased fuel use is distributed between the turbines and the exhaust jet momentum, both contributing to the thrust gain. However, this shift in operating conditions, moving away from the engine design point, should have a negative rather than a neutral impact on its lifespan. As a result, a redesign of the compressor block is advised to accommodate the higher thermal and energy demands when using LNG or MeOH under a set-TIT operating regime. Therefore, although the set-TIT operation offers a notable yet diminished overall emission reduction potential, it should not be regarded as a viable integration option for the fuels in question. Instead, it highlights the potential benefits of aero-engine redesign in the context of alternative fuels. Engine models such as the one proposed, coupled with multidisciplinary engine design methods, could be crucial for the development of fuel-specific engine re-sizing solutions or even novel configurations, as emphasized by the authors in [43].

Finally, for the aircraft mission analysis, the proposed engine model is integrated with an in-house flight evaluation tool. Assuming an A320 is retrofitted to run on LNG or MeOH, the analysis demonstrates that using these fuels for an average European mission range is feasible without exceeding the aircraft's weight restrictions. This is achievable because, although $\dot{m}_f$ shows a linear relationship with changes in total FW (a slight decrease for LNG and a twofold increase for MeOH), the maximum Take-Off Weight (TOW) remains largely unchanged for LNG and only slightly increases for MeOH. As a result, the net-thrust profile of the mission remains nearly identical for both fuels, confirming that the set-F_N operation is a valid assumption for evaluating alternative fuels, if assumed that no changes are made to the aircraft's aerodynamic design and that the mission range is within certain limits. Finally, regarding all evaluated emissions, again, both fuels display reduction capabilities, with LNG having an advantage in reducing CO_2 and CO, with MeOH showing greater potential for NO_x.

6. Conclusions

To conclude, a 0-D GTF model with an integrated detailed chemistry isobaric burner was developed, tested and applied to evaluate two non drop-in fuels, LNG and MeOH. A robust simulation framework combining the PROOSIS v6.2.0 software and a detailed CRN model was successfully coupled to assess the thermodynamic and emission performance of the fuels. The integration of advanced CRN techniques allowed for more detailed analysis of the combustion process, pollutant formation mechanisms and the impact of alternative fuel properties on engine performance. The proposed model provided a detailed and accurate analysis of LNG and MeOH as alternative aviation fuels while even being capable of producing cycle and emission predictions during a simulation of real flight scenario.

LNG stands out as a near-term solution with manageable integration efforts and enhanced emission mitigation and minor cycle improvement abilities in every evaluated scenario. Emission-wise, its LTO reduction potential compared to Jet-A, ranges from 19–21% for CO_2, 45–48% for NO_x and constant 39% for CO, depending on the thrust or TIT requirements assumed. Despite its challenges, MeOH demonstrates strong potential as a long-term alternative fuel, offering exceptional post-combustion temperature reduction and NO_x reduction capabilities, albeit with increased fuel storage requirements. Cycle-wise MeOH causes fuel mass consumption to increase by 105% while also reducing TIT by an average of 3% under set-F_N operation. Additionally, under set-TIT operation, MeOH holds the ability to increase thrust by 10%. Emission-wise, it is capable of reducing CO_2, NO_x and CO LTO emissions by 2–10%, 55–64% and 39%, respectively. Finally, both fuels are capable of carrying out an average European mission, representing a major area of application for the examined engine while also mitigating the overall emission footprint of the aircraft, in accordance with aforementioned numerical results. Yet no conclusions should be drawn regarding the current profitability of the evaluated alternative fuels in aviation.

Future research on alternative fuel integration in aviation should focus on expanding this work by integrating the proposed models within multidisciplinary preliminary engine design methods, enabling optimization with respect to performance, engine life and emissions. The focus should be on optimizing aero-engines by addressing the unique challenges associated with each fuel type while leveraging their respective advantages. This approach requires the integration of state-of-the-art 0-D engine models which incorporate detailed chemistry for combustor modeling, such as the one proposed. This would allow for more tailored solutions, enhancing the performance and efficiency of engines powered by alternative fuels.

Last but not least, to fully evaluate the viability of LNG and MeOH integration in the aviation industry, future research should include a more in-depth critical analysis of logistical and economic issues, such as fuel supply infrastructure, required aircraft and engine modifications and cost and risk assessment of these non-drop-in fuels. For example, the impact of LNG or methanol on mission range limits should be addressed in future work concerning this topic. Nevertheless, the computational tools developed in this work should serve as a valuable asset in undertaking such analyses. Demonstrating the advantages of these alternative fuels over the baseline performance of current engines is essential to justify the industry's investment in their costly integration.

Author Contributions: Conceptualization, K.M., S.V. and A.A.; methodology, S.V. and A.A.; software, S.V., N.A. and A.A.; validation, S.V.; formal analysis, S.V.; investigation, A.A. and S.V.; data curation, S.V.; writing—original draft preparation, S.V.; writing—review and editing, S.V., A.A. and K.M.; visualization, S.V.; supervision, A.A., N.A. and K.M.; project administration, N.A. and K.M.; funding acquisition, A.A., N.A. and K.M. All authors have read and agreed to the published version of the manuscript.

Funding: This research has been co-financed by the European Regional Development Fund of the European Union and Greek national funds through the Operational Program Competitiveness, Entrepreneurship, and Innovation under the call RESEARCH–CREATE–INNOVATE (project code: T2EDK-00034—Advanced alternative ground and air transport fuels fromresidual lipids—Lipid4fuel).

Data Availability Statement: The data contained within this article are available upon request from the corresponding author.

Conflicts of Interest: The authors declare no conflicts of interest.

Nomenclature and Symbols

The following abbreviations are used in this manuscript, while the symbols referred to in this paper are presented below :

a.k.a.	also known as
w.r.t.	with respect to
eq.	equation
Alt	Altitude
App	Approach
BADA	Base of Aircraft Data
BPR	By-pass Ratio
CAMACM	Commercial Aircraft Mission Analysis Computational Model
Cl	Climb
CRN	Chemical Reactor Network
DZ	Dilution Zone
FAR	Fuel to Air Ratio
FW	Fuel Weight
GHG	Green House Gas
GTF	Geared Turbofan
HC	Hydrocarbon
ICAO	International Civil Aviation Organization
Id	Idle
LHV	Lower Heating Value
LNG	Liquefied Natural Gas
LTO	Landing & Take-Off
LTT	Laboratory of Thermal Turbomachines
MeOH	Methanol
NTUA	National Technical University of Athens
OF	Objective Function
OPR	Overall Pressure Ratio
PROOSIS	PRopulsive Object-Oriented SImulation Software
PS	Power Setting
PZ	Primary Zone
RQL	Rich-Quench-Lean
SAF	Sustainable Aviation Fuel
SZ	Secondary Zone
TIT	Turbine Inlet Temperature
TO	Take-Off
TOW	Take-Off Weight
TSFC	Thrust Specific Fuel Consumption
UHC	Unburnt Hydrocarbons

Symbol	Description	Units
A_{ref}	Burner Reference Area	m^2
AR_i	Air Ratio per Zone$-i$	-
EI_k	Emission Index of species$-k$	g/kg of fuel
FAR	Fuel-to-Air Ratio	-
F_N	Net Thrust	N
L	Liner Length	m
LR_i	Length Ration per Zone$-i$	-
$\dot{m}_{core}$	Core Massflow	kg/s
$\dot{m}_f$	Fuel flow	kg/s
P_t	Total Pressure	bar
T_{ad}	Adiabatic Post-combustion Temperature	Kelvin
T_t	Total Temperature	Kelvin
t	Aircraft Mission time	min
ϕ	Equivalence Ratio	-
ω	Adaptation Weight	-

Appendix A. Fuel Combustion Mechanisms and Compositions

For the simulation of the Jet-A fuel, the detailed kerosene combustion mechanism of J. Luche [44] is employed, as in the initial work concerning the CRN model [30]. This mechanism consists of 91 species and 991 corresponding reactions, developed specifically for Jet-A combustion simulation, being validated over 0.5–10 atm pressure, 300–1800 K gas temperature and 0.5–2.0 equivalence ratio ranges, while also including nitrous oxides formation and reduction reactions. The parameter ranges clearly overlap the respective ranges covered within this work. Given that real jet fuel can consist of over 300 different hydrocarbons (HCs), their quality and quantity depending on season and oil-well location [40], they cannot be all accounted for in 0D–1D models. Consequently, it is common practice [40,45] to group these hydrocarbons into fractions based on HC families (e.g., paraffins) and represent each family with a single species (e.g., n-decane). This type of fuel surrogate is typically tailored to each specific mechanism. For the Luche mechanism, the fuel surrogate composition is outlined in Table A1, indicating that $n - C_{10}H_{22}$ represents the paraffins, $cy - C_9H_{18}$ the naphthenes and $ph - C_3H_7$ the aromatics [44].

For the alternative fuels, LNG and MeOH, the GRI 3.0 [46] is utilized, a mechanism optimized for conditions of 1000–2500 K, pressures of 10 Torr–10 atm, and equivalence ratios of 0.1–5 in premixed systems. This mechanism is mostly applied in natural gas combustion applications but also includes MeOH in its species pool. Typical LNG composition is assumed containing mostly methane, while the MeOH is assumed 100% pure, as seen in Table A1.

Table A1. Fuel parameters for the CRN model.

Parameter	**Jet-A**	**LNG**	**MeOH**
Mechanism	Luche [44]	GRI 3.0 [46]	
Composition (as molar ratios)	74% $n - C_{10}H_{22}$ 11% $cy - C_9H_{18}$ 15% $ph - C_3H_7$	90% CH_4 6% C_2H_6 4% C_3H_8	100% CH_3OH

Appendix B. PW1127G Performance Reference Data

Table A2. PW1127G reference ICAO data [37].

PW1127G	Take-Off	Climb	Approach	Idle
PS [%]	100	85	30	7
F_n [KN]	120.430	102.365	36.129	8.430
$\dot{m}_f$ [kg/s]	0.8	0.67	0.232	0.08
EI_{NOx} [g/kg]	20.814	16.954	10.593	5.203
EI_{CO} [g/kg]	0.268	0.362	4.885	24.526
EI_{UHC} [g/kg]	0.070	0.067	0.058	0.458
OPR [-]	31.7		-	
BPR [-]	12.3		-	

Appendix C. CRN Model: Adaptation of Geometric and Flow Input Data

Table A3. Adjusted parameters for the CRN burner model of the GTF.

Parameter	Initial	Adapted
A_{ref} [m^2]	0.126	0.126
L [m]	0.161	0.161
LR_{PZ} * [%]	29.90	34.05
LR_{SZ} [%]	33.60	33.60
LR_{DZ} [%]	36.50	32.35
ϕ_{PZ} * [-]	1.1	1.276
ϕ_{SZ} [-]	0.6	0.722

* active DVs during adaptation process.

Appendix D. A320 London–Barcelona Mission Data and Assumptions

The software used for the mission analysis of the engine model is CAMACM (Commercial Aircraft Mission Analysis Computational Model), a tool developed by LTT/NTUA [47]. The aircraft is simulated as a point mass upon which all forces act, such as lift, drag, thrust and weight. Environmental conditions are based on the International Standard Atmosphere [48], while air conditions are assumed stationary, homogeneous, ideal, and free of moisture.

The Airbus A320 aircraft is chosen for being an actual area of application for the PW1000. It is configured with 150 passengers and an additional 15% reserve fuel. Therefore, the payload corresponds to 14.25 tons, assuming that each passenger and their luggage have a combined mass of 95 kg. According to BADA v3.9 file [42], the A320 has the following specifications:

- Maximum TOW: 77 tons
- Maximum Payload: 21.5 tons
- Maximum FW: 20 tons
- Operating Empty Weight: 41.345 tons

Due to its cryogenic nature, LNG cannot be stored in the existing fuel tanks located in the wings. The A320 has four storage Compartments on the lower deck, shown in Figure A1, with compartments 1 and 5 easily accommodating the luggage needs of 150 passengers. Therefore, Compartments 3 and 4 can be used for LNG tanks, providing a total available volume of 18.1 m^3.

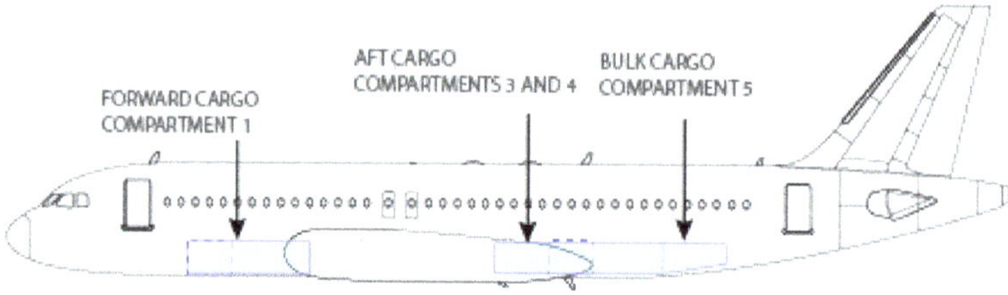

Figure A1. A320 cargo-hold compartments.

Due to the required characteristics of cryogenic fuel tanks (cylindrical with insulation) and the necessary space for piping and wiring, it is assumed that the usable fuel volume is 15 m^3. This means that 6960 kg of LNG can be stored in the tanks (with LNG density ρ_{LNG} = 464 kg/m^3). Considering the gravitational index (gravitational index = fuel mass/(fuel mass + tank mass)) for LNG is assumed to be 90% [19], the mass of the tanks is approximately 770 kg. Therefore, this additional mass is added to the OEW of the aircraft only when LNG is used to power the aircraft.

Finally, the mission flight route along with the aircraft's altitude and flight speed w.r.t. mission time are shown in Figure A2.

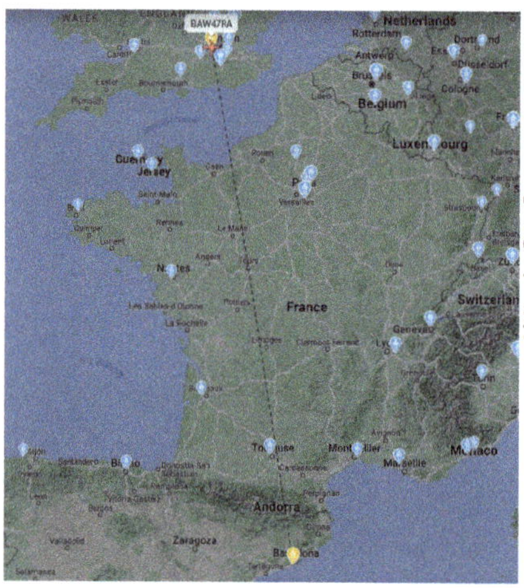

(a)

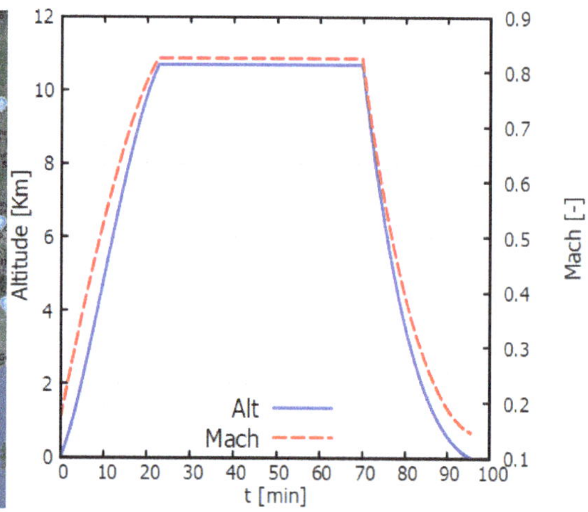

(b)

Figure A2. Additional mission information. (**a**) Mission air-path over map [49]. (**b**) A320 altitude and speed time profiles, computed using CAMACM.

References

1. Ansell, P.J. Review of sustainable energy carriers for aviation: Benefits, challenges, and future viability. *Prog. Aerosp. Sci.* **2023**, *141*, 100919. [CrossRef]
2. Lai, Y.Y.; Christley, E.; Kulanovic, A.; Teng, C.C.; Björklund, A.; Nordensvärd, J.; Karakaya, E.; Urban, F. Analysing the opportunities and challenges for mitigating the climate impact of aviation: A narrative review. *Renew. Sustain. Energy Rev.* **2022**, *156*, 111972.
3. Marszałek, N.; Lis, T. The future of sustainable aviation fuels. *Combust. Engines* **2022**, *191*, 29–40. [CrossRef]
4. Grewe, V.; Rao, A.G.; Grönstedt, T.; Xisto, C.; Linke, F.; Melkert, J.; Middel, J.; Ohlenforst, B.; Blakey, S.; Christie, S.; et al. Evaluating the climate impact of aviation emission scenarios towards the Paris agreement including COVID-19 effects. *Nat. Commun.* **2021**, *12*, 3841. [CrossRef]

5. Lee, D.S.; Fahey, D.W.; Skowron, A.; Allen, M.R.; Burkhardt, U.; Chen, Q.; Doherty, S.J.; Freeman, S.; Forster, P.M.; Fuglestvedt, J.; et al. The contribution of global aviation to anthropogenic climate forcing for 2000 to 2018. *Atmos. Environ.* **2020**, *244*, 117834. [CrossRef]
6. Yin, F.; Rao, A.G. Climate impact of aviation emissions and the implications on aircraft engine design. In Proceedings of the Tutorial Handout of the ASME Turbo Expo 2022 Turbomachinery Technical Conference and Exposition GT2022, Rotterdam, The Netherlands, 13–17 June 2022; Volume 206.
7. Herzog, T. *World Greenhouse Gas Emissions in 2005*; WRI Working Paper; World Resources Institute: Washington, DC, USA, 2009.
8. International Energy Agency. Global Energy Review: CO_2 Emissions in 2021. 2022. Available online: https://www.iea.org/reports/global-energy-review-co2-emissions-in-2021-2 (accessed on 10 December 2024).
9. Christodoulakis, J.; Karinou, F.; Kelemen, M.; Kouremadas, G.; Fotaki, E.F.; Varotsos, C.A. Assessment of air pollution from Athens International Airport and suggestions for adaptation to new aviation emissions restrictions. *Atmos. Pollut. Res.***2022**, *13*, 101441. [CrossRef]
10. Price, R.O. Methanol: An Environmentally Attractive Alternative Commercial Aviation Fuel. In Proceedings of the IEEE 25th Intersociety Energy Conversion Engineering Conference, Reno, NV, USA, 12–17 August 1990; Volume 4, pp. 331–336.
11. Altarazi, Y.S.M.; Talib, A.R.A.; Yusaf, T.; Gires, J.Y.E.; Ghar, M.F.A.; Lucas, J. A review of engine performance and emissions using single and dual biodiesel fuels: Research paths, challenges, motivations and recommendations. *Fuel* **2022**, *326*, 125072. [CrossRef]
12. European Partnership. Clean Aviation: Strategic Research and Innovation Agenda. *EU Counc. Regul.* **2021**, *L 427*, 17–119.
13. Airbus. Cities, Airports & Aircraft. In *Global Market Forecast*; Airbus: Leiden, The Netherlands, 2019.
14. Liu, Z.; Deng, Z.; Davis, S.J.; Giron, C.; Ciais, P. Monitoring global carbon emissions in 2021. *Nat. Rev. Earth Environ.* **2022**, *3*, 217–219. [CrossRef]
15. Schripp, T.; Anderson, B.E.; Bauder, U.; Rauch, B.; Corbin, J.C.; Smallwood, G.J.; Lobo, P.; Crosbie, E.C.; Shook, M.A.; Miake-Lye, R.C.; et al. LeClercq. Aircraft engine particulate matter emissions from sustainable aviation fuels: Results from ground-based measurements during the NASA/DLR campaign ECLIF2/ND-MAX. *Fuel* **2022**, *325*, 124764. [CrossRef]
16. Sethi, V. *ENABLing CryogEnic Hydrogen Based CO_2 free Air Transport (ENABLEH2)*; Technical Report 8146886; Publications Office of the European Union: Luxembourg, 2022.
17. Przysowa, R.; Gawron, B.; Lecki, T.B.; Legowik, A.; Merkisz, J.; Jasinski, R. Performance and emissions of a microturbine and turbofan powered by alternative fuels. *Aerospace* **2021**, *8*, 25. [CrossRef]
18. Chiong, M.C.; Chong, C.T.; Ng, J.; Lam, S.S.; Tran, M.; Chong, W.W.F.; Jaafar, M.N.M.; Valera-Medina, A. Liquid biofuels production and emissions performance in gas turbines: A review. *Energy Convers. Manag.* **2018**, *173*, 640–658. [CrossRef]
19. Rompokos, P.; Kissoon, S.; Roumeliotis, I.; Nalianda, D.; Nikolaidis, T.; Rolt, A. Liquefied Natural Gas for Civil Aviation. *Energies* **2020**, *13*, 5925. [CrossRef]
20. Sasi, S.; Mourouzidis, C.; Roumeliotis, I.; Nikolaidis, T.; Pachidis, V.; Norman, J. Impacts of Alternative Aviation Fuels on Engine Cycle Design and Aircraft Mission Capability. In Proceedings of the ASME Turbo Expo 2023: Turbomachinery Technical Conference and Exposition. Volume 2: Ceramics and Ceramic Composites; Coal, Biomass, Hydrogen, and Alternative Fuels, Boston, MA, USA, 26–30 June 2023; ASME: New York, NY, USA, 2023; p. V002T03A005.
21. Seyam, S.; Dincer, I.; Agelin-Chaab, M. Novel hybrid aircraft propulsion systems using hydrogen, methane, methanol, ethanol and dimethyl ether as alternative fuels. *Energy Convers. Manag.* **2021**, *238*, 114172. [CrossRef]
22. Santasalo-Aarnio, A.; Nyari, J.; Wojcieszyk, M.; Kaario, O.; Kroyan, Y.; Magdeldin, M.; Larmi, M.; Järvinen, M. *Application of Synthetic Renewable Methanol to Power the Future Propulsion*; SAE Technical Paper 2020-01-2151; SAE: Warrendale, PA, USA, 2020.
23. Verhelst, S.; Turner, J.W.G.; Sileghem, L.; Vancoillie, J. Methanol as a fuel for internal combustion engines. *Prog. Energy Combust. Sci.* **2019**, *70*, 43–88. [CrossRef]
24. Dodds, W.J.; Ekstedt, E.E.; Bah, D.W. Methanol combustion in a CF6l-80A engine combustor. In Proceedings of the AIAA/SAE/ASME, 19th Joint Propulsion Conference, Seattle, WA, USA, 27–29 June 1983.
25. Cican, G.; Mirea, R.; Rimbu, G. Experimental Evaluation of Methanol/Jet-A Blends as Sustainable Aviation Fuels for Turbo-Engines: Performance and Environmental Impact Analysis. *Fire* **2024**, *7*, 155. [CrossRef]
26. Liu, Y.; Sun, X.; Sethi, V.; Li, Y.; Nalianda, D.; Abbott, D.; Gauthier, P.; Xiao, B.; Wang, L. Development and application of a preliminary design methodology for modern low emissions aero combustors. *Proc. Inst. Mech. Eng. Part A J. Power Energy* **2021**, *235*, 783–806. [CrossRef]
27. Nozari, M.; Eidiattarzade, M.; Tabejamaat, S.; Kankashvar, B. Emission and performance of a micro gas turbine combustor fueled with ammonia-natural gas. *Int. J. Engine Res.* **2022**, *23*, 1012–1026. [CrossRef]
28. Valles, M.A.; Kessler, I.; Windom, B.C. Review and Development of Natural Gas/Hydrogen Fuel Flexible Reduced Chemical Mechanism for High-Order Modeling of Gas Turbines. In Proceedings of the ASME Turbo Expo 2023: Turbomachinery Technical Conference and Exposition. Volume 3A: Combustion, Fuels, and Emissions, Boston, MA, USA, 26–30 June 2023; ASME: New York, NY, USA, 2023; p. V03AT04A030.

29. Vogel, M.; Kaufmann, J.; Völkl, V.; Hirsch, C.; Sattelmayer, T. Comparison of Equivalence Ratio Fluctuations in a Lean Premixed Combustor for Kerosene and Natural Gas. In Proceedings of the ASME Turbo Expo 2023: Turbomachinery Technical Conference and Exposition. Volume 3A: Combustion, Fuels, and Emissions, Boston, MA, USA, 26–30 June 2023; ASME: New York, NY, USA, 2023; p. V03AT04A011.
30. Villette, S.; Adam, D.; Alexiou, A.; Aretakis, N.; Mathioudakis, K. A Simplified Chemical Reactor Network Approach for Aeroengine Combustion Chamber Modeling and Preliminary Design. *Aerospace* **2024**, *11*, 22. [CrossRef]
31. Alexiou, A.; Aretakis, N.; Kolias, I.; Mathioudakis, K. Novel Aero-Engine Multi-Disciplinary Preliminary Design Optimization Framework Accounting for Dynamic System Operation and Aircraft Mission Performance. *Aerospace* **2021**, *8*, 49. [CrossRef]
32. Kurzke, J. Fundamental Differences Between Conventional and Geared Turbofans. In Proceedings of the ASME Turbo Expo 2009: Power for Land, Sea, and Air. Volume 1: Aircraft Engine; Ceramics; Coal, Biomass and Alternative Fuels; Controls, Diagnostics and Instrumentation; Education; Electric Power; Awards and Honors, Orlando, FL, USA, 8–12 June 2009; ASME: New York, NY, USA, 2023; pp. 145–153.
33. Alexiou, A. *Introduction to Gas Turbine Modelling with PROOSIS*, 4th ed.; Empresarios Agrupados International (EAI) S.A.: Madrid, Spain, 2020.
34. McKinney, R.; Cheung, A.; Sowa, W.; Sepulveda, D. The Pratt & Whitney TALON X Low Emissions Combustor: Revolutionary Results with Evolutionary Technology. In Proceedings of the 45th AIAA Aerospace Sciences Meeting and Exhibit, Reno, NV, USA, 1–8 January 2007.
35. Cantera Co. Cantera Is an Open-Source Suite of Tools for Problems Involving Chemical Kinetics, Thermodynamics, and Transport Processes. Available online: https://cantera.org (accessed on 10 December 2024).
36. Gordon, S.; McBride, B.J. *Computer Program for Calculation of Complex Chemical Equilibrium Composition and Applications*; NASA RP1311; NASA: Washington, DC, USA, 1994.
37. ICAO Aircraft Engine Emissions Databank. Available online: https://www.easa.europa.eu/en/domains/environment/icao-aircraft-engine-emissions-databank (accessed on 10 December 2024).
38. Lefebvre, A.H.; Ballal, D.R. *GAS Turbine Combustion: Alternative Fuels and Emissions*; CRC Press: Boca Raton, FL, USA, 2010.
39. Ranjith P. V, M. Shivashankar, G. Sivaramakrishna, Vimala Narayanan. Reduction of Gas Turbine Combustor Pattern Factors using CFD. *Int. J. Eng. Res. Technol.* **2014**, *3*, 1786–1791.
40. Braun-Unkhoff, M.; Riedel, U. Alternative fuels in aviation. *CEAS Aeronaut J.* **2015**, *6*, 83–93. [CrossRef]
41. EUROCONTROL Think Paper #21–22 August 2023. Available online: https://www.eurocontrol.int/sites/default/files/2023-08/eurocontrol-think-paper-21-long-haul-decarb.pdf (accessed on 10 December 2024).
42. Nuic, A. *User Manual for the Base of Aircraft Data (BADA)*, Revision 3.9; EUROCONTROL Experimental Centre Project BADA Report 11/03/08-08; Supporting European Aviation: Brussels, Belgium, 2011.
43. Alexiou, A.; Kolias, I.; Aretakis, N.; Mathioudakis, K. Aero-Engine Preliminary Design Optimization and Operability Studies Supported by a Compressor Mean-Line Design Module. *Aerospace* **2023**, *10*, 726. [CrossRef]
44. Luche, J. Obtention de Modeles Cinetiques Reduits de Combustion: Application a un Mecanisme du Kerosene. Ph.D. Thesis, Universite D'Orleans, Orléans, France, 2003.
45. Liu, J.; Zhang, D.; Hou, L.; Yang, J.; Xu, G. Laminar burning speed of aviation kerosene at low pressures. *Energies* **2022**, *15*, 2191; ISSN 1996-1073. [CrossRef]
46. Smith, G.P.; Golden, D.M.; Frenklach, M.; Moriarty, N.W.; Eiteneer, B.; Goldenberg, M.; Bowman, C.T.; Hanson, R.K.; Song, S.; Gardiner, W.C.; et al. GRI-Mech 3.0. 1999. Available online: http://www.me.berkeley.edu/gri_mech (accessed on 10 December 2024).
47. Kelaidis, M.; Aretakis, N.; Tsalavoutas, A.; Mathioudakis, K. Optimal Mission Analysis Accounting for Engine Aging and Emissions. *ASME J. Eng. Gas Turbines Power* **2009**, *131*, 011201. [CrossRef]
48. *ISO 2533:1975*; Standard Atmosphere; Edition 1. International Organization for Standardization: Geneva, Switzerland, 1975. Available online: https://www.iso.org/standard/7472.html (accessed on 10 December 2024).
49. Flight Radar 24. Available online: https://www.flightradar24.com (accessed on 10 December 2024).

aerospace

MDPI

Article

Surrogate Models and Related Combustion Reaction Mechanisms for a Coal-Derived Alternative Jet Fuel and Its Blends with a Traditional RP-3

Quan-De Wang [1,*], Lan Du [2], Bi-Yao Wang [2,*], Qian Yao [3], Jinhu Liang [4], Ping Zeng [2] and Zu-Xi Xia [2]

[1] Jiangsu Key Laboratory of Coal-Based Greenhouse Gas Control and Utilization, Carbon Neutrality Institute and School of Chemical Engineering, China University of Mining and Technology, Xuzhou 221008, China
[2] Aviation Fuel and Chemical Airworthiness Certification Centre of CAAC, The Second Research Institute of CAAC, Chengdu 610041, China; dulan@fccc.org.cn (L.D.); zengping@fccc.org.cn (P.Z.); xiazuxi@fccc.org.cn (Z.-X.X.)
[3] College of Energy and Power Engineering, Nanjing University of Aeronautics and Astronautics, Nanjing 210016, China; yaoqian2008@foxmail.com
[4] School of Environmental and Safety Engineering, North University of China, Taiyuan 030051, China; jhliang@nuc.edu.cn
* Correspondence: quandewang@cumt.edu.cn (Q.-D.W.); wangbiyao@fccc.org.cn (B.-Y.W.)

Abstract: Jet fuel from direct coal liquefaction (DCL) is an important alternative kerosene and represents a high-performance fuel for specific applications in civil applications. The study on its chemical positions and combustion properties is critical for the development of surrogate models and related combustion reaction mechanisms, which is valuable for promoting its usage in aeroengines. However, research on DCL-derived jet fuel is rather scarce. Herein, this work reports a systematic study on a DCL-derived jet fuel and its blends with traditional RP-3 jet fuel in two different ratios. Specifically, major physicochemical properties related to the aviation fuel airworthiness certification process are measured. Advanced two-dimensional gas chromatography (GC × GC) analysis is used to analyze the detailed chemical compositions on the DCL derived jet fuel and its blend with RP-3, which is then employed for surrogate model development. Moreover, ignition delay times (IDTs) are measured by using a heated shock-tube (ST) facility for the blended fuels over a wide range of conditions. Combustion reaction mechanisms based on the surrogate models are developed to predict the experimental measured IDTs. Finally, sensitivity analysis and rate-of-production analysis are carried out to identify the key chemical kinetics controlling the ignition characteristics. This work extends the understanding of the physicochemical properties and ignition characteristics of alternative jet fuels and should be valuable for the practical usage of DCL derived jet fuels.

Keywords: direct coal liquefaction; jet fuel; two-dimensional gas chromatography; shock tube; ignition delay time; combustion reaction mechanism

1. Introduction

Petroleum-derived jet fuels are almost the only commercial aviation fuels that are used worldwide. However, considering the contiguous consumption and the uneven distribution of petroleum resources, alternative jet fuels, including sustainable jet fuels (SAFs), receive significant attention due to the rapid increase in aviation passengers and the demand for energy security. Alternative jet fuels can be synthesized via different techniques depending on the feedstock selection and production pathway [1,2]. Coal, natural gas, and biomass are representative feedstocks. As another major energy resource, the clean utilization of coal resources has become a major issue for many countries that are rich in coal but lack of petroleum/natural gas. The transformation of coal to liquid fuels has a long history and has also been widely used in many countries, i.e., Germany, South Africa, and China [3–5]. The conversion of coal to liquid fuels is generally performed through two production technical routes, i.e., indirect coal liquefaction (ICL) and direct coal liquefaction (DCL). The ICL technique is mainly achieved via the Fischer–Tropsch (FT) synthesis process, which converts the coal into syngas first, and then, the syngas is converted into long-chain hydrocarbons via catalytic process [5]. On the other hand, the DCL process uses the high-pressure hydrogenation process to break the chemical structure of coal and convert the unsaturated chemical bond to alkanes [3]. Due to the different technique routes, the chemical compositions of the liquid fuels from ICL and DCL are quite different. Specifically, iso-alkanes are major products from ICL [5,6], while polycyclic naphthenes with two or three rings are major products from DCL [3]. The usage of these fuels requires a comprehensive understanding of their combustion properties and related combustion chemistry, which is of critical importance for their airworthiness certification process.

During the past several decades, a series of research has been conducted on alternative jet fuels from different production routes. The corresponding chemical compositions and physiochemical and combustion properties of these fuels are studied and then used to develop and validate combustion reaction mechanisms that can be used for engine design [7]. For FT synthetic jet fuels, extensive efforts have been performed. Chemical kinetic mechanisms for two different FT fuels, i.e., Syntroleum S-8 and Shell GTL produced from natural gas, were constructed based on chemical composition analysis [8]. Dagaut et al. [9,10] also performed experimental and kinetic modeling study on a gas-to-liquid FT jet fuel. The chemical composition effects of FT jet fuels from Shell and Sasol on the autoignition behaviors were also experimentally studied [11]. In principle, the physiochemical and combustion properties of fuels are fundamentally determined by the chemical compositions and structures. Thus, advanced analytical methods [12,13] have been introduced to uncover the chemical compositions of jet fuels, which can promote more accurate surrogate models [14,15]. And this has been extended to study on alternative jet fuels [9,11,16–22]. Specifically, accurate two-dimensional gas chromatography (GC $\times$ GC) analysis has been employed for the determination of chemical compositions of FT jet fuels and SAF recently [23–25]. Compared with the extensive studies on FT jet fuels, fewer studies have been performed on DCL jet fuels [19,26,27]. The major reason may be due to the existence of few industrial projects because of the high cost of the high-pressure hydrogenation process. However, as the green hydrogen technology and catalytic technique continue to develop, the cost and technique problems of DCL have decreased. Further, due to the special process of DCL, the derived jet fuel represents a performance fuel with high energy density due to the ring structures [28]. Thus, a better understanding of the physicochemical and combustion properties of DCL-derived jet fuels is necessary to streamline the airworthiness certification process and design more efficient combustion chambers.

Based on the above considerations, this work contributes to the literature studies on DCL-derived jet fuels. Specifically, the detailed chemical compositions of a DCL-derived jet fuel and its blends with a traditional RP-3 jet fuel with two different ratios are systematically analyzed via two-dimensional gas chromatography (GC $\times$ GC) analysis. The major physicochemical properties related to airworthiness certification process are studied. Based on this, surrogate models for DCL-derived jet fuel and the blended fuels are developed, and two different kinds of combustion reaction mechanisms are constructed via detailed and decoupling methods. The ignition properties of the DCL-derived jet fuel with a traditional RP-3 jet fuel are experimentally studied and used to validate the developed mechanisms. Finally, sensitivity analysis and rate-of-production analysis are conducted to identify the key chemical kinetics controlling the ignition characteristics.

2. Experimental Methods

2.1. GC × GC Analysis

Surrogate model approach usually selects several representative hydrocarbons with large fractions in real fuels to mimic their physicochemical properties. Thus, it strongly depends on the chemical composition analysis of the real fuels. But the real fuels usually consist of hundreds of different hydrocarbons; hence, accurate analysis of the fuel's chemical compositions remain a great challenge for traditional established standardized techniques [12]. In recent years, a multidimensional separation technique to identify volatile and semi-volatile compounds in complex mixtures, GC × GC analysis, which is a universal method for the complete group-type to component-by-component analysis, has been introduced to perform chemical composition analysis of real fuels [12]. Herein, to gain insight into the chemical compositions of the DCL-derived jet fuel and its blends with RP-3, GC × GC analysis is employed to reveal their chemical compositions. The Shimadzu QP2010 GC × GC–MS/FID facility with two chromatographic columns, DB-17 (2.5 m × 0.10 mm × 0.10 μm) and DB-1 (30 m × 0.25 mm × 0.25 μm), is employed to separate the samples. The temperature at the injection port is set to be 548 K with a sample size of 1.0 μL. The initial temperature is 358 K and then gradually increased to 558 K with a rate of 1.5 K/min. The detection ranges of mass spectra (MS) in this work are 40–500 m/z (mass to charge ratio). Finally, the experimental results are processed by using GC × GC images, which are then matched with the NIST Mass Spectral Database to determine the quantitative compositions of the fuels.

2.2. Shock-Tube Experiment

Although the ignition characteristics of pure DCL-derived jet fuel have been experimentally studied, further work remains because it can only be used by blending with traditional jet fuels according to current airworthiness certification standards. Hence, this work performs experimental study on the ignition characteristics of DCL-derived jet fuel with traditional RP-3 with volume ratios of 30%/70% and 70%/30%. A high-pressure shock tube (HPST) at North University of China (NUC) is adopted to measure the IDTs. The experimental system has been described and validated previously [19,29–31]. Thus, only a brief description is introduced. The HPST contains a 3.0 m driver section and a 6.8 m driven section with inner diameter of 0.1 m, and they are connected through a 0.3 m double diaphragm section. Figure 1 shows the scheme of the shock-tube facility. Before the experiments, the reactant mixture is prepared in a gas distribution tank based on Dalton's law of partial pressures, and it is maintained as static for 12 h for homogeneity. The detailed combustion conditions are listed in Table 1. Specifically, the descriptors, namely φ, x, Avg. P_5, and T_5, denote the equivalence ratio, concentrations of the corresponding species in the reactant mixtures, the reflected pressure and reflected temperature, respectively. The DCL-derived jet fuel is synthesized by CHN Energy Investment Group, while the RP-3 jet fuel is produced in Dalian West Pacific Petrochemical Company. They are provided by the Aviation Fuel and Chemical Airworthiness Certification Centre of CAAC. Helium with purity of 99.99% is used as the driven gas in HPST. The incident shock velocity is measured through five PCB113B26 piezoelectric transducers and is computed through a linear extrapolation of the four known velocities to the endwall. The pressure behind the reflected shock wave is recorded through a PCB 113B26 piezoelectric pressure sensor. The temperature and pressure behind the reflected shock wave is estimated by using the Gaseq program [32]. The definition of IDT is the time interval between the pressure increasing due to the arrival of the incident shock at the endwall and the maximum rate of increase in the pressure signal.

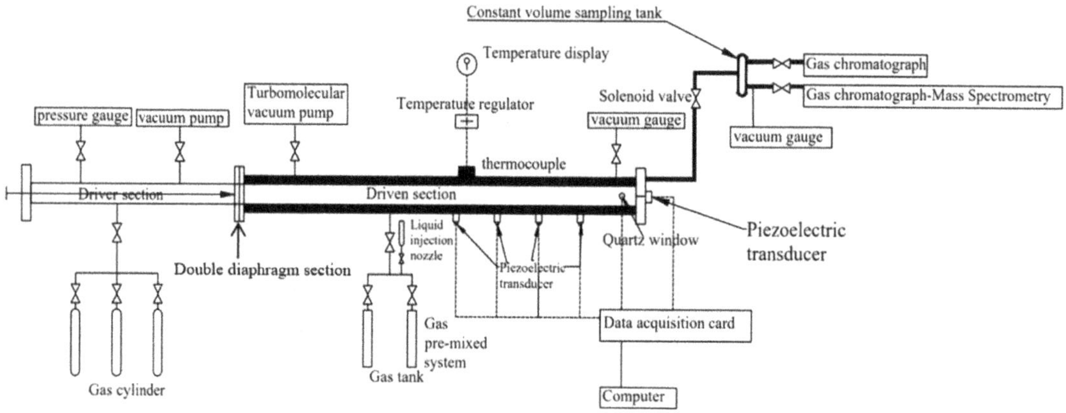

Figure 1. Scheme of the shock-tube experimental facility used in the present work.

Table 1. HPST experimental conditions in this work.

Fuel (DCL/RP-3 Blends)	φ	x_{Fuel} (mol%)	x_{O2} (mol%)	x_{N2} (mol%)	Avg. P_5 (bar)	T_5 Range (K)
30/70 blends in volume	0.5	0.64	20.87	78.50	10.02	1110–1530
	1.0	1.27	20.73	78.00	9.94	1060–1450
	2.0	2.50	20.48	77.03	10.23	1030–1490
70/30 blends in volume	0.5	0.64	20.87	78.50	10.03	1090–1510
	1.0	1.27	20.73	78.00	10.16	1070–1520
	2.0	2.50	20.48	77.03	10.13	1080–1500

For the uncertainty of the experimental IDTs, it changes with the combustion conditions, i.e., pressure, temperature, equivalence ratio, and also the facility. Generally, the results from the present HPST facility are in good accordance with other related facilities [29,33,34], indicating the uncertainty should be consistent. Based on the literature analysis of the uncertainty analysis [33,35], the uncertainty of the measured IDTs can be controlled within 20%.

2.3. Kinetic Modeling Approach

The experimentally measured IDTs are simulated by using the Cantera software [36] by using the closed homogeneous batch reactor at constant volume, which has been confirmed to be accurate for shock-tube facility-derived IDTs from short test time experiments [37,38]. Figure 2 shows a conceptual diagram of the simulation procedure. Varying volume profiles are not considered because the shock attenuation ranges in the present experiments are minor. To obtain consistent results that can be compared with experiments, the IDT from the simulation is also defined as the extrapolation of the maximum dp/dt to the zero point. For the chemical kinetic analysis, sensitivity analysis is performed through the brute-force procedure [39], while rate-of-production (ROP) analysis is conducted by determining the contribution of each reaction to the net production or destruction rates of a species transiently during ignition simulations [24,25], which allows to quickly identify dominant reaction paths.

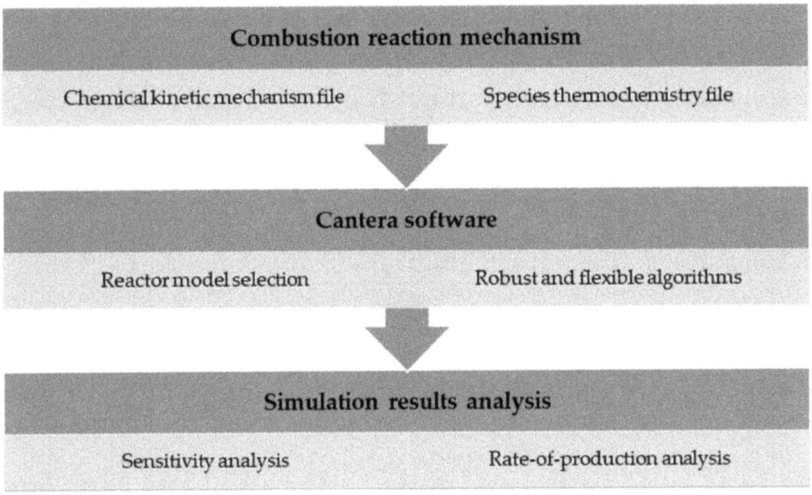

Figure 2. The conceptual diagram of the Cantera software used in this work.

3. Results and Discussion

3.1. Physicochemical Properties

Although the aerospace engine operability is fundamentally influenced by the fuel's ability to ignite and sustain a flame under demanding conditions [28], several critical physicochemical properties must be fulfilled according to airworthiness certification to ensure the fuel's compatibility with existing engines and airport infrastructure. Table 2 lists the measured major physicochemical properties of the DCL-derived jet fuel and also the traditional RP-3 jet fuel associated with the specifications. It can be seen that all the measured physicochemical properties are comparable with the traditional RP-3 jet fuel and satisfy the specifications. Compared with the traditional RP-3, the DCL jet fuel is cleaner, as shown from the sulfur and naphthalene contents due to the restricted production routes. The other important physicochemical properties are very close between the DCL and RP-3 jet fuels, but larger differences exist for the aromatics and freezing point, which are significantly influenced by their chemical compositions, as discussed below. For some of the measured physicochemical properties of the blended fuels, it can be seen that they are also close to the pure DCL-derived jet fuel and RP-3.

Figure 3 shows the detailed chemical compositions, including the mass fractions of the different alkane types and the corresponding hydrocarbon species of the studied DCL jet fuel and its blends with RP-3 with volume ratios of 70/30 and 30/70 from GC $\times$ GC analysis. From Figure 3, the DCL-derived jet fuel is mainly composed of cyclic alkanes, which contributes 87.21% mass fraction of the total compositions. Of these, polycyclic alkanes account for nearly half of the total mass fraction of DCL-derived jet fuel, while n-alkanes and iso-alkanes account for only 11.59% of the compositions. It is also worth noting that aromatics hardly exist in the DCL-derived jet fuel due to the hydrogenation production process. On the contrary, the hydrocarbon types in the RP-3 fuel exhibit a more even distribution. The iso-alkanes contributes most to RP-3, but the mass fraction is only 31.47%, which is just slightly larger than that of n-alkanes and monocyclic alkanes. It is also noted that the polycyclic alkanes in RP-3 are significantly lower than that of monocyclic alkanes, which is quite different compared with the DCL-derived jet fuels. Moreover, the present RP-3 jet fuel contains a large mass fraction of aromatics. For the two blends of DCL-derived jet fuel and RP-3, it can be seen that the compositions of the hydrocarbon types fall generally between the two pure jet fuels, and the specific mass fractions tend to be close to the pure fuel with larger volume fractions.

Table 2. Physicochemical properties of the DCL and comparisons with RP-3. The standard refers to the airworthiness certification standards for jet fuels enacted by the American Society for Testing Materials (ASTM).

Property	DCL	DCL/RP-3		RP-3	Standard
		30/70	70/30		
Acidity (mg KOH/g)	0.002	—	—	0.002	ASTM D3242, ≤0.015
aromatics (% vol)	0.9	—	—	16.8	ASTM D1319, ≤25
sulfur (mass %)	<0.0001	—	—	0.066	ASTM D5453, ≤0.3
sulfur, mercaptan (mass %)	<0.0003	—	—	0.0012	ASTM D3227, ≤0.003
flash point (°C)	45.5	—	—	47.5	ASTM D1655, ≥38
density 15 °C (kg/m^3)	827.4	—	—	799.7	ASTM D4052, 775–840
freezing point (°C)	−66.0	—	—	−52.5	ASTM D5972, ≤−47
viscosity, −20 °C (mm^2/s)	4.528	4.395	4.242	4.230	ASTM D445, ≤8.0
Net heat of combustion (MJ/kg)	42.992	43.257	43.362	43.296	ASTM D3338, ≥42.8
smoke point (mm)	25.2	26.1	27.6	26.6	ASTM D1332, ≥25.0
naphthalene (vol %)	0.0091	0.72	0.34	0.81	ASTM D1840, ≤3.0

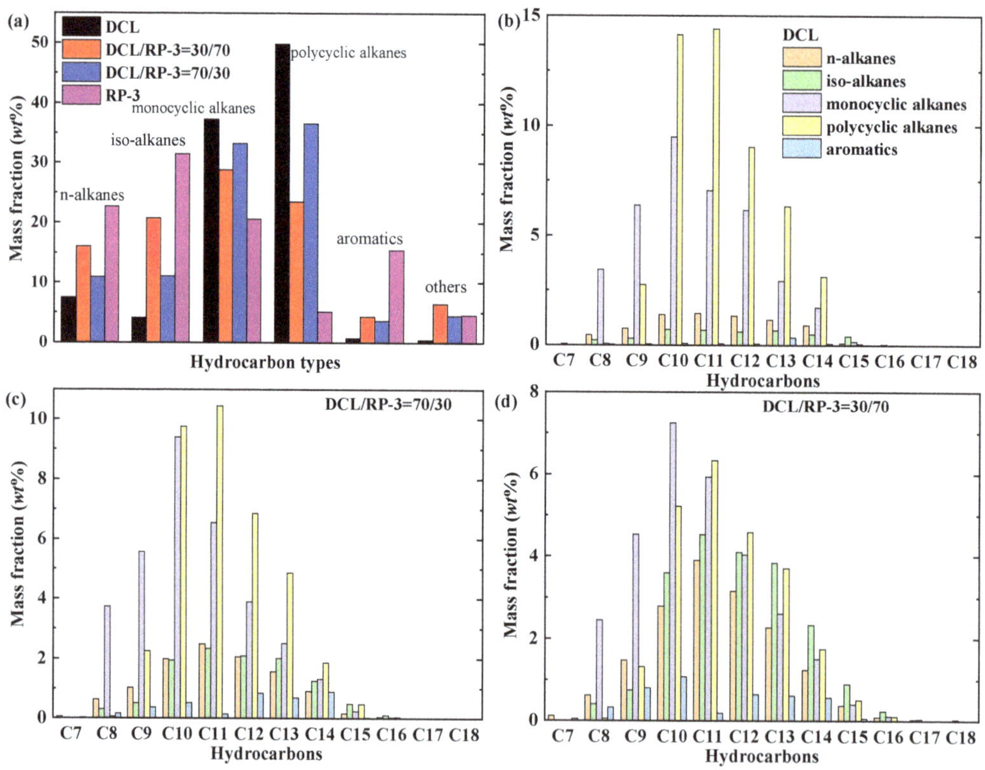

Figure 3. Detailed compositions of the studied DCL-derived jet fuel and its blends with RP-3 with volume ratios of 70/30 and 30/70 from GC × GC analysis. The results in (**a**) for RP-3 are adopted from [24]. Figures (**b**–**d**) display the carbon number distributions of the studied DCL, and its blends with RP-3 with volume ratios of 70/30 and 30/70, respectively.

For the carbon number distributions, the three fuels are primarily distributed within the C8–C15 range. Specifically, for the DCL-derived jet fuel, it can be seen that the main hydrocarbons of the polycyclic alkanes are C10–C12 hydrocarbons, while the main hydrocarbons of the monocyclic alkanes are C9–C12 hydrocarbons. The detailed compositions of the blended fuel with 70% DCL jet fuel also tend to show similar distributions. As the volume ratio of RP-3 increases to 70%, it can be seen that the hydrocarbon distributions become close to pure RP-3 [24]. The detailed compositions and related carbon number distributions provide fundamental information for surrogate model development.

3.2. *Ignition Properties of the Blends*

Figure 4 shows the experimentally measured IDTs of the studied blended jet fuel of DCL with RP-3 in volume ratios of 30/70 and 70/30, and the effect from temperature and equivalence ratio is clarified. From Figure 4, the temperature demonstrates a typical Arrhenius relationship on the IDTs under the studied high-temperature combustion conditions, and the results can be linearly fitted. Under the studied temperature ranges, the effect of the equivalence ratio for the two blends with different ratios is also similar. Generally, the IDTs decrease as equivalence ratio increases, and this is more distinct at a temperature below 1250 K. As the temperature further increases, the IDTs become convergent, as shown from the fitted results. The controlling combustion chemical kinetics should be the major reason for the effect of the equivalence ratio on the IDTs. At the relatively lower temperature, the IDTs of the reactant mixture can be shortened due to the increase in fuel concentration because the chemistry of the fuel radicals dominates and has the largest effect on the ignition process. To be more specific, ignition at this stage is mainly affected by the fuel chemistry related to the two reactions, i.e., fuel + HO_2 = radical + H_2O_2 and the subsequent H_2O_2 (+M) = OH + OH (+M) that will generate two reactive OH radicals, which can decrease the IDTs [40,41]. As temperature increases, the chain branching reaction, H + O_2 = OH + O, begins to control the ignition process. But the reaction of the fuel with H atoms will still compete with this reaction. Consequently, the effect of the equivalence ratio will converge and then demonstrate an opposite trend compared with the lower temperature condition. From Figure 4, the IDTs of the two blended fuels with different volume ratios do not exhibit a large difference. At lower temperature conditions (lower than 1200 K), the IDTs of the two blended fuels have very close equivalence ratios of 0.5 and 1.0, and the blended jet fuel with the small RP-3 component can ignite a little faster. However, at a temperature larger than 1200 K, the blended jet fuel with a large RP-3 component generally shows longer IDTs irrespective of the equivalence ratio. Overall, the difference between the two volumes is small.

Figure 5 further compares the IDTs of pure DCL [19] and RP-3 [24] and their blends with different ratios at 10 bar under different equivalence ratio conditions. Generally, the IDTs of the four fuels show similar reactivity trends at different equivalence ratio conditions. Under a typical equivalence ratio condition, the IDTs of the pure DCL fuel are shorter compared with the other three fuels at temperatures approximately above 1100 K, and the IDTs of the four fuels become convergent at around 1100 K. The two blends of DCL with RP-3 at different volume ratios are close to the IDTs of RP-3 jet fuel, and the increment of RP-3 composition (the DCL/RP-3 mixture with volume ratio of 30/70) makes the IDTs closer to the RP-3 jet fuel. It is also worth noting that the DCL/RP-3 in a volume ratio of 70/30 with a large DCL component demonstrates the longest IDTs at high temperature conditions, which should be caused by the composition changes due to the small amounts of RP-3 additions.

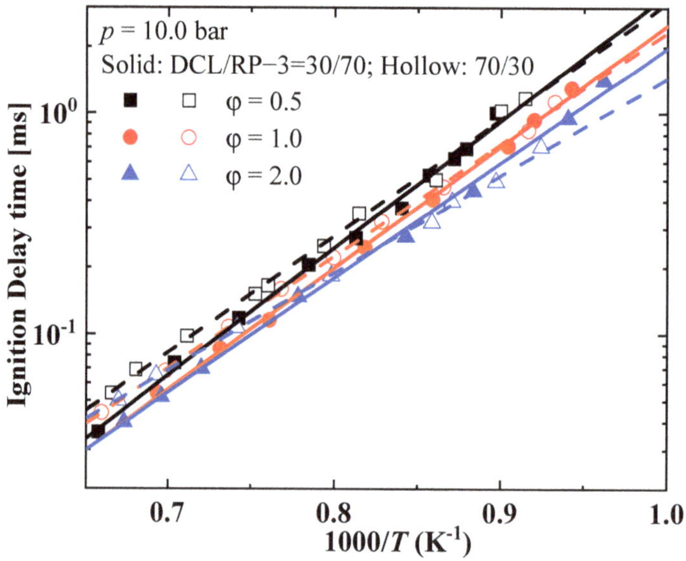

Figure 4. IDTs from ST experimental measurements of the studied blend jet fuel of DCL with RP-3 in volume ratios of 30/70 and 70/30.

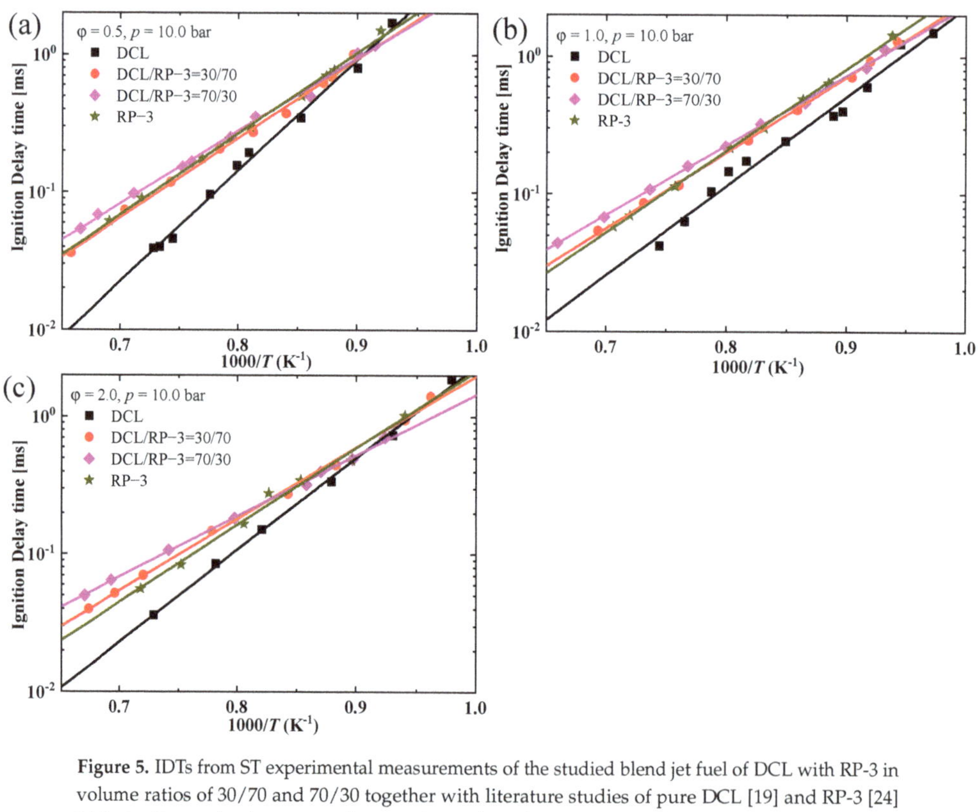

Figure 5. IDTs from ST experimental measurements of the studied blend jet fuel of DCL with RP-3 in volume ratios of 30/70 and 70/30 together with literature studies of pure DCL [19] and RP-3 [24] jet fuels. Figures (**a**–**c**) represent the experimental results at pressure of ~ 10.0 bar with equivalence ratios of 0.5, 1.0, and 2.0, respectively.

3.3. Surrogate Models and Combustion Reaction Mechanism

Based on GC × GC analysis, surrogate models for DCL-derived jet fuel, RP-3, and their blends are proposed. Table 3 shows the surrogate models and the corresponding mole fractions of the surrogate components. Herein, a new version that is different from the literature model of RP-3 [24] is proposed. Specifically, decalin is also considered as a surrogate component considering the existence of about 5% polycyclic alkanes. For n-alkanes, n-undecane is selected as the surrogate component due to the largest amount of mass fractions, and the combustion reaction mechanisms are also widely studied. For iso-alkanes, it can be seen that the carbon number distributions are rather evenly distributed in the range from C9 to C13. However, studies on iso-alkanes are few and mainly focused on iso-dodecane. Hence, this study represents the iso-alkanes. For monocyclic alkanes, the most abundant species is C10 hydrocarbons, followed by the C11 and C9 hydrocarbons. Herein, n-butyl cyclohexane is proposed to mimic the monocyclic alkanes in the jet fuels. Decalin is employed to represent polycyclic alkanes because it is the simplest polycyclic alkanes and has been studied extensively. Larger polycyclic alkanes are studied very rarely. Moreover, it can be seen that polycyclic alkanes are mainly composed of C10 and C11 species, indicating decalin should be a major component in the studied fuels. Aromatic hydrocarbons in the studied fuels are usually distributed from C9 to C14. Herein, n-propyl benzene is adopted because previous studied revealed that the alkyl length effect on the ignition properties of aromatics is small, especially at high-temperature conditions [42]. Through a transformation and normalization process of the measured mass fraction of the species distributions, the mole fractions of the surrogate components are derived as shown in Table 3. The detailed mechanism is mainly based on previous studies on different jet fuels [24,25], comprising a sub-mechanism of these surrogate components that has been extensively validated. To be more specific, a detailed mechanism provided as supplementary materials is developed based on the comprehensively validated hierarchical NUIG mechanism [24,25,43–45] with the sub-mechanisms of decalin [46], iso-dodecane [47], n-butyl cyclohexane [48], and n-propyl benzene [49]. Using the surrogate models shown in Table 3, the adiabatic flame temperature, with initial temperature of 300 K and pressure of 1 bar with an equivalence ratio of 1.0, is estimated from chemical equilibrium calculations. The computed temperatures of DCL, DCL/RP-3 = 30/70, DCL/RP-3 = 70/30, and RP-3 are 2286, 2286, 2287, and 2289 K, respectively. Thus, the differences between the four jet fuels are very minor. The flame temperature can significantly affect the laminar flame properties, while it has small effect on the IDTs. From the computed flame temperature, it can be anticipated that the flame speed between these fuels could be small.

Table 3. Surrogate models for DCL derived jet fuel, RP-3, and the blends. The compositions are in mole fraction (%).

Surrogate Component	DCL	DCL/RP-3 = 30/70	DCL/RP-3 = 70/30	RP-3
n-Undecane	6.85	16.11	10.47	22.60
Iso-dodecane	3.46	19.11	9.72	28.76
n-Butyl cyclohexane	38.08	32.21	35.46	22.90
Decalin	51.61	26.70	39.65	5.28
n-Propyl benzene	0.00	5.87	4.70	20.46

3.4. Kinetic Modeling Results

Figure 6 compares the simulated IDTs with the experimentally measured results of the two blended fuels with different ratios from DCL and RP-3 in the present work together with the literature IDTs for DCL-derived jet fuel [19] and RP-3 [24]. From Figure 6, it can be seen that the detailed mechanism can predict the experimental results for the four fuels with reasonable accuracy. The variations of IDTs as a function of temperature and equivalence ratio can be reflected. Additionally, the relative reactivity from the IDTs of the four fuels can also be predicted by the detailed mechanism. Compared with experimental results, the predicted IDTs of DCL are also faster than the other three fuels, while the IDTs of the other three fuels are very close. However, it is also worth noting that at high-temperature conditions, the predicted IDTs are generally faster than the experimental results at an equivalence ratio of 0.5. Overall, the reactivity can be well reflected by the detailed mechanism and the surrogate models. To further reveal the combustion chemical kinetics of the four fuels, rate-of-production (ROP) and sensitivity analysis are conducted.

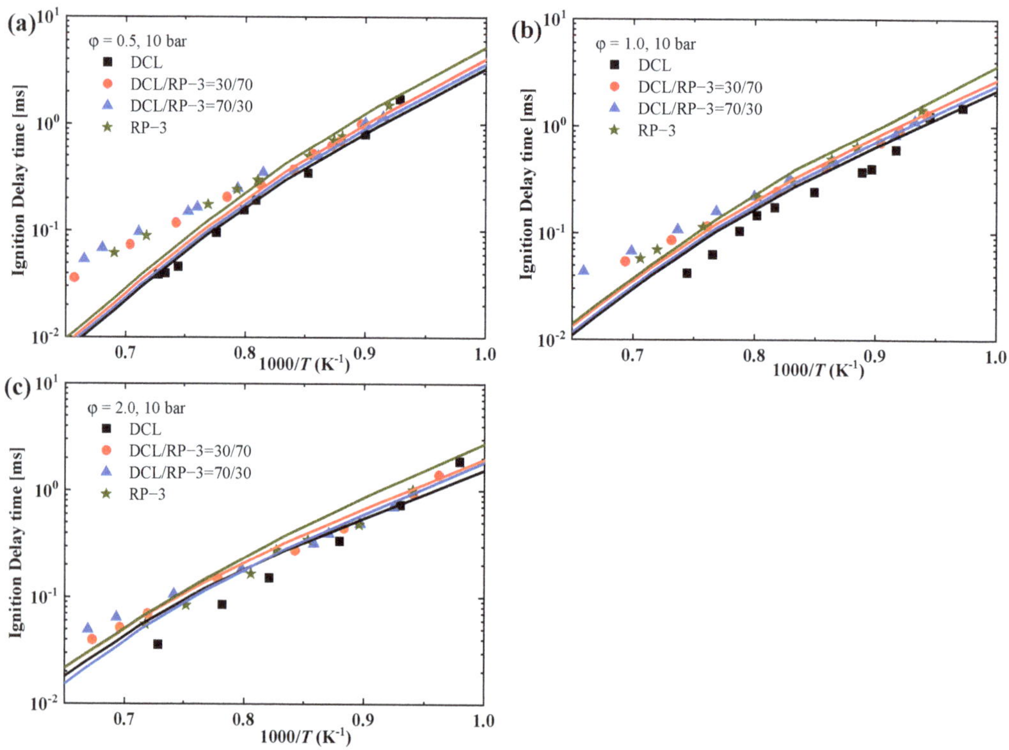

Figure 6. Predicted IDTs of the present experimental results by using the detailed (solid line) and decoupling (dashed line) mechanisms for the four fuels at pressure of ~10.0 bar with equivalence ratios of 0.5 (**a**), 1.0 (**b**), and 2.0 (**c**), respectively.

It is widely recognized that the high-temperature combustion properties of large hydrocarbons are mainly controlled by the small C0–C4 intermediates formed through the pyrolysis process [39,50]. Thus, to reveal the different combustion properties of the four jet fuels, ROP is conducted, and Figures 7 and 8 show the time histories of major intermediates as a function of time during the ignition processes of the four fuels at 10 bar and an equivalence ratio of 1.0, with temperature at 1100 K and 1450 K, respectively. From Figures 7 and 8, the major intermediates formed during the ignition processes of the four fuels are quite different, indicating the chemical compositions and molecular structures show significant influence on the intermediates that further affects the IDTs of the fuels. Additionally, it can also be seen that the temperature demonstrates an important effect on the evolution of the species concentrations as a function of time, especially on the alkenes, i.e., ethylene (C_2H_4), propene (C_3H_6), 1,3-butadiene (C_4H_6), and iso-butene (IC_4H_8). This could be induced by the temperature effect on the reactions leading to the formation of these species. At high-temperature conditions, the β-scission reactions leading to the formation of these intermediates can be facilitated and can continuously produce these species. In addition, the greater importance of hydrogen chemistry at high-temperature conditions decreases the consumption of these species, resulting in more gradual tendencies of these species compared with conditions at lower temperatures whose concentrations demonstrate sudden changes approaching zero.

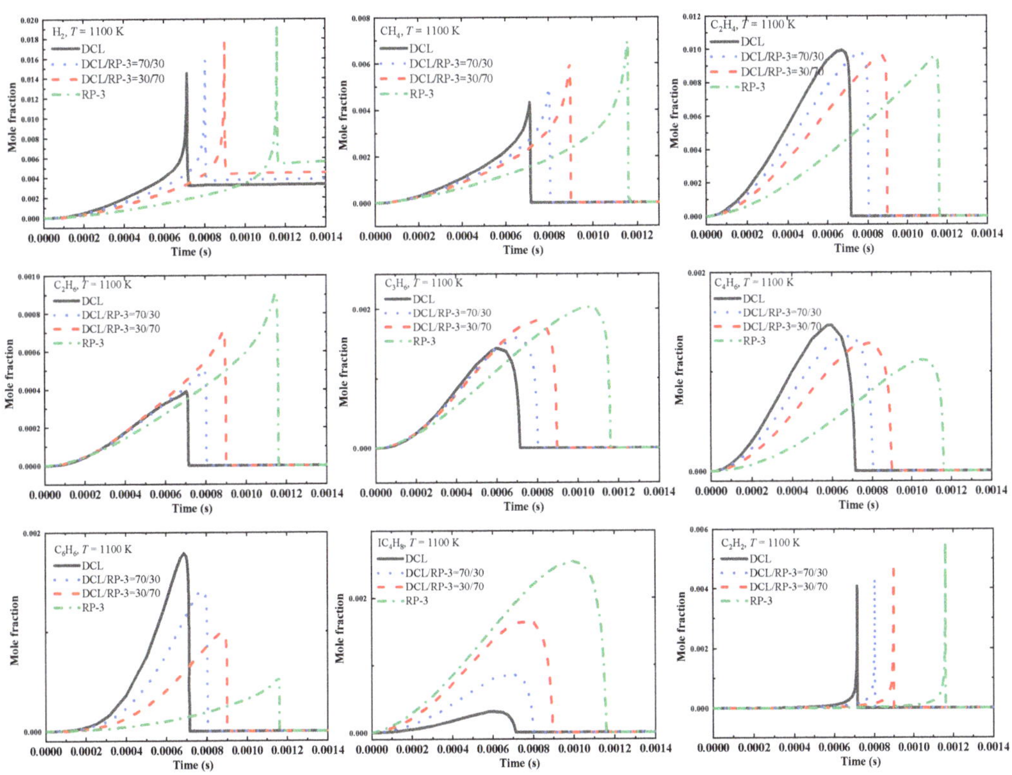

Figure 7. Species profiles of the major intermediates from ROP analysis during the ignition processes of the studied four fuels at 10 bar, 1100 K, and an equivalence ratio of 1.0.

For the formed intermediates from different fuels, it can be seen that the relative quantities of the intermediates from the four fuels at 1100 and 1450 K are quite similar. Specifically, the quantities of H_2 and CH_4 from RP-3 are larger than that from DCL, and the two blends with different ratios lie between them. This could be induced by the lower H/C ratio of DCL compared with RP-3 due to the large amounts of cyclic alkanes. For C_2H_4, the quantities from the four fuels do not exhibit a large difference because it is mainly formed through the first-step β-scissions of the fuel radicals from the abstraction reactions, and the total molecular mass of the fuels is very close for the four fuels. For C_2H_6 and C_3H_6, the existence of linear alkanes in RP-3 tends to form these intermediates compared with cyclic alkanes, which leads to the quantity tendencies of the two intermediates. The large amounts of cyclic alkanes in DCL makes the formed 1,3-butadiene and benzene (C_6H_6) larger than that from RP-3 due to the dehydrogenation reactions and ring-opening reactions of cyclic alkanes. For iso-butene, the formation is mainly affected through the iso-alkanes in the fuels; thus, the quantity from RP-3 is larger than that from DCL. For acetylene (C_2H_2), it can be seen that its production from RP-3 is larger than that from DCL. Generally, acetylene is usually formed through the reactions of C_2H_3 with H, i.e., C_2H_2 + H (+M) = C_2H_3 (+M) and C_2H_3 + H = C_2H_2 + H_2. Considering the similar production tendency of C_2H_4, the formation of acetylene could be related to the decomposition reaction of propene to the formation of C_2H_3 and CH_3 radicals. Overall, from the above-detailed analysis, the major intermediates formed from different fuels are directly affected by the chemical compositions and molecular structures of the fuels.

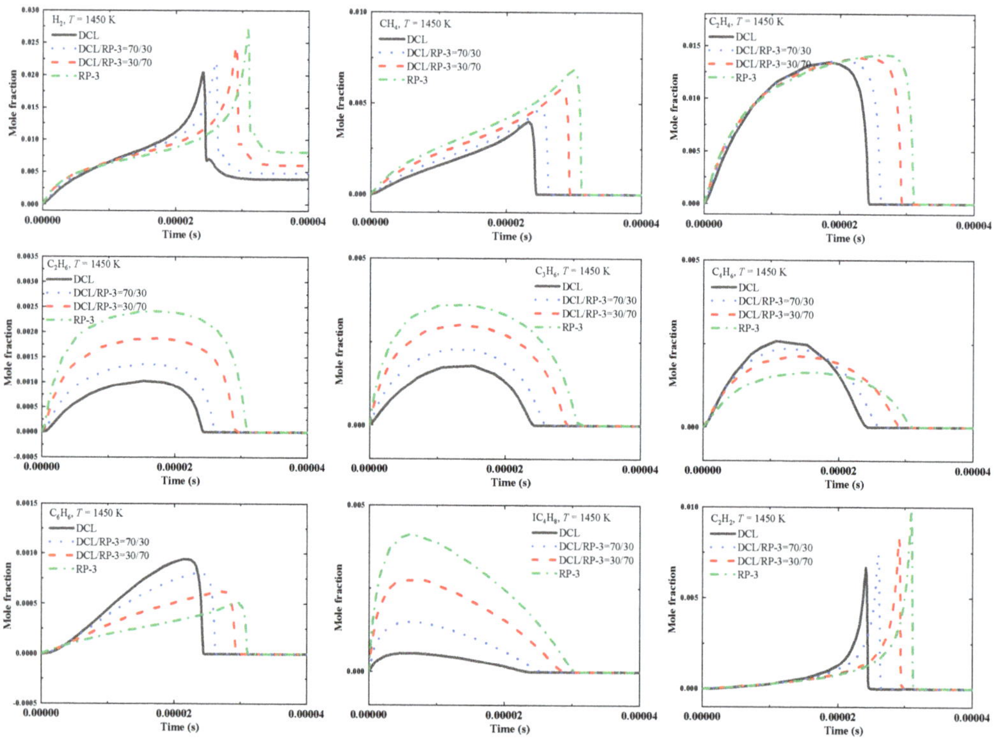

Figure 8. Species profiles of the major intermediates from ROP analysis during the ignition processes of the studied four fuels at 10 bar, 1450 K, and an equivalence ratio of 1.0.

Figure 9 shows the top 10 reactions with corresponding positive and negative sensitivity coefficients for the IDTs of the studied four fuels at 10 bar, 1100, and 1450 K with an equivalence ratio of 1.0. From the definition of the brute-force sensitivity coefficients, a negative sensitivity coefficient denotes that the IDTs are shortened when the reaction rate constant is doubled. Hence, the reactivity of the reactant system is increased and vice versa. From Figure 9, both the chemical compositions and temperature exhibit a large effect on the sensitivity analysis results of the four fuels. Firstly, the most sensitive reactions of the four fuels are strongly affected by their compositions, more specifically the surrogate models. The most influential reactions of the four fuels are rather different. Specifically, for DCL compared with the other three fuels, it can be seen that the most important reaction with a negative sensitivity coefficient is the decomposition reaction of hydrogen peroxide (H_2O_2) to the formation of two reactive hydroxy radicals from Figure 9. The major reason for this difference can be attributed to the large number of polycyclic alkanes in the real DCL. In the mechanism, decalin is employed to represent this kind of hydrocarbon. The results for DCL are in good accordance with the literature studies on decalin [51,52]: the chain-branching reaction $H_2O_2(M) = 2OH + (M)$ demonstrates most important promoting effect on the IDTs in high-temperature combustion conditions because this reaction can generate two reactive OH radicals by consuming only one relatively unreactive HO_2 radical. Furthermore, besides reactions involving small C0–C4 species, the reactions of decalin with H_2O_2 also demonstrate a reactivity-promoting effect on the IDTs because the H-atom abstraction reaction by HO_2 radical plays an increasingly important role in the high-temperature combustion conditions [51]. As the DCL is blended with RP-3, the effect from decalin-related reactions and the chain-branching reaction $H_2O_2(M) = 2OH + (M)$ on the IDTs becomes less important, especially at high-temperature conditions. At 1100 K for RP-3, it can be seen that the reactions related to decalin hardly affect the IDTs due to its composition in the surrogate model, and the reactions related to toluene and benzyl radical show a large effect, which is in good agreement with previous studies on traditional jet fuels [39].

Secondly, the temperature effect on the sensitivity analysis results shows a large dependence on the fuels. For DCL, it can be seen that the sensitive reactions at 1100 K and 1450 K do not exhibit a large difference. However, the sensitive reactions for the blended fuels are significantly affected from temperature 1100 K to 1450 K. For the blended fuels, the reactions related to decalin show little effect, while the decomposition reactions of n-butyl cyclohexane to cyclohexyl and n-butyl show some effect on the IDTs. Besides this, the dominant reactions with large, sensitive coefficients are mainly related to C0–C4 species. However, for RP-3, the sensitivity results are generally consistent at 1100 K and 1450 K, and a major difference is that some reactions related to propene and small species become important at 1450 K. Except for these differences in the sensitivity analysis results of the four fuels, there are still some consistent observations from the sensitivity analysis results. Generally, the most influential reactions with an ignition-promoting effect are the chain-branching reactions that can generate reactive radicals $H_2O_2(M) = 2OH + (M)$ and $H + O_2 = OH + H$. The relative importance of the two reactions depends on the fuel types. On the other hand, the chain-termination reaction, $CH_3 + HO_2 = CH_4 + O_2$, which is competitive with another reactive channel, $CH_3 + HO_2 = CH_3O + OH$, demonstrates the greatest ignition-inhibiting effect for most cases. Further, it can be concluded that under the studied high-temperature conditions, the abstraction and decomposition reactions directly related to the polycyclic alkanes, monocyclic alkanes, and aromatic hydrocarbons in the fuels can demonstrate a large effect on the IDTs compared with the n-alkanes and iso-alkanes.

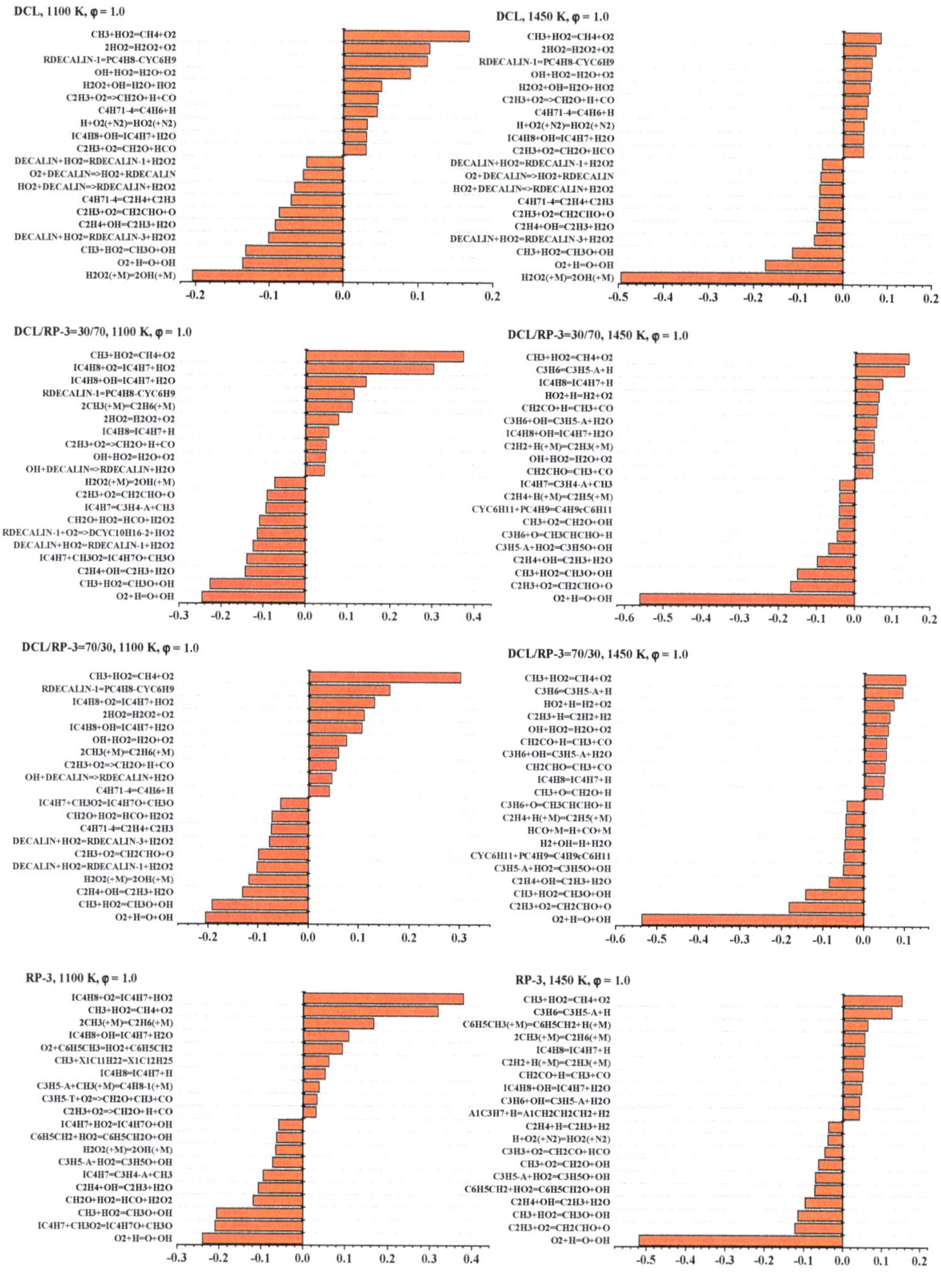

Figure 9. The most influential reactions on the IDT of the four fuels at 10 bar, 1100, and 1450 K with an equivalence ratio of 1.0.

4. Conclusions

This work reports extensive experimental and modeling studies on an alternative jet fuel from DCL and its blends with traditional RP-3 jet fuel. The major conclusions are summarized as follows.

- Advanced physicochemical analysis and chemical compositions were carried out for the DCL jet fuel and its blends with RP-3. Generally, the major physicochemical properties of the DCL jet fuel satisfied most of the airworthiness certification standards, and it is cleaner, with a lower freezing point (−66.0 °C compared with that of −52.5 °C of RP-3);
- Advanced GC × GC analysis revealed that the DCL jet fuel is mainly composed of cyclic alkanes (87.21% mass fraction) with a small number of n/iso-alkanes (11.59%), and the carbon numbers are primarily distributed within the C9–C14 range;
- The IDTs of the blended fuels of DCL and RP-3 with different volume ratios do not exhibit a large difference, and they tend to be close to the IDTs of RP-3 within the studied combustion conditions, i.e., equivalence ratios of 0.5, 1.0, and 2.0 and temperature ranges 1030–1530 K, with pressure at approximately 10 bar;
- ROP analysis indicated that major intermediates, including 1,3-butadiene and benzene, can be largely formed due to the dehydrogenation and ring-opening reactions of cyclic alkanes during the ignition process of DCL, while the quantities of other intermediates, including hydrogen, methane, propene, acetylene, iso-butene, ethylene, and ethane, increase as the volume ratios of RP-3 increase;
- Sensitivity analysis results highlighted that the ignition properties of the studied jet fuels are significantly affected by the chemical compositions and molecular structures of the jet fuels. Future studies on more accurate chemical composition analysis and other combustion properties, including laminar flame speeds and soot, should be conducted for surrogate model and mechanism optimization, which is critical for airworthiness certification and large-scale commercial applications of the DCL jet fuels.

Supplementary Materials: The following supporting information can be downloaded at https://www.mdpi.com/article/10.3390/aerospace12060505/s1. Detailed_mech.THERM.txt: The thermochemical data file; Detailed_mech.MECH.txt: The combustion mechanism file.

Author Contributions: Q.-D.W., conceptualization, formal analysis, funding acquisition, supervision, and writing—review and editing; L.D., data curation, formal analysis, and investigation; B.-Y.W., methodology, investigation, and writing—original draft; Q.Y., data curation, formal analysis, and validation; J.L., data curation, formal analysis, investigation, and methodology; P.Z., data curation and validation; Z.-X.X., conceptualization, funding acquisition, and resources. All authors have read and agreed to the published version of the manuscript.

Funding: This research was funded by the Joint Research Fund from National Natural Science Foundation of China and Civil Aviation Administration of China, grant number U2133215; National Natural Science Foundation of China, grant number 12172335; Advanced Aerospace Power Innovation Workstation, grant number HKCX2024-01-021; and Sichuan Science and Technology Program, grant number 2023ZYD0138.

Data Availability Statement: Data will be made available on request.

Conflicts of Interest: The authors declare no conflicts of interest.

Abbreviations

The following abbreviations are used in this manuscript:

DCL	Direct coal liquefaction
GC × GC	Two-dimensional gas chromatography
HPST	High-pressure shock tube
IDT	Ignition delay time

References

1. Pires, A.P.P.; Han, Y.L.; Kramlich, J.; Garcia-Perez, M. Chemical Composition and Fuel Properties of Alternative Jet Fuels. *Bioresources* **2018**, *13*, 2632–2657. [CrossRef]
2. Zhang, C.; Hui, X.; Lin, Y.Z.; Sung, C.J. Recent development in studies of alternative jet fuel combustion: Progress, challenges, and opportunities. *Renew. Sustain. Energy Rev.* **2016**, *54*, 120–138. [CrossRef]
3. Mochida, I.; Okuma, O.; Yoon, S.H. Chemicals from Direct Coal Liquefaction. *Chem. Rev.* **2014**, *114*, 1637–1672. [CrossRef] [PubMed]
4. Huang, Y.; Rolfe, A.; Rezvani, S.; Herrador, J.M.H.; Franco, F.; Pinto, F.; Snape, C.; Hewitt, N. Converting brown coal to synthetic liquid fuels through direct coal liquefaction technology:Techno-economicevaluation. *Int. J. Energy Res.* **2020**, *44*, 11827–11839. [CrossRef]
5. Khodakov, A.Y.; Chu, W.; Fongarland, P. Advances in the development of novel cobalt Fischer-Tropsch catalysts for synthesis of long-chain hydrocarbons and clean fuels. *Chem. Rev.* **2007**, *107*, 1692–1744. [CrossRef] [PubMed]
6. Li, J.; He, Y.; Tan, L.; Zhang, P.; Peng, X.; Oruganti, A.; Yang, G.; Abe, H.; Wang, Y.; Tsubaki, N. Integrated tuneable synthesis of liquid fuels via Fischer–Tropsch technology. *Nat. Catal.* **2018**, *1*, 787–793. [CrossRef]
7. Sankaran, V. Research and development needs in combustion modeling. *Appl. Energy Combust. Sci.* **2025**, *21*, 100307. [CrossRef]
8. Naik, C.V.; Puduppakkam, K.V.; Modak, A.; Meeks, E.; Wang, Y.L.; Feng, Q.Y.; Tsotsis, T.T. Detailed chemical kinetic mechanism for surrogates of alternative jet fuels. *Combust. Flame* **2011**, *158*, 434–445. [CrossRef]
9. Dagaut, P.; Karsenty, F.; Dayma, G.; Dievart, P.; Hadj-Ali, K.; Mze-Ahmed, A.; Braun-Unkhoff, M.; Herzler, J.; Kathrotia, T.; Kick, T.; et al. Experimental and detailed kinetic model for the oxidation of a Gas to Liquid (GtL) jet fuel. *Combust. Flame* **2014**, *161*, 835–847. [CrossRef]
10. Dagaut, P.; Dievart, P. Combustion of synthetic jet fuels: Naphthenic cut and blend with a gas-to-liquid (GtL) jet fuel. *Proc. Combust. Inst.* **2017**, *36*, 433–440. [CrossRef]
11. Valco, D.; Gentz, G.; Allen, C.; Colket, M.; Edwards, T.; Gowdagiri, S.; Oehlschlaeger, M.A.; Toulson, E.; Lee, T. Autoignition behavior of synthetic alternative jet fuels: An examination of chemical composition effects on ignition delays at low to intermediate temperatures. *Proc. Combust. Inst.* **2015**, *35*, 2983–2991. [CrossRef]
12. Melder, J.; Zinsmeister, J.; Grein, T.; Jürgens, S.; Köhler, M.; Oßwald, P. Comprehensive Two-Dimensional Gas Chromatography: A Universal Method for Composition-Based Prediction of Emission Characteristics of Complex Fuels. *Energy Fuels* **2023**, *37*, 4580–4595. [CrossRef]
13. Heyne, J.; Bell, D.; Feldhausen, J.; Yang, Z.; Boehm, R. Towards fuel composition and properties from Two-dimensional gas chromatography with flame ionization and vacuum ultraviolet spectroscopy. *Fuel* **2022**, *312*, 122709. [CrossRef]
14. Wu, Z.; Mao, Y.; Raza, M.; Zhu, J.; Feng, Y.; Wang, S.; Qian, Y.; Yu, L.; Lu, X. Surrogate fuels for RP-3 kerosene formulated by emulating molecular structures, functional groups, physical and chemical properties. *Combust. Flame* **2019**, *208*, 388–401. [CrossRef]
15. Dooley, S.; Won, S.H.; Chaos, M.; Heyne, J.; Ju, Y.; Dryer, F.L.; Kumar, K.; Sung, C.-J.; Wang, H.; Oehlschlaeger, M.A.; et al. A jet fuel surrogate formulated by real fuel properties. *Combust. Flame* **2010**, *157*, 2333–2339. [CrossRef]
16. Kim, D.; Martz, J.; Abdul-Nour, A.; Yu, X.; Jansons, M.; Violi, A. A six-component surrogate for emulating the physical and chemical characteristics of conventional and alternative jet fuels and their blends. *Combust. Flame* **2017**, *179*, 86–94. [CrossRef]
17. Kathrotia, T.; Oßwald, P.; Naumann, C.; Richter, S.; Köhler, M. Combustion kinetics of alternative jet fuels, Part-II: Reaction model for fuel surrogate. *Fuel* **2021**, *302*, 120736. [CrossRef]
18. Kathrotia, T.; Oßwald, P.; Zinsmeister, J.; Methling, T.; Köhler, M. Combustion kinetics of alternative jet fuels, Part-III: Fuel modeling and surrogate strategy. *Fuel* **2021**, *302*, 120737. [CrossRef]
19. Yang, Z.-Y.; Zeng, P.; Wang, B.-Y.; Jia, W.; Xia, Z.-X.; Liang, J.; Wang, Q.-D. Ignition characteristics of an alternative kerosene from direct coal liquefaction and its blends with conventional RP-3 jet fuel. *Fuel* **2021**, *291*, 120258. [CrossRef]
20. Richter, S.; Kukkadapu, G.; Westbrook, C.K.; Braun-Unkhoff, M.; Naumann, C.; Köhler, M.; Riedel, U. A combined experimental and modeling study of combustion properties of an isoparaffinic alcohol-to-jet fuel. *Combust. Flame* **2022**, *240*, 111994. [CrossRef]
21. Abi Nurazaq, W.; Wang, W.-C.; Lin, J.-K. The properties of sustainable aviation fuel II: Laminar flame speed. *Energy* **2024**, *294*, 130946. [CrossRef]
22. Guzman, J.; Brezinsky, K. Experimental and modeling study of the oxidation of F-24 jet fuel, and its mixture with an iso-paraffinic synthetic jet fuel, ATJ. *Combust. Flame* **2021**, *224*, 108–125. [CrossRef]
23. Yang, Z.; Kosir, S.; Stachler, R.; Shafer, L.; Anderson, C.; Heyne, J.S. A GC $\times$ GC Tier α combustor operability prescreening method for sustainable aviation fuel candidates. *Fuel* **2021**, *292*, 120345. [CrossRef]
24. Wang, Q.-D.; Wang, B.-Y.; Yao, Q.; Liang, J.; Zeng, P.; Liu, J.-G.; Xia, Z.-X. An experimental and kinetic modeling study on the ignition property of an alternative gas to liquid jet fuel. *Combust. Flame* **2025**, *271*, 113805. [CrossRef]
25. Wang, Q.-D.; Zeng, P.; Yao, Q.; Liang, J.; Wang, B.-Y.; Huang, F.; Liu, J.-G.; Xia, Z.-X. An experimental and kinetic modeling study on the ignition kinetics of a sustainable aviation fuel and its blends with a traditional RP-3 jet fuel. *Fuel* **2025**, *380*, 133191. [CrossRef]

26. Wang, B.-Y.; Zeng, P.; He, R.; Li, F.; Yang, Z.-Y.; Xia, Z.-X.; Liang, J.; Wang, Q.-D. Single-Pulse Shock Tube Experimental and Kinetic Modeling Study on Pyrolysis of a Direct Coal Liquefaction-Derived Jet Fuel and Its Blends with the Traditional RP-3 Jet Fuel. *ACS Omega* **2021**, *6*, 18442–18450. [CrossRef]
27. Fang, X.Y.; Huang, X.Y.; Chen, W.K.; Qiao, X.Q.; Ju, D.H. Development of a skeletal surrogate mechanism for emulating combustion characteristics of diesel from direct coal liquefaction. *Combust. Flame* **2020**, *218*, 84–97. [CrossRef]
28. Wang, F.; Rijal, D. Sustainable Aviation Fuels for Clean Skies: Exploring the Potential and Perspectives of Strained Hydrocarbons. *Energy Fuels* **2024**, *38*, 4904–4920. [CrossRef]
29. Panigrahy, S.; Liang, J.; Ghosh, M.K.; Wang, Q.-D.; Zuo, Z.; Nagaraja, S.; Mohamed, A.A.E.-S.; Kim, G.; Vasu, S.S.; Curran, H.J. An experimental and detailed kinetic modeling study of the pyrolysis and oxidation of allene and propyne over a wide range of conditions. *Combust. Flame* **2021**, *233*, 111578. [CrossRef]
30. Liang, J.; Zhao, C.; Zhao, Z.; Wang, X.; Jia, M.-X.; Wang, Q.-D.; Zhang, Y.; Zhao, F. An experimental and kinetic modeling study on the high-temperature ignition and pyrolysis characteristics of cyclohexylamine. *Combust. Flame* **2023**, *252*, 112769. [CrossRef]
31. Wang, Q.-D.; Sun, Y.; Zhao, Z.; Zhang, Y.; Zhao, F.; Li, Y.; Liang, J. Ignition kinetics of nitrocyclohexane behind reflected shock waves in inert and air environments. *Combust. Flame* **2023**, *255*, 112865. [CrossRef]
32. Morley, C. Gaseq: A Chemical Equilibrium Program for Windows, 0.79. 2005. Available online: http://www.gaseq.co.uk/ (accessed on 17 September 2010).
33. Petersen, E.L.; Rickard, M.J.A.; Crofton, M.W.; Abbey, E.D.; Traum, M.J.; Kalitan, D.M. A facility for gas- and condensed-phase measurements behind shock waves. *Meas. Sci. Technol.* **2005**, *16*, 1716–1729. [CrossRef]
34. Guzman, J.; Kukkadapu, G.; Brezinsky, K.; Westbrook, C. Experimental and modeling study of the pyrolysis and oxidation of an iso-paraffinic alcohol-to-jet fuel. *Combust. Flame* **2019**, *201*, 57–64. [CrossRef]
35. Baigmohammadi, M.; Patel, V.; Nagaraja, S.; Ramalingam, A.; Martinez, S.; Panigrahy, S.; Mohamed, A.A.E.-S.; Somers, K.P.; Burke, U.; Heufer, K.A.; et al. Comprehensive Experimental and Simulation Study of the Ignition Delay Time Characteristics of Binary Blended Methane, Ethane, and Ethylene over a Wide Range of Temperature, Pressure, Equivalence Ratio, and Dilution. *Energy Fuels* **2020**, *34*, 8808–8823. [CrossRef]
36. Goodwin, D.G.; Speth, R.L.; Moffat, H.K.; Weber, B.W. *Cantera: An Object-oriented Software Toolkit for Chemical Kinetics, Thermodynamics, and Transport Processes*, Version 2.4.0; Cantera: Tulsa, OK, USA, 2018.
37. Davidson, D.F.; Hanson, R.K. Recent advances in shock tube/laser diagnostic methods for improved chemical kinetics measurements. *Shock Waves* **2009**, *19*, 271–283. [CrossRef]
38. Sung, C.J.; Curran, H.J. Using rapid compression machines for chemical kinetics studies. *Prog. Energy Combust. Sci.* **2014**, *44*, 1–18. [CrossRef]
39. Wang, Q.-D.; Fang, Y.-M.; Wang, F.; Li, X.-Y. Systematic analysis and reduction of combustion mechanisms for ignition of multi-component kerosene surrogate. *Proc. Combust. Inst.* **2013**, *34*, 187–195. [CrossRef]
40. Curran, H.J. Developing detailed chemical kinetic mechanisms for fuel combustion. *Proc. Combust. Inst.* **2019**, *37*, 57–81. [CrossRef]
41. Bugler, J.; Marks, B.; Mathieu, O.; Archuleta, R.; Camou, A.; Gregoire, C.; Heufer, K.A.; Petersen, E.L.; Curran, H.J. An ignition delay time and chemical kinetic modeling study of the pentane isomers. *Combust. Flame* **2016**, *163*, 138–156. [CrossRef]
42. Darcy, D.; Tobin, C.J.; Yasunaga, K.; Simmie, J.M.; Wurmel, J.; Metcalfe, W.K.; Niass, T.; Ahmed, S.S.; Westbrook, C.K.; Curran, H.J. A high pressure shock tube study of n-propylbenzene oxidation and its comparison with n-butylbenzene. *Combust. Flame* **2012**, *159*, 2219–2232. [CrossRef]
43. El-Sabor Mohamed, A.A.; Sahu, A.B.; Panigrahy, S.; Baigmohammadi, M.; Bourque, G.; Curran, H. The effect of the addition of nitrogen oxides on the oxidation of propane: An experimental and modeling study. *Combust. Flame* **2022**, *245*, 112306. [CrossRef]
44. Wang, Q.-D.; Panigrahy, S.; Yang, S.; Martinez, S.; Liang, J.; Curran, H.J. Development of Multipurpose Skeletal Core Combustion Chemical Kinetic Mechanisms. *Energy Fuels* **2021**, *35*, 6921–6927. [CrossRef]
45. Dong, S.; Wagnon, S.W.; Pratali Maffei, L.; Kukkadapu, G.; Nobili, A.; Mao, Q.; Pelucchi, M.; Cai, L.; Zhang, K.; Raju, M.; et al. A new detailed kinetic model for surrogate fuels: C3MechV3.3. *Appl. Energy Combust. Sci.* **2022**, *9*, 100043. [CrossRef]
46. Zeng, M.R.; Li, Y.Y.; Yuan, W.H.; Li, T.Y.; Wang, Y.Z.; Zhou, Z.Y.; Zhang, L.D.; Qi, F. Experimental and kinetic modeling study of laminar premixed decalin flames. *Proc. Combust. Inst.* **2017**, *36*, 1193–1202. [CrossRef]
47. Fang, R.; Kukkadapu, G.; Wang, M.; Wagnon, S.W.; Zhang, K.; Mehl, M.; Westbrook, C.K.; Pitz, W.J.; Sung, C.-J. Fuel molecular structure effect on autoignition of highly branched iso-alkanes at low-to-intermediate temperatures: Iso-octane versus iso-dodecane. *Combust. Flame* **2020**, *214*, 152–166. [CrossRef]
48. Wang, H.; Dames, E.; Sirjean, B.; Sheen, D.A.; Tangko, R.; Violi, A.; Lai, J.Y.W.; Egolfopoulos, F.N.; Davidson, D.F.; Hanson, R.K.; et al. JetSurF II: A High Temperature Chemical Kinetic Model of n-Alkane (up to n-Dodecane), Cyclohexane, and Methyl-, Ethyl-, n-Propyl and n-Butyl-Cyclohexane Oxidation at High Temperatures. Available online: https://web.stanford.edu/group/haiwanglab/JetSurF/JetSurF2.0/ (accessed on 17 September 2010).

49. Liu, Y.-X.; Richter, S.; Naumann, C.; Braun-Unkhoff, M.; Tian, Z.-Y. Combustion study of a surrogate jet fuel. *Combust. Flame* **2019**, *202*, 252–261. [CrossRef]
50. Wang, H.; Xu, R.; Wang, K.; Bowman, C.T.; Hanson, R.K.; Davidson, D.F.; Brezinsky, K.; Egolfopoulos, F.N. A physics-based approach to modeling real-fuel combustion chemistry—I. Evidence from experiments, and thermodynamic, chemical kinetic and statistical considerations. *Combust. Flame* **2018**, *193*, 502–519. [CrossRef]
51. Yu, L.; Wu, Z.Y.; Qiu, Y.; Qian, Y.; Mao, Y.B.; Lu, X.C. Ignition delay times of decalin over low-to-intermediate temperature ranges: Rapid compression machine measurement and modeling study. *Combust. Flame* **2018**, *196*, 160–173. [CrossRef]
52. Kukkadapu, G.; Whitesides, R.; Wang, M.; Wagnon, S.W.; Mehl, M.; Westbrook, C.K.; McCormick, R.; Sung, C.-J.; Pitz, W.J. Development of a diesel surrogate for improved autoignition prediction: Methodology and detailed chemical kinetic modeling. *Appl. Energy Combust. Sci.* **2023**, *16*, 100216. [CrossRef]

MDPI AG
Grosspeteranlage 5
4052 Basel
Switzerland
Tel.: +41 61 683 77 34

MDPI Books Editorial Office
E-mail: books@mdpi.com
www.mdpi.com/books

www.ingramcontent.com/pod-product-compliance
Lightning Source LLC
Chambersburg PA
CBHW061555220926
51850CB00005B/346
* 9 7 8 3 7 2 5 8 4 9 5 7 4 *